T0270893

COMPOSITION AND ANALYSIS OF HEAVY PETROLEUM FRACTIONS

CHEMICAL INDUSTRIES

A Series of Reference Books and Textbooks

Consulting Editor

HEINZ HEINEMANN
Berkeley, California

1. *Fluid Catalytic Cracking with Zeolite Catalysts,* Paul B. Venuto and E. Thomas Habib, Jr.
2. *Ethylene: Keystone to the Petrochemical Industry,* Ludwig Kniel, Olaf Winter, and Karl Stork
3. *The Chemistry and Technology of Petroleum,* James G. Speight
4. *The Desulfurization of Heavy Oils and Residua,* James G. Speight
5. *Catalysis of Organic Reactions,* edited by William R. Moser
6. *Acetylene-Based Chemicals from Coal and Other Natural Resources,* Robert J. Tedeschi
7. *Chemically Resistant Masonry,* Walter Lee Sheppard, Jr.
8. *Compressors and Expanders: Selection and Application for the Process Industry,* Heinz P. Bloch, Joseph A. Cameron, Frank M. Danowski, Jr., Ralph James, Jr., Judson S. Swearingen, and Marilyn E. Weightman
9. *Metering Pumps: Selection and Application,* James P. Poynton
10. *Hydrocarbons from Methanol,* Clarence D. Chang
11. *Form Flotation: Theory and Applications,* Ann N. Clarke and David J. Wilson
12. *The Chemistry and Technology of Coal,* James G. Speight
13. *Pneumatic and Hydraulic Conveying of Solids,* O. A. Williams
14. *Catalyst Manufacture: Laboratory and Commercial Preparations,* Alvin B. Stiles
15. *Characterization of Heterogeneous Catalysts,* edited by Francis Delannay
16. *BASIC Programs for Chemical Engineering Design,* James H. Weber
17. *Catalyst Poisoning,* L. Louis Hegedus and Robert W. McCabe
18. *Catalysis of Organic Reactions,* edited by John R. Kosak
19. *Adsorption Technology: A Step-by-Step Approach to Process Evaluation and Application,* edited by Frank L. Slejko
20. *Deactivation and Poisoning of Catalysts,* edited by Jacques Oudar and Henry Wise

21. *Catalysis and Surface Science: Developments in Chemicals from Methanol, Hydrotreating of Hydrocarbons, Catalyst Preparation, Monomers and Polymers, Photocatalysis and Photovoltaics,* edited by Heinz Heinemann and Gabor A. Somorjai
22. *Catalysis of Organic Reactions,* edited by Robert L. Augustine
23. *Modern Control Techniques for the Processing Industries,* T. H. Tsai, J. W. Lane, and C. S. Lin
24. *Temperature-Programmed Reduction for Solid Materials Characterization,* Alan Jones and Brian McNichol
25. *Catalytic Cracking: Catalysts, Chemistry, and Kinetics,* Bohdan W. Wojciechowski and Avelino Corma
26. *Chemical Reaction and Reactor Engineering,* edited by J. J. Carberry and A. Varma
27. *Filtration: Principles and Practices: Second Edition,* edited by Michael J. Matteson and Clyde Orr
28. *Corrosion Mechanisms,* edited by Florian Mansfeld
29. *Catalysis and Surface Properties of Liquid Metals and Alloys,* Yoshisada Ogino
30. *Catalyst Deactivation,* edited by Eugene E. Petersen and Alexis T. Bell
31. *Hydrogen Effects in Catalysis: Fundamentals and Practical Applications,* edited by Zoltán Paál and P. G. Menon
32. *Flow Management for Engineers and Scientists,* Nicholas P. Cheremisinoff and Paul N. Cheremisinoff
33. *Catalysis of Organic Reactions,* edited by Paul N. Rylander, Harold Greenfield, and Robert L. Augustine
34. *Powder and Bulk Solids Handling Processes: Instrumentation and Control,* Koichi Iinoya, Hiroaki Masuda, and Kinnosuke Watanabe
35. *Reverse Osmosis Technology: Applications for High-Purity-Water Production,* edited by Bipin S. Parekh
36. *Shape Selective Catalysis in Industrial Applications,* N. Y. Chen, William E. Garwood, and Frank G. Dwyer
37. *Alpha Olefins Applications Handbook,* edited by George R. Lappin and Joseph L. Sauer
38. *Process Modeling and Control in Chemical Industries,* edited by Kaddour Najim
39. *Clathrate Hydrates of Natural Gases,* E. Dendy Sloan, Jr.
40. *Catalysis of Organic Reactions,* edited by Dale W. Blackburn
41. *Fuel Science and Technology Handbook,* edited by James G. Speight
42. *Octane-Enhancing Zeolitic FCC Catalysts,* Julius Scherzer
43. *Oxygen in Catalysis,* Adam Bielański and Jerzy Haber
44. *The Chemistry and Technology of Petroleum: Second Edition, Revised and Expanded,* James G. Speight

45. *Industrial Drying Equipment: Selection and Application,* C. M. van't Land
46. *Novel Production Methods for Ethylene, Light Hydrocarbons, and Aromatics,* edited by Lyle F. Albright, Billy L. Crynes, and Siegfried Nowak
47. *Catalysis of Organic Reactions,* edited by William E. Pascoe
48. *Synthetic Lubricants and High-Performance Functional Fluids,* edited by Ronald L. Shubkin
49. *Acetic Acid and Its Derivatives,* edited by Victor H. Agreda and Joseph R. Zoeller
50. *Properties and Applications of Perovskite-Type Oxides,* edited by L. G. Tejuca and J. L. G. Fierro
51. *Computer-Aided Design of Catalysts,* edited by E. Robert Becker and Carmo J. Pereira
52. *Models for Thermodynamic and Phase Equilibria Calculations,* edited by Stanley I. Sandler
53. *Catalysis of Organic Reactions,* edited by John R. Kosak and Thomas A. Johnson
54. *Composition and Analysis of Heavy Petroleum Fractions,* Klaus H. Altgelt and Mieczyslaw M. Boduszynski

ADDITIONAL VOLUMES IN PREPARATION

NMR Techniques in Catalysis, edited by Alexis T. Bell and Alexander Pines

The Chemistry and Technology of Coal: Second Edition, Revised and Expanded, James G. Speight

Upgrading Petroleum Residues and Heavy Oils, Murray R. Gray

COMPOSITION AND ANALYSIS OF HEAVY PETROLEUM FRACTIONS

Klaus H. Altgelt
Consultant
San Rafael, California

Mieczyslaw M. Boduszynski
Chevron Research and Technology Company
Richmond, California

CRC Press
Taylor & Francis Group
Boca Raton London New York

CRC Press is an imprint of the
Taylor & Francis Group, an **informa** business

Library of Congress Cataloging-in-Publication Data

Altgelt, Klaus H.
Composition and analysis of heavy petroleum fractions / Klaus H. Altgelt. Mieczyslaw M. Boduszynski.
p. cm. — (Chemical industries ; v. 54)
Includes bibliographical references and index.
ISBN 0-8247-8946-6
1. Petroleum—Analysis. I. Boduszynski, Mieczyslaw M. II. Title. III. Series.
TP691.A45 1994 93-39037
665.5—dc20 CIP

The publisher offers discounts on this book when ordered in bulk quantities. For more information, write to Special Sales/Professional Marketing at the address below.

This book is printed on acid-free paper.

Marcel Dekker
270 Madison Avenue, New York, New York 10016

Current printing (last digit):
10 9 8 7 6 5 4 3 2

PRINTED IN THE UNITED STATES OF AMERICA

Preface

Almost five years ago, one of us (KHA) was approached by Marcel Dekker, Inc. to write a new edition of an earlier work, *Chromatography in Petroleum Analysis*. Instead we chose the topic of the present book, for two reasons: (1) the rising economic significance of heavy crude oils and their residues and (2) the recent progress in heavy crude oil analysis. This is a "hot" field, with great promise and not without controversy.

This field is economically important because most of the new crude oil produced is heavy, with relatively small amounts of light (low-boiling) components. On the other hand, the demand for light distillates is growing rapidly. Thus, increasing amounts of low-boiling fuel must be produced from high-boiling feed. Furthermore, sulfur, nitrogen, and even aromatic rings must be drastically reduced. Compositional analysis of heavy (high-boiling) petroleum fractions is an indispensible tool for the petroleum chemist in the search for improved methods for the conversion of messy heavy material to clean, low-boiling fuels.

Great progress has been made in instrumentation and methodology. Sensitivity and speed of analytical instruments (e.g., mass, nuclear magnetic resonance, and infrared spectrometers) have greatly improved during the last 10–15 years. New approaches in the use and combination of analytical methods have also been important. We strongly feel that careful separation of heavy petroleum fractions before spectroscopic and other measurements is mandatory for reliable results. Even the type and sequence of the diverse separation methods are crucial. The first step should always be distillation, as discussed in the book. Depending on the task, several chromatographic techniques in specific sequences may be necessary. The right combination of separation and measuring techniques will give optimal results at reasonable cost.

Another recent innovation is the introduction of the atmospheric equivalent boiling point (AEBP) scale, which spans the entire boiling range of a crude oil, including nondistillable residue. This concept enables us to recognize the gradual change in composition from one boiling range to the next. It helps the analytical chemist choose the appropriate methods for a certain fraction based on the knowledge of the lower-boiling fractions. It also allows the refining engineer to visualize the composition of an entire crude oil in uniform terms. The AEBP concept is a major theme in our book and its application and appeal will be explored.

We have tried to present an in-depth view of the current analytical methodology. Also included is a survey of our present knowledge of the composition of heavy petroleum fractions. For some readers, it may be amazing to see how much detail is now accessible for certain fractions. For others, it may be disappointing to recognize how limited our understanding is of the highest-boiling fractions, especially the nondistillable residue, despite the enormous efforts expended over the last 25 years.

We could not have succeeded in our endeavor without the support from many people. First and foremost, we thank our families—especially our spouses—for their heavily taxed (yet seemingly unlimited) patience despite our extended neglect of their rightful title to our companionship in daily life, joys, and chores.

We are also deeply indebted to our colleagues and friends from Chevron Research and Technology Company, who have so willingly and ably assisted us with reading our drafts and providing us with helpful comments, corrections, and even written contributions. Dr. John Shinn read the entire manuscript, a momentous task, and gave us innumerable, valuable suggestions. Dr. T. H. Gouw rewrote major parts of our discussion concerning the practice and theory of distillation (Chapter 3). Dr. Carl Rechsteiner made many important improvements to our chapter on mass spectrometry (Chapter 7). Dr. Don Wilson critically read our chapter on NMR techniques (Chapter 8), and Dr. Don Young reviewed the section on IR (Chapter 9).

Our thanks also to the publishers and authors who gave us permission to reproduce figures and tables from their papers, and especially to those authors who kindly sent us high-quality copies of their original figures for our book: Drs. Biggs, Cookson, Farrall, Green, Snape, and Swain. Finally, Mr. David Grudoski introduced one of us (KHA) to the art of making illustrative diagrams by Macintosh computer, which was both useful and enjoyable.

We appreciate the patience and courtesy extended by our publisher, Marcel Dekker, Inc. Finally, we thank the management of Chevron Reserch and Technology Company for permission to publish some of the material developed under their sponsorship and for allowing one of us (KHA) the use of their library and other facilities.

Klaus H. Altgelt
Mieczyslaw M. Boduszynski

Contents

1

Introduction

I. INTENT AND SCOPE OF THIS BOOK

We believe that compositional analysis of heavy petroleum fractions will play a decisive role in improving refinery operations and will, thus, be a significant aid in saving energy and resources and in mitigating pollution problems. This book is an advocacy for its application as well as a guide to its implementation. Here are some reasons for our belief in the importance of this issue.

According to recent reports, about 50% of the petroleum products consumed in the United States at the present time are gasoline and an additional 40% are other distillate fuels boiling below about 650°F (345°C). The Committee on Production Technologies for Liquid Transportation Fuels (1) states that "the US transportation sector . . . will depend almost completely for the foreseeable future on liquid fuels . . . and . . . any transition from the use of liquid transportation fuels (to compressed natural gas or electricity) is likely to be slow." Therefore, we can expect a continued high percentage of the petroleum products to be light distillate fuels boiling below 650°F. On the other hand, an increasing amount of the petroleum *produced* currently is of the "heavy" variety, containing large amounts of material boiling above 650°F (2,3). At least 85% of the world proven reserves of petroleum belong to this category (4). Figure 1.1 shows how the average API gravity of all the crudes refined in the United States decreased during the 10 years between 1980 and 1990. Related to this drop in API gravity is a constant rise in the sulfur content as illustrated in Fig. 1.2. These trends mean that most of the crude oil produced now and in the future must be converted from heavy material (>650°F) to light distillate products in multiple, complex refining steps (5).

Major conversion processes include catalytic cracking, hydrocracking, and coking. The yield and quality of the products resulting from these processes are quite variable and depend on feed type, process type, and processing conditions. Sulfur-, nitrogen-, and metal-containing compounds make the upgrading process difficult because of their propensity to poison catalysts. Heteroatoms must be removed in several steps with different catalysts. The ease of their removal depends on their chemical environment and functionality [e.g., aliphatic versus aromatic, thiophenic sulfur versus

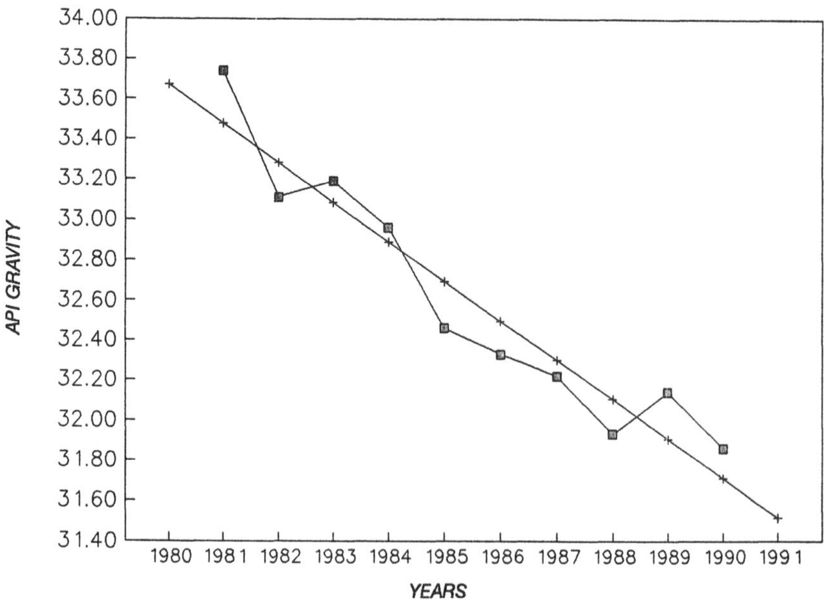

Figure 1.1 Change in API gravity of crude oils refined in the United States between 1980 and 1990. (From Ref. 5. Reproduced with permission of the publisher. Original data from the U.S. Department of Energy.)

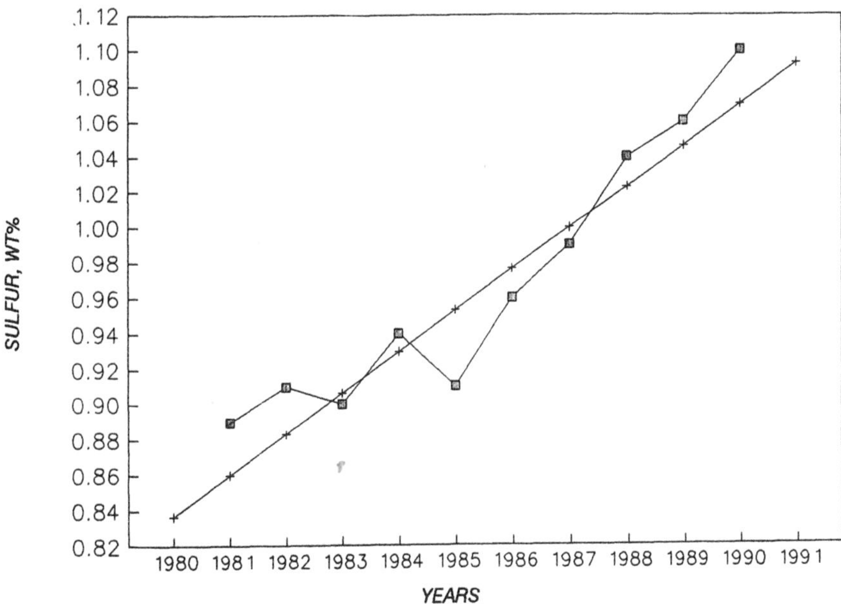

Figure 1.2 Change in sulfur content of crude oils refined in the United States between 1980 and 1990.(From Ref. 5. Reproduced with permission of the publisher. Original data from the U.S. Department of Energy.)

sulfide sulfur, neutral versus basic nitrogen, metals in low-molecular-weight (low-MW) free porphyrins versus high-MW structures]. The heteroatom content of the crudes to be processed is expected to increase dramatically, whereas the Conradson carbon residue content in the fluid catalytic cracker (FCC) feed also is expected to double or triple (see Fig. 1.3).

Measurements of compositional changes from feed to product to assess the effectiveness of an upgrading step for process optimization will involve much more detail than is presently common. The application of new technology together with new computational tools will offer a more fundamental approach to unraveling the chemical reactions that occur in catalytic processing of complex petroleum mixtures. Detailed compositional analysis will be vital in developing reaction networks and kinetic models of refining processes. Naber et al. (6) emphasize the need for detailed understanding on the molecular level of the product composition and its effect on performance properties in our efforts to improve product quality and process integration.

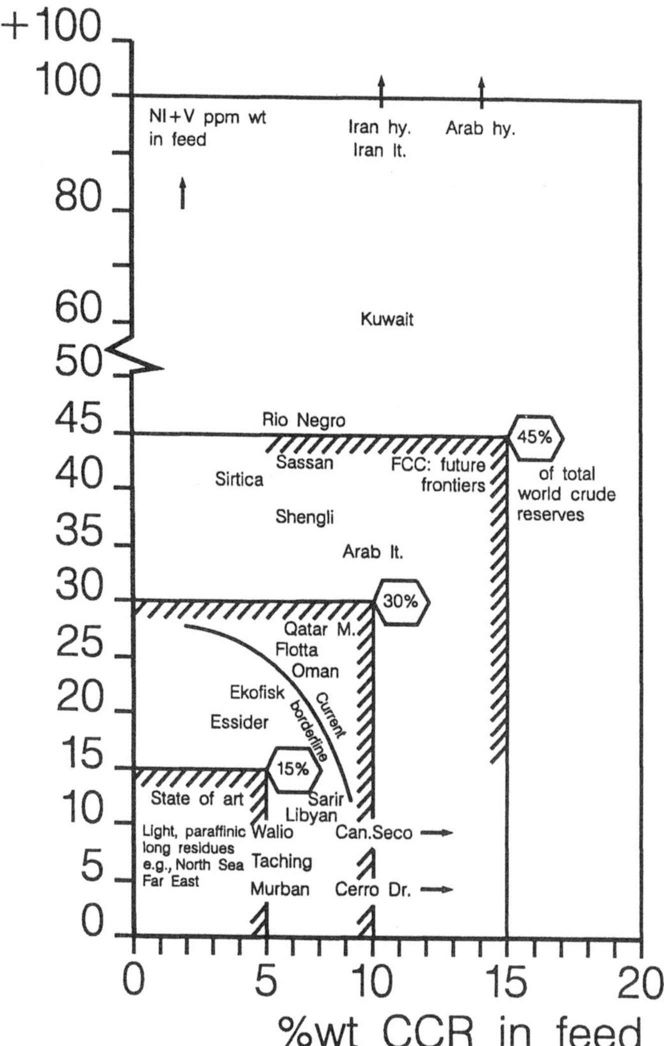

Figure 1.3 FCC feedstock "heaviness" diagram. Resid (>370°C) properties in relation to FCC processability. (From Ref. 6. Reproduced with permission of the publisher.)

With the help of compositional analysis, catalytic processing of petroleum feedstocks can be approached as "molecular tailoring" where feedstock molecules are converted into new compounds designed to meet the requirements of specific products. This usually involves a reduction of molecular weight and alteration of molecular structure; in other words, "tailoring" the size and shape of molecules to make products with the most desirable composition. The higher the boiling point of the feed, the more complex is its composition in terms of both the variety and the structure of its molecules. In addition, as the composition of the sample becomes more complex, compositional analysis becomes more difficult, but its potential benefits become even greater.

Compounding the problems arising from the need to process heavier crude oils is the fact that quality requirements for petroleum products are constantly tightened. Sulfur and nitrogen contents as well as the concentration of aromatic hydrocarbons are limited by law or by customer demand, and the downward trend continues. Naber et al. (6) write "future specifications for middle distillates call for deep desulfurization (often to 95% or more) and imply an aromatics content of less than 30%, with cetane index above 50." In California, low aromatic diesel (LAD) with less than 10% aromatics will be required after 1993. We contend that, in this climate of increasing difficulties, proper compositional analysis of petroleum and its fractions, particularly the heavy ones, becomes increasingly attractive. Compositional analysis, judiciously applied, can contribute to the preservation of precious resources while at the same time helping the oil industry to improve their operations.

With this book we also hope to provide a better insight into the composition of petroleum, especially their heavy ends, and into the importance of understanding its complexity. We consider the complexity of heavy ends, the significance of their various components and their interactions as the key to their analysis and also to their processability. We place special emphasis on the increase in complexity of petroleum fractions with increasing boiling point, increased complexity in terms of the variety of components as well as the structure of the individual molecules. We strive to illustrate both the opportunities and the limitations of analytical technology in unraveling this complexity. With this understanding, the rules of the game of compositional analysis will then, hopefully, become clearer.

An important innovation, underlying much of the discussion presented here, is the introduction of the atmospheric equivalent boiling point (AEBP) concept (7,8). This concept extends the "boiling range" of petroleum to very high-boiling and even to nonvolatile fractions, increasing it about threefold (from 1000°F to about 2700°F) and permits the description of an entire crude oil in uniform terms. With the AEBP concept and with the understanding of the huge and continuously increasing complexity of heavy crude oil fractions, the rules of the game of compositional analysis will, hopefully, become clearer.

Probably the greatest benefit of the AEBP concept to the analytical chemist is its use as a road map, enabling one to build on the known composition of low-boiling fractions for the analysis of high-boiling ones. This approach facilitates the most efficient use of our modern analytical methods.

Impressive improvements have been made during the last 10 years to the point where the analytical chemist can now evaluate the changes occurring in a processing step in great detail. Yet, we still face tremendous challenges. With high-boiling distillates and, in particular, with residues, the variety of components and their molecular complexity is so large that some of our instrumental methods do not work at all and others do not give us simple, unequivocal answers. In many instances, we are forced to rely on our experience with simpler systems for the interpretation of some of our measurements. Fortunately, the changes in composition from lighter to heavier cuts of a given crude oil are only gradual with no abrupt, unexpected jumps. We assume, and in certain cases we observe, these gradual changes to persist throughout the entire boiling point and molecular-weight range of the crude oils. This helps us in the interpretation of measurements in ranges which are not yet fully accessible to us. On the other hand, the extension of research into such ranges also accounts for the differences of interpretation we see in the literature. These thoughts will be more fully explored in the next chapter.

Beyond giving insight into the composition of petroleum, we try to give our readers an overview of what we believe are the most important modern methods in petroleum analysis that cover all major aspects of our field. In this way, we hope to help them in the selection of practical approaches to solving certain problems they are likely to encounter in petroleum analysis and processing. In describing the various analytical methods, we try to give sufficient detail to explain the underlying concepts and the jargon so that the petroleum chemist or engineer knows enough about them to be able to talk to and understand the analytical experts with whom they may cooperate.

II. DEFINITION OF HEAVY PETROLEUMS

In our book, we use the word "petroleum" interchangeably with crude oil although, to some extent, we have included papers on bitumens in our discussion. The term "heavy" is meant to indicate material boiling higher than about 650°F (345°C), including distillation residues. Our primary definition of "heavy petroleums" refers then to the absence, more or less, of low-boiling components. As far as we discuss petroleum *fractions*, this definition is straightforward in that we consider only high-boiling distillates or residues. In our discussion of whole crude oils, we cannot be quite so exclusive and refer to those which contain only minor amounts of the lighter fractions.

By using the term "heavy," we made a concession to general usage of it as a convenient though not strictly accurate abbreviation. We will have more to say about this in Chapter 2, Section II.A and Chapter 4, Section IV.C.

REFERENCES

1. Committee on Production Technologies for Liquid Transportation Fuels (CPTLTF) (1990a). *Fuels to Drive Our Future*. National Acadamy Press, Washington, DC, p. 13.
2. Meyer, R. F., Fulton, P. A., and Dietzman, W. D. (1982). A Preliminary Estimate of World Heavy Crude Oil and Bitumen Resources. *The Future of Heavy Crude Oil and Tar Sands*, edited by R. F. Meyer, J. C. Wyan, and J. C. Olson, McGraw-Hill, New York, pp. 97–158.
3. Speight, J. G. (1986). Upgrading heavy feedstocks. *Ann. Rev. Energy*, **11**:253.
4. Logwinuk, A. K., and Caldwell, D. L. (1983). Application of ART to heavy oil processing. *Chem. Econ. Eng. Rev.*, **15**:31.
5. Swain, E. J. (1991). U.S. crude slate gets heavier, higher in sulfur. *Oil Gas J.*, September, 59–61.
6. Naber, J. E., Stork, W. H. J., Blauwhoff, P. M. M., and Groeneveld, K. J. W. (1991). Technological response to environmental concerns, product quality requirements and changes in demand patterns. *Erdoel Kohle*, **107**:124–132.
7. Boduszynski, M. M. (1987b). Composition of heavy petroleums. 1. Molecular weight, hydrogen deficiency, and heteroatom concentration as a function of atmospheric equivalent boiling point up to 1400°F (760°C). *Energy Fuels*, **1**:2–11.
8. Boduszynski, M. M., and Altgelt, K. H. (1992). Composition of heavy petroleums. 4. Significance of the extended atmospheric equivalent boiling point (AEBP) scale. *Energy and Fuels*, **6**:72–76.

2

Compositional Analysis

Dream and Reality

In this chapter, we give the reader a brief account of what we call "heavy petroleum fractions," their geometrically rising complexity with increasing boiling point, the limitations encountered in their analysis and even in their description, and the approaches to overcome or circumvent some of these limitations. This chapter may be seen as an overview of the subject matter dealt with in greater detail later and as an introduction to our philosophy and our conceptual approach to the compositional analysis of heavy petroleum fractions.

I. OPPORTUNITIES AND LIMITATIONS

If we knew the entire composition, every detail, of a feedstock going into a refinery process and that of the product coming out, we could certainly make better use of our crude oils and of our processing facilities. Better process models could be devised; problems of catalyst fouling and off-spec products would be more easily and quickly recognized, understood, and solved; new catalysts could be developed more rapidly; and more efficient refineries could be built with readily optimized "downstream" operations.

Should we then make an all-out effort and try to find out every compositional detail of a feedstock to be processed? Sometimes yes, but usually not. With heavy petroleum fractions, knowing every detail is generally unaffordable, often impossible, and, fortunately, usually unnecessary. It is often impossible because of two obstacles: (1) the enormous complexity of heavy petroleum fractions and (2) the limitations of our measuring techniques. It is usually unnecessary because of the great similarity between members of a given group or subgroup of compounds if we define, select, and separate these properly.

With our present effort, we hope to help the research chemist and the refinery engineer in three ways to analyze their feedstocks and products more efficiently: first, by giving them a firm grasp of this complexity and showing them what to expect from an analysis; second, by guiding them in the process of defining, selecting, and separating appropriate compound groups; and third, by describing modern analytical techniques for the molecular characterization of these complex mixtures.

II. TERMINOLOGY

A. Basic Terminology

In this book, we try to use logical and chemical terminology wherever possible, instead of traditional terms. The latter are defined but used in the text only where necessary to avoid misunderstandings.

Petroleum components can be classified—although analytically often not resolved—into two major compound groups, namely, hydrocarbons and heterocompounds. The term hydrocarbons is used for molecules made up only of carbon and hydrogen atoms. Heterocompounds are compounds which, in addition to carbon and hydrogen, also contain one or more heteroatoms such as sulfur, nitrogen, oxygen, vanadium, nickel, or iron.

Hydrocarbons and heterocompounds can be described by a general molecular formula:

$$C_nH_{2n+Z}X$$

where C = carbon, H = hydrogen, n = number of carbon atoms in the molecule, Z = hydrogen deficiency value, and X = heteroatoms. The hydrogen deficiency value Z of hydrocarbons can be calculated from the number of double bonds, DB, and rings, R, in the molecule by the equation:

$$Z = -2(R + DB - 1)$$

Hydrocarbons are comprised of the following classes of compounds:

1. Acyclic alkanes, which in petroleum chemistry are usually called paraffins. Paraffins include normal and isoparaffins, i.e., straight chain and branched ones, respectively. Both normal (*n*-paraffins) and isoparaffins have the same molecular formula C_nH_{2n+2}.

2. Cycloalkanes, which in petroleum chemistry are called naphthenes. Although most cycloalkanes in petroleum have paraffinic side chains attached to their saturated rings, they are defined by their number of rings, i.e., mononaphthenes (or monocyclic alkanes), dinaphthenes (or dicyclic alkanes), trinaphthenes (or tricyclic alkanes), etc. Mononaphthenes have the molecular formula C_nH_{2n}; dinaphthenes, C_nH_{2n-2}; trinaphthenes, C_nH_{2n-4}; etc. The Z value decreases by 2 with every additional naphthenic ring. Naphthenes can have five-membered rings (e.g., cyclopentane) or six-membered rings (e.g., cyclohexane). The rings can be catafused or perifused or nonfused. Another expression is catacondensed or pericondensed. Perhydrochrysene (catacondensed) and perhydropyrene (pericondensed) are examples of the two fused ring types; bicyclohexyl is an example of nonfused ring systems. There are also diamondoid (similar to diamond structure) trinaphthenes, e.g., adamantane. Figure 2.1 shows a few examples of various naphthenes and explains the terminology.

3. Olefins, which are also called alkenes and have at least one nonaromatic double bond in their structure. Mono-olefins have the same molecular formula, C_nH_{2n}, as mononaphthenes. Olefins are so scarce in crude oils that they may be neglected there. However, they may occur in large amounts in cracked refinery streams, especially in light and middle distillates.

4. Aromatics, which are compounds containing at least one benzene ring. Most of the aromatic hydrocarbons in petroleum consist of aromatic and naphthenic rings and bear normal and branched alkane side chains. Molecules containing one aromatic ring are classified as monoaromatic; those with two aromatic rings, as diaromatic; those with three aromatic rings, as triaromatic; and so on. Naphthenic rings and paraffinic side chains, which may also be part of their structure, do not affect this classification. Alkylbenzenes, the simplest type of monoaromatic hydrocarbons in petroleum, have the formula C_nH_{2n-6}. Monoaromatics with one naphthenic ring have the formula C_nH_{2n-8}; those with two naphthenic rings, C_nH_{2n-10}; and so on. There are many combinations of "alkylnaphthenoaromatics," some of which may have identical molecular formulas. Figure 2.2 shows a few examples of aromatic hydrocarbons, including some with different structures but the same formula.

The complexity of aromatic hydrocarbons rises immensely as the number of aromatic rings increases. Not only does the number of aromatic ring

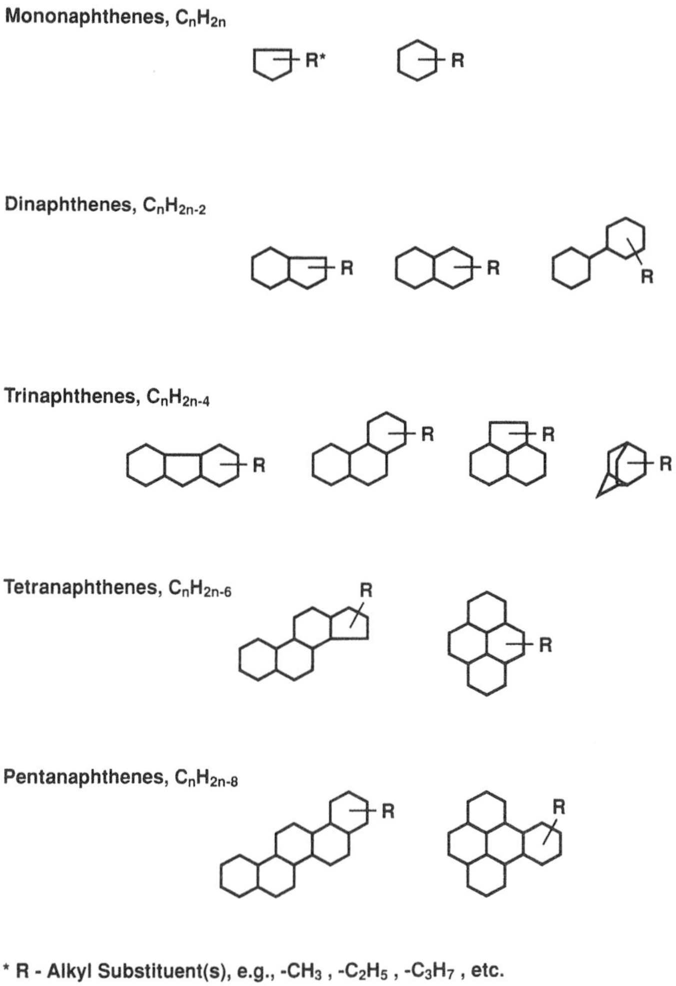

Mononaphthenes, C_nH_{2n}

R* R

Dinaphthenes, C_nH_{2n-2}

R R R

Trinaphthenes, C_nH_{2n-4}

R R R R

Tetranaphthenes, C_nH_{2n-6}

R R

Pentanaphthenes, C_nH_{2n-8}

R R

* R - Alkyl Substituent(s), e.g., $-CH_3$, $-C_2H_5$, $-C_3H_7$, etc.

Figure 2.1 Examples of various naphthene types.

isomers increase rapidly but the possible arrangements of naphthenic and aromatic rings relative to each other in a molecule grow at an even greater rate. The different rings may be clustered or interspersed. Either set of rings may be arranged in a straight line (e.g., tetracene) or at angles (e.g., chrysene), and they may be catacondensed (e.g., tetracene and chrysene) or pericondensed (e.g., fluoranthene, pyrene, and perylene). In addition

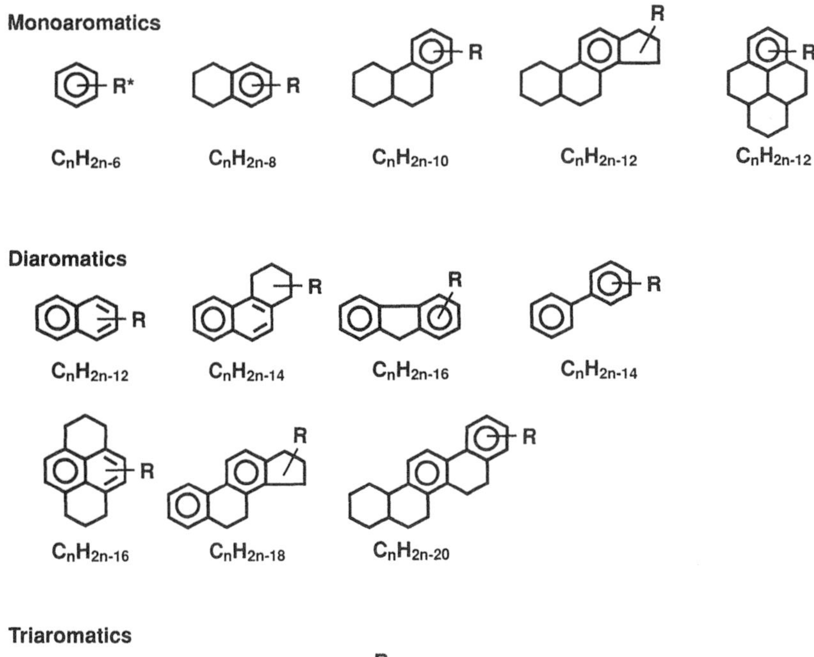

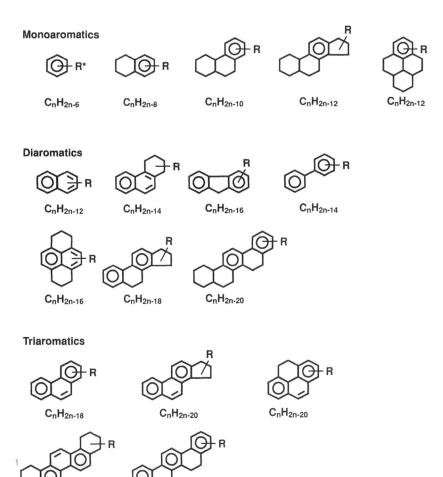

*** R - Alkyl Substituent(s), e.g., $-CH_3$, $-C_2H_5$, $-C_3H_7$, etc.**

Figure 2.2 Examples of various aromatic types.

to these permutations, we have all the variations in number and type of alkyl side chains. Figure 2.3 shows a few examples of polycyclic aromatic hydrocarbons (PAHs).

In heterocompounds, the presence of one or more heteroatoms per molecule is added to the multitude of hydrocarbon structural arrangements. The type of heteroatom (S, N, O, V, Ni, Fe) and its functionality (>S, −SH, ≥N, >NH, −NH2, −OH, > C = O, etc.) further complicate the issue. A few simple examples of heterocompound types found in heavy petroleum fractions are shown in Fig. 2.4. The combination of multiple heteroatoms and different functionalities in the same molecule add yet another dimension. In some high-molecular-weight crude oil fractions, average values of several heteroatoms per molecule have been found. One of these had the general formula

$$C(ar)_{80.0}C(al)_{99.3}H_{218.3}S_{1.14}N_{3.96}O_{1.89}V_{0.009}Ni_{0.019}Fe_{0.013}$$

i.e., it contained 80 aromatic C, 99.3 aliphatic C, 1.14 S, 3.96 N, 1.89 O, and together 0.041 metal atoms. Another one had 110.6 aromatic C, 171.5 aliphatic C, 8.25 S, 5.18 N, 2.18 O, and 0.105 metal atoms:

$$C(ar)_{110.6}C(al)_{171.5}H_{361.5}S_{8.25}N_{5.18}O_{2.18}V_{0.076}Ni_{0.017}Fe_{0.012}$$

(see Chapter 4, Tables 4.19 and 4.21). These are only average numbers. Whereas about half of the molecules in these fractions would contain fewer heteroatoms, the other half must contain even more. Such heteroatom concentrations and combinations make the complexity of these heavy petroleum fractions truly immense.

B. Inherited Terminology

Terminology is an important topic, especially in so complex a subject as petroleum chemistry. Unfortunately, terminology usually evolves with the field and is, therefore, rarely as logical and precise as it should be. This is certainly the case in our field. To some extent, we must live with the inherited terms, even though many are based on operational procedures, most are vague, and some are even misleading. One such inherited vague term is heavy as in our title heavy petroleum fractions. Heavy, in petroleum chemistry, originally referred to materials of high density. Because most oils or oil fractions of high density are also high boiling and because heavy is such a convenient, short word, this term came to be used interchangably with high boiling. However, some highly paraffinic oils or oil fractions may have significantly higher boiling points than much heavier, i.e., denser, aromatic oils or oil fractions. This point will be elaborated on with examples in Chapter 4, Section I.

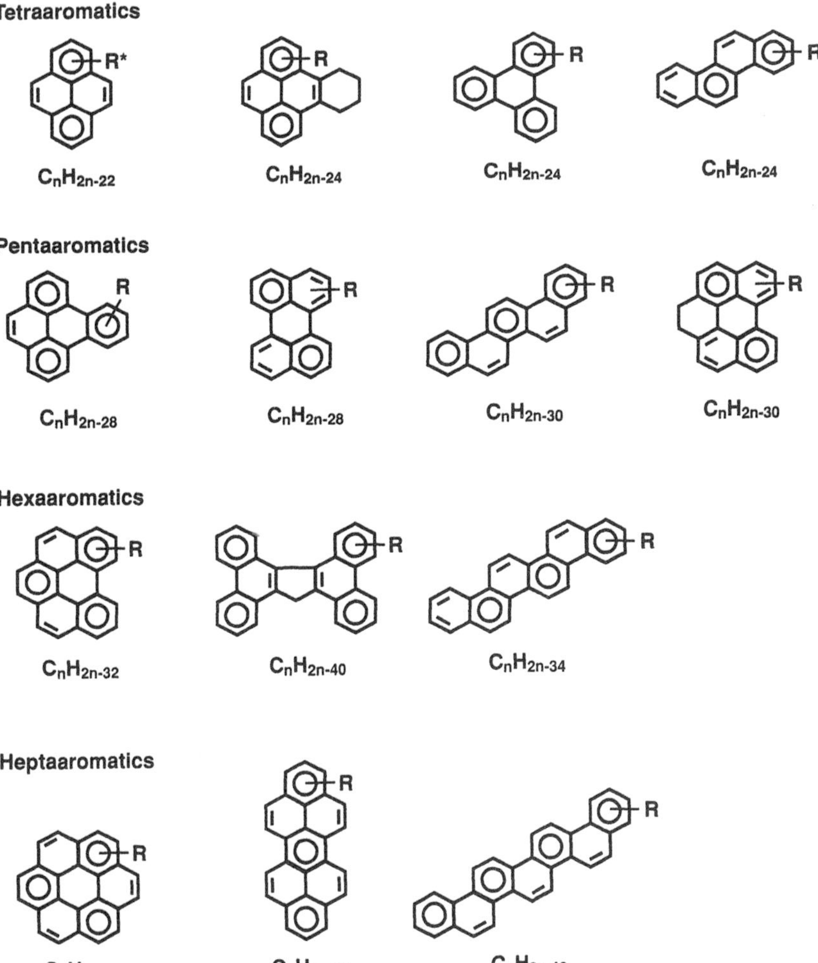

* R - Alkyl Substituent(s), e.g., $-CH_3$, $-C_2H_5$, $-C_3H_7$, etc.

Figure 2.3 Examples of various polycyclic aromatic hydrocarbons.

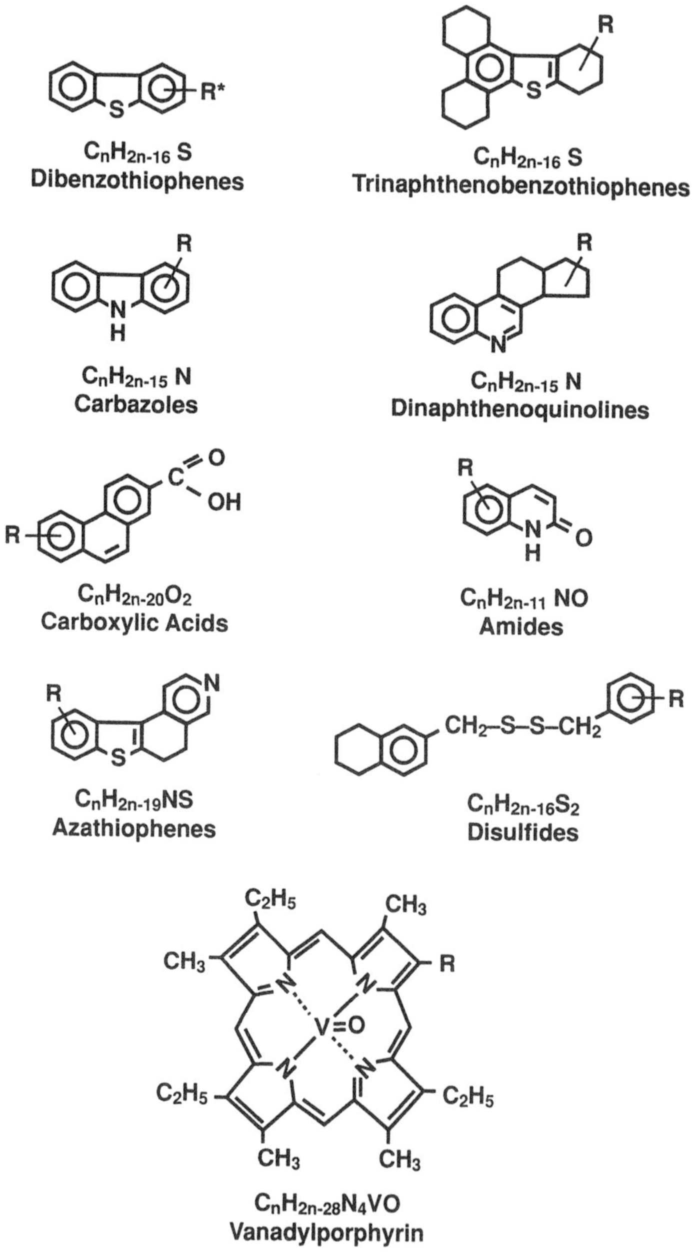

$C_nH_{2n-16}S$
Dibenzothiophenes

$C_nH_{2n-16}S$
Trinaphthenobenzothiophenes

$C_nH_{2n-15}N$
Carbazoles

$C_nH_{2n-15}N$
Dinaphthenoquinolines

$C_nH_{2n-20}O_2$
Carboxylic Acids

$C_nH_{2n-11}NO$
Amides

$C_nH_{2n-19}NS$
Azathiophenes

$C_nH_{2n-16}S_2$
Disulfides

$C_nH_{2n-28}N_4VO$
Vanadylporphyrin

* R - Alkyl Substituent(s), e.g., $-CH_3$, $-C_2H_5$, $-C_3H_7$, etc.

Figure 2.4 Some heterocompound types found in petroleum fractions.

Some distillation cuts of petroleum are named and categorized by their origin (refinery process) or by their intended use. For example, there are the straight-run fractions (distilled directly from crude oil) and cracked ones (produced by catalytic cracking or hydrocracking). There are the light and heavy naphthas, also called gasolines, from straight-run and cracked streams. Middle distillates, from both straight-run and cracked stocks, are also referred to as kerosene, diesel, and jet cuts. Then there are the light, medium, and heavy vacuum gas oils, again straight run or process derived (e.g., FCC heavy cycle oil or heavy coker gas oil), and the light, medium, and heavy lube base stocks, which are highly refined and dewaxed vacuum gas oil fractions.

Residues, also called residua or resids, are those fractions which are nondistillable under given conditions and remain at the bottom of a distillation tower. They are, therefore, referred to as tower bottoms and also as the bottom of the barrel. But one bottom may be quite different from another, depending on the distillation conditions. The term atmospheric residue describes the material remaining at the bottom of the atmospheric distillation column which has an upper boiling limit of approximately 650°F (345°C). For deeper cuts, distillations are performed at lower pressures to minimize thermal degradation of the fractions. What stays at the bottom of a vacuum distillation column is then called vacuum residue, or sometimes vacuum bottoms, and has an equivalent boiling point above approximately 1000°F (540°C). Still deeper cuts can be achieved (although not in refinery columns) with high-vacuum short-path ("molecular") distillation which yields distillates up to approximately 1300°F (700°C) and leaves behind truly "nondistillable residue."

Vacuum residues can often be used without further treatment as asphalts, i.e., as binder for road pavements or as roofing material, pipe coatings, and similar applications. Whereas the term residue designates a material by its nature, namely, as the nondistillable portion of a crude oil (under given conditions), the term asphalt might designate the same material by its intended use for paving or coating. Asphalt is also the name given to one of the products made from vacuum residue in a process known as propane/butane deasphalting. The other product of this process is called deasphalted oil (DAO). By further upgrading and dewaxing, DAO may be converted to bright stock, a lube base oil of high viscosity and light color.

The terms pitch and tar have occasionally, and incorrectly, been used interchangably with asphalt, for the propane/butane deasphalting product as well as for the finished coating material. Although their end use may be the same, their origins are not. Pitch is a product of petroleum or coal pyrrolysis, and it is solid at room temperature. Originally, the term tar designated an oily product of the destructive distillation of coal, wood, or peat. Thus, originally it referred to an artificial product unrelated to petroleum. Later, the term was also used for heavy oils as, for instance, in tar sands, i.e., now for natural materials related to petroleum. Thus, tar is a particularly striking example of inconsistent and potentially confusing terminology.

Wax in the terminology of petroleum chemists is a mixture of *n*-paraffins (*n*-alkanes) and other hydrocarbon types having long straight alkane chains. This is another operationally defined term, namely, by its ability to crystallize. Slack wax, a direct product of lube oil dewaxing processes, has a greater proportion of hydrocarbons other than *n*-paraffins in the mixture and is, therefore, only partly crystallized.

Early efforts, aimed at unraveling the composition of heavy petroleum fractions, also introduced operational and often confusing terminology which is used to this day. Examples of such traditional terms are asphaltenes and maltenes which are used to describe the insoluble and soluble fractions of a vacuum residue or an asphalt. They are defined by the respective insolubility and solubility of these fractions in light hydrocarbons such as *n*-pentane, *n*-hexane, or *n*-heptane. The proportion of the insoluble asphaltenes to soluble maltenes depends, of course, on the type of hydrocarbon (or other liquid) used for precipitation. Thus, "pentane-asphaltenes" are different from "hexane-asphaltenes" and from "heptane-asphaltenes." A variety of other conditions also affect the yield and composition of asphaltenes (see Chapter 8). To distinguish asphaltenes from other *n*-alkane-insoluble admixtures, their definition included solubility in toluene. Components insoluble in toluene but soluble in carbon disulfide were called carbenes, and those insoluble in carbon disulfide were called carboids. Maltenes were typically separated further by extraction from various adsorbents into so-called oils (a fraction which was readily desorbed with *n*-alkane solvent) and resins (a strongly adsorbed fraction which was desorbed with polar solvents). The definition of all these "components" is strictly in operational terms and not in terms of chemical structure.

Available separation methods cannot completely separate or distinguish the overwhelming variety of compounds present in heavy petroleum fractions. However, for many years it has been recognized that it is desirable, and usually possible, to separate certain *groups* of compounds. Numerous separation methods and complex separation schemes have been developed over the years. We will discuss some of these in Chapter 6. All these efforts certainly contributed to a better understanding of the composition of heavy petroleum fractions; however, they also introduced many potentially confusing terms in the attempt to describe the various chromatographic fractions.

Sometimes, similar terms were used to describe dissimilar fractions; in other cases, similar fractions were given different names. For example, a number of chromatographic methods were developed to separate a group of components called aromatics. Different methods, however, produced fractions comprising different types of aromatics; but all of these were usually called the same, just aromatics. Furthermore, the separation of "aromatics" from heterocompounds becomes increasingly difficult with higher-boiling fractions. Depending on the chromatographic method as well as the boiling range, aromatics may contain significant amounts of sulfur and nitrogen compounds. As long as the term aromatics is used only to distinguish a particular fraction from other fractions of different compound types, e.g., from saturates or polar heterocompounds, it is acceptable though vague. However, often the aromatics content of one distillate cut is weighed against that of another without regard for their composition. Such comparisons may be quite misleading.

Terminology used to categorize heterocompounds is even more confusing. Here, the earlier separation methods introduced such terms as resins or polars to describe heteroatom-rich fractions, which were strongly retained on chromatographic columns and required the use of polar solvents for their desorption and recovery. Occasionally resin-1, resin-2, and resin-3 fractions were postulated when different solvents were used for their desorption. The authors of the API Research Project 60, an extensive study of the heavy ends of petroleum, introduced several new, more chemically oriented terms such as acids, bases, and neutral nitrogen compounds. These were used to describe the various heteroatom-containing fractions separated by nonaqueous chromatography on ion-exchange resins and by charge-transfer chromatography. The API-60 separation scheme was applied to both very high-boiling distillates and truly nondistillable residues without prior precipitation of asphaltenes from the latter. Now, what used to be traditionally described as asphaltenes was separated along with the soluble portion into acids, bases, and neutrals. These terms are better than the older ones, especially when they are related to boiling ranges or to molecular-weight ranges.

More recently, the API-60 separation scheme was modified and extended by Green et al. (1) at the National Institute for Petroleum and Energy Research (NIPER). They separated petroleum heavy ends into over 30 fractions, introducing more specific chemical terms such as, strong acids, weak acids, and very weak acids, strong bases, medium bases, and very weak bases to describe some of the fractions. This was a good innovation over the older resins 1, 2, and 3. Other more general terms such as neutrals and polars are occasionally unavoidable but may produce situations where polars are actually isolated from a neutrals fraction.

Finally, a fundamental question arises: Should an amphoteric molecule such as hydroxy pyridine be considered an acid or a base? Some analysts say the answer depends on the analytical method by which the compound is determined. We believe the answer should depend on the environment of the compound in the original mixture. An amphoteric compound in a predominantly acidic environment, for example, should be considered basic even if it was found and isolated by chromatography on an anion exchange resin. This issue again points to the need to think carefully about the objective of an analysis and to devise an appropriate classification system even before starting the experimental work.

C. Suggestions for an Improved Terminology

Ideally, the nomenclature should be based on strictly chemical terms. However, the immense number of molecular species in petroleum makes some grouping and certain compromises unavoidable. We have already mentioned the basic terminology of such groups as paraffins, naphthenes, aromatics, and polars, and the various ring types (mono, di-, tri-, etc.). Here we will try to develop a more orderly classification scheme with groupings on the basis of separations, identifications, and chemical differentiations.

As the broadest category, we accept the traditional *compound groups*: saturates, aromatics, and polars. These can be isolated within certain limitations by simple chromatographic methods. With more refined chromatographic columns and instrumentation, much narrower fractions can be separated, the *compound classes*. Examples are paraffins, naphthenes, monoaromatics, diaromatics, triaromatics, acid and base concentrates, and neutral heterocompounds. Within these, we can distinguish *compound types*, such as mononaphthenes, dinaphthenes, trinaphthenes, benzothiophenes, sulfides, pyrroles, carbazoles, and azacompounds. These can no longer be separated, but they can be identified by spectroscopic methods (MS, IR, NMR) and often by their *Z*-number. The next step toward increased differentiation leads to the *homologous series*, which are groups of compounds with alkyl substituents of increasing chain length. *Carbon number members* or *homologues* are individual compounds identified by their carbon number. They may belong to several homologous series.

Figure 2.5 gives some concrete examples of these categories and how they interrelate. Obviously, our scheme is a compromise rather than a clean solution to the terminology problem. Its advantage is the close relation to chemical composition, augmented by experimental means of differentiation, namely, by chromatographic and other separations and by MS analysis, i.e., by methods which again stress basic chemical properties. We see this scheme as a first step toward a better terminology and hope for constructive input from other petroleum chemists.

Compound Group:	e.g., saturates (aliphatic hydrocarbons)
Compound Class:	e.g., naphthenes (cycloalkanes)
Compound Type:	e.g., tetranaphthenes (tetracyclic alkanes)
Z Series:	Z = -6H, C_nH_{2n-6}, e.g.,

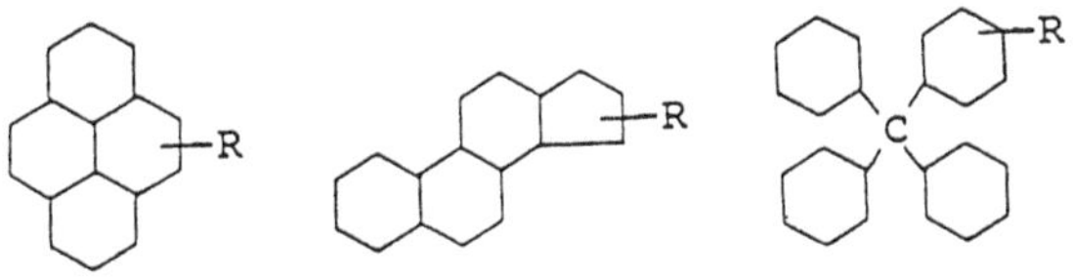

Here, R may stand for several substituents.

Carbon No. Member: e.g., $C_{27}H_{48}$

Homologous Series: e.g.,

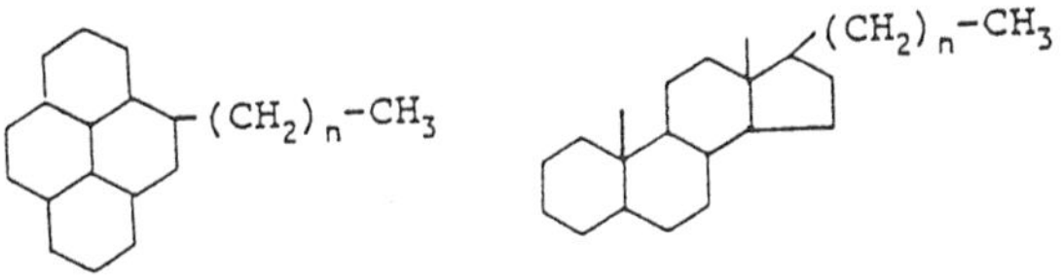

$(CH_2)_n-CH_3$ $(CH_2)_n-CH_3$

Carbon No. Homolog: e.g.,

n-undecylperhydropyrene or cholestane (sterane)

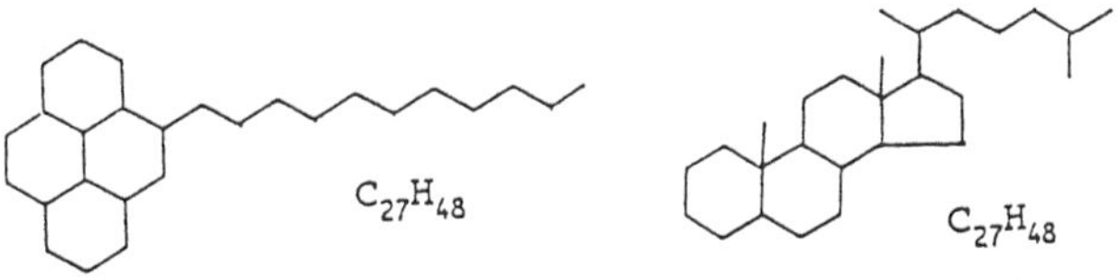

$C_{27}H_{48}$ $C_{27}H_{48}$

Figure 2.5 Examples of grouping according to our scheme.

II. MOLECULAR-WEIGHT RANGE AND NUMBER OF COMPONENTS IN PETROLEUM

Beyond the limitations imposed by inherited terminology, we face limitations caused by the immense number of compounds in petroleum and by their often great complexity.

The true molecular-weight distribution of petroleum compounds cannot be readily measured. It is generally accepted that molecular weights of petroleum compounds range from less than one hundred to several thousand daltons. However, molecular weights higher than about 2000 daltons are still uncertain and probably inflated because of aggregation effects which are not yet fully resolved. This issue is discussed in greater detail in Chapter 4, Section I. In Fig. 2.6, we show the molecular-weight distribution of a typical heavy crude oil. It is instructive also to see the numbers in Table 2.1.

Even if we could determine the presence and the amount of every molecular species in a heavy petroleum fraction, let alone those of a whole crude oil, we would be swamped with so much data that we could hardly handle them, even now in the computer age. Table 2.2 gives an idea of the number of different compounds with which we would be dealing. It shows the number of possible isomers only in the simplest type of compounds found in petroleum, the paraffins (acyclic alkanes). Even though

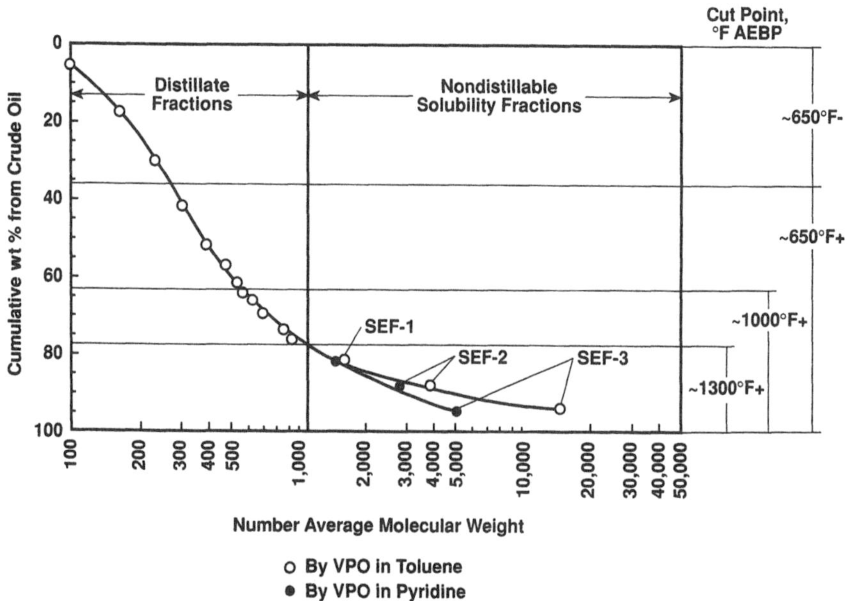

Figure 2.6 Molecular-weight distribution of a heavy crude oil (Maya crude oil, 22.2°API).

Table 2.1 Molecular-Weight Distribution of Maya Crude Oil

MW range	Approximate carbon no. of limiting MW	Cumulative weight %
<400	<30	50
>400	>30	50
<1000	<75	75
>1000	>75	25
<2000	<150	85
>2000	>150	15
>3000	>220	10
>5000	>370	5

Table 2.2 Carbon Number, Boiling Point, and the Number of Paraffin Isomers

Carbon no.	Boiling point[a] (°C)	Boiling point[a] (°F)	Isomers	Examples of petrol. distill. cuts
5	36	97	3	
8	126	258	18	Gasoline
10	174	345	75	
12	216	421	355	Diesel and jet fuels, middle distillates
15	271	519	4347	
20	344	651	$3.66 \cdot 10^{5}$	Vacuum gas oil
25	402	755	$3.67 \cdot 10^{7}$	
30	449	840	$4.11 \cdot 10^{9}$	
35	489	912	$4.93 \cdot 10^{11}$	Atmospheric residue
40	522	972	$6.24 \cdot 10^{13}$	
45	550	1022	$8.22 \cdot 10^{15}$	
60	615	1139	$2.21 \cdot 10^{22}$	Vacuum residue, asphalt
80	672	1242	$1.06 \cdot 10^{31}$	
100	708	1306	$5.92 \cdot 10^{39}$	Nondistillable residue

[a]Atmospheric equivalent boiling point (AEBP) of *n*-alkanes.
(From Ref. 2. Reproduced with permission of the publisher.)

not all of the structurally possible isomers will actually exist, the numbers are overwhelming.

The number of isomers increases rapidly with the number of carbon atoms in a molecule because of the rapidly rising number of their possible structural arrangements. Even for the paraffins in the C_5–C_{12} range (a typical carbon number range for naphthas), the number of possible isomers is large (>600). The number of isomers, *which have actually been experimentally observed* (*though not all identified*) in this range (see, e.g., Refs. 3–5), is in the order of 200–400. For any of the higher-boiling cuts, the numbers quickly become unmanageable. If we add to the paraffins the other common petroleum components, the naphthenes, aromatic hydrocarbons, and the various heterocompounds, the numbers become larger yet by several orders of magnitude for the higher-boiling fractions. This makes a complete compositional analysis of heavy petroleum fractions utterly impossible. We must settle for simplifications.

IV. THE VIRTUES OF DISTILLATION

The first step to simplification is the distillation of a crude oil into a number of fractions (or cuts). It reduces the enormous number of different molecules by limiting their sizes (carbon number) as well as their structural diversity in each fraction. Figure 2.7 illustrates this issue in a plot of carbon number versus boiling point.

Rather than on a common curve, the data points fall into a wedge-shaped area. The upper left boundary is defined by the normal paraffins which have the highest carbon number and molecular weight for a given boiling point.* Looking at it the other way, for a given carbon number, the paraffins have the lowest boiling points. Naphthenes boil at somewhat higher temperatures and are, therefore, located to the right of the paraffins. Aromatic hydrocarbons and, especially polar heterocompounds, have higher boiling points yet and are positioned even farther to the right. The most polar, unsubstituted aromatic heterocompounds fall on the bottom curve (which is hypothetical beyond the point belonging to 2-quinolone). These are the compounds of lowest mass for a given boiling point; or, seen from the other perspective again, they are the compounds with the highest boiling points for a given molar mass.

Figure 2.7 shows an essential fact of petroleum composition: *Diverse compounds with similar molar masses cover a broad boiling range*; *and, conversely*, *a narrow boiling point cut can contain a wide molar mass range*.

---

*Strictly speaking, some isoparaffins have the highest carbon number, but a consistent data set is not available.

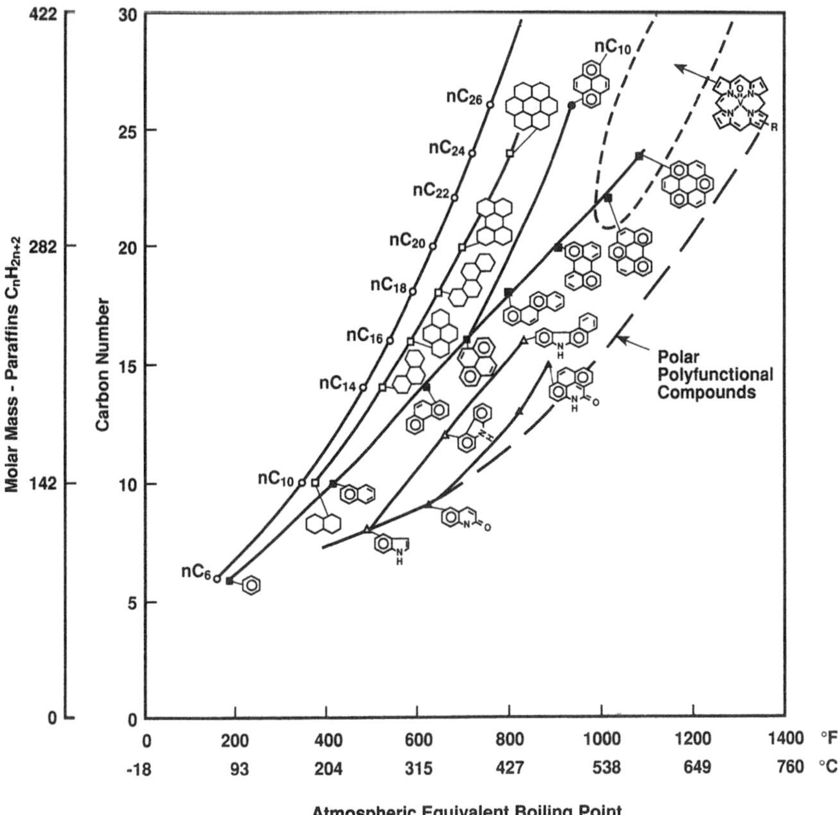

Figure 2.7 The effect of molecular weight and structure on boiling point.

Distillation limits the molecular-weight range of each compound type in a cut, but the molecular-weight range of one compound type in the mixture, say that of *n*-paraffins, can be quite different from that of another, e.g., *n*-alkylnaphthalenes. A 650–800°F (345–425°C) cut, for instance, may contain *n*-paraffins between 282 and 380 daltons and *n*-alkylnaphthalenes of considerably lower MW, namely, between 240 and 296 daltons. As Fig. 2.7 demonstrates, a narrow distillation cut can contain only a limited selection of different chemical species. It is easier to distinguish and separate the different compound classes in a narrow distillation cut than in a broad one or in whole crude oil because each of its compound types comprises only a small range of molar masses, and the molar mass range of each compound type is different from that of the others. Another advantage of distillation is the mitigation of interactions between large and small molecules which could interfere with subsequent separations by chemical nature.

Last but not least, distillation is the main separation process in refineries. The boiling point and boiling range of refinery streams and products are the basis for costing and for planning their use. They are, therefore, of prime importance for refining and marketing decisions. Distillation as a laboratory operation ties in directly with refinery distillations. Laboratory crude oil distillation, the so-called distillation assay (see Chapter 3 for more details), has long been the first step in the assessment of crude oils. It provides information on yields of various boiling-range fractions. As the first step in the comprehensive compositional analysis of petroleum, it produces fractions of known boiling ranges for further compositional analysis.

V. COMPOSITIONAL ANALYSIS OF PETROLEUM FRACTIONS: THE HIGHER YOU GO, THE HARDER IT GETS

The higher the boiling point of a fraction, the more difficult is its analysis. The molecular composition of light naphthas, which have a nominal boiling range of start to 265°F (130°C), can be readily determined by a single analysis such as gas chromatography (GC). The composition of heavy naphthas (nominal boiling range 265–430°F = 130–220°C) is more difficult to measure. The GC PIONA analysis (paraffins, isoparaffins, olefins, naphthenes, aromatics) already reflects this difficulty by "lumping" components into the five compound classes rather than resolving individual compounds. Yet the detailed compositional analysis of heavy naphthas is still accessible by a single analysis such as GC/MS (gas chromatography/mass spectrometry).

Middle distillates (nominal boiling range 430–650°F = 220–345°C) already represent a challenge to the otherwise very powerful GC/MS analysis. Incomplete resolution of individual compounds makes data interpretation and quantitation very difficult if not impossible. In this range, GC can no longer separate all the components, and the mass spectrometer may see and analyze mixtures of species belonging to different compound types rather than single compounds. Group-type MS techniques are often preferred here (see Chapter 5). Liquid chromatography (LC), and recently supercritical fluid chromatography (SFC), are often used to determine the composition of middle distillates in terms of the compound groups saturates, aromatics, and polars.

The conventional vacuum gas oils (nominal boiling range 650–1000°F = 345–540°C) are too complex for a single GC/MS analysis, and group-type MS methods are generally used instead. These, however, are based on a major assumption, namely, that the sample components match the compound types used in the calibration matrix. If a sample contains compounds other than those in a matrix, they will be either ignored or misidentified. A direct analysis of vacuum gas oils by group-type MS methods is frequently impaired by a high concentration of sulfur or nitrogen compounds. LC methods are usually required to separate fractions of saturates and aromatics which, separately, are amenable to MS group-type analyses (e.g., ASTM D2687 and D3239, respectively). Special high-performance liquid chromatographic (HPLC) methods are also used to separate specific, well-defined compound-class fractions for further analysis by field ionization mass spectrometry (FIMS), low-voltage high-resolution mass spectrometry (LVHR-MS), and other spectroscopic techniques (e.g. ^{1}H-NMR, ^{13}C-NMR, FTIR, etc.).

The conventional vacuum residues (>~1000°F = >540°C) require the use of extensive separation and characterization schemes. Further fractionation by high-vacuum short-path ("molecular") distillation greatly facilitates their analysis by producing very high-boiling distillate cuts in the nominal boiling range of ~1000–1300°F (~540–700°C) which are amenable to various LC separations into compound classes. The compound-class fractions can then be subjected to molecular characterization by spectroscopic methods. Their composition cannot be measured nearly with the same degree of detail and rigor as that of the lower-boiling cuts. None of the existing MS group-type methods are applicable to analysis of these very high-boiling fractions.

Truly nondistillable residues (>~1300°F = >~700°C) are by far the most difficult to analyze. Their limited volatility and frequently limited solubility complicate or impair the use of various analytical techniques for their molecular characterization. Furthermore, their broad molecular-weight range and very high concentration of heteroatoms make the analysis of these fractions particularly difficult. Despite great efforts in this area, knowledge of their composition is still far from complete. Their amount in heavy crude oils can be substantial, 20–45%.

The straight-run distillates (fractions derived directly from crude oil by distillation) would be insufficient for filling the demand for present-day transportation fuels and other products. Their volume and also their quality would be unacceptable. Cracking of high-boiling fractions to smaller molecules furnishes about 70% of all the gasoline made today in the United States. Similarly, jet and diesel fuels are made largely by conversion of vacuum gas oils to lower-boiling fractions. Vacuum gas oils, in turn, are in large part produced by conversion of vacuum residues. In the process, the complexity of the higher-boiling components is, in part, passed on to the resulting lower-boiling products. Thus, compositional analysis of "heavy petroleum fractions" must cover both straight-run and process-derived fractions. Some of the latter ones (e.g., residue-derived hydrocracked vacuum gas oils, heavy coker gas oils, or FCC heavy cycle oils) usually pose a greater analytical challenge than their straight-run counterparts.

The cracked products require still further upgrading. The trend toward deep desulfurization, denitrogenation, and continued lowering of aromatics calls for ever deeper hydrotreating and hydrofinishing. Compositional analysis of feeds and products not only will help the refiner but will become an indispensible tool for process optimization, for increasing the products with desirable properties and mitigating those with undesirable ones. Naber et al. (6) write "Detailed understanding of the relationships between molecular product composition and (performance) properties is crucial for obtaining improved product quality. A consistent molecular approach results in (process) models that are coherent with respect to physical principles and are more powerful for use in extrapolation than conventional ones. Moreover, it facilitates the understanding of process integration."

VI. LIMITATIONS OF OUR MEASURING TECHNIQUES

We can distinguish at least four types of such limitations:

1. Inadequate resolution of separation techniques
2. Insufficient instrumental resolution
3. Insufficient instrumental sensitivity
4. The range for which an analytical method was developed and is valid

A. Resolution of Separation Techniques

The first limitation, inadequate resolution of separation techniques, is apparent by the overlap between consecutive fractions. In distillation, separations without considerable overlap are impossible with mixtures as complex and rich in minute differentiations as petroleum. Even with modern, high-plate-count columns and with optimally chosen cut points, each fraction will always contain some components belonging to the previous and the following cuts. With higher-boiling fractions, the separations become increasingly difficult. High-vacuum short-path distillation, which allows the separation of heavy crude fractions up to about 1300°F (700°C), is a low-efficiency, nonequilibrium process and produces broad fractions with substantial overlap.

Even with gas and liquid chromatography, baseline separations of heavy petroleum fractions are impossible. Perhaps the best and simplest example of this problem is the isolation of "saturates." Even this group of aliphatic hydrocarbons (paraffins and naphthenes) is not easily separated from other components in heavy petroleum fractions. The separation becomes increasingly difficult with increasing boiling point as length and number of alkyl substituents on the other nonaliphatic compounds increases, and saturates and certain aromatics and heterocompounds become structurally more similar. Distillation of heavy petroleum fractions into narrow boiling-range cuts can greatly facilitate subsequent chromatographic separations. But even with prior distillation, high-performance chromatographic columns are usually necessary to obtain "clean" saturates.

B. Instrumental Resolution

The second limitation is insufficient instrumental resolution. In mass spectrometry, it manifests itself in the inability to resolve species by their exact molar mass and formula. Field ionization (FI) and field desorption (FD) mass spectrometry (MS) have been successfully applied to the analysis of very high-boiling petroleum fractions, producing spectra of unfragmented molecular ions. FIMS or FDMS are particularly useful for the analysis of saturates, which cannot be analyzed by other MS techniques without severe fragmentation. Both techniques allow the determination of the following aliphatic hydrocarbon types and their carbon number distributions:

paraffins (C_nH_{2n+2})
mononaphthenes (C_nH_{2n})
dinaphthenes (C_nH_{2n-2})
trinaphthenes (C_nH_{2n-4})
tetranaphthenes (C_nH_{2n-6})
pentanaphthenes (C_nH_{2n-8})
hexanaphthenes (C_nH_{2n-10})

The low nominal mass* resolution of FIMS and FDMS is an obvious limitation. Presently, these methods often do not distinguish between many molecules of equal nominal mass belonging to different Z series, for example, between heptanaphthenes (C_nH_{2n-12}) and paraffins (C_nH_{2n+2}). At a molar mass (M) of 324 amu (atomic mass units), a resolving power of about 3500 ($M/\Delta M$ = 3454) would be necessary to distinguish these two compound types ($C_{23}H_{48}$ and $C_{24}H_{36}$, ΔM = 0.0938 amu).

Low-voltage MS (LV-MS) can provide a parent ion spectrum, i.e., a spectrum of the unfragmented molecular ions, for most petroleum molecules except for saturates which fragment easily. Low-voltage high-resolution

*Nominal (integer) mass as opposed to exact mass, which may be measured with sufficient resolution.

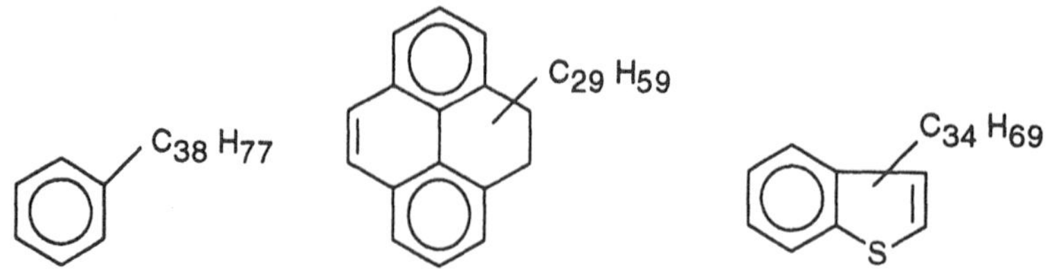

$C_{38}H_{77}$ $C_{29}H_{59}$ $C_{34}H_{69}$

C_{38}-alkylbenzene C_{29}-alkyldihydropyrene C_{34}-alkylbenzothiophene

Figure 2.8 Examples of three compound types with equal nominal mass.

mass spectrometry (LVHR-MS), with its resolving power of about 50,000, also produces parent ion spectra. It can easily distinguish the hydrocarbons shown in Fig. 2.8, a C_{38}-alkylbenzene, belonging to the series C_nH_{2n-6}, and C_{29}-alkyldihydropyrene, belonging to the series C_nH_{2n-20}, both with a nominal mass of 610 amu and a ΔM of 0.0938 amu. They can be differentiated with a resolving power of about 6500 which is within easy reach of LVHR-MS, but outside that of current FIMS. However, the mass of the C_{34}-alkylbenzothiophene from Fig. 2.8 is different from that of the C_{29}-dihydropyrene by only 0.0034 amu and would require a resolving power of 180,000. This is beyond that of most instruments. Whereas certain N compounds can be resolved from hydrocarbons, molecules containing several heteroatoms are generally too complex. Other problems are caused by the presence of ^{13}C isotopes.

Interestingly, the HPLC aromatic ring-type separation can readily resolve the latter two series of compounds. This is an example of how proper fractionation can, to a great extent, overcome the limitations of MS and other characterization methods.

On the other hand, fractionation, even by HPLC, cannot solve all problems. For instance, it is completely impossible by MS alone to distinguish, in complex mixtures, between different species having the same chemical formula, e.g.,

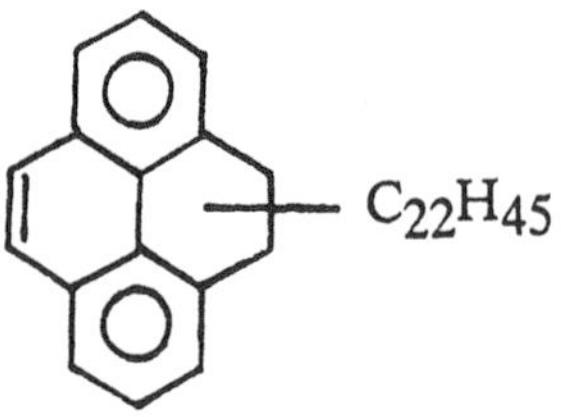

$C_{22}H_{45}$ and $C_{16}H_{33}$

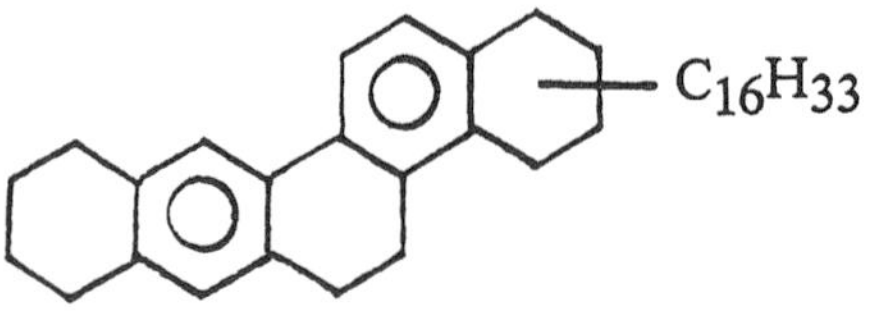

The two compounds have the same composition $C_{38}H_{56}$, i.e., they have the same general formula C_nH_{2n-20} and the same exact molar mass of 512.438. Even chromatography can have great difficulty separating these two compound types.

IR and NMR spectroscopy are both limited by the overlap between frequency ranges. Fourier-transform techniques have greatly improved both sensitivity and resolution of IR spectroscopy, yet overlap problems, with few exceptions, still severely hamper quantitative measurements. Even more progress has been made in NMR. Aside from improvements in instrumental resolution, new experimental techniques were developed, or applied for the first time to petroleum fractions, in the last 10 years, e.g., spectral editing and the so-called two-dimensional (2D) analyses (see Chapter 7, Section III). Spectral editing permits the measurement of subspectra which display only one of the four possible CH_n carbons (n = 0, 1, 2, or 3), thus disentangling the regular NMR spectrum in which the peaks of these carbons overlap or mix with each other. In 2D NMR spectroscopy, the various molecular components are separated either in heteronuclear coupled (^1H–^{13}C) or in homonuclear coupled (^1H–^1H) spectra. Because different independent parameters are measured, the components are observed in different groupings. By drawing the concentration contour lines in plots with the two parameters as the coordinates, it is now possible to see, i.e., to visually isolate, different atomic groups that were previously hidden because their NMR frequency ranges overlapped with those of other groups. Despite these improvements, many NMR peaks still overlap to such a degree that a complete characterization of high-boiling hydrocarbon mixtures by NMR alone is not possible, even without the additional complication by heteroatoms.

In one instance, good use is made of the overlap of NMR peaks. Petroleum fractions are so complex and the variety of naphthenic configurations so great that the corresponding low-resolution NMR peaks are often so closely spaced as to be unresolved and form a broad hump rather than a distinct pattern. In contrast, the paraffinic carbon peaks are generally more widely spaced and sufficiently resolved for evaluation. In these cases, the naphthenic carbon can be measured or estimated from the unresolved hump under the resolved paraffinic carbon peaks. This is the only way to measure or estimate the total naphthenic carbon of high-boiling petroleum fractions. Indirect ways to calculate it from NMR spectra and elemental analysis data are still in the experimental stage (7).

C. Instrumental Sensitivity

The third limitation of our measuring techniques is instrumental sensitivity or, more precisely, the signal-to-noise ratio. In many cases, instrumental sensitivity has been improved by several orders of magnitude during the last 10–20 years. No matter how much better it becomes, we will always find it limited because our demands and applications rise right along with the improvements. But with increased sensitivity, new and often unexpected problems arise as pointed out with many examples by Rogers (8). Slight changes in the ambient pressure or temperature, e.g., by opening doors, may affect baselines or even create peaks. Special precautions, regular testing, and even simulations may become necessary. Simulations not only guard against potential problems in the hardware of modern computerized instruments but also against bias in the software.

D. Range of Permissible Application

The fourth limitation is the range for which a method was developed and our tendency to overstep it. The various group-type MS methods are a good example. These methods, by calibration, certain assumptions, and mathematical manipulation, allow the quantitative determination of selected compound types (defined by the formula C_nH_{2n+Z}) in a sample. For instance, the ASTM method D2786 is applied to a chromatographic fraction of saturates and distinguishes seven types of aliphatic hydrocarbons and alkylbenzenes. More recent versions of the group-type MS method (9,10) do not require chromatographic separation of a sample and can handle up to 22 and 33 compound types, respectively, ranging from paraffins and naphthenes to polycyclic aromatic hydrocarbons, and some heterocompounds.

These group-type MS methods use only milligram amounts of sample and are fast and very cost-effective. Thus, they are employed extensively in industry and research organizations. They were developed for specific and stated ranges of molar mass and chemical composition, primarily for middle distillates and light vacuum gas oils. But the temptation is great to apply them also to other fractions which fall outside the prescribed ranges. If we overstep these ranges, we cannot be sure of the results and must check them with other, independent methods. Otherwise, very erroneous conclusions may be drawn.

VII. TYPES OF INFORMATION

A. Average Versus Detailed Information

Most compositional analyses done on petroleum fractions give data which are averaged over all the molecules in the sample. Examples are the results from elemental analysis, molecular weights by vapor phase osmometry (VPO), and structural information from IR and NMR measurements. These data give no hint at how broad the molecular and chemical distribution in the sample might be.

In contrast, GC/MS analysis allows the identification of individual components and the determination of their concentration in the sample. This kind of data, however, is available only for relatively low-boiling distillates. For many of the high-boiling samples, GC/MS cannot be used. HPLC separation followed by FIMS, FDMS, and LVHR-MS analysis, which will provide the formula and carbon number distribution, are preferred in these cases. Whereas this, too, is very useful information, it has its own limitations. One is that a molecular formula may represent more than one compound type as we have seen earlier. Additional information, say from NMR, could complement the MS analysis, but the NMR data would now be averaged over all the molecules.

The situation becomes more complicated yet with increasing molecular weight, hydrogen deficiency, and heteroatom concentration. Chromatographic separations, although very useful, become more difficult and less efficient. Interferences in MS analysis of the various chromatographic fractions increase rapidly with an increasing number of compounds having the same molecular formula but different molecular structures. The point we are trying to make is that even MS, the method giving the most detailed data, cannot always, with or without fractionation, save us from averages when it comes to structural information, regardless of how high and impressive its resolving power.

With some exceptions, a certain degree of averaging is inescapable in structural measurements and is tacitly accepted. But problems occur when we begin to draw "average structures." Average structures have great appeal because pictures are so much easier to comprehend than columns of numbers. Seeing an average structure, we feel we understand the molecular nature of the species in that mixture; not only what they look like but also how they might behave under certain circumstances, e.g., under processing conditions in a reactor.

But usually this feeling is deceptive. This is particularly true when average structures are drawn for complex samples containing mixtures of compounds which differ not only in chemical composition but also in molecular weight. Numerous average structures of asphaltenes and other operationally defined petroleum fractions, including even unfractionated residues, have been published. Average structures of samples as heterogeneous as those are next to meaningless. Average structures are not the same as representative structures. Average structures are artificial constructs created by the averaging process, often with no resemblance to any real molecules in the sample. Representative structures, in contrast, are the structures of molecules which are present in the sample in rather large amounts either as indicated or in only slightly modified form, e.g., certain ring structures with different degrees of alkyl substitution.

Fractionation by distillation, followed by multistep selective chromatographic separation, can isolate narrow, quite homogeneous, and chemically well-defined fractions from the complex sample matrix. Mass spectrometry, when applied to such narrow fractions, often gives unequivocal formulas and carbon number (molecular weight) distributions, at least for those of <1000°F (540°C) boiling point. Other spectral techniques, e.g., NMR, IR, and UV, provide complementary structural information. Whereas, on the one hand, average data on broad samples may be nearly meaningless, greatly detailed data, on the other hand, are usually time-consuming and expensive to generate. Compromises often give good information at reasonable cost.

For appropriate simplification, we must make reasonable choices about what to measure, bulk constituents, e.g., compound groups (e.g., aromatics), chemically meaningful compound classes (e.g., monoaromatics), compound types (e.g., dinaphtheno-monoaromatics, C_nH_{2n-10}), a specific compound type (e.g., 1,2,3,4,5,6,7,8-octahydrophenenthrenes (C_nH_{2n-10}), carbon-number members (homologues, e.g., $C_{16}H_{22}$, C_2-alkyl-1,2,3,4,5,6,7,8-octahydrophenenthrenes), or individual compounds (e.g., 9,10-dimethyl-1,2,3,4,5,6,7,8-octahydrophenanthrene, $C_{16}H_{22}$).

B. The Right Information

What is the right information and how much information is right? That depends on what we need it for. The grouping of components even into such relatively broad fractions as saturates, aromatics, and polars by chromatographic separation is a great simplification, especially if applied to a reasonably narrow distillation cut. Knowledge of the composition of a petroleum fraction in terms of such chemical groups is sometimes sufficient information for a given task. However, merely measuring the concentration of total aromatics, e.g., in hydrocracking feeds and intermediate process streams, may not suffice for monitoring process performance, particularly if a buildup of polycyclic aromatic hydrocarbons (PAHs) is of concern. In this case, identification of specific PAHs and determination of their concentration down to a few parts per million (ppm) may be necessary to monitor reaction pathways leading to their formation.

Another example may be hydrodenitrogenation (HDN) of feedstocks for fluid catalytic cracking (FCC) to avoid poisoning of the FCC catalyst. Here, the determination of polars (which are usually a concentrate of nitrogen compounds) in the HDN feedstock and product may not be adequate. In this case, we want to know the amount and also the composition of the offending species. We might then isolate the nitrogen compounds to determine their nature (e.g., in terms of basic, acidic, and neutral nitrogen compounds). In addition, we may want to identify different nitrogen compound types (e.g., pyridines, quinolines, indols, carbazoles, etc.), and to determine their molecular formula and carbon number distribution.

In other cases, we may have to prove the presence or absence of specific compounds, for example, of some of those that are restricted by environmental laws and regulations. This may require measuring their concentration down to a few parts per billion (ppb). The analytical approach may involve separation and preconcentration of the offending species prior to their identification while the bulk of the sample can be disregarded.

In choosing an analytical method, it is vital to establish what specific information is needed and how much more information additional data would contribute. Recently, Liguras and Allen (11) presented a pertinent discussion of this point which we outline in Chapter 10, Section IV.

C. A Matter of Interpretation

Mostly, we take for granted that the information we receive and, particularly, that which we pass on is exact and correct. If that were always true, there would be less disagreement between researchers. In fact though, our experimental data are often incomplete, and we tend to extrapolate or overinterpret them on the basis of our experience and intuition. This may be useful, but only if we make it clear when we begin to speculate.

More difficult is the case in which a method only appears to give exact data when, in reality, it does not. Typically, this happens at the forefront of research. Startling new results call for caution and verification, preferably by other independent methods.

The reason for our disagreements is usually our attempt to understand and explain our world in spite of insufficient data, of incomplete information. For the dearth of data, we try to compensate with rational thought and intuition. The more we work in the forefront of research and the scarcer established, certain knowledge is in a particular field, the more we must use our intuition and the likelier we will interpret data differently from other workers. This is good. This is how we progress.

However, if we disagree in our interpretations because we do not use the same basic definitions, progress is impeded. In petroleum chemistry, with its immense complexity, we will be better off using exact definitions, precise and common terminology, and appropriate categorization. In this book, we attempt to use clear terms and logical categorizations of petroleum constituents (based on chemical composition and boiling range), and hope thereby to help avoid disagreements based on poor definitions. Then any differences in interpretation persisting in spite of clear terminology will, hopefully, lead to further progress.

We also attempt to clearly indicate what is truly known and what, at this point, has to be viewed as hypothetical. This too, we hope, will contribute to a greater understanding and to faster progress in our field.

VIII. OUR STRATEGY AND BIAS

Our strategy (12–14) is based on the view that compositional trends in fractions of increasing boiling point are continuous and that this continuity extends even to nondistillable residues. It incorporates separation by complementary methods and selective molecular characterization. Recognition of this continuity led one of us (12) to propose the atmospheric equivalent boiling point (AEBP) concept which is discussed at some length in the following chapter. It includes not only the customary boiling-point range of distillates, but also a hypothetical AEBP range of nondistillable residues. The latter is calculated from molecular weights and densities or from MW and H/C ratios by means of modified Winn equations (14).

The continuity in composition and AEBP then extends across the entire range of a crude oil, from the lowest boiling species to the most complex high-MW residue component. As the boiling point of the petroleum fractions increases, so does their complexity in terms of molecular-weight range, molecular-type range, and molecular structure. The increase in complexity is gradual, not abrupt. Thus, by studying simpler fractions, we can anticipate and plan for the analysis of the more difficult fractions.

The continuity concept in conjunction with the AEBP concept has two advantages. First, a plot of fraction mass versus AEBP is an important piece of information in itself. It allows comparing different crude oils on a common and rational basis as shown in Chapter 4. Second, it emphasizes the importance of determining the AEBP range of petroleum fractions before deciding on the molecular characterization approach.

Figure 2.9, together with Table 2.3, illustrates how column and short-path distillation and solubility fractionation [by sequential elution fractionation (SEF)] fit into the primary fractionation scheme for an entire crude oil. It also outlines further steps for molecular characterization of the fractions.

This, then, is our bias. It is reflected in our approach to compositional analysis.

1. The primary objective of our analytical efforts is a better understanding of the chemistry of petroleum refining processes.

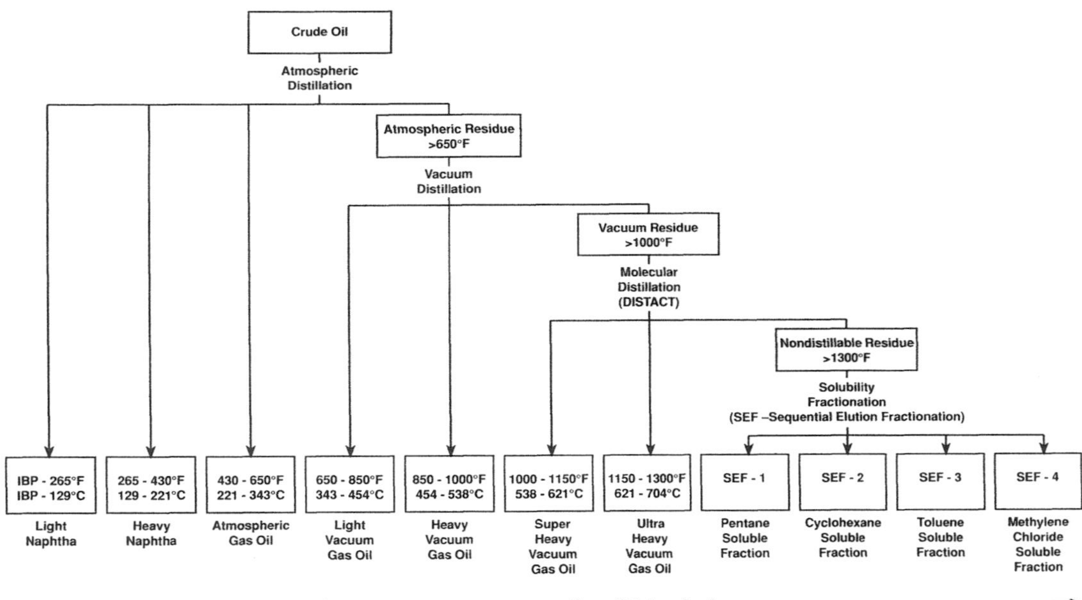

Figure 2.9 Comprehensive compositional analysis of heavy petroleum fractions. Major steps in comprehensive compositional analysis include (1) atmospheric equivalent boiling-point (AEBP) distribution; (2) elemental analysis (C, H, S, N, O, V, Ni, Fe); (3) HPLC compound-class separations; and (4) spectroscopic characterization.

Table 2.3 Overview of the Main Characterization Methods for Heavy Petroleum Fractions

Main objective	Methods
1. The AEBP distribution	
Distillate fractions	Simulated distillation (e.g., HTGC, SFC, VTGA)
2. Elemental analysis	
Distillate fractions	(C, H, S, N, O, V, and Ni)
Nondistillable fractions	(C, H, S, N, O, V, Ni, Fe) and SEC-ICP analysis (V, Ni, S)
3. HPLC compound-class separations	
Distillate fractions	Various HPLC methods (choice depends on task and AEPB range)
Nondistillable fractions	Very few applicable methods, e.g., API-60 ABN separation (most HPLC methods give poorly defined fractions and low sample recovery)
4. Characterization of the HPLC compound-class fractions in terms of *Z* series and their carbon-number distribution	
Distillate fractions	FIMS and LVHR-MS [routine for the 650–1000°F (345–540°C) AEBP distillates, experimental for the 1000–1300°F (540–700°C) AEBP distillates]
Nondistillable fractions	Beyond the range of current MS methods
5. Complementary molecular characterization by spectroscopic methods	
Distillate fractions	^{1}H-NMR, ^{13}C-NMR, FTIR, UV-VIS, fluorescence, GC/MS
Nondistillable fractions	^{1}H-NMR, ^{13}C-NMR, FTIR, XAS, XDS, ESCA, EPR (some limitations due to limited sample solubility)

2. We recognize the great complexity of petroleum fractions and the continuity in its increase with increasing boiling point (AEBP).
3. We stress the importance of crude oil fractionation by column and short-path distillation and, finally, by solubility (SEF) fractionation to produce fractions which are categorized by their AEBP ranges, and which are easier to analyze than the unfractionated crude oil.
4. We emphasize the importance of chromatographic (HPLC) separations to produce chemically meaningful fractions for further molecular characterization by spectroscopic methods.
5. We recognize that knowledge of the composition of lower-boiling petroleum fractions facilitates the analysis and data interpretation of higher-boiling fractions.
6. We agree with Aczel et al. (15) who, reporting on the characterization of coal liquids, remark: "Selection of the best analytical strategy for the analysis . . . depends on the nature of the information needed, the availability of analytical tools, funding . . . , and time." Indeed.
7. Last but not least, we try to mitigate the confusion caused by traditional terminology. This includes the two temperature scales presently in use. Because it is unlikely that the Celsius scale will be fully adopted by the petroleum industry in this country within the next 10 years, we have reluctantly chosen the Fahrenheit scale as the primary one for our book, but we also give the temperature readings in degrees Celsius wherever possible.

REFERENCES

1. Green, J. B., Hoff, R. J., Woodward, P. W., and Stevens, L. L. 1984. Separation of liquid fossil fuels into acid, base, and neutral concentrates. 1. An improved nonaqueous ion exchange method. *Fuel*, **63**:1290–1301.
2. Boduszynski, M. M. 1988. Composition of heavy petroleums. 2. Molecular characterization. *Energy Fuels*, **2**:597.
3. Szakasits, J. J., and Robinson, R. E. 1991. Hydrocarbon type determination of napthas and catalytically reformed products by automated multidimensional gas chromatography. *Anal. Chem.*, **63**:114–120.
4. Johansen, N. G., Ettre, L. S., and Miller, R. L. 1983. Quantitative analysis of hydrocarbons by structural group type in gasolines and distillates. 1. Gas chromatography. *J. Chromatog.*, **25**:293–417.
5. Whittemore, I. M. 1979. High resolution gas chromatography of the gasolines and naphthas. In *Chromatography in Petroleum Analysis*, edited by K. H. Altgelt and T. H. Gouw, Marcel Dekker, New York, Chap. 3, pp. 41–74.
6. Naber, J. E., Stork, W. H. J., Blauwhoff, P. M. M., and Groeneveld, K. J. W. 1991. Technological response to environmental concerns, product quality requirements and changes in demand patterns. *Erdoel Kohle*, **107**:124–132.

7. Altgelt, K. H. 1992. Manuscript in preparation.
8. Rogers, L. B. 1990. The new generation of measurement. *Anal. Chem.*, **62**:703A–711A.
9. Teeter, R. M. 1985. High resolution mass spectrometry for type analysis of complex hydrocarbon mixtures. *Mass Spectrom. Rev.*, **4**:123.
10. Bouquet, M., and Brument, J. 1990. Characterization of heavy hydrocarbon cuts by mass spectrometry. Routine and quantitative measurements. *Fuel Sci. Tech. Int.*, **8**(9):961.
11. Liguras, D. K., and Allen, D. T. 1989. Structural models for catalytic cracking. 2. Reactions of simulated oil mixtures. *Ind. Eng. Chem. Res.*, **28**:674.
12. Boduszynski, M. M. 1987. Composition of heavy petroleums. 1. Molecular weight, hydrogen deficiency, and heteroatom concentration as a function of atmospheric equivalent boiling point up to 1400°F (760°C). *Energy Fuels*, **1**:2.
13. Boduszynski, M. M., and Altgelt, K. H. 1992. Composition of heavy petroleums. 4. Significance of the extended atmospheric equivalent boiling point. *Energy Fuels*, **6**:72–76.
14. Altgelt, K. H., and Boduszynski, M. M. 1992. Composition of heavy petroleums. 3. An improved boiling point–molecular weight relation. *Energy Fuels*, **6**:68–72.
15. Aczel, T., Colgrove, S. G., and Reynolds, S. D. 1985. High resolution mass spectrometric analysis of coal liquids. *Amer. Chem. Soc. Div. Fuel Chem.*, **30**:209.

3

Crude Oil Distillation and Significance of AEBP

With a contribution by T. H. Gouw

I. SIGNIFICANCE OF DISTILLATION

A. Distillation as a Principal Separation Method

In this chapter, we discuss in greater detail some of the issues touched on in the previous chapter. Restating our premise, distillation is the most basic and important separation process in refineries. Crude oils, virtually all the intermediate streams, and most of the products—from gasoline to lube oil base stocks—go through distillation columns. We, as many others, think distillation should have a similar place in the laboratory of the analytical petroleum chemist in that it should precede any other separation methods in most cases. There are several reasons for this position.

1. The separation of a crude oil by distillation into various cuts provides a very important set of data in its own right.
2. It reduces the enormous number of different molecules and, thus, simplifies the interpretation of compositional analyses performed on them.
3. The division of a crude oil into a number of narrow cuts eases their comparison with other crudes and with coal liquids or shale oils. With-

out prior distillation into specific cuts, comparisons of such diverse samples have much less meaning.

4. When looking for the changes in composition from feed to product in refinery processes, the subdivision of both into several cuts makes the comparison more detailed and meaningful.
5. The study of low-boiling cuts increases our understanding of the less tractable high-boiling fractions and helps in planning their analysis.

Distillation reduces both the molecular-weight range and the variety of chemical groups. Take naphtha as an example with its boiling range between 85 and 430°F (30 and 220°C). Its carbon numbers range from 4 to 11 or 12, and it contains paraffins, naphthenes with up to two or three rings, and monoaromatics with up to one naphthenic ring. Its upper boiling limit prohibits it from containing naphthalenes, let alone any aromatics with higher ring numbers.

Middle distillates have a boiling range from about 430 to 650°F (220 to 345°C). Their carbon numbers range from about 10 to 20, excluding not only molecules with higher carbon numbers but also those with lower ones. Their diversity is greater than that of naphthas, but still quite limited. With rising boiling point, the variety of the high-boiling end, especially that of heterocompounds, increases rapidly, but more and more of the low-boiling species, which make up the bulk of a crude oil, are excluded.

B. Boiling Point and Molecular Weight

The relation among carbon number (or molecular weight), boiling point, and chemical composition was already introduced briefly in Chapter 2, Section IV. Here we elaborate on the concept. The graph is shown—once more—in Fig. 3.1. We recognize the wedge-shaped area with the upper left boundary defined by the straight chain alkanes and the lower right curve by the—mostly hypothetical—unsubstituted polar compounds. The *n*-paraffins have the highest carbon number and molecular weight for a given boiling point.* Or, looking at it the other way, for a given carbon number, they have the lowest boiling points. (Strictly speaking, some isoparaffins have even lower boiling points, and these should bound the area on the left-hand side. But for the sake of simplicity, we chose the *n*-paraffins.) Naphthenes boil at somewhat higher temperatures and are,

*Only the boiling points up to C_{17} are experimental values. The higher ones, up to C_{100}, were calculated on the basis of empirical relations (API Research Project 44 Loose Leaf Data Sheets, 1972; see also Ref. 2). In our figures, we used the commonly accepted API values for the higher boiling points although the values obtained by Glinzer's equation seem to be better.

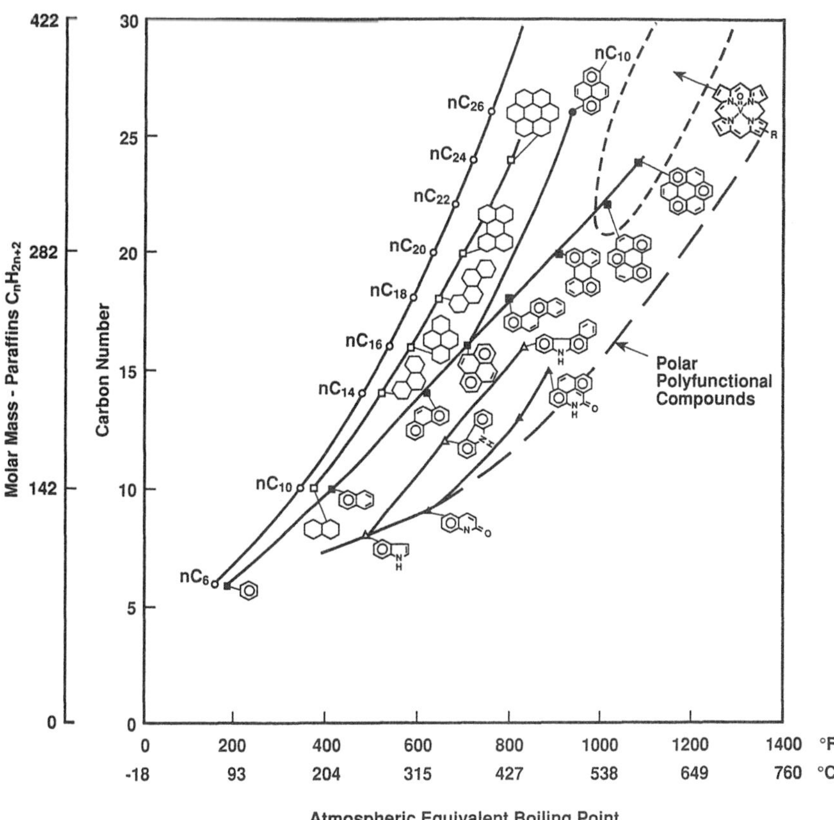

Figure 3.1 The effect of molecular structure on boiling point. (From Ref. 1.)

therefore, located to the right of the *n*-paraffins in the plot. Aromatic hydrocarbons have higher boiling points yet and are positioned even farther to the right. The parent aromatic hydrocarbons, i.e., those without alkane substituents, neatly fall on a common curve in our plot. Alkane substitution, as in the example of *n*-decylpyrene, raises the molecular weight more than the boiling point and, thus, elevates the corresponding point above the parent curve. Aromatic polar species have their own curves to the right of the aromatic hydrocarbon curve. The shift is greater, the higher the aromaticity and the degree of polarity. Condensed polyaromatic molecules with several polar groups and no alkyl chains have the highest boiling points for a given molecular weight. Their points are the farthest to the right and form the bottom curve in Fig. 3.1. Because of insufficient boiling point data, this curve is hypothetical beyond the point belonging to 2-quinolone.

It is worth repeating also this statement: As Fig. 3.1 shows, diverse compounds with similar molecular weights cover a broad boiling range; conversely, a narrow distillation cut can contain a wide range of molecular weights. This important fact is caused by the different intermolecular forces which affect the heat of vaporization and, thus, the boiling point of a compound. For a homologous series of alkanes, the weak Van der Waals dispersion forces prevailing between the molecules become greater in proportion with increasing carbon number. Compounds with aromatic rings have additional attractive intermolecular forces acting on them and, therefore, have higher boiling points than aliphatic molecules of similar molecular weight and structure. The intermolecular forces are even stronger in polar compounds capable of hydrogen bonding or other types of polar interactions.

The effect of distillation can be visualized by cutting the plot into vertical slices as we have done in Fig. 3.2. In reality, of course, the distillation cuts are not separated nearly as cleanly as in these slices. Indeed, the overlap between cuts is substantial. However, the principles of our discussion are still correct. Distillation obviously limits the molecular-weight range of each compound type in a slice, but the molecular-weight range of one compound type in the mixture, say that of *n*-alkanes, can be quite different from that of another, e.g., the aromatics. On the other hand, a narrow distillation cut—a slice in Fig. 3.2—contains only a limited selection of different chemical species. Because each of these species comprises a small range of molar masses, each of which is different from that of other species, it is easier to distinguish the different compound classes in a narrow distillation cut than in a wider one or, certainly, than in a whole petroleum. Another advantage of distillation is that it eliminates or mitigates the interactions between large and small molecules which could interfere with separations by chemical nature at a later stage.

II. DISTILLATION METHODS

A. Atmospheric and Vacuum Distillation

1. Principles of Distillation

Distillation is a tool to separate molecules primarily by the differences in their vapor pressure. The vapor pressure decreases with increasing MW, aromaticity, and polarity. It is inversely proportional to the boiling point.

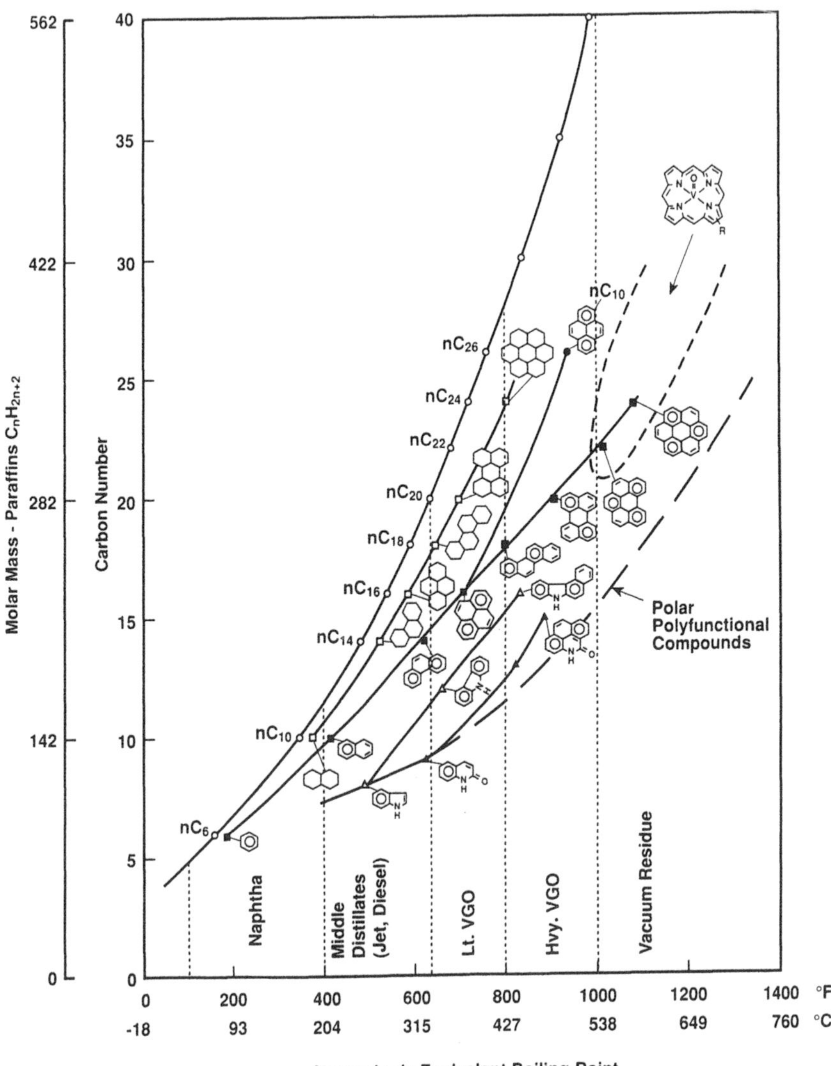

Figure 3.2 Distillation cuts in the MM–AEBP plot.

Among the prevalent distillation methods, we can distinguish between two major categories: column distillation and short-path distillation. Column distillation is performed at high pressures, at atmospheric pressure, and at reduced pressures. High pressures are used mainly in large-scale refinery distillations with low-boiling distillates. They result in higher distillation temperatures and save the cost of using refrigerants for condensor cooling. This technique is not of interest for our book. Here we restrict our discussion to laboratory distillations.

In short-path distillation, also known as molecular distillation, a very high vacuum ($\leq 10^{-3}$ Torr) is applied. The sample passes rapidly as a very thin film over a heated surface. The lighter molecules evaporate and are condensed on a cooled surface which is located within 2–3 cm of the condensing surface. The vacuum must be high enough to ensure that the mean-free-path length of a distillate molecule is shorter than the distance between the heated and the cooled surfaces.

Column distillations at atmospheric pressure and under medium or high vacuum are normally conducted in the batch mode. Before the distillation, the feed is placed in a reboiler attached to the bottom of the column, and an overhead product is obtained which becomes progressively higher boiling and more complex in composition. Short-path distillation, on the other hand, is carried out in the continuous mode, and the composition of the "overhead" remains the same.

The bulk of the distillations performed in the laboratory are carried out in packed columns. The efficiency of a distillation column is measured in terms of its number of theoretical trays or plates. The higher this number, the higher its efficiency. Generally, the number of theoretical plates is proportional to the column length. For most packings in laboratory distillations, the height of a theoretical plate is roughly equal to the diameter of the column. The standard distillation column has a diameter of about 25–60 mm, a height of 1–1.5 m, and an efficiency of 15–50 theoretical plates. Occasionally, one may need taller columns or columns with a special high-efficiency packing delivering as many as 70–100 theoretical plates. These are rather unusual, though.

Of special interest in the distillation laboratory are the spinning band columns. These are constructed of precision-bore glass tubing 1–20 mm in diameter and 30–100 cm in length, in which a closely fitted twisted Teflon or metal band rotates at high speed. The centrifugal force exerted on both the ascending vapors and the descending reflux brings both into intimate contact on the column walls, achieving the high mass and heat transfer required for good efficiencies. The major advantage of these columns is that they have very low pressure drops. They can be used for high-vacuum distillations of oils (with low or zero amounts of nondistillables) up to boiling points of about 800–900°F (430–480°C).

In comparison to separation efficiencies of tens of thousands of plates in chromatography, the 20–100 theoretical plates in distillation do not appear impressive. However, these efficiencies cannot be compared on a one-to-one basis because the theoretical-plate concept in distillation is somewhat different from that in chromatography. In chromatography, the sample solution and eluent move in the same direction (concurrent process), whereas in distillation, the liquid reflux goes in the opposite direction of the ascending vapors (countercurrent process). Obviously, the countercurrent process is much more efficient. As a rule of thumb, a distillation column of n theoretical plates has about the same efficiency as a chromatographic column of n^2 plates. Thus, a 100-plate distillation column has a resolution comparable to that of a 10,000-plate chromatographic column.

A theoretical plate in distillation is a hypothetical section of a column which produces the same difference in composition of the ascending distillate as exists at equilibrium between a liquid mixture and its vapor. It acts as an ideal bubble-cap tray would. An easy way to understand its principle is to see what happens to a binary mixture in a packed column. Assume the column is composed of sections, each representing a theoretical plate (TP). The packing provides a large surface area for the descending reflux. As the rising vapor comes in contact with the descending liquid, some of its higher-boiling component transfers to the reflux, and some of the lower-boiling component transfers from the liquid to the vapor. Therefore, the vapor arriving at the top of a TP section contains a lower amount of high-boiling material than it had when it came in at the bottom, and its lighter component is correspondingly enriched. On the other hand, the reflux leaving the theoretical plate at the bottom contains a higher amount of high-boiling material than when it came in at the top. Now we compare the composition of the liquid phase at the top with that of the liquid phase at the bottom. If the section indeed has the length of a theoretical plate, then the composition of the liquid phase at its top is equal to the (theoretical) vapor composition in equilibrium with the liquid at the bottom.

If the ratio of the lighter to the heavier component in the vapor is equal to R_v, and the same ratio in the liquid phase is R_l, then we can define an enrichment factor of the process, R_v/R_l. This factor is equal to the relative volatility of these two components, α, which, in turn, is roughly equal to the ratio of their partial pressures. Thus, under ideal conditions, the molar ratio of these two components in the vapor phase to that in the liquid phase changes by α when the vapor passes through one theoretical plate. It changes by α^2 after passage through two theoretical plates and by α^n after n theoretical plates.

Another common measure of column efficiency is the height equivalent to a theoretical plate (HETP). High efficiency corresponds to a high number of theoretical plates and a low plate height. In the petroleum laboratory, typical distillation columns have HETPs of around 5 cm (2 in.), although specially packed columns with HETPs of as little as 1–2 cm are available. Here, it may be useful to question why low HETPs are desirable because one can construct a column with any desired number of theoretical plates by making it tall enough. The main problem here is that larger columns require more stock because the column holdup is also so much larger. Large column holdups do not only cause sample losses but they also have a negative effect on overall column efficiency and performance.

Another undesirable feature of very tall columns is a considerable pressure drop from bottom to top. High pressure drops are undesirable because they increase the pressure in the reboiler and, hence, its temperature. This can be of serious concern in the distillation of thermally labile compounds or high-boiling feeds. The pressure drop limits the depth to which one can cut into the feedstock.

One may now conclude that low-HETP columns must be the right choice. However, low-HETP columns often have high pressure drops per theoretical plate. The ideal packing has both a low pressure drop and a low HETP. Such packings are hard to make and very expensive.

An important factor affecting the operational efficiency of a column is the reflux ratio employed. In general, HETPs and the corresponding number of theoretical plates are reported for the performance of a distillation column at total reflux, i.e., at zero throughput. As distillate is collected, the efficiency of the column may drop dramatically. A column with an infinite number of plates will only show a separation efficiency of slightly more than two plates when operated at a reflux to distillate ratio of 1:1 (i.e., 50% takeoff). A 10-plate column operated at a reflux of 10:1 shows a separation efficiency of 6–7 theoretical plates.

In refinery operation, there is a tendency to construct columns with a large number of theoretical plates and to operate them at a low reflux ratio to reduce heating and cooling costs. In the laboratory, one would tend to use less-efficient columns to reduce column holdup and to compensate by running the unit at a higher reflux ratio. This, however, reduces the throughput. At a reflux to distillate ratio of $R:1$, each distillate molecule travels the column $R + 1$ times up and R times down on average before it is collected. Hence, a balance must be found between sufficiently high reflux ratios for good efficiency on the one hand and low enough reflux ratios for reasonable throughput rates on the other. Reflux ratios of 5:1 to 20:1 are commonly used in the laboratory.

Let us assume a column operating at an efficiency of 10 theoretical plates. With this column, we could expect separations close to α^{10}. Let us assume we collected a distillation cut boiling between 450 and 550°F (235–290°C) from this column. Let us further assume that the midpoint component of this cut has a partial pressure (volatility) of 375 Torr and its highest-boiling component has one of 200 Torr. For this mixture, we can expect a final molar ratio of these two components (midpoint and high-boiling) close to $(375/200)^{10} = 537$. This is a decent though not a great separation. Or assume an equimolar mixture of four *n*-alkanes: tridecane (bp = 235.5°C = 456°F), tetradecane, pentadecane, and hexadecane (bp = 287°C = 549°F), which comes close to the boiling range of our first example. In a column of 10 theoretical plates, the separation between each of the 4 components would be $\alpha^{10} = \sim 70$. In a column of only 5 theoretical plates, the most common type, the separation between these components would be only about 8, i.e., the molar ratio of tridecane to tetradecane at the nominal boiling point of tridecane would be only 8:1. This result explains the substantial overlaps we observe in distillations.

2. Vacuum Distillation

For practical purposes, one should refrain from exceeding 650°F (345°C) in the reboiler. At this temperature, some of the petroleum components begin to decompose because the energy needed for the evaporation of a molecule exceeds the energy required to break C–S, C–N, or C–C bonds (about 60–80 kcal/mole, depending on structure). Cracking or coking may result at these higher temperatures, especially with prolonged heating. The temperature of prime interest is the cutpoint temperature of the distillate. There is no direct relationship between the reboiler and the actual distillate temperature, except that the latter is lower than the former. In distillations at atmospheric pressure, distillate cutpoints at 650°F (345°C) reboiler temperature can range from about 210°F to 570°F (100 to 300°C) depending on the stock and the distillation column employed. In general, the difference between distillate and reboiler temperature increases with increasing MW, aromaticity, and heteroatom content of the material in the reboiler and with increasing pressure drop in the column.

Vacuum distillation allows us to obtain higher-boiling distillates and to cut deeper into the stock without decomposition. Operational pressures in vacuum distillation with packed columns generally range from 5 to 200 mm Hg. At lower pressures, the vapor velocity increases, resulting in greater pressure drops. As a consequence, eventually the pressure in the reboiler will be reduced only marginally so that there is little advantage in going to pressures below 5 mm Hg.

For lower pressures yet, one needs an empty column or a spinning band column. With these, the pressure drop is small enough that low pressures can be maintained in the reboiler. For instance, with a distillate pressure of 0.1 mm Hg, one may have a pressure of 0.5–1 mm Hg in the reboiler which, theoretically, allows the collection of distillates with an atmospheric equivalent temperature cutpoint of as high as 1050°F (560°C). The actual observed boiling points during this distillation are, of course, much lower. Despite the low pressure, the reboiler may have to be heated as high as 700°F (370°C) for such high-boiling distillates.

From the actual boiling points, obtained at reduced pressure, the so-called atmospheric equivalent temperatures (AET) are calculated at which the material would boil under atmospheric pressure if it were stable and would not decompose. The concept of AET is an ingenious technique which extends the boiling point range of atmospheric distillation beyond the decomposition limit. It provides a common basis for the categorization and direct comparison of petroleum components across the entire volatility range accessible by atmospheric and vacuum distillation.

Vacuum distillations, at maximum throughput in a given still, are slower than atmospheric distillations. As a rule of thumb, distillation rates are inversely proportional to the square root of the vacuum measured in Torr. Thus, a distillation performed at a vacuum of 1 Torr takes roughly 25–30 times as long as one made at atmospheric pressure (760 Torr) with the same reflux ratio. On the other hand, relative volatilities are higher, and separations therefore better, at reduced pressure (3). AETs can be estimated from the actual boiling points obtained under vacuum by the formula (4):

$$\text{AET}(^\circ\text{C}) = \frac{(748.1 \times A)}{1/(\text{VT, K}) + (0.3861 \times A) - 0.00051606} - 273$$

$$A = \frac{5.9991972 - (0.9774472 \times \log P)}{2663.129 - (95.76 \times \log P)}$$

where VT, K is the observed vapor temperature in degrees Kelvin, and P is the pressure of the system in mm Hg observed when the vapor temperature was read. AETs can also be looked up in tables of the Annual Book of ASTM Standards (5).

Of interest is the "true boiling point" (TBP) distribution of crude oil as determined by ASTM D 2892.

B. Short-Path Distillation

As an alternative to empty column or spinning band distillations or to cut even deeper into the stock, high-vacuum short-path distillation can be used. Other names for this technique are molecular and sometimes wiped-film distillation although the latter is often performed with insufficient vacuum and belongs then to a different category. Short-path stills are of a completely different design than the standard distillation columns (30). The differences are listed in Table 3.1.

The most important feature of the short-path still is a very high vacuum of at least 10^{-3} mm Hg. It ensures that the mean-free-path length of the

Table 3.1 Differences Between High-Vacuum Short-Path and Conventional Packed Column Distillation

	High-vacuum short-path still	Packed column
Operational mode	Continuous	Batch
Pressure	0.001 Torr	0.5–760 Torr
Feed heating	Thin film	Reboiler
Heating duration	10–60 s	1–10 h
Distance between heated and condensing surfaces	1.5–3 cm	50–150 cm
Column packing	None	Packing
Maximum cutpoint, AET	1300°F (700°C)	1040°F (560°C)
Separation principle	Kinetic	Phase equilibrium

molecules in the gas phase is approximately 2–3 cm, which is the distance between the evaporator and the condensor surfaces. These conditions afford much lower distillation temperatures than possible in regular open columns at the same surface temperature. The sample is spread into a very thin film on the evaporating surface for quick evaporation and a short residence time. The combination of high vacuum, short distance, and short residence time allow very deep distillation without decomposition. Modern versions of these short-path stills can fractionate oils up to atmospheric euivalent boiling points of 1300°F (700°C) with fairly high throughput rates. The small DISTACT laboratory short-path still, for example, has a rate of about 100–800 mL/h with a residence time of less than a minute. Large production plants can operate with throughputs of up to 300 L/m^2/h.

In contrast to column distillation, short-path distillation is a nonequilibrium process. Boiling points cannot be measured during the process. They must be determined afterward by simulated distillation of the product.

In short-path stills, the preheated sample is pumped at a precisely regulated rate onto the surface of a heated vertical cylinder in the evaporation chamber where it is spread by wipers into a uniform, thin, falling (downward flowing) film. Figure 3.3 shows a diagram of such a short-path still. The lower-boiling components are flash-evaporated and collected on a condenser, typically a coil inside the feed cylinder. Only two fractions result, the flash distillate and the residue.

C. Simulated Distillation

Regular distillation requires relatively large sample sizes and long operation times. For example, the ASTM "true boiling point" method D2892 is a 100-h procedure. Simulated distillation (SIMDIS) can greatly save on both

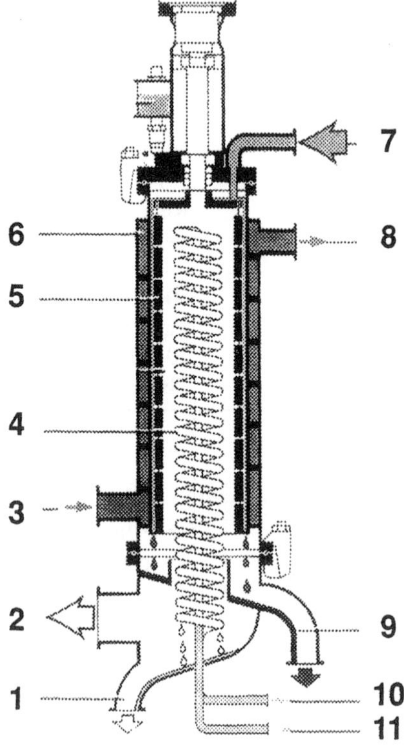

1 Distillate Discharge
2 Exhaust Vacuum Port
3 Heating Inlet Fluid
4 Internal Condensor
5 Roller-Wiper System
6 Heater Jacket
7 Feed
8 Heating Outlet Fluid
9 Residue Discharge
10 Cooling Water, Inlet
11 Cooling Water, Outlet

Figure 3.3 Diagram of a short-path still.

accounts, and it has the additional advantage of being more precise and covering even higher temperatures than vacuum distillation (in columns). SIMDIS is ordinarily performed by gas chromatography (GC) which can handle samples up to about 1000°F (540°C) (6). For very high-boiling samples, vacuum thermal gravimetric analysis (VTGA), high-temperature GC, and supercritical fluid chromatography (SFC) have been used.

GC-SIMDIS is a well-established method and has been adopted by the ASTM since 1973 (D2887). It is based on the observation that hydrocarbons are eluted from a nonpolar column in order of their boiling points. The nonpolar packing in this case is coated with silicone gum. The retention times can be calibrated in terms of boiling points with an upper limit of about 1000°F (540°C) AEBP. As Butler (6) stated, GC-SIMDIS uses the chromatographic column as a micro still. Minor irregularities, such as shifts in retention time values of aromatics versus paraffins of the same boiling points, are smoothed out by use of low-resolution columns. Occasionally, adjustments are applied for samples with high concentrations of aromatics and heterocompounds, e.g., for some heavy atmospheric and vacuum gas oils. The application of GC-SIMDIS to petroleum samples have been reviewed by Butler (6).

Glinzer (2) derived a highly accurate general relation for the calibration of GC columns with *n*-alkanes of up to 36 C atoms, with deviations from experimental values between 0.1 in the lower ranges and 1.6°C in the upper ones (400–500°C). The calibration is different for pure aromatic hydrocarbons, particularly for unsubstituted condensed polyaromatic compounds. However, with increasing side chain number and length the deviation diminishes rapidly to the point where polycyclic aromatic hydrocarbons (PAHs) such as those prevalent in crude oils follow about the same calibration as the paraffins. Figure 3.4 compares GC-SIMDIS boiling points of various unsubstituted PAHs with the API calibration curve for *n*-alkanes.

Higher boiling ranges can be covered by short, thin-film capillary columns (7–9). Upper AETs of 1470°F (800°C) were reached, but only with column temperatures of 800°F (430°C), i.e., under severe conditions likely to induce decomposition.

Another SIMDIS technique for high-AEBP fractions, vacuum thermal gravimetric analysis (VTGA), was applied to petroleum samples by Southern et al. (10), Mondragon et al. (11), Boduszynski (12), and Schwartz et al. (13). Here, a sample of 15 mg is placed on a pan in the sample chamber, which is then pumped down to a vacuum of 0.2–0.5 Torr. The pan is heated from room temperature to 400°C (752°F) at a rate of 5°C/min (9°F/min), and the weight loss is recorded as a function of temperature. The sample is exposed to temperatures above 345°C (650°F) for only 11 min to minimize decomposition. Calibration with *n*-alkanes and polyethylene fractions gave a smooth, almost linear calibration curve (13). However, in our experience, VTGA tends to give low values (by about 20°F) when applied to real samples such as short-path distillates of vacuum residues.

SIMDIS by supercritical fluid chromatography (SFC) is a third promising method for high-boiling fractions. It was introduced by Schwartz et al. (13–15a). Raynie et al. (16) tested various experimental conditions including two surface coatings and found conditions which gave much improved results for aromatics even over the accepted GC method. Distillation curves

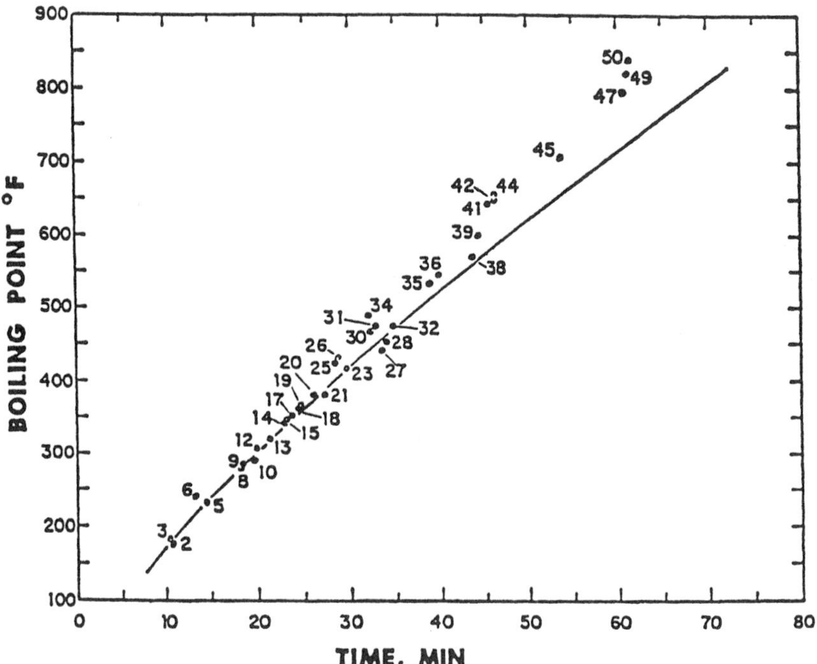

Figure 3.4 GC-SIMDIS boiling points of PAHs in relation to the *n*-alkane calibration curve (ASTM D 2887-84). (From Ref. 5. Reproduced with permission of the publisher.) A few of the compounds are identified as follows: 25—naphthalene; 30/31—methylnaphthalenes; 35—acenaphthene; 41—phenanthrene; 42—anthracene; 45—pyrene; 50—chrysene. The remainder can be looked up in the ASTM reference (5).

of high-boiling petroleum distillates by SFC matched quite well those determined by GC (ASTM D-2887) in the common boiling range except for the end point, but their temperature range extends up to almost 1400°F (750°C) versus 1100°F (600°C) for GC. Despite such promising results, SIMDIS by SFC still has to be considered experimental rather than routine, as must VTGA.

For SFC, an even smaller sample than 15 mg is sufficient. CO_2 serves as the mobile phase at pressures of 2000–5000 psi and at temperatures as low as 100–150°C. Capillary columns of 10 m length and 50–80 μm inner diameter coated with a 0.2–1.0-μm film are used which allow flow rates of about 5 cm/s. Data, acquired with a flame ionization detector, are processed by computer. Calibration of samples other than paraffins still seems to require some more work.

Figure 3.5 shows two thermogravimetric profiles of the same sample, an Arabian Heavy atmospheric residue. One (VTGA) was obtained under vacuum, the other (FTGA) under nitrogen flow at atmospheric pressure. Both curves have two maxima. The first maximum of each is caused by evaporation (distillation), the second one by decomposition. The different causes of the maxima are indicated by their temperature behavior. As expected for evaporation, the first part of the vacuum (VTGA) curve begins at a lower temperature than that of the nitrogen flow (FTGA) curve. However, the second maximum of the two curves occurs at the same temperature, indicating a cause independent of vacuum or flow conditions and suggesting decomposition as a logical explanation. Schwartz et al. (13) could prove that decomposition was indeed taking place by monitoring the gas phase with an on-line TGA-GC/MS technique. This experiment clearly demonstrates that, under the conditions of these tests, measurable decomposition sets in at the latest at a temperature of 370°C (700°F) and that it rapidly increases at higher temperatures.

An example will demonstrate how useful SIMDIS can be, especially when combined with a new evaluation technique (15). A heavy crude oil

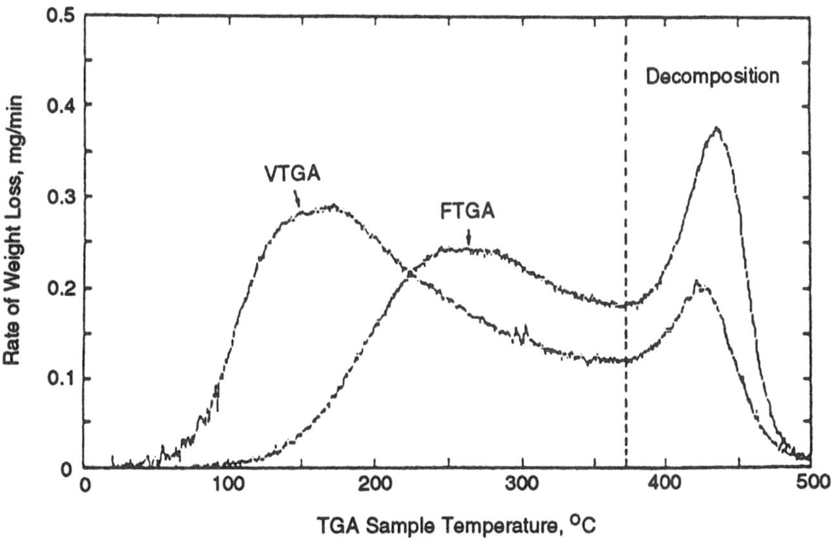

Figure 3.5 Comparison of vacuum and flow thermogravimetric profiles of an atmospheric residue. (From Ref. 13. Reproduced with permission of the publisher.)

was distilled into nine cuts and a residue according to the scheme shown in Fig. 3.6. The boiling curves of cuts 2–7 were measured by ASTM D 2887 (GC), those of cuts 8 and 9, by high temperature GC. The SIMDIS results were plotted in Fig. 3.7 as differential volume % (in the crude) versus boiling point. The large overlap revealed in this plot may seem discouraging. However, plotting the same data in a different way shown in Fig. 3.8, namely, the boiling point of all the SIMDIS results versus the cumulative liquid volume % (LV%) of crude, we see how a common, almost linear distillation curve evolves from the very nonlinear single curves of the nine original distillation cuts. This composite distillation curve (15) is much more precise than any one constructed from column distillation alone.

Interestingly, SIMDIS techniques not only have upper temperature limits but also lower ones. In GC-SIMDIS, the lower limit is determined by the resolution of the solvent peak from the first eluting sample peak. For this reason, the ASTM method D 2887 is restricted to samples of initial boiling points of at least 100°F (38°C). In the SFC method, the lower boiling limit is dependent on the initial column temperature and pressure. The lowest possible pressure of CO_2 is about 800 psi, namely, the vapor pressure in the supply cylinder at room temperature. However, for many high-boiling samples lacking low-boiling components, a higher initial pressure,

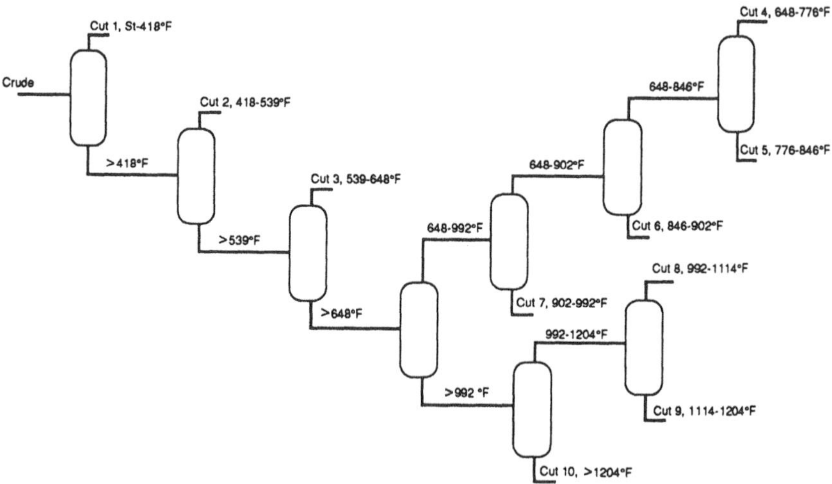

Figure 3.6 Distillation scheme for the separation of a heavy crude oil into nine distillation cuts and one residue.

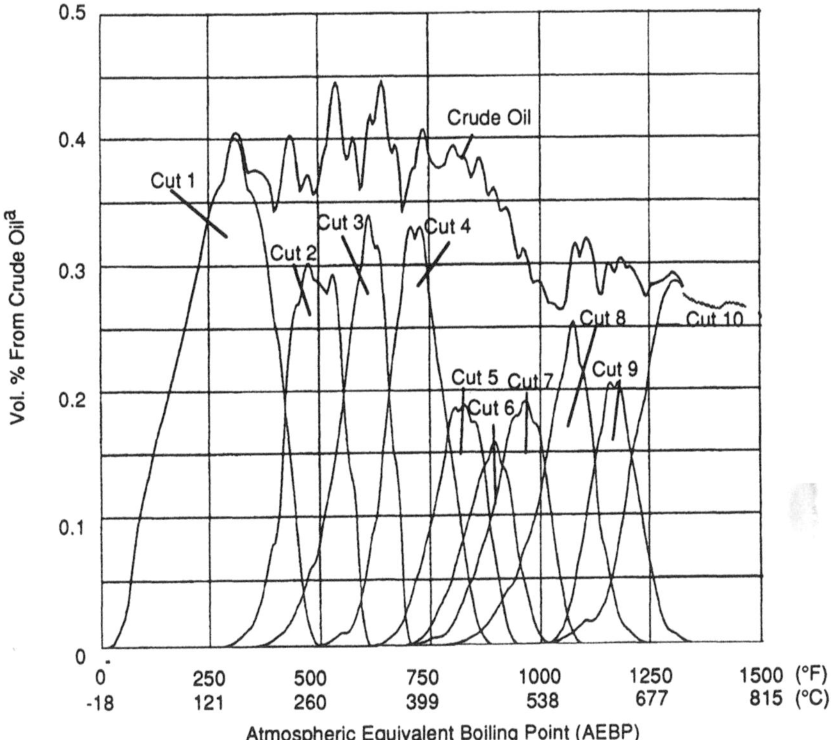

[a]5°F AEBP pseudocomponent.

Figure 3.7 Differential plot of LV% versus mid-boiling point of 10 fractions.

e.g., 850 psi, avoids unnecessarily long analysis times while providing the desirable resolution of the lower members. This restricts the lower boiling temperature of samples to be analyzed by SFC to about 250°F (120°C).

III. EQUIVALENT DISTILLATION

A. What Is Equivalent Distillation?

Earlier we discussed the central importance distillation holds in our view for the compositional analysis of petroleum, and the relations of boiling point with molecular weight on the one hand and with chemical structure on the other hand. We illustrated how the introduction of atmospheric equivalent temperature (AET) greatly expanded the basis for these rela-

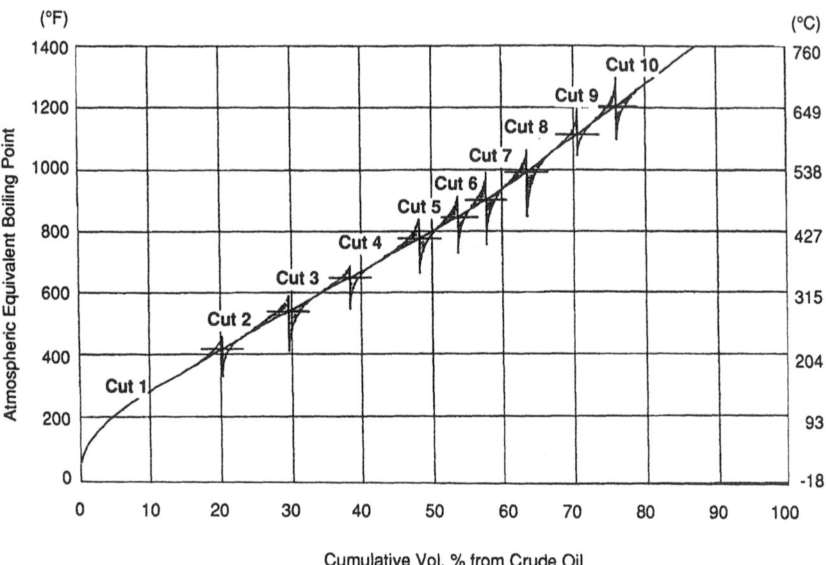

Figure 3.8 Common distillation curve as composite of 10 single (simulated) distillation curves.

tions by extending the upper limit of the boiling range from 650°F (345°C) to about 1300°F (700°C), the present limit of short-path distillation. But the sometimes large portion of nondistillables was excluded from this scheme.

Within the distillates, the physical and chemical properties are known to change only gradually with the boiling point, allowing interpolations and even extrapolations with reasonable certainty. Whenever the range of distillates was expanded, first by vacuum distillation, then by short-path distillation, it turned out that the new distillate portions of the previously "nondistillable" residues followed the same patterns as the previous distillates. Plots of C-number or molecular weight, or of sulfur and nitrogen concentrations, could be extended smoothly into the new regions. Boduszynski (12) argued that it would not only be highly desirable to find these gradual changes extended beyond the realm of distillates into that of residues, but that such continuity was also quite likely. If this continuity could be proven, we could then interpolate findings across the dividing line between distillates and residues and, hopefully, even extrapolate into those ranges that are still beyond exact measurements. Just as vacuum distillation extended the range or atmospheric distillates and led to the original concept of AET, a method of "equivalent distillation" could lead to an expanded concept of atmospheric equivalent temperature. Boduszynski (12) developed such a method. To make it very clear that this was a new concept, he gave the expanded boiling point scale a new name, "atmospheric equivalent boiling point" (AEBP).

B. Sequential Elution Fractionation, a Way to Perform Equivalent Distillation

Keeping in mind that distillation separates by molecular weight for a given class of compounds, Boduszynski (17) developed a separation method that does the same but is based on solubility instead of volatility. In his sequential elution fractionation (SEF), a nondistillable DISTACT residue is extracted in a column with four solvents of increasing solvent power.

First the residue is dissolved in methylene chloride and deposited on a 50/50 mixture of Teflon particles (30/60 mesh Chromosorb-T) and glass beads (40 mesh). (The glass beads keep the packing from collapsing during solvent flow.) A 1-g sample deposited on 200 g of this substrate leaves a layer of about 0.7 μm thickness on the particles. After drying under vacuum at room temperature, the coated particles are packed into an LC glass column (25 mm i.d. × 50 cm). In chromatographic apparatus, four solvents [*n*-pentane, cyclohexane, toluene, and a mixture of methylene chloride and methanol (4 : 1 v/v)] are pumped in sequence upward through the column at the fairly fast flow rate of 20 ml/min. Each of these solvents extracts one of the four fractions from the packing, beginning with the pentane-soluble SEF-1 to the methylene chloride-soluble/toluene-insoluble SEF-4 fraction. Usually, $<1\%$ of the sample remains on the column. As an illustration, the yields of SEF fractions from five nondistillable residues are shown in Table 3.2.

The underlying principle of both methods, distillation and SEF, are the various molecular interactions, called the Van der Waals forces. These consist of the dispersion (London) forces, permanent dipole–permanent dipole forces, and hydrogen-bonding forces. Whereas the latter two of these reflect the different chemical nature of the molecules in the mixture, the dispersion forces are roughly proportional to molecular weight. In reality, they are proportional to the number of interacting units, e.g., CH_2 groups, and, thus, for *n*-alkanes at least, to MW. For each homologous series of compounds, or for each type of compound, the dispersion forces at least directionally increase with molecular weight. As the dispersion forces increase, so does the boiling point and so does the solubility param-

Table 3.2 Yield of Nondistillable SEF Fractions from Atmospheric Residues of Five Crude Oils

	Yield from atmospheric residues (wt%)				
Fraction	Kern River	Arabian Heavy	Offshore Calif.	Maya	Boscan
SEF-1	16.0	15.2	15.7	14.7	24.3
SEF-2	5.7	8.2	9.5	6.6	13.1
SEF-3	3.1	3.1	10.7	13.0	10.7
SEF-4	0.5	0.5	1.1	1.3	2.0
Total	25.3	27.0	37.0	35.6	50.1

Source: Ref. 12. Reproduced with permission of the publisher.

eter (18,19), which governs the separation in liquid fractionations such as SEF. Therefore, *for each compound type*, the basis of both separation methods, distillation and solubility fractionation, is the molecular weight.

Although, in practical terms, for nondistillable residues it is no longer possible to distinguish and separate simple compound types as it is in distillates, the principle of equivalency in the two separation methods is maintained. This means, *if* these materials could be distilled, their distillation cuts would show the same compositional trends as the fractions obtained by SEF. Actually, much of SEF-1 can be "distilled" by SIMDIS. Figure 3.9 shows the VTGA-SIMDIS distillation curves of several fractions of an Arabian Heavy atmospheric residue (12), including that of a SEF-1 fraction. Despite substantial overlap between the cuts, there is a clear progression in AEBP from cut to cut, extending also to the SEF fraction. Furthermore, the similarity of the SEF curve with the distillates curves is striking.

In solubility fractionation methods, the molecular-weight discrimination hinges to some extent on the gradual rather than the abrupt change of solvent power and other experimental parameters. In extraction, for instance, it is important not to soak the sample, but constantly to remove the extract so as to maintain the proper solvent strength. For this reason, Boduszynski used a relatively high elution rate for his SEF method.

SEF is certainly not the only way to approach equivalent distillation and it is not the only extraction method proposed (see, for instance, Ref. 20), but for now it seems to be the best one because of its effectiveness and reproducibility. Supercritical fluid extraction might soon evolve as a good alternative.

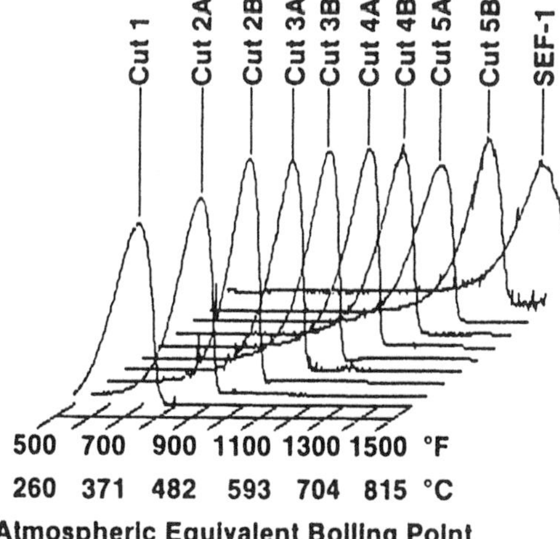

Figure 3.9 AEBP distribution curves measured by VTGA-SIMDIS on distillate cuts and a SEF fraction derived from Arabian Heavy atmospheric residue. (From Ref. 12. Reproduced with permission of the publisher.)

IV. THE CONCEPT OF AN ATMOSPHERIC EQUIVALENT BOILING POINT SCALE

A. Definition and Significance of AEBP

In the last section we described how the idea of equivalent distillation led to the concept of the atmospheric equivalent boiling point (AEBP). The AEBP scale encompasses the entire boiling range of a petroleum, starting with that of atmospheric pressure, continuing with those accessible by reduced pressures, and including the *equivalent boiling ranges* of nondistillable residue fractions. The plot of carbon number versus boiling point of Fig. 3.1 can now be enlarged to that of Fig. 3.10. By this stratagem of including nondistillable residue, an entire crude oil can now be described in terms of its various physical and chemical properties as they change with increasing boiling point, true and equivalent, on a common basis. This has at least two major advantages. First, it delineates the continuity of the changing composition from those boiling ranges that are easy to analyze—the light distillates—to those which are less easily accessible—middle distillates—to those which are only incompletely tractable—such as VGOs—and, finally, the most difficult ones—the vacuum residues. Such continuity of change is often a decisive clue in the interpretation of analytical measure-

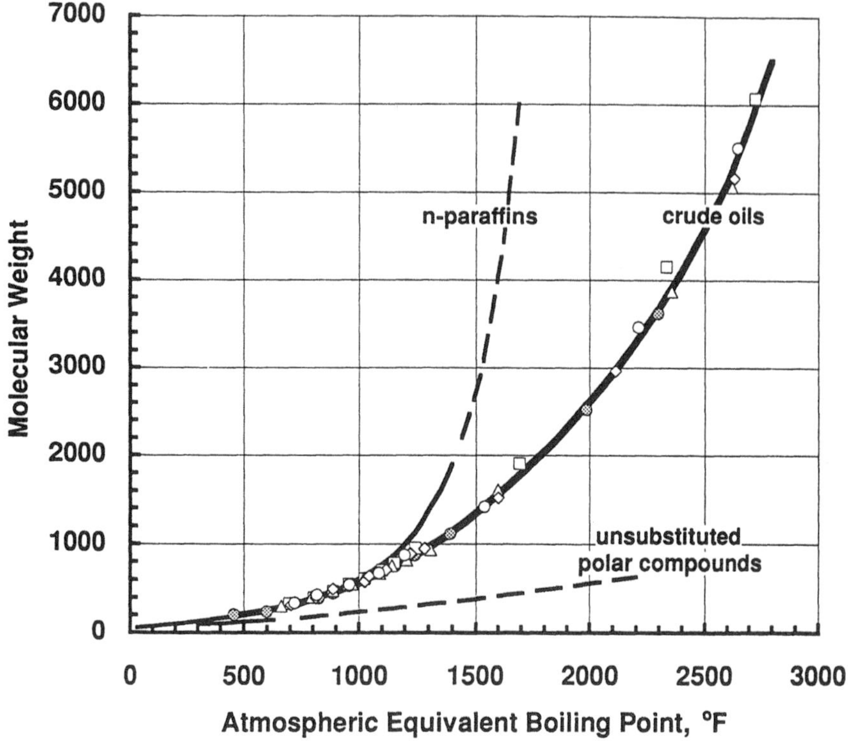

Figure 3.10 MW–AEBP plot for the entire Kern River crude oil in relation to *n*-paraffins and a largely hypothetical curve of unsubstituted, highly polar aromatic compounds (as in Fig. 3.1).

ments performed on high-AEBP fractions. To be able to clearly display and see the continuous change of a property is a distinct advantage provided by the concept of AEBP. The second great advantage of AEBP is the possibility to compare crude oils and crude oil fractions on a common, rational basis.

Even though each petroleum distillation cut consists of a mixture of compound types, each having a series of homologues with different substituents, any higher-boiling cut contains higher-molecular-weight members of each series than the previous lower-boiling cuts. Figure 3.11 illustrates this shift of MW with increasing AEBP. Showing the parent ion mass spectra of several Boscan crude fractions, five distillates and three SEF fractions, it demonstrates how the molecular weight, and even more so the upper MW limit, increases with the AEBP.

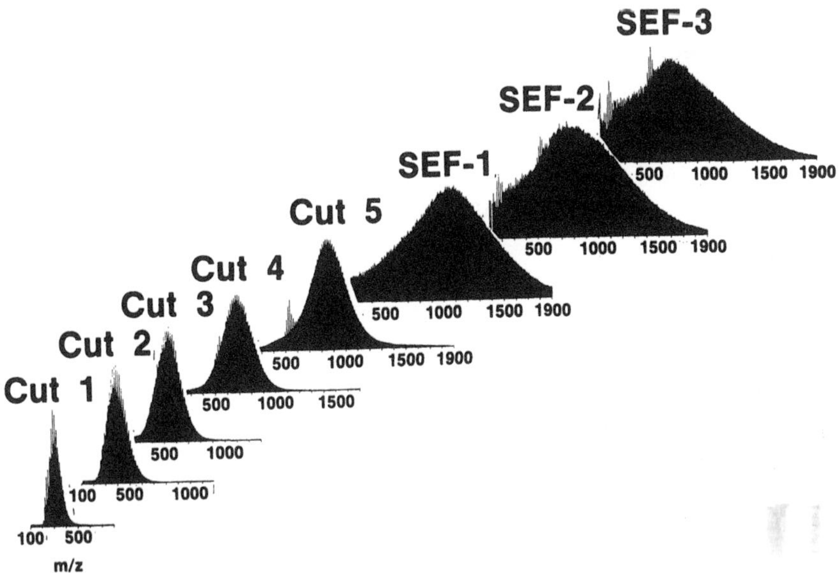

Figure 3.11 Molecular weight distributions by FIMS of short-path distillation cuts and nondistillable SEF fractions of Kern River crude oil. (From Ref. 12. Reproduced with permission of the publisher.)

Although for many purposes the boiling *range* of a cut is of primary interest, for others, especially for detecting and demonstrating trends, a single data point is preferable. A convenient single representative number for the boiling range of such broad mixtures is the mid-boiling point or, in our more general terms, the mid-AEBP. The mid-AEBP of a distillate is the temperature of the 50% mass point on a distillation curve which is best established by simulated distillation.

B. AEBP of Nondistillable Petroleum Fractions

Two questions arise at this point. What separation method is the best substitute for distillation? How are the mid-AEBPs of its fractions determined? We have already pointed out that Boduszynski's sequential elution fractionation method, SEF, comes very close to the requirement of separating by the same principles as distillation does. Being an extraction method, it does not suffer from the occlusion problems which asphaltene precipitation has. It is simple, reproducible, and can handle reasonably large samples (up to 1 g in a 20 × 1-in. column). On the other hand, we saw that the molecular weight is an important basis for the separation of both distillation and solubility fractionations. Thus, the second question above can be asked more specifically: How can we convert the MWs* of the SEF fractions into mid-AEBPs?

Several equations relating mid-boiling point with MW and specific gravity of petroleum or synfuel fractions exist (21–25). Of these, Altgelt and Boduszynski showed their combination of Eqs. (3.1) and (3.2),

$$\mathrm{MW} = 140 + 3.40 \times 10^{-7}\left(\frac{\mathrm{AEBP}^3}{\mathrm{SG}^{2.5}}\right) \quad \text{for AEBP} > 500°\text{F } (260°\text{C}) \tag{3.1}$$

$$\mathrm{MW}_{<500°\mathrm{F}} = \mathrm{MW}_{>500°\mathrm{F}}\,[1 - (600 - \mathrm{AEBP})1.4 \times 10^{-4}] \tag{3.2}$$

to be the best one. When tested with a set of model compound hydrocarbons published by the Thermodynamic Research Center (TRC) at the Texas A&M University (1989), it showed a linear response and little sensitivity to hydrocarbon type. Its application to lube oil fractions (29) and coal oil fractions (Ref. 23 and White, personal communication) gave data points which fall on a straight line with very little scatter as shown in Fig. 3.12. In Figs. 3.12–3.18, the calculated MWs are plotted versus the measured (or true) ones. Ideally, all data points should fall on the 45° lines; in reality, they are somewhat scattered around them. Values calculated from synfuel data (24) lie slightly above the line (see Fig. 3.13). Because we had noted that heteroatoms cause the calculated MWs of model compounds to come out on the high side (Figs. 3.14a and 3.14b), we assume that the higher synfuels MWs may result from a higher heteroatom content compared to that of the other samples.

Sometimes the H/C ratio of samples is known but not the specific gravity. For those cases, Altgelt and Boduszynski (25) developed another correlation, Eq. (3.3) [Eq. (3.2) still applies for AEBP < 500°F, 260°C], to calculate MWs from

$$\mathrm{MW} = 170 + 2.67 \times 10^{-7}(\mathrm{AEBP}^3\,(\mathrm{H/C})^{0.9}) \quad \text{for AEBP} > 500°\text{F} \tag{3.3}$$

boiling points. Its application to petroleum samples is shown in Fig. 3.15. Two aspects of this plot are noteworthy. First, the data points from the

---

*Because petroleum fractions are mixtures, strictly speaking we should refer to *average* molecular weights. In addition, because molecular weights of petroleum samples are usually measured by vapor phase osmometry or other colligative methods, the correct designation would be the *number average* molecular weight, M_n. We assume this is understood and, for the sake of simplicity, continue to use the terms molecular weight and MW in most places.

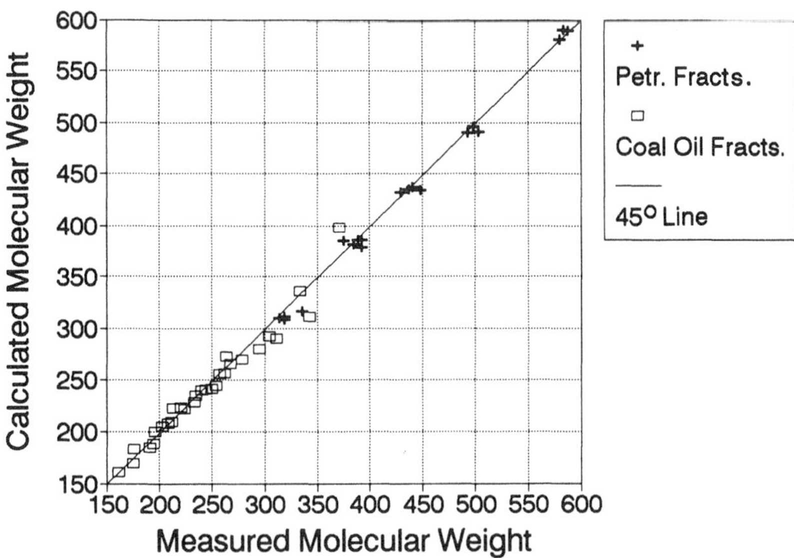

Figure 3.12 Molecular weights of petroleum and coal oil fractions calculated by Altgelt and Boduszynski's AEBP–SpGr–MW equation versus measured MWs.

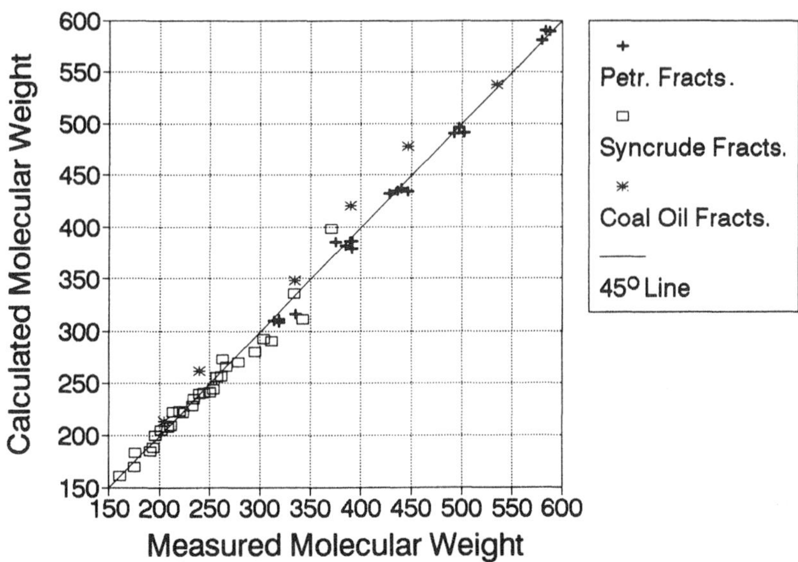

Figure 3.13 Altgelt and Boduszynski's AEBP–SpGr–MW equation applied to petroleum, coal oil, and synfuel fractions.

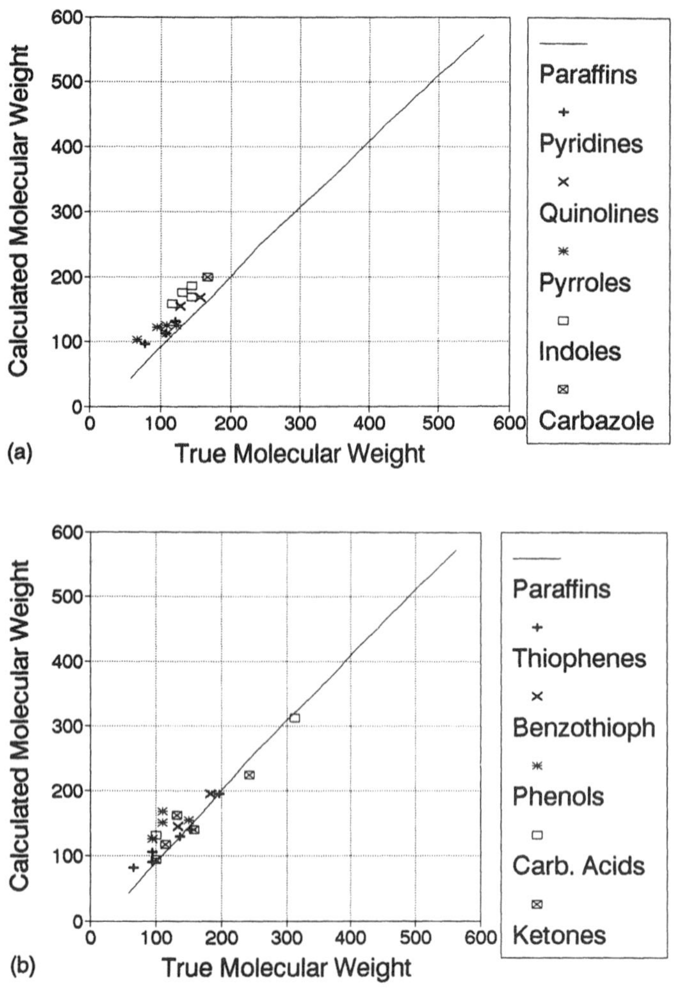

Figure 3.14 Altgelt and Boduszynski's AEBP–SpGr–MW equation applied to model heterocompounds: (a) N-compounds (TRC Data; SpGr), (b) S- and O-compounds (TRC Data, SpGr).

two laboratories (1,26), covering six different crude oils and an extraordinarily wide MW span, ranging from 100 to 1100 daltons, fall remarkably close to the 45° line. Second, one of the data sets consists of short-path distillates derived from atmospheric residues, which extend the boiling range from the usual limit for distillates of 650°F to almost 1300°F, i.e., by more than twice. In Fig. 3.16 we take a closer look at these values, now

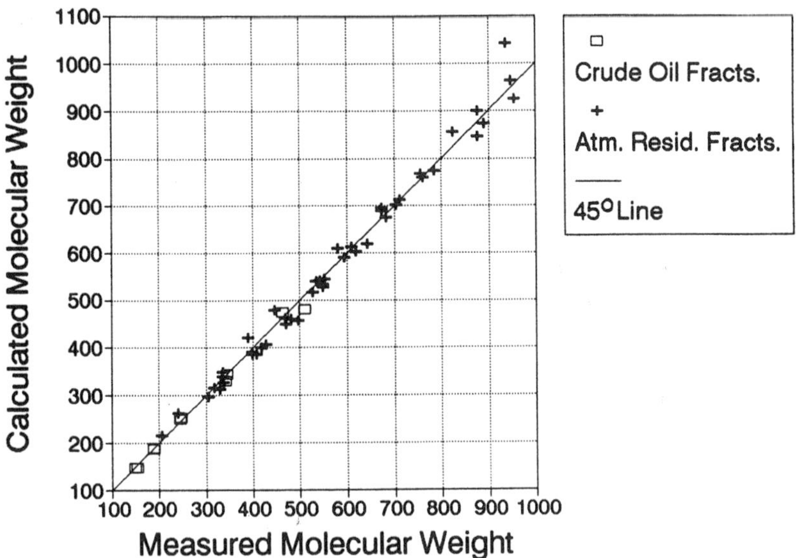

Figure 3.15 Altgelt and Boduszynski's AEBP–H/C–MW equation applied to petroleum samples.

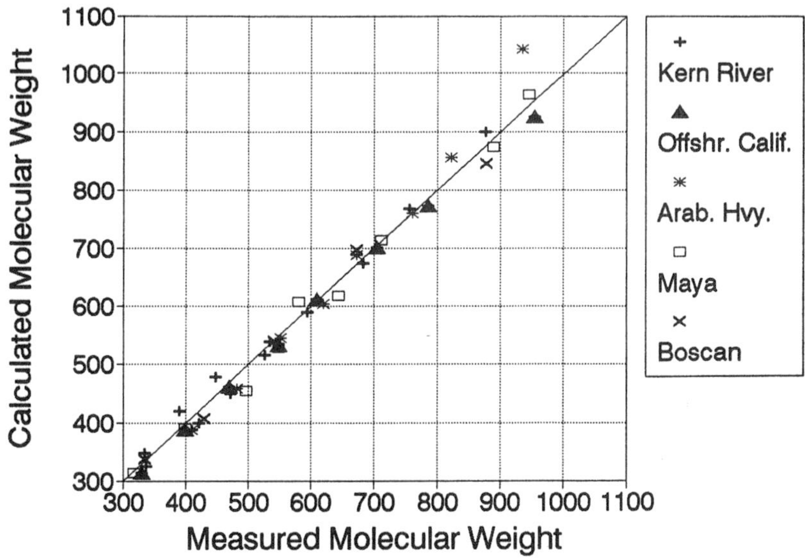

Figure 3.16 Altgelt and Boduszynski's AEBP–H/C–MW equation applied to model short-path distillation cuts from atm. residue fractions of five crude oils.

distinguishing between the fractions obtained from five different crude oils. Any compositional effect certainly is small for these samples even though they range from the paraffinic Altamont crude to the asphaltic Boscan and the biodegraded Kern River crudes. Only if we include coal oil data (Ref. 23 and White, personal communication) do we see these points lying above the band of the petroleum points, Fig. 3.17. This offset again suggests the possibility of a compositional difference between the coal oil data and the petroleum data. Indeed, the Altgelt–Boduszynski–H/C correlation applied to TRC heterocompound data shows most of these to be above the 45° line, just as with the corresponding plots using specific gravity in Figs. 3.14a and 3.14b (see Ref. 23).

We believe our Eqs. (3.1)–(3.3) provide a reliable way to calculate MWs of petroleum distillates from their mid-AEBPs and specific gravities or H/C ratios. Furthermore, we deem it safe to assume that this correlation also holds for SEF fractions of nondistillable residues because the SEF method is to similar to distillation in its separation mode. With this assumption, we can now determine the mid-AEBPs of SEF fractions simply by measuring the number average molecular weights and using Eq. (3.1) or (3.3), solved for AEBP. Equation (3.2) can be ignored for this purpose

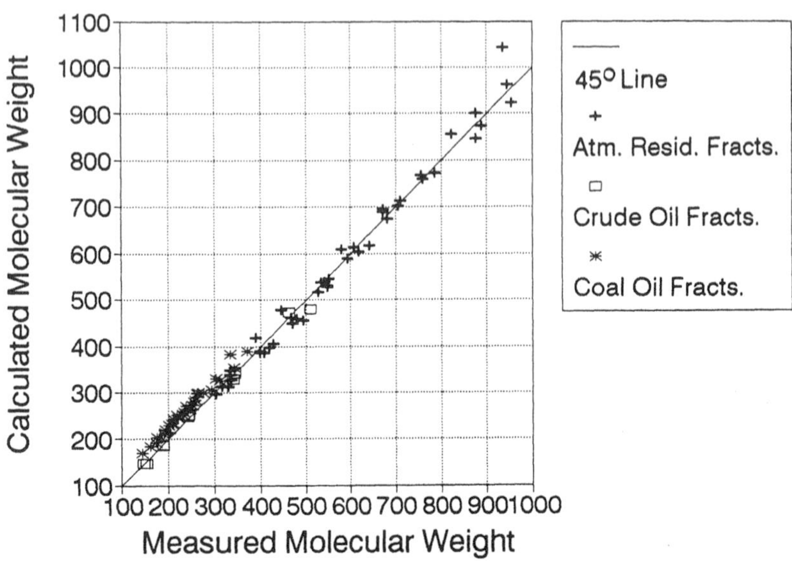

Figure 3.17 Altgelt and Boduszynski's AEBP–H/C–MW equation applied to petroleum samples and coal oil fractions.

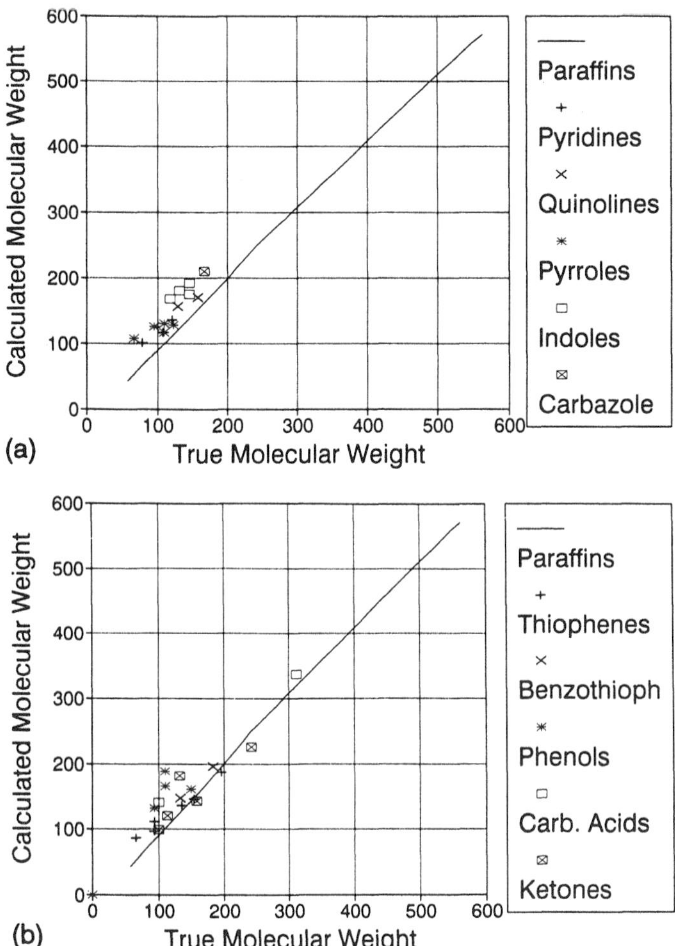

Figure 3.18 Altgelt and Boduszynski's AEBP–H/C–MW equation applied to model heterocompounds: (a) N-compounds (TRC Data; H/C), (b) S- and O-compounds (TRC Data, H/C).

as the MWs of the nondistillable residue fractions are much higher than the minimum of 500 daltons required by Eqs. (3.1) and (3.3).

This approach seems to conflict with the observation (see, e.g., Ref. 27) of a limit to the actual AEBP at 1670°F (910°C). Plots of boiling point versus a function of reciprocal carbon number ($a + b/\mathrm{C}$) for different types of compounds converge at this temperature for infinite C. However, this is the limiting temperature for liquid *n*-paraffins only because it is based on molecules with an infinitely high *n*-alkane content (25,27). It is not a limiting value for any other compound type. Furthermore, our "equivalent" distillation is only a concept, a thought construct, for the sake of continuity and not to be taken literally as a physical extension of the actual boiling range as are vacuum and short-path distillation.

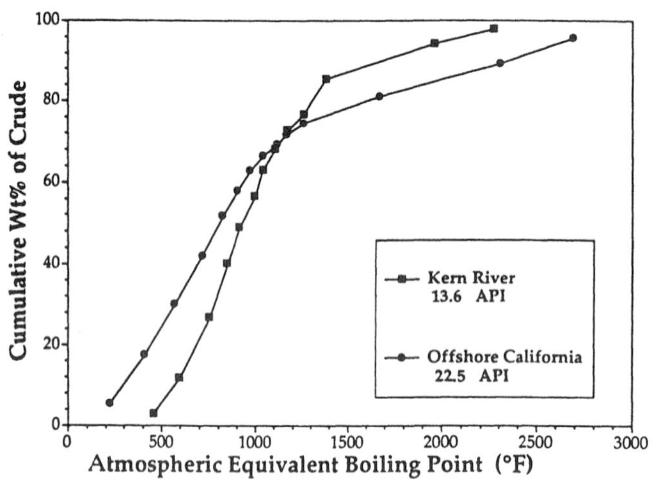

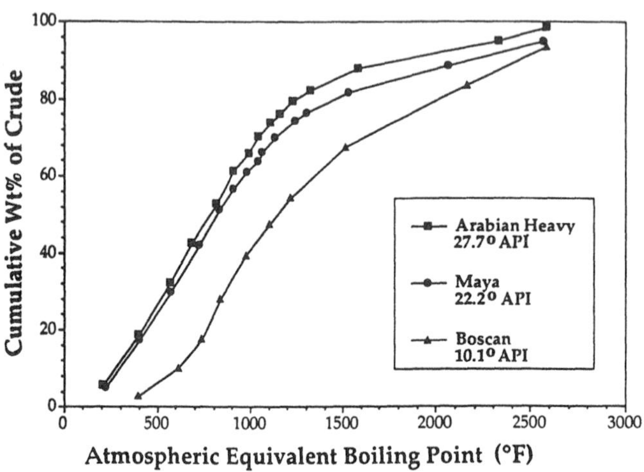

Figure 3.19 Cumulative assays. AEBP distribution curves for ARs from five crude oils.

One could argue that if we accept the MW as the basis of interconversion between distillate and SEF fractions, why not use a MW scale directly instead of the more complicated AEBP scale? The answer is yes; a MW scale would be scientifically cleaner than a boiling-point scale in that it would completely exclude variations arising from differences in chemical composition. However, we think distillation is such a basic, important, and widespread technique in petroleum refining that it should have a similar place in petroleum analysis, in general, and for the categorization and comparison of petroleum samples in particular. Ultimately, our analyses are used (and paid for) by the petroleum chemist or engineer charged with improving the refinery operations. Thus, we should present the results of our research in the terms closest to their requirements.

We think the principle of AEBP is sound, but some of the underlying measuring techniques can stand improvement. One problem needing attention is that of measuring the correct molecular weight. As will be discussed later (Chapter 4) in detail, VPO tends to give high numbers for polar fractions because of their strong tendency to associate and form aggregates. So far, VPO measurements of such fractions in the polar solvent pyridine give results two- to threefold lower than those obtained by measurements in toluene. However, even these may still be too high. Proper calibration of the VPO instrument also is not trivial.

Figure 3.19 gives several examples of AEBP distribution curves (28). It shows plots of cumulative weight percent for the atmospheric residue (AR) fractions from five different crude oils versus their mid-AEBPs. Again, note the smoothness of the curves, especially between the distillate data points and those of the SEF fractions. The pattern is very similar for all the crudes, from the waxy Altamont to the heavy Maya and Boscan crudes. We will come back to this figure later for more extensive discussions in the following chapter. Here we only want to point out the consistency of the data across the entire AEBP range.

REFERENCES

1. Boduszynski, M. M. 1988. Composition of Heavy Petroleums. 2. Molecular Characterization. *Energy Fuels*, **2**, 597–613.
2. Glinzer, O. 1985. A General Vapor Pressure Relation for *n*-Alkanes. *Erdoel Kohle*, **38**, 213.
3. Tipson, R. S. 1965. Distillation Under Moderate Vacuum. *Distillation*, edited by E. S. Perry and A. Weissberger, Interscience Publishers, New York, p. 512.
4. Annual Book of ASTM Standards. 1991. Vol. 05.01, p. 439.
5. Annual Book of ASTM Standards. 1991. Vol. 05.01, pp. 422–434.

6. Butler, R. D. 1979. Simulated Distillation by Gas Chromatography, *Chromatography in Petroleum Analysis*, edited by K. H. Algelt and T. H. Gouw, Marcel Dekker, New York.
7. Jackson, B. W., Judges, R. W., and Powell, J. L. 1976. Boiling Range Distribution of Petroleum with a Short Capillary Column. *J. Chromatogr. Sci.*, **14**, 49.
8. Luke, L. A., and Ray, J. E. 1985. Simulated Distillation of Atmospheric Residues Using Short Pyrex Capillary Columns. *J. High. Res. Chromatogr. Chromatogr. Commun.*, **8**, 193.
9. Trestianu, S., Zilioli, G., Sironi, A., Saravalle, C., and Munari, F. 1985. Automatic Simulated Distillation of Heavy Petroleum Fractions up to 800°C TBP by Capillary Chromatography. Part I. Possibilities and Limits of the method. *J. High Res. Chromatogr. Chromatogr. Commun.*, **8**, 771.
10. Southern, T. G., Iacchelli, A., Cuthiell, D., and Selucky, M. L. 1985. Simulated Distillation of Coal-Derived Liquids Using Combined Gas Chromatography-Vacuum Thermogravimetry. *Anal. Chem.*, **57**, 303–308.
11. Mondragon, F., and Ouchi, K. 1984. New Method for Obtaining the Distillation Curves of Petroleum Products and Coal-Derived Liquids Using a Small Amount of Sample. *Fuel*. **63**, 61–65.
12. Boduszynski, M. M. 1987. Composition of Heavy Petroleums. 1. Molecular Weight, Hydrogen Deficiency, and Heteroatom Concentration as a Function of Atmospheric Equivalent Boiling Point up to 1400°F (760°C). *Energy Fuels*, **1**, 2–11.
13. Schwartz, H. E., Brownlee, R. G., Boduszynski, M. M., and Fu Su. 1987. Simulated Distillation of High-Boiling Petroleum Fractions by Capillary Supercritical Fluid Chromatography and Vacuum Thermal Gravimetric Analysis. *Anal. Chem.*, **59**, 1393.
14. Schwartz, H. E., Higgins, J. W., and Brownlee, R. G. 1986. Simulated Distillation of High-Boiling Petroleum Fractions by Capillary Supercritical Fluid Chromatography and Vacuum Thermal Gravimetric Analysis. *LC–GC*, **4**, 639.
15. Farrell, T. R. 1990. Chevron Research and Technology Company. Unpublished results.
15a. Schwartz, H. E. 1988. Simulated Distillation by Packed Column Supercritical Column Chromatography. *J. Chromatog. Sci.*, **26**, 275–279.
16. Raynie, D. E., Markides, K. E., and Lee, M. L. 1991. Boiling Range Distribution of Petroleum and Coal-Derived Heavy Ends by Supercritical Fluid Chromatography. *J. Microcol. Sep*. **3**, 423–433.
17. Boduszynski, M. M. 1987. Comprehensive Analysis of Petroleum. II. Sequential Elution Fractionation (SEF) of Petroleum Residua. *Chevron Company Report*, unpublished.
18. Hildebrand, J., and Scott, R. 1949. *Solubility of Non-Electrolytes*. Reinhold, New York.
19. Hanson, C. M. 1969. The Universality of Solubility Parameter. *Ind. Eng. Chem., Prod. Res. Develop.*, **8**, 2.

20. Rodgers, P. A., Creagh, A. L., Prange, M. M., and Prausnitz, J. M. 1987. Molecular Weight Distributions for Heavy Fossil Fuels from Gel Permeation Chromatography and Characterization Data. *AICHE 1987 Spring National Meeting*, Houston, March 29–April 2, Prepr. N. 20B.
21. Winn, F. W. 1957. Physical Properties by Nomogram. *Petroleum Refiner*, **36**, 157.
22. Sim, W. J., and Daubert, T. E. 1980. Prediction of Vapor-Liquid Equilibria of Undefined Mixtures. *Ind. Eng. Chem. Process Des. Dev.*, **19**, 386–393.
23. White, C. M., Perry, M. B., Schmidt, C. E., and Douglas, M. J. 1987. Relationship Between Refractive Indices and Other Properties of Coal Hydrogenation Distillates. *Energy Fuels*, **1**, 99.
24. Trytten, L. C., and Gray, M. R. 1990. Estimation of Hydrocracking of C–C Bonds During Hydroprocessing of Oils. *Fuel*, **69**, 397.
25. Altgelt, K. H., and Boduszynski, M. M. 1992. Composition of Heavy Petroleums. 3. An Improved Boiling Point–Molecular Weight Relation. *Energy Fuels*, **6**, 68–72.
26. Alexander, G. L., Creagh, A. L., and Prausnitz, J. M. 1985. Phase Equilibria for High-Boiling Fossil-Fuel Distillates. 1. Characterization. *Ind. Eng. Chem. Fundam.*, **24**, 301–310.
27. Van Nes, K., and van Weston, H. A. 1951. *Aspects of the Constitution of Mineral Oils*, Elsevier Publishing Company, New York.
28. Boduszynski, M. M., and Altgelt, K. H. 1992. Composition of Heavy Petroleums. 4. Significance of the Extended Atmospheric Equivalent Boiling Point (AEBP) Scale. *Energy and Fuels*, **6**, 72–76.
29. Chevron data, 1991, unpublished.
30. Vercier, P., and Mouton, M. 1982. Preparation de distillats a hauts points-d'ebullion par distillation a court trajet. *Analusis*, **10**, 57–70.

4

Properties of Heavy Petroleum Fractions

I. OVERVIEW

Several different approaches to the compositional analysis of a petroleum and its fractions have been taken which vary in terms of expense and information content. Some of these will be discussed in the following chapters. The simplest way is the measurement of certain physical and chemical properties. More information is provided by the group-type characterization methods. Fractionation into compound groups and compound classes (Chapter 6), especially combined with molecular characterization by MS and other spectroscopic methods, not only yields the most detailed and comprehensive compositional data but also carries the greatest expense.

Measurements of bulk properties are generally easy to perform and, therefore, quick and cheap. Several correlate well with certain compositional characteristics and are widely used as a quick and inexpensive means to determine those. The most important properties of a whole crude oil are its boiling-point distribution, its density (or API gravity), and its viscosity. The boiling-point distribution, or "distillation assay," gives the yield of the various distillation cuts. It is a prime property in its own right which indicates how much gasoline and other transportation fuels can be made from an oil without costly conversion. Density and viscosity are measured for secondary reasons. The former helps to estimate the paraffinicity of the oil, and the latter permits the assessment of its undesirable residual material. Boiling-point distribution, density, and viscosity are easily measured and give a quick first evaluation of an oil. Sulfur content, another crucial and primary property of a crude oil, is also readily determined. Certain composite characterization values, e.g., the UOP K factor, calculated from density and mid-boiling-point as described later correlate better with molecular composition than density alone.

For atmospheric and vacuum residues, density and viscosity still are of great interest. But for such materials, hydrogen, nitrogen, and metal contents and carbon residue values, besides sulfur content, become even more important. Table 4.1 shows three examples of common inspections.

In the narrower context of our theme, particularly in the advanced compositional analysis of heavy petroleum fractions, the measurement of certain bulk properties is only the first and easiest step. Beyond the immediate utility of knowing their individual values, such properties, when used together, give a more complete picture than each by itself. Take, for instance, the molecular weight (MW) and its distribution. The average MW or carbon number of a heavy petroleum fraction, and especially its distribution, are basic properties of interest in their own right. Secondary applications of the MW allow us to extend the AEBP scale into the range of nondistillables (discussed in Chapter 3) and to translate H-, C-, or heteroatom contents from weight percent to the more meaningful average number of atoms per molecule (as will be demonstrated in Section IV).

Density and viscosity have been employed in combination with MW and refractive index for rather precise determinations of paraffinicity, aromaticity, and naphthenicity of feeds and products. More on this will be said later in this chapter and in Chapter 5.

II. MOLECULAR WEIGHT

The MW and MW distribution have multiple applications in the compositional analysis of heavy petroleum fractions as mentioned earlier. Usually, the carbon number is quoted in lieu of the MW, but only for paraffins are the two equivalent; for cyclic hydrocarbons and for heterocompounds, they are not. As described in Chapter 3, the MW of a petroleum fraction can be estimated with good precision from its mid-boiling point in combination with either specific gravity [Eq. (3.1)] or the H/C ratio [Eq. (3.3)]. It may be determined by a number of methods including nonfragmenting mass spectrometry, also called parent ion MS; vapor phase osmometry (VPO), often called vapor pressure osmometry; and size exclusion chromatography

Table 4.1 Inspections for Crude Oils, Atmospheric Residues, and Vacuum Residues

Crude oils	Atmospheric residues	Vacuum residues
Gravity, API	Yield, vol%	Yield, vol%
Sulfur, wt%	Gravity, API	Gravity, API
Pour point, °F	Sulfur, wt%	Sulfur, wt%
Acid, mg KOH/g	Pour point, °F	Nitrogen, ppm
Viscosity @ 40°C (104°F), cst	Viscosity @ 50°C (122°F), cst	Hydrogen, wt%
Viscosity @ 50°C (122°F), cst	Viscosity @ 100°C (212°F), cst	MCRT,[a] wt%
Characterization, UOP K		Ramsbottom carbon, wt%
TBP yields, vol%		Asphaltenes, wt%
Butanes and lighter		Nickel, ppm
Light gasoline (55–175°F)		Vanadium, ppm
Light naphtha (175–300°F)		Iron, ppm
Hvy. naphtha (300–400°F)		Pour point, °F
Kerosine (400–500°F)		Viscosity @ 50°C (122°F), cst
Atm. gas oil (500–650°F)		Viscosity @ 100°C (212°F), cst
Lt. vac gas oil (650–800°F)		Viscosity @ 135°C (275°F), cst
Hvy. vac gas oil (800–1000°F)		Cutter, vol% in fuel oil
Vacuum residue (>1000°F)		Fuel oil yield, vol%
		Characterization, UOP K

[a]Micro carbon residue.

(SEC), which is still often referred to by its older name, gel permeation chromatography (GPC). Freezing-point depression (cryoscopy), another colligative technique and commonly used in the past, has essentially been superseded by VPO primarily because the latter is more sensitive and, therefore, covers a wider MW range. It is also easier to operate. In modern laboratories, cryoscopy is employed only for the measurement of samples with molecular weights of < 300–500 daltons.

A. Measuring Techniques

1. Vapor Phase Osmometry

Vapor phase osmometry measurements are relatively simple and cheap, but they give only the number average MW, not the MW distribution. Vapor phase osmometry is one of a group of techniques which measure colligative properties, i.e., properties based on the number concentration of molecules. The original name, vapor pressure osmometry, suggests it measures the degree to which the vapor pressure of a solvent is lowered by the presence of a solute. Actually, it measures a temperature effect which is related to this reduction and proportional to it.

The measuring elements are two temperature-sensitive thermistors placed in a closed, heat-insulated chamber as shown in Fig. 4.1. By means of syringes, a small drop of solvent is placed on one of these, and a small drop of sample solution (in the same solvent) is placed on the other one. The chamber is saturated with solvent vapor and carefully temperature controlled. Because the solution has a lower vapor pressure than the solvent, solvent from the chamber atmosphere will condense on it. The difference in vapor pressure is proportional to its sample concentration. The heat of condensation warms the solution drop until its vapor pressure is the same as that of the surroundings. From then on, a steady state of condensation and warming is established. The temperature increase is measured and recorded. The solvent drop on the other thermistor is in equilibrium with the solvent in the chamber and, theoretically, gives no temperature effect. In reality, convection and other effects cause minute disturbances in the system, the effect of which can be minimized by subtracting the voltage of the solvent thermistor from that of the solution thermistor.

The temperature difference ΔT, is related to concentration and MW at infinite dilution by the equation

$$\Delta T = \frac{K_1 c}{\text{MW}} \tag{4.1}$$

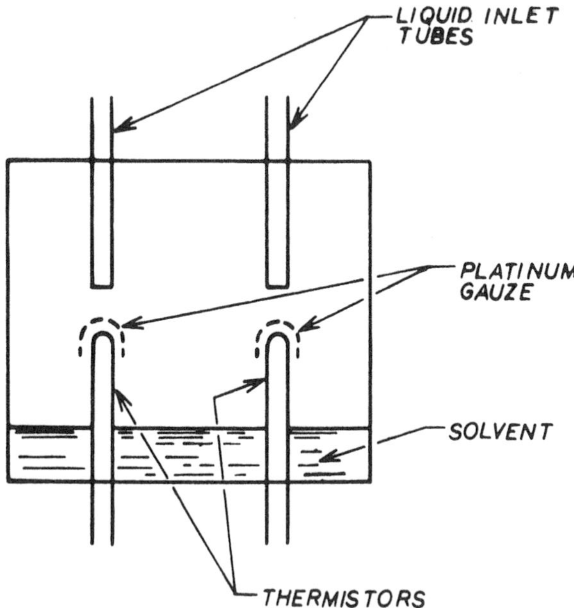

Figure 4.1 Schematic of a vapor phase osmometer. (Reproduced with the kind permission of Alltech and Dr. Burge.)

where K_1 is a constant determined by calibration and c is the solute concentration. The effect is measured at several concentrations, and the results are plotted versus 1/MW and extrapolated to $c = 0$.

Modern instruments, e.g., the new Wescan model, have reduced experimental errors about 10-fold over early models by allowing the application of very small and precisely repeatable sample and solvent drops on the thermistor beads. Thus, measurements of MWs up to about 50,000 daltons are now routinely possible. In addition, samples of much lower concentrations can now be measured allowing linear extrapolation to zero concentration, in contrast to those from more concentrated solutions.

However, systematic errors can easily occur. A potential problem with the number average MW is that it may be shifted to the low side by low-MW contaminants, e.g., by traces of solvent other than that used for the VPO measurement or, if the sample is an oil fraction, by excessive overlap from the nearest lower-boiling fraction(s). Residual toluene would not affect the MWs measured in toluene, but it certainly would distinctly lower those measured in pyridine or in any other solvent.

Kyriacou et al. (1) found in their fractions substantial amounts of stabilizer that were left behind when they evaporated the THF from their size exclusion chromatography cuts. No doubt, these lowered the MWs, particularly those of the fractions with the highest MW and those with the lowest sample concentration. Indeed, some of Kyriacou's first eluting fractions, which should have had the highest MW of a series, were actually lower than the second ones. These fractions had both the highest MW and the lowest sample concentration. They required large amounts of solvent to be evaporated, which left large amounts of contaminant behind.

One definition of the number average MW is given by the formula

$$M_n = \frac{100}{\sum(\text{wt}\%_i/M_i)} \tag{4.2}$$

Because of this relation, the error introduced by a contaminant increases with increasing sample MW and decreasing contaminant MW. Take, for example, a sample of 3000 MW and a contaminant of 100 MW. An amount of 1 wt% contaminant would lower the number average MW (M_n) to 2325 daltons, i.e., by 22.5%. An amount of only 0.1 wt% would still bring the M_n down to 2915 daltons, i.e., by almost 3%. The same amounts (1 and 0.1 wt%) of contaminant in a sample of 5000 daltons would reduce the M_n by 33% and close to 5%, respectively.

Another type of problem is caused by aggregation effects. These are worst with the polar components which are most concentrated in the nondistillables. A common solvent for VPO is toluene which works well for hydrocarbons and lightly to moderately polar compounds. For instance, the VPO MWs of all the distillates (and the pentane-soluble fractions of nondistillables) are in good agreement with FIMS data obtained on the same samples (1,2). However, for the highly polar residues and particularly for their nondistillable fractions, more polar solvents such as pyridine are required. MWs of such materials measured by VPO in pyridine were found to be distinctly lower than those measured in toluene (3–5) indicating less aggregation here (if contamination with trace amounts of previously used solvent in these samples can truly be excluded). Examples of this effect can be seen in Fig. 4.4.

VPO measurements of pentane insolubles in nitrobenzene (3) gave even lower MWs than those in pyridine, with differences ranging from 5 to over 50%. Raising the temperature of the VPO sample chamber to 100°C (210°F) and above lowered the results by another few percent. Such behavior strongly suggests that even nitrobenzene at the regular VPO temperature (37.5°C, 100°F) was not sufficient to overcome all the intermolecular aggregation. It is not an isolated event but seems to be a common occurrence as it was observed with pentane insolubles (asphaltenes) from a wide variety of crude oils.

Al-Jarrah and Al-Dujaili (5) made a systematic study of the effects of solvent and temperature on the degree of aggregation of hexane insolubles of a northern Iraqi crude oil. Although they used only this one source, they covered a wide range for both solvent type and temperature and measured not only MWs (by VPO) but also intrinsic viscosities. Their MW results, presented in Table 4.2, exhibit very high degrees of aggregation in benzene and somewhat lower ones in chlorobenzene and THF. Even in nitrobenzene at 90°C (194°F), there appeared to be some aggregation as to judge from the MWs at 90 and 120°C (248°F). The intrinsic viscosities, shown in Table 4.3, give the same picture. From the data, it is unclear whether even the most severe conditions, nitrobenzene at 120°C (248°F), gave zero aggregation as the authors assume. On the other hand, we wonder about possible C–S bond breakage with nitrobenzene at these high temperatures. Clearly, more work needs to be done before this issue is resolved.

Another potential problem with any MW measurements of insolubles is their tendency to become less soluble after isolation. In crude oil and its heavy fractions, alkane-insoluble polar molecules (asphaltenes) are kept in solution by peptization, i.e., by interaction with soluble polar molecules. In this environment, they seem to form only loose aggregates (6). However, once they are isolated by precipitation, they agglomerate into much tighter

Table 4.2 Effect of Solvent and Temperature on the MW of Asphaltenes Measured by VPO

Solvent	Temperature (°C)	M_n (g/mol)	Degree of association
Benzene	37	16,840	18.0
	45	16,440	17.6
	60	14,920	15.95
Chlorobenzene	37	6,375	6.82
	45	2,750	2.94
	60	2,080	2.22
	90	2,025	2.16
THF	37	13,900	14.86
	45	7,550	8.07
	60	6,330	6.77
Nitrobenzene	90	1,400	1.50
	120	935	1.00

Source: Data adapted from Ref. 17.

Table 4.3 Effect of Solvent and Temperature on the Intrinsic Viscosity of Asphaltenes

Solvent	Temperature (°C)	[η] (g/100 ml)	Degree of association
Benzene	37	0.301	17.70
	45	0.244	14.35
	60	0.232	13.65
Chlorobenzene	37	0.144	8.47
	45	0.067	3.94
	60	0.045	2.65
THF	37	0.220	12.94
	45	0.131	7.70
	60	0.105	6.17
Cyclohexanone	37	0.050	2.94
	45	0.047	2.76
	60	0.045	2.65
Nitrobenzene	37	0.023	1.35
	45	0.022	1.29
	60	0.021	1.235
	120	0.017	1.00

Source: Data adapted from Ref. 17.

aggregates, which are difficult to redisperse even in their own maltenes (pentane solubles) (7). Obviously, they cannot be molecularly dissolved even in such "good" single solvents as toluene.

Chung et al. (8) pointed out the importance of using sufficiently low concentrations for VPO measurements. For benzil, the standard material commonly used for calibrating the instrument, they found an upper limit of 0.6×10^{-3} moles per mole of chloroform (about 1 g/kg chloroform or 0.1 wt%), above which the measured voltage, ΔV, or the temperature difference, ΔT, is no longer proportional to the concentration. This means that, at higher concentrations, extrapolation to infinite dilution would be nonlinear. According to Chung et al., at higher concentrations the relation between ΔV and the number of moles of solute, N_2, can be described by

$$\Delta V = K\phi \frac{N_2}{N_1 + \eta N_2} \tag{4.3}$$

where K is a constant related to instrument design, experimental conditions, and solvent; ϕ is a constant for a given solute-solvent system; η is a "concentration parameter" dependent on solvent and concentration; and

N_1 and N_2 are the mole numbers of solvent and solute in the mixture. The ratio $N_2/(N_1 + N_2)$ would correspond to the concentration, c, in Eq. (4.1), although it would not be equal to it. (Chung et al. found η values of 40 for chloroform, 15 for toluene, and 10 for acetone. However, these were given only for general reference because they could change with the concentration, presumably increasing with increasing concentration.) For linear extrapolation to zero concentration, the solute concentration must be low enough for $\eta N_2 << N_1$.

Chung et al. imply that many of the differences in VPO molecular weights found with different solvents could be attributed to high concentrations rather than to aggregation effects. This claim has not yet been tested to our knowledge. The upper concentration limits prescribed by Chung et al. do not seem to be prohibitively low. Assuming a residue fraction of 1000 average MW and toluene as solvent, η would be 15, $N_1 = 1000/92 = 11$. If we require $\eta N_2 = 10^{-2} N_1$, $\eta N_2 = 11 \times 10^{-2} = 0.11$ mol/kg. Assuming Chung's value of 15 for η, this would make $N_2 = 0.007$ mol/kg and the concentration 7 g/kg or 0.7 wt%. For samples with higher MWs, proportionately higher concentrations would be allowed, e.g., 1.5 wt% for a sample of 2000 MW. Most reported measurements were made in the range of 2–6 wt% (e.g., Ref. 3), not all that much higher.

Selucky et al. (9) and Brulé (10) found a time effect in the aggregation of pentane insolubles. Although their method was size exclusion chromatography (SEC), their results have a direct bearing on VPO measurements, which is why we mention them here. They observed much more detailed SEC curves at very low concentrations (0.01–0.05%) than at higher ones. Later, Selucky et al. (11) reported slow changes in the SEC pattern of freshly prepared dilute asphaltene solutions on standing under exclusion of air and light. In the course of a few weeks, apparent MWs of several thousand daltons dropped to roughly 1000 daltons. The authors concluded that these changes had to be caused by the dissociation of larger aggregates to smaller ones or, quite possibly, to the basic unaggregated molecular units. This phenomenon calls into question any claims of having measured true MWs of such materials by regular SEC and, by implication, also those determined by VPO at low and moderate temperatures. This does not mean such measurements would be worthless. However, they give particle weights rather than reliable, true molecular weights. This finding also reemphasizes the need to spell out the exact conditions of the measurements when reporting VPO (and SEC) data.

An aspect of the number average MW of VPO seldom considered is the possibility of some, or even preferential, aggregation of several small polar molecules with a much larger polar molecule during sample collection (precipitation) or during fractionation. The aggregates dissociate in the "good" solvents used for MW measurement, releasing some or all of the low-MW contaminants. Such aggregation is favored by precipitation of asphaltenes where occlusion of polar molecules from the maltene fraction is common. However, this effect does not even require contamination with foreign material but is possible in any sample containing highly polar molecules of widely different MWs. It slants the resulting number average MW toward low values because the more numerous small molecules dominate over the fewer large ones even though their amounts by weight may be the same or smaller. A number average MW of 5000, thus, does not preclude the presence in that fraction of several weight percent of molecules at least 10 times larger, i.e., of ≥ 50,000.

This feature is not specific to the VPO method. It applies to any method producing number average results. Only further fractionation or other suitable measurements can provide the MW *distribution* (actually the size distribution, see Section II) of these polar fractions. Other suitable measurements could be those of SEC performed with a sufficiently polar solvent, or intrinsic viscosity, $[\eta]$, which gives an average close to the weight average. SEC, of course, has the advantage of giving the entire MW distribution. The different changes of $[\eta]$ and M_n with fraction number allow certain conclusions about the MW distribution.

Reprecipitation or further fractionation of insolubles mitigates the preferential solvation of large polar molecules by small ones. The higher-MW asphaltene fractions obtained by subfractionation or by extraction methods contain fewer, if any, small molecules or contaminants; consequently, there is more high-MW–high-MW aggregation in these fractions than high-MW–contaminant or high-MW–low-MW interaction. But the true MWs of non-aggregated insolubles in such high-MW fractions may also be higher than expected from the measured average MWs of the whole, unfractionated insolubles sample. As a consequence of these considerations and of other effects, MWs of petroleum fractions >2000 daltons measured by VPO are dubious. The true MWs are likely to be lower, but they could also be higher in some cases.

Nevertheless, it is safe to say that only a small fraction of atmospheric residues (<4 wt%) was found to have average MWs larger than 5000 daltons, even in such heavy crude oils as Boscan and Arabian Heavy (4). These results are in keeping with the nature of the number average molecular-weight methods used. Earlier reports (e.g., Ref. 12) of *large amounts* of very high-MW components in pentane insolubles were either based on methods giving higher averages than the number average MW, or they were compromised by aggregation effects. Aggregation effects can lead to even greater errors with methods such as ultrafiltration where experimental artifacts are extremely difficult to suppress.

One more possible reason for finding higher MWs for asphaltenes after their isolation is their susceptibility to oxidation in the solid state. Such oxidation increases their polarity and, thus, their aggregation.

2. Size Exclusion Chromatography

Yau et al.'s (13) authoritative book on modern high-performance SEC was written for the practitioner and is still pertinent today. A recent, much briefer review was provided by Hagel and Janson (14). The practice of SEC as an HPLC tool for the analysis of asphalts and asphaltenes has been discussed by Brulé (10). Here we restrict our discussion to a brief overview and to some points of specific interest for the analysis of heavy petroleum fractions.

We prefer the more descriptive name "size exclusion chromatography" to the older one "gel permeation chromatography," both of which refer to the same procedure. The principle of SEC is the exclusion of larger sample molecules from smaller pores in the packing. Because of this exclusion, larger sample molecules cannot reside in the entire column volume but are restricted to smaller regions. In the extreme, the largest ones are restricted to the interstitial volume, i.e., the space between particles, whereas the smallest ones can permeate the entire open column volume, i.e., the interstitial and all the pore volume. As a consequence, the large molecules elute first and the smallest ones last. In reality, especially with polar samples, two complications can arise:

1. Because of solute–solute or solute–solvent interaction (larger particles are formed), part or all of the sample elutes earlier than otherwise.
2. Because of solute–packing interaction (adsorption), part or all of the sample elutes later than normally.

Adsorption, though initially reversible, may cause permanent damage to the packing if neglected. Even with all the superb equipment now available for SEC, these old problems which have plagued its application to fossil fuel samples (see, e.g., Ref. 15) must not be forgotten. Foremost among these is the proclivity of PAHs and polar samples for adsorption. Mixed solvents containing small amounts of a strongly polar component, e.g., pyridine and/or *o*-cresole (16), can overcome much of the adsorption problem. Buchanan et al. (17) used pyridine as solvent with good success. With certain precautions, their column train lasted for over a year. Another means of limiting adsorption is high-temperature SEC (18). Removal of finely dispersed insolubles by filtering the sample solution through a dense (0.4-μm) filter and thorough flushing after a run helps to keep the columns clean. Establishing the material balance enables one to find out whether the entire sample has been measured or how much was lost. This is important for the interpretation of results and for column maintenance.

For heavy petroleum fractions, column sets with packings of nominal pore sizes between 50–100 and 5000–10,000 Å are used for complete resolution. Properly chosen bimodal column combinations give good linear calibration curves in terms of ln MW versus elution volume, often better ones than obtained with sets of more than two pore sizes (13). Well-characterized polymer samples as well as narrow polymer fractions, e.g., polystyrenes, are available for calibrating column sets across the entire MW range. However, these polymers are generally quite different from petroleum components and cannot be directly used for calibration. In fact, petroleum (and other fossil fuel) fractions differ from polymers in three ways: (1) in their basic hydrocarbon structure; (2) in their often polar nature; and (3) in their varying composition with MW. Therefore, it is preferable to employ a set of *narrow* subfractions obtained by preparative SEC from the same or a similar petroleum fraction for calibration (see, e.g., Ref. 19). Calibration curves obtained by polymers should at least be checked with such subfractions and adjusted as necessary.

By a combination of dialysis and chromatography, Acevedo et al. (20) prepared a set of narrow octylated, and thus solubilized, asphaltene fractions as calibration standards. With a train of three styragel columns (pore diameters 10^2, 10^3, and 10^4) and THF as solvent, they obtained a linear calibration curve over the entire MW range (500–12,000 daltons). This calibration curve had a slightly steeper slope, namely, by a factor of 1.24, than that obtained with polystyrene fractions which was also linear. It crossed the PS curve near 10,000 MW.

Reerink and Lijzenga (21) pointed out some pitfalls in the attempt to establish the correct calibration curve for heavy petroleum fractions. The use of petroleum subfractions can be misleading because they are usually too broad; and in samples with broad distributions, the number average MW may be quite different from the peak MW. From Reerink and Lijzenga's theoretical considerations and experimental data it can be concluded that a good approximation is a calibration curve based on narrow polystyrene fractions (or, better yet, single compounds) and multiplied by a factor of 1.5 (or 1.24 according to Acevedo et al.) across the entire MW range. An additional vertical adjustment is needed to have the polystyrene and petroleum MWs come out equal at 10,000 (see the above paragraph).

It is important to remember that SEC separates by molar volume rather than MW. Size exclusion chromatography will, therefore, differentiate by structure in addition to MW. Rodgers et al. (22) seem to have greatly improved the calibration of SEC columns for petroleum fractions by taking this effect into account. They propose using the traditional equation relating MW to elution volume, V_{el}:

$$\ln \text{MW} = A + BV_{el} \tag{4.4}$$

with new constants defined by the equations

$$A = A_0 + A_1(H/C) + A_2H_\alpha + A_3H_\beta + A_4H_\gamma \tag{4.5}$$

and

$$B = B_0 + B_1(H/C) + B_2 \ln (H/C) + B_3H_\alpha \tag{4.6}$$

where H/C is the molar hydrogen to carbon ratio, H_α is the fraction of H atoms attached to carbons alpha to an aromatic ring, H_β is the fraction of H atoms attached to carbons beta to an aromatic ring or farther away, but nonterminal, and H_γ is the fraction of H atoms attached to methyl groups in γ position (to "terminal carbons," according to the authors).

Rodgers et al. ignored the effect of heteroatoms in their study. They obtained the H/C ratio from elemental analysis and the hydrogen distribution from ^{1}H-NMR data, which they normalized to include only alpha (1.7–4.0 ppm), beta (0.9–1.7 ppm), gamma (0.5–0.9 ppm), and aromatic (6.0–9.0 ppm) H atoms. The 4.0–6.0-ppm range, which covers olefinic and phenolic H, was not taken into account. Rodgers et al. tested their equation with 45 hydrocarbon compounds, run one at a time and ranging from *n*-paraffins to PAHs including eicosane, squalene, cyclohexane, decacyclene, and coronene. The linearity between ln MW and the elution volume, V_{el}, was maintained and only the slope and intercept were affected by the characterization parameters. The average deviation between experimental and true MWs was ±3.6% and the largest deviation was 11.9%. In a statistical assessment, they found the greatest errors to occur in their terms with A_2 and B_3, both involving H_α. The relatively large deviations there seem to have been caused by compounds whose hydrogens are not only alpha but also olefinic. Among the compounds tested, cyclohexane, cyclooctane, and bicyclohexyl gave the greatest deviations (11.9, −11.6, and −11.4, respectively), and coronene, 2-methylnaphthalene, and heptadecane gave the smallest ones (0.0, 0.0, and 0.4, respectively).

The response factors of different samples and even of the different MW fractions of the same sample usually vary and must be determined for quantitative measurements. This holds true for refractive index (RI) detectors and even more so for those using UV, IR, or visible light. For detection by RI, the response factor is proportional to the difference in RI between the sample molecules of a given fraction and the solvent. Some examples of RIs are presented in Table 4.4, which also shows the difference

Table 4.4 Refractive Indices of THF and of Various Model Compounds

Compound	MW	RI[a]	RI − RI_{THF}
THF	72	1.4070	—
Alkanes			
n-Decane	142	1.4102	0.0032
n-Pentadecane	212	1.4315	0.0245
n-Eicosane	282	1.4425	0.0355
n-Pentacosane	352	1.4491	0.0421
Cyclohexane	84	1.4262	0.0192
Decyl cyclohexane	224	1.4534	0.0464
Aromatics			
Benzene	78	1.5011	0.0941
Toluene	92	1.4961	0.0891
Decyl benzene	316	1.4832	0.0762
Eicosyl benzene	456	1.4805	0.0735
Naphthalene	128	1.7 ± 0.2[b]	0.3
Methyl naphthalene	142	1.6176	0.2106
Decyl naphthalene	268	1.5434	0.1364
Phenanthrene	178	1.5943	0.1873
Pyrene	202	1.7700[c]	0.3630
Heterocompounds			
Pyrrole	67	1.5085(78)[c]	0.1015(08)
Indole	117	1.6300	0.2330
Carbazole	167	—	—
Pyridine	79	1.5075[c]	0.1005
Quinoline	129	1.6091[c]	0.2021
		1.6235[b]	0.2165
Thiophene	84	1.5289	0.1219
2-Propylthiophene	126	1.5049	0.0979
Benzothiophene	134	1.6374	0.2304

[a]All data from TRC Thermodynamic Tables except as noted.
[b]*CRC Handbook of Physical and Chemical Properties*, 1990.
[c]From Ref. 23.

in RI between various model compounds and THF, the most common solvent employed with SEC. In samples containing alkanes and aromatics in roughly equal amounts, the RI effect between fractions tends to cancel out, whereas predominantly paraffinic samples give high MWs (22).

In SEC, errors may arise from changes in the response factor with increasing elution volume. The reason for this change is that petroleum components of different MW also have a different composition. Ultraviolet detectors respond primarily to differences in aromaticity and degree of aromatic ring condensation. Refractive index detectors are affected by color intensity, light scattering by very high-MW aggregates, and also by the change in composition, e.g., the alkane content, with MW. Brulé (10) described a method for determining the RI response curve of a sample in a given chromatographic system. Figure 4.2 shows his RI response curves for four asphalts. They are fairly flat in the low and medium MW range (high to medium elution volumes) and increase steeply toward high particle weights (low elution volumes). See also Bartle et al.'s (24) discussion of the effect of composition on the response factors for coal liquid fractions.

In solutions of highly polar samples, such as most insolubles, the molecules are aggregated in clusters as discussed earlier. In very dilute methylene chloride solutions, ≤0.05%, these clusters come apart, and true

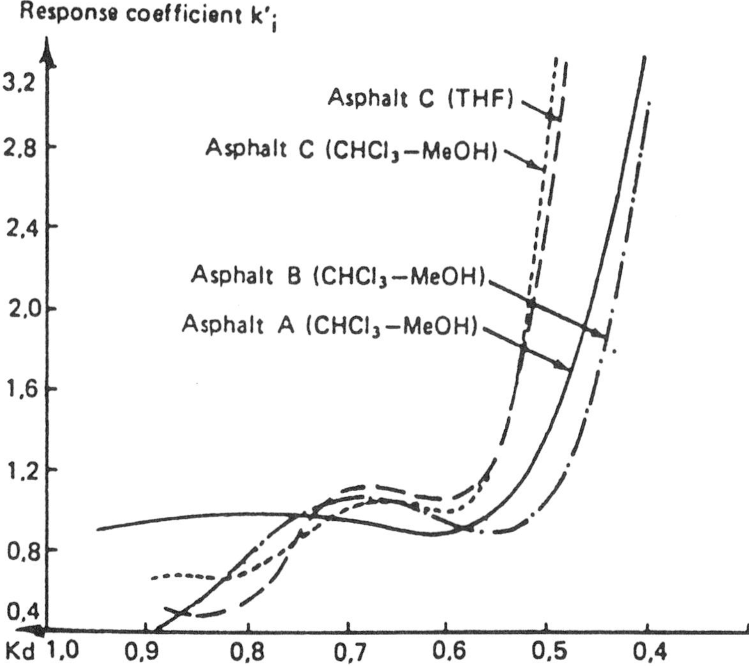

Figure 4.2 Differential refractometer response curves for four asphalts in high-performance SEC. (From Ref. 10. Reproduced with permission of the publisher.)

molecular solutions seem to result (Ref. 11; see also Section II.A.1). The dissociation process, however, is very slow, taking several days to several weeks for completion. Presumably, it could be accelerated by raising the temperature, but no systematic study has been published.

Bartle et al. (24) discussed several errors encountered in the measurement of MW distributions of fossil fuels by SEC. Aside from those already mentioned, they point to potential problems caused by differences in flow rate and temperature and some other ones, such as column loading, band spreading, and baseline drift. Standardization of experimental conditions is important—as usual.

In principle, SEC is very powerful. It is the separation method which comes closest to differentiating by MW only (actually by molecular size), almost unaffected by chemical composition. In contrast to VPO, it gives the entire MW distribution and, therefore, any MW average desired. In contrast to MS, it is quantitative for any, even the highest, molecular or particle weights found in heavy petroleum samples. For these reasons, SEC is still used frequently in petroleum analysis despite the tendency of heavy components toward adsorption and aggregation, and other potential problems. As a preparative tool, compared to distillation, it has the disadvantage of limited sample size and the need to remove the solvent from the fractions.

3. Mass Spectrometry

Nonfragmenting MS is nearly ideal for petroleum distillates. It provides a wealth of detail by giving the formulas and concentrations of practically all the components in a fraction. From these data, the MW *distribution* and the number average MW can be easily obtained. The weight average MW, M_W, should only be calculated from the mass spectra of narrow fractions. For broad fractions, M_W would be underestimated because of the molar mass discrimination.

The main mass spectrometric (MS) methods for measuring the MW of heavy petroleum fractions are FIMS (field ionization MS) and FDMS (field desorption MS). These are nonfragmenting methods which give the so-called parent ion spectra, i.e. spectra showing only the peaks belonging to unfragmented molecules. The techniques will be described in Chapter 7. Aczel (25) considered MS uncertain for coal liquid fractions boiling higher than about 1100°F (~600°C) because of their low volatility. In our experience, the volatility of petroleum fractions is sufficient for FIMS at least up to the limit of short-path distillation, i.e., to 1300°F (~700°C) AEBP. In fact, we observed the SEF-1 fraction of several crude oils to be almost completely volatile in the mass spectrometer.* [SEF-1 is the first of four

*The reason for the different volatilities of the same material is the higher vacuum in the mass spectrometer than in the short-path still.

solubility fractions obtained from nondistillable residue by sequential elution fractionation (SEF); see Chapter 3.]

The SEF fractions typically gave symmetrical MS molar mass profiles (MW distributions) up to a limit of about 2000 daltons. However, the ionization sensitivity changes with molar mass and molecular structure. This "molar mass discrimination" can severely impair the application of MS to broad petroleum fractions. It depends only on the width of the MW distribution and its compositional complexity, but not on the average MW. Broad fractions of any average MW will suffer from this effect, whereas narrow ones, even of high MW, may not. For this reason, MS is ordinarily applied only to compound-class fractions rather than to entire distillation cuts. For such narrow fractions, MW measurements by FIMS usually agree quite well with those by other methods for number average MWs below about 1000 daltons.

Figure 4.3 shows the mass spectra of five narrow boiling cuts and of three solubility fractions of Boscan crude oil. The symmetrical profiles are typical for petroleum fractions. Because of the compositional diversity of these cuts and the differences in ion sensitivity among alkanes, aromatics, and polar compounds, the actual distributions may be different from the mass profiles as shown. Therefore, we cannot be certain of their symmetry,

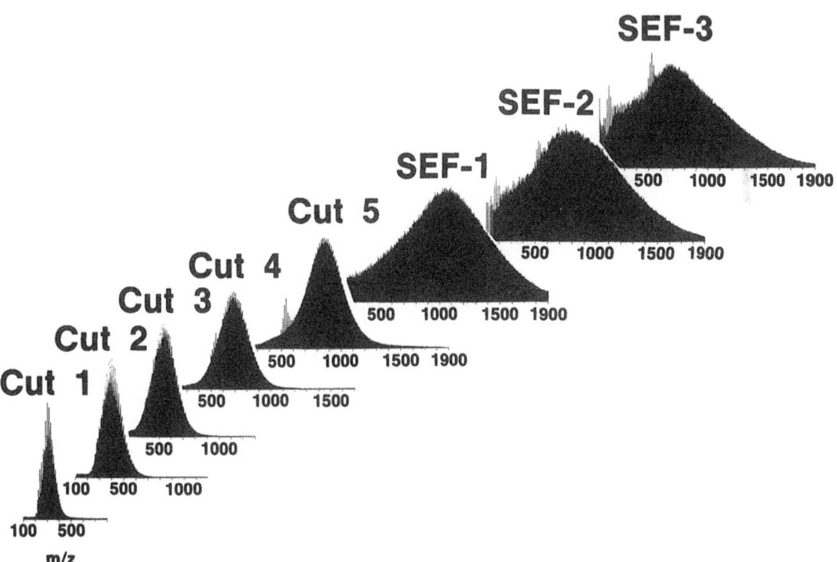

Figure 4.3 FIMS molecular weight profiles of Boscan AR fractions. (From Ref. 2. Reproduced with permission of the publisher.)

and a quantitative evaluation would be hazardous. On the other hand, a cursory look reveals an amazing agreement with VPO measurements for the distillates as indicated by the data in Table 4.5. Here we have listed the VPO MWs of the fractions of Fig. 4.3 together with the upper and lower MW limits taken from the mass spectra. In the last column, we also show the peak MWs of the first six fractions—five distillates and a solubility fraction (SEF-1).

We chose the peak MWs because, for truly symmetrical distributions of the *number* of ions per molar mass, they would be identical with the number average MWs (and the median MWs). The agreement with the VPO results for the distillates is remarkable. It suggests that several errors must compensate each other in the mass distributions displayed by the FIMS spectra. Others (1) have reported similar correspondence between FIMS and VPO molecular weights of petroleum distillates. The discrepancy observed with the SEF-1 fraction may be caused either by the VPO or the MS measurements or by both.

4. NMR Spectroscopy

NMR spectroscopy has been suggested and used by some groups (26–29) for measuring MWs without the limitations incurred by the other methods discussed above. Their approach is based on the determination of substi-

Table 4.5 Mid-AEBPs and MW Data of Boscan AR Fractions

		Molecular weight			
Fraction	Mid-AEBP	VPO	MS L limit[a]	MS U limit[b]	MS peak
Cut 1	720[c]	335[d]	150	500	310
Cut 2	815[c]	430[d]	200	700	380
Cut 3	955[c]	540[d]	240	900	540
Cut 4	1090[c]	675[d]	260	1200	670
Cut 5	1195[c]	875[d]	280	1500	950
SEF-1	1495[e]	1420[f]	300	~2000[g]	1040
SEF-2	2140[e]	3460[f]	320	>2000[g]	—
SEF-3	2570[e]	5510[f]	350	>2000[g]	—

[a]Lower MW limit, estimated from FIMS.
[b]Upper MW limit, estimated from FIMS.
[c]By VTGA SIMDIS.
[d]In toluene.
[e]Calculated from MW.
[f]In pyridine.
[g]Upper limit uncertain because the sample was incompletely vaporized in the mass spectrometer.

tuted and total aromatic carbon atoms per molecule (on average) by ^{13}C and ^{1}H-NMR spectroscopy. However, in our evaluation, it too suffers from shortcomings. It involves several assumptions which, for more complex samples, causes it to render only maximum and minimum values. For several distillates (up to 1033°F), it gave results closely matching those obtained by VPO (see Refs. 26–29). But in this range, of course, VPO works reliably and need not be replaced. With more polar and higher-MW samples, the upper limit calculations often did not converge; and when they did, the spread between lower and upper limits was very large, e.g., 920 versus 2120 and 2110 versus 4400 (28,29). Furthermore, in this range, the assumptions become very tenuous. Thus, in the range where VPO is unreliable, this method does not work either.

B. Molecular-Weight Distributions

Four representative examples of heavy crude oil MW distributions are shown in Figs. 4.4a–4.4d. The corresponding data can be found in Table 4.6. The molecular weights of the distillates were measured by VPO in toluene; those of the SEF fractions, in both toluene and pyridine. In view of our discussion of the MWs determined by parent ion MS and of the possible errors of the VPO method, we must assume that the high-MW values in toluene, and quite possibly even those in pyridine, are too high.

The selected crude oils cover a wide range of composition. For instance, Maya crude (Fig. 4.4a) is a medium heavy paraffinic oil; Boscan (Fig. 4.4b), a heavy asphaltic one; and Kern River (Fig. 4.4c), a heavy biodegraded oil. The crudes displayed in Fig. 4.4 are the same whose AEBP distributions were shown in Fig. 3.9. Note the small amounts of material with MWs higher than 5000 daltons in any of the oils, especially because the MWs measured in pyridine may still be too high. The distillates were also subjected to FIMS. The agreement in MW between the two methods was very good, even for Cut 5, the highest boiling distillate; see Table 4.5.

Figure 4.5 shows how atmospheric distillates, the diverse residues, and some nondistillable solubility fractions obtained from six crude oils fit into the MW–AEBP plot. The vertical lines indicate the various AEBP limits of atmospheric, vacuum, and short-path distillations and also the upper MW limit of SEF-1, the first solubility fraction of the nondistillables. The wedge segments to the left of these lines represent the components of the respective distillates (atmospheric, vacuum, and short-path); and those to the right, the corresponding residues. The last vertical line to the right indicates the AEBP limit between the pentane soluble SEF-1 fraction and the pentane insoluble SEF-2–4 fractions (not the same as asphaltenes; see Chapter 10).

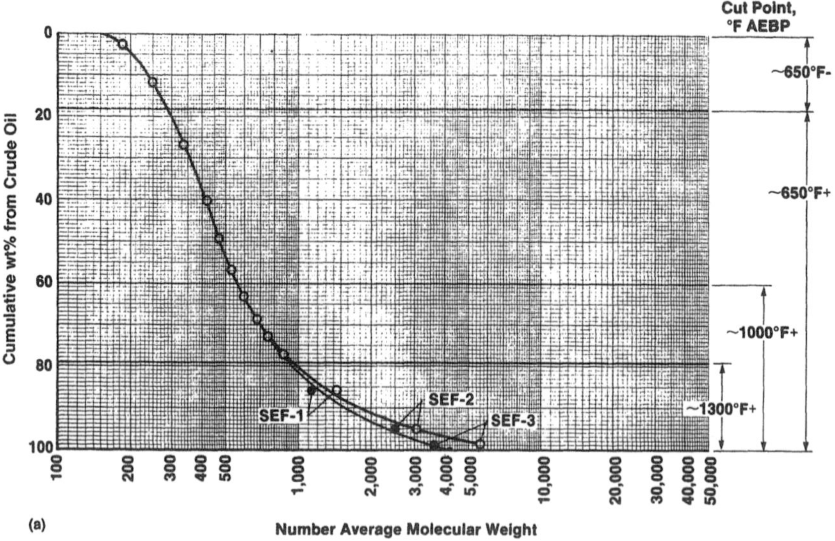

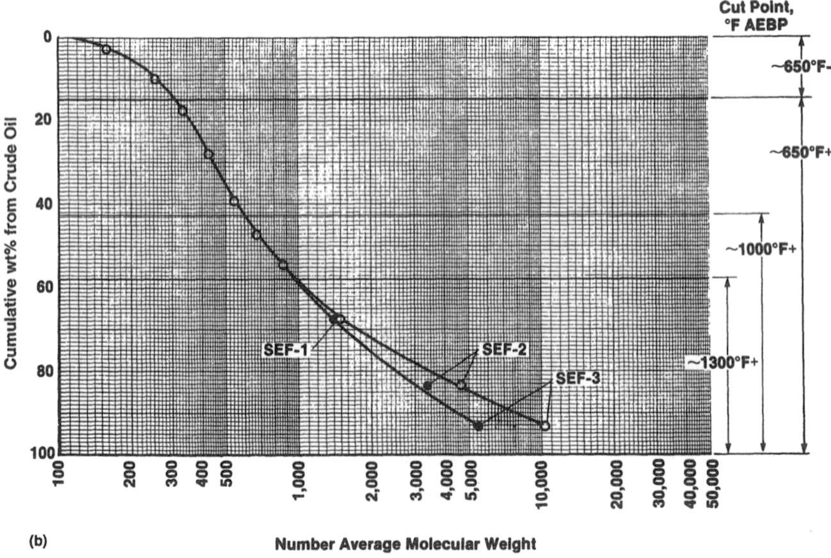

Figure 4.4 Molecular-weight distributions of four crude oils: (a) Kern River; (b) Boscan; (c) offshore California; (d) Maya. These charts also demonstrate the effect of solvent on the apparent MW measured by VPO: ○—measurements in toluene; ●—measurements in pyridine. (Parts a and b from Ref. 4. Reproduced with the permission of the publisher.)

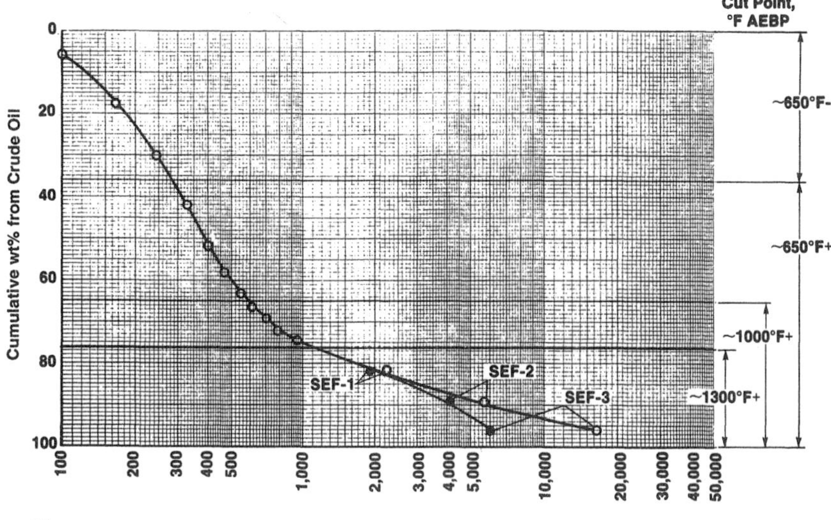
Cut Point,
°F AEBP
~650°F-
~650°F+
~1000°F+
~1300°F+
SEF-1
SEF-2
SEF-3
Cumulative wt% from Crude Oil
0
20
40
60
80
100
100
200
300
400
500
1,000
2,000
3,000
4,000
5,000
10,000
20,000
30,000
40,000
50,000
(c)
Number Average Molecular Weight

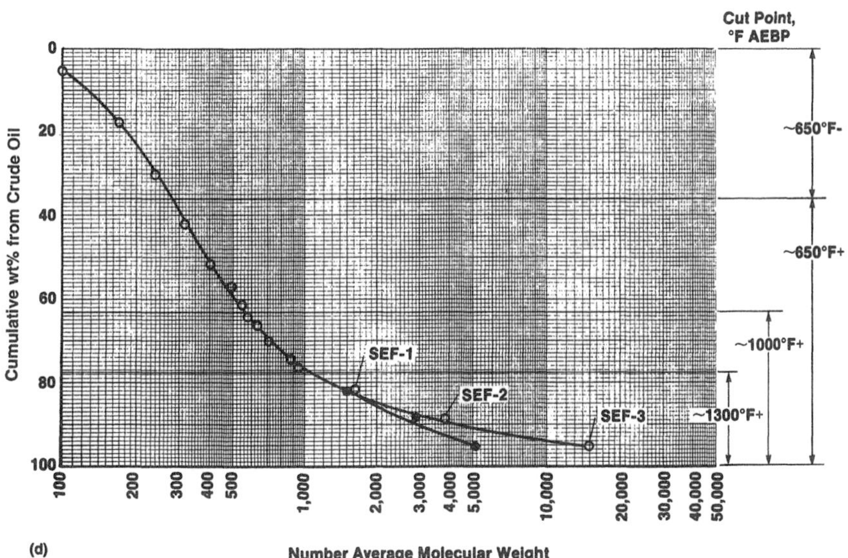
Cut Point,
°F AEBP
~650°F-
~650°F+
~1000°F+
~1300°F+
SEF-1
SEF-2
SEF-3
Cumulative wt% from Crude Oil
0
20
40
60
80
100
100
200
300
400
500
1,000
2,000
3,000
4,000
5,000
10,000
20,000
30,000
40,000
50,000
(d)
Number Average Molecular Weight

Table 4.6 Molecular Weights of Distillates and SEF Fractions of Five Crude Oils

Fraction	Avg. Mid-AEBP (°F)	Molecular weight				
		AH[a]	MA	OC	KR	BO
Cut 1	700	305	317	330	337	335
Cut 2A	800	400	400	400	420	430
Cut 2B	890	480	495	470	470	—
Cut 3A	1000	550	550	550	527	542
Cut 3B	1025	620	580	610	595	—
Cut 4A	1100	673	643	705	682	673
Cut 4B	1150	760	710	785	755	—
Cut 5A	1240	823	890	955	875	877
Cut 5B	1280	936	947	—	—	—
SEF-1	b	1615	1520	1910	1110	1420
SEF-2	b	3870	2965	4145	2520	3460
SEF-3	b	5065	5165	6070	3620	5512

[a]AH = Arabian Heavy; MA = Maya; OC = offshore California; KR = Kern River; BO = Boscan.

[b]The AEBP of the SEF fractions varies with their MWs.

Source: Data adapted from Ref. 30.

The curves for the *n*-paraffins and the hypothetical unsubstituted polar compounds provide a reference for the crude oil curve (marked by the heavy line and data points). The crude oil curve is remarkably close to that of the *n*-paraffins in the range up to about 1000°F (540°C). This means that most of the structure in these fractions is paraffinic (it may also mean that our MWs are somewhat too high or our mid-AEBPs are too low). As will be shown later, the higher boiling fractions, especially those beyond 1300°F (~700°C) contain substantial amounts of aromatic C and heteroatoms. In that range, the oil curve deviates distinctly from the paraffin curve.

Surprisingly, the data range of the oils, i.e., the MW spread at any AEBP, is very narrow up to the limit of distillable fractions (1300°F = ~700°C). Despite the reasonable diversity of the six crude oils used for establishing the MW–AEBP relation (see Chapter 3), the standard deviation of MW was only in the order of ±5% up to the highest measured AEBP (1300°F). We might expect greater deviations beyond this limit, but that would be difficult or impossible to establish. Although the mid-AEBP of SEF-1 fractions may be accessible by VTGA or possibly SFC SIMDIS, at this time we have no such data. Thus, we drew the oil curve in the upper

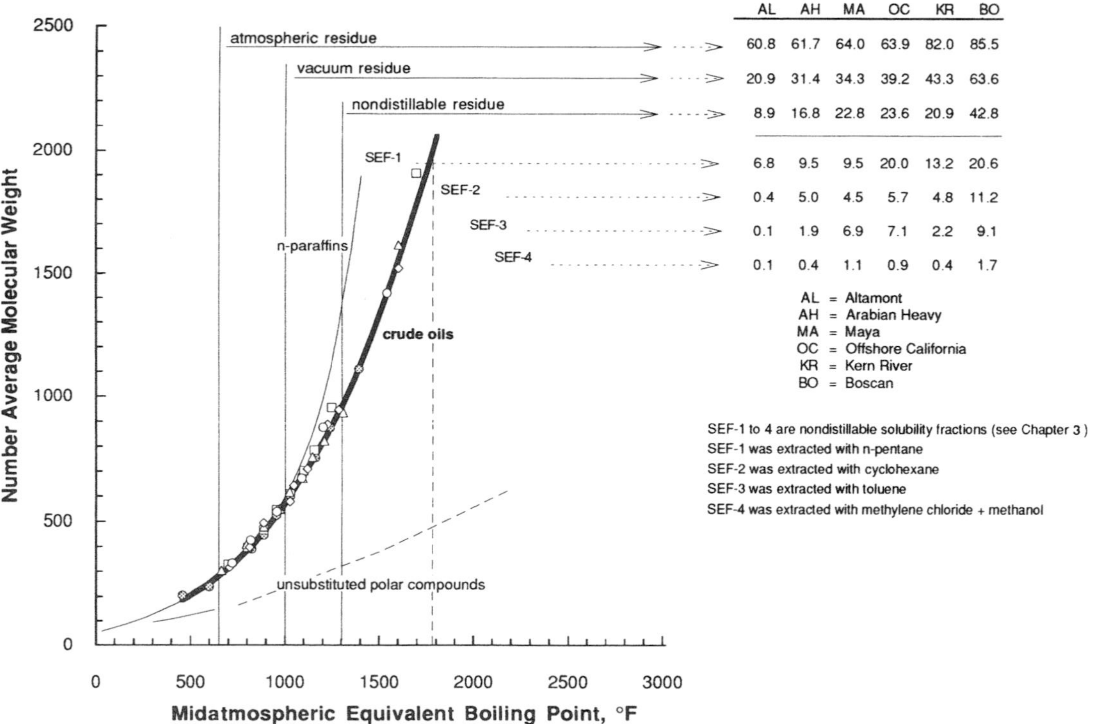

Figure 4.5 Extended MW–AEBP plot and data for six crude oils.

AEBP region provisionally as an extension of that in the lower one, with AEBPs calculated from the MWs by means of the *H/C* equation by Altgelt and Boduszynski; see also Chapter 3, Section IV.B.

Because the MWs of the SEF-2–4 fractions may be too high, the crude oil curve shown in the extended plot of Fig. 3.10 may extend too far, i.e., it may be too long. The slope of the curve in this range is unaffected by the MW because the AEBPs of these fractions were calculated from the MWs. Lower MWs would simply reduce both values (MW and AEBP) and, thus, only shorten the curve but not otherwise change it.

Returning to Fig. 4.5 we see the yields of the atmospheric, vacuum, and nondistillable residues of the six crude oils listed in the upper right corner together with those of the SEF fractions. The atmospheric residue accounts for >60 wt% of these oils, the vacuum residue, for 20–64 wt%, and the nondistillables, for 9–43 wt%. Most of the latter are pentane soluble (SEF-1); the remainder is roughly evenly distributed between SEF-2 and SEF-3, with very little material in the SEF-4 fraction.

Note how low the molecular weights of the oils are at the limits between the distillates and their respective residues (the vertical lines): about 280 daltons for 650°F (345°C), 600 for 1000°F (540°C), and 950 for 1300°F (700°C), the limit between distillable and nondistillable material. For the crude oil curve, the boiling points represent mid-AEBPs rather than the sharp, unequivocal boiling points of single compounds belonging to the *n*-paraffins and the unsubstituted polar compounds of the wedge boundaries.

The height between the limiting curves of the plot indicates the MW spread between the extreme components of an oil fraction, namely, the paraffins and the unsubstituted polar molecules. The corresponding molecular weights of the latter are much lower than those of the crude oil fractions: about 140, 230, and 300 daltons, respectively. Obviously, to judge from the oil curve, there are very few of these highly polar, low-MW compounds in the distillates. (An appreciable amount of low-MW material would depress the number average MW more than observed.)

The MW distributions (determined by FIMS) of five distillation cuts and three SEF fractions from a Boscan atmospheric residue (AR) were already presented in Fig. 4.3. Even though these distributions may be somewhat distorted because of the mass discrimination effect (see above and Chapter 7), the figure gives a good overall impression of the MW range for these fractions. Note the steady increase in both MW and width of the MW distribution with increasing AEBP. The broadening of the molecular-weight distribution reflects the increasing complexity of the samples with rising boiling point. For the same reason, the lower MW limit remains almost constant.

The lower and upper MW limits cannot be judged from the spectra as they appear in Fig. 4.3, but were obtained from the original spectra. The spectra shown in the figure were adjusted to equal maximum height rather than to equal total mass (equal area). This resulted in a more uniform display which helps the overall comparison, but it distorted the size relations between the fractions. The area under the spectrum of Cut 1, for example, is much smaller than any of the other ones.

Apparent inconsistencies in the last two SEF fractions, e.g., the regression in the lower limit MWs and the peak MWs in comparison to those of earlier fractions in Fig. 4.3 is caused by another effect, namely, the limited volatility of these fractions. For instance, only about 70 wt% of SEF-2 and 40 wt% of SEF-3 were volatile, leaving 30% and 60%, respectively, of these fractions behind on the probe. Regardless of what happened to the upper end of the distribution, remembering that SEF-2 and -3 are equivalent to pentane insolubles, we see here that these, contrary to widespread opinion, contain substantial amounts of low-MW (though highly polar) constituents.

These distributions are representative, i.e., the corresponding fractions from other crude oils would have about the same width and the same average MW although details of the distributions might be different. Figure 4.6 illustrates this point with two examples, the FIMS profiles of equivalent DISTAC cuts of Kern River (Cut 5) and Boscan AR (Cut 5). The number average MWs (M_n) of both are very close, and so are the widths of the distribution. Otherwise, the Boscan cut has a narrower profile than the Kern River cut. Of interest also are the anomalous peaks around M = 541 in the Boscan spectrum which represent vanadyl porphysins.

Going back to Fig. 4.5 once again, the third row of its table gives the weight percent nondistillables in each crude oil (<1300°F or about <700°C). The distillable fractions (<1300°F AEBP), which make up most of each crude oil, between 57 and 90 wt%, contain material of up to roughly 1400 MW. The remainder, the nondistillables, contain some larger molecules but also many small ones (<1400 daltons). SEF-1 is the first solubility fraction of the nondistillables and almost completely volatile in the mass spectrometer. In Fig. 4.3 and Table 4.5 we see that SEF-1 of Boscan crude has a MW span ranging from about 300 to slightly more than 2000 daltons. Thus, Boscan crude, one of the heaviest crude oils known, contains nearly 80 wt% (100–20.6 as indicated in Fig. 4.5) of material with <2000 MW.

SEF-2 was about 70 wt% volatile in the mass spectrometer. Most of this material had MWs <2000 daltons (Fig. 4.5). Likewise, about 40 wt% of SEF-3 contains molecules in this range. Adding up all the portions of <2000 MW, we arrive at 79 + ~7.5 + ~3.5 = ~90 wt%, which leaves only

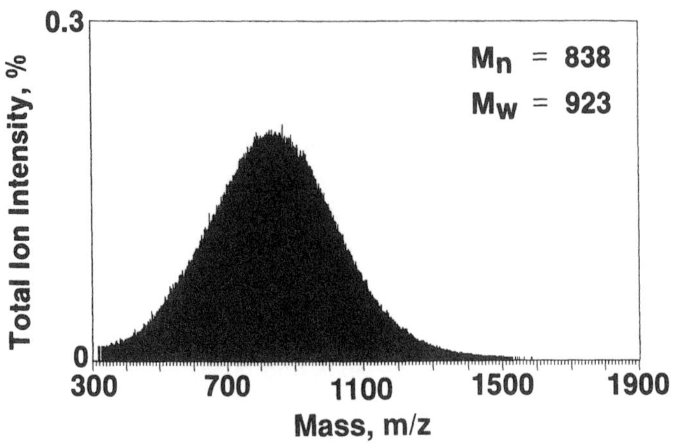

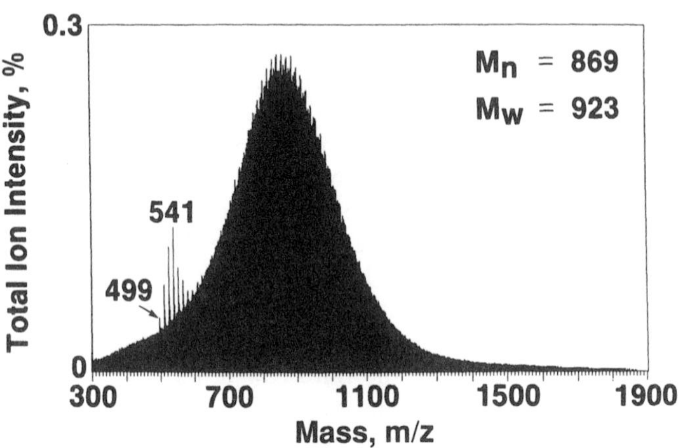

Figure 4.6 FIMS molecular-weight profiles of DISTAC Cut 5 fractions from Kern River (top) and Boscan (bottom) atmospheric residues. (From Refs. 2 and 30. Reproduced with permission of the publisher.)

about 10 wt% of higher-MW material. This result is based on yields and FIMS data, in contrast to the curve in Fig. 4.4b, which is based on VPO data (which are less reliable for petroleum fractions of >2000 daltons).

By the same reasoning, Altamont, the lightest crude oil listed in the table of Fig. 4.5, has <1% material of >2000 MW, if any, and the other four crude oils have <5 wt% in this category.

We mentioned the difficulty in determining the true MWs of nonaggregated asphaltenes and other insoluble petroleum fractions. Acevedo et al. (32) tried to shed some light on this issue. By chromatography on washed silica, they found that their Cerro Negro pentane insolubles could be divided into a "neutral" and several polar fractions. When they treated the polar part by mild reductive methylation (with K and naphthalene in THF at room temperature), they obtained a product which, on the silica column, behaved like the "neutral" insolubles fraction. Furthermore, its average MW had been reduced from >3000 to 1600, as measured by SEC, or to 1800, as measured by VPO, almost down to the MW of the neutral fraction (1200 and 1400, respectively). They concluded that the difference between the polar and the neutral fraction was that the former contained hydrogen-bonding groups which caused the molecules to aggregate. The true number average MW of their insolubles in their nonaggregated state, they claim, is under 2000 mass units and close to that of their neutral fraction.

This is an interesting observation, but it seems to be contradicted in a paper by Shaw (33). Shaw reduced several asphaltene samples from different crude oils by reaction with CH_3I-NaI. When he used refluxing at 75°C, he observed a drop in MW after the reaction. However, when he treated his samples with the same reagents at room temperature, he found no change in MW. The mild treatment should have been sufficient to alkylate H-bonding groups (34). Thus, Shaw concluded that the high-temperature MW reduction of his samples was caused by C–S bond breakage rather than by elimination of hydrogen bonding. If there was any hydrogen bonding, he argued, it did not contribute to aggregation because low-temperature treatment had no effect.

Rose and Francisco (35) obtained a similar result when they methylated two vacuum residues and the pentane insolubles of one of these by a phase-transfer–catalysis method. They found no effect on the residues and only a minor one on the insolubles, namely, a reduction of the MW measured by VPO in toluene from 8000 to 7250 daltons. The MW measured by VPO in orthodichlorobenzene at 130°C, in contrast, was only 4130 daltons. Thus, the methylation did not suppress all the aggregation that existed. There may have been two reasons for this failure. Either the methylation was incomplete, leaving enough of the acid groups intact for continued aggregation by hydrogen bonding, or else forces other than hydrogen bonding (such as charge-transfer effects) were in large part responsible for the aggregation. It is open at this time which one pertains if not both.

In this context, a paper by Del Paggio et al. (18) is of interest. This group looked at the effect of temperature on the SEC of two atmospheric residues. When they ran these samples at 130°C in 1-methylnaphthalene, they found a distinct shift of the high-temperature elution curves relative to the curves obtained at room temperature. Unfortunately, the shift as shown in their figures suggests that the curves were mislabeled. They claim, with good reason, that high temperature improved the separation because of better sample solubility, reduced polar interactions, and lower viscosity. On the other hand, their curves labeled for the higher temperature have a narrower range and are shifted toward higher rather than lower MWs. If, indeed, the curves were mislabeled, these results would make sense. However, without further clarification we can only point out that there was an effect, but not what it means.

Only a few MW distributions of petroleum heavy ends including the nondistillable residues have been published, but all of the recently reported ones fit the pattern of the five crudes discussed earlier. The data from three groups are shown in Fig. 4.7 and in Tables 4.7–4.10. Rodgers et al. (22) measured the MW of an Alaskan North Slope residue using analytical SEC. Their data are listed in Table 4.7. Champagne et al. (36) fractionated Athabasca bitumen by preparative SEC and measured the MWs of their fractions by three methods, namely, by VPO in two solvents (THF and benzene), by freezing-point depression, and by GC-MS (see Table 4.8). The third group, Khan and Hussain (37), also used preparative SEC to

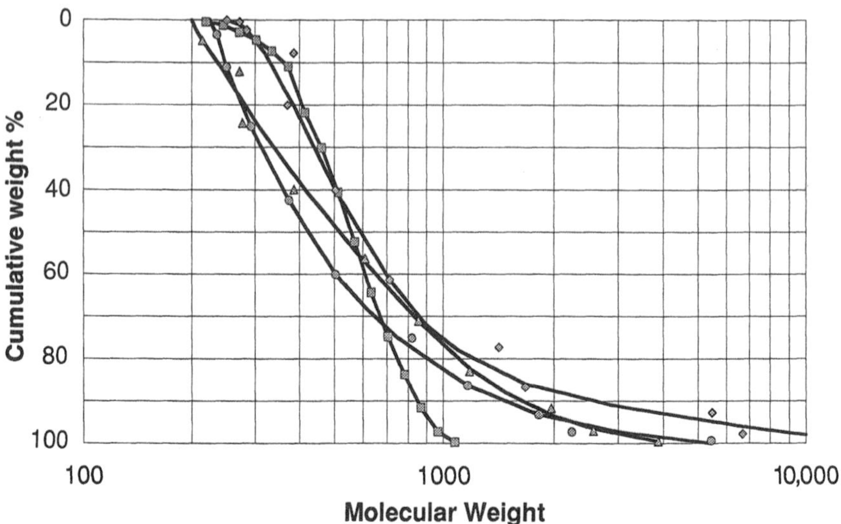

Figure 4.7 Molecular-weight distributions of various heavy oil samples: ■—Alaskan North Slope residue (Ref. 22); ▲—light Middle East crude (Ref. 37); ●—medium Middle East crude (Ref. 37); ◆—Athabasca bitumen (Ref. 36).

Table 4.7 Weight Percent and Molecular Weight of SEC Fractions Obtained from an Alaska North Slope Nondistillable Residue

Wt%	Cum wt%[a]	MW (SEC)
0.81	0.41	219
1.16	1.16	244
1.64	2.79	271
2.21	4.72	302
2.97	7.31	335
4.01	10.80	373
5.42	15.51	415
7.20	21.82	461
9.35	30.10	512
11.38	40.46	570
12.38	52.34	633
11.48	64.27	704
9.44	74.73	783
8.44	83.67	870
7.18	91.48	968
4.25	97.20	1076
9.68	99.66	1196

[a]Calculated according to Eq. (4.7).
Source: Data adapted from Ref. 22.

fractionate their samples, two Middle Eastern crude oil residues (no boiling-point range given). They determined the MWs by analytical SEC and by VPO but did not reveal any details on the latter method. As seen from their data, Tables 4.9 and 4.10, the agreement between the SEC and VPO MWs was good except for the three highest fractions, for which VPO gave higher values than SEC. We calculated the cumulative weight percent values, Cum w_i, of the fractions of these authors by a formula from polymer chemistry:

$$\text{Cum } w_i = \frac{w_1}{2} + \sum_i w_{i-1} + \frac{w_i}{2} \tag{4.7}$$

where i is the fraction number. This formula reflects the common practice of plotting weight percent of the fractions versus MW in a bar graph and connecting the midpoints of the bars to construct the distribution curve.

Another MW distribution, this one of a Cold Lake vacuum residue (AEBP >525°C, 977°F), was reported by Kyriacou et al. (1) and is listed in Table 4.11. They, too, obtained their data by preparative SEC separation

Table 4.8 Weight Percent and Molecular Weight of Athabasca Bitumen SEC Fractions

Fraction number	Wt%	Cum wt%[a]	Molecular weights by various methods			
			$VPO_{(THF)}$[b]	$VPO_{(B)}$[c]	$FPD_{(B)}$[d]	GC-MS
10	0.5	97.8	6690	—	—	—
11	5.6	92.7	5470	3020	3290	—
12	6.9	86.6	1680	1920	1890	—
13	12.2	77.2	1420	1390	1370	—
14	19.0	61.3	713	772	704	—
15	23.1	40.0	504	476	434	—
16	16.8	20.0	371	456	356	—
17	7.8	7.7	368	414	376	—
18	3.0	2.3	284	419	401	280
19	0.8	0.4	271	444	421	270
20–25	0.3	0.2	250	556	544	250

[a]See Table 7.
[b]Vapor phase osmometry with THF as solvent.
[c]Vapor phase osmometry with benzene as solvent.
[d]Freezing point depression with benzene as solvent.
Source: Data adapted from Ref. 36.

Table 4.9 Weight Percent and Molecular Weight of Fractions from a Light Middle Eastern Crude Determined by SEC

Fraction number	Wt%	Cum wt%[a]	Molecular weights by various methods	
			SEC	VPO
1	1.35	99.33	3236	5443
2	2.91	97.20	2108	2240
3	4.98	93.25	1417	1821
4	8.83	86.34	1171	1165
5	13.71	75.07	616	821
6	16.72	59.87	466	504
7	18.17	42.42	403	375
8	16.93	24.87	270	292
9	11.01	10.90	256	250
10	3.95	3.43	217	235
11	1.04	0.93	227	—
12	0.42	0.21	—	—

[a]See Table 7.
Source: Data adapted from Ref. 37.

Table 4.10 Weight Percent and Molecular Weight of Fractions from a Medium Middle Eastern Crude Determined by SEC

Fraction number	Wt%	Cum wt%[a]	Molecular weights by various methods	
			SEC	VPO
1	1.15	99.42	—	3870
2	3.66	97.02	—	2568
3	7.11	91.63	—	1968
4	10.25	82.95	1710	1178
5	13.70	70.97	937	856
6	15.79	56.22	598	608
7	16.84	39.91	435	387
8	14.54	24.22	337	276
9	9.83	12.03	291	271
10	4.81	4.71	—	241
11	1.78	1.41	—	—
12	0.42	0.31	—	—
13	0.10	0.05	—	—

[a]See Table 7.
Source: Data adapted from Ref. 37.

Table 4.11 Molecular Weight Distribution of Cold Lake Vacuum Bottom Fractions

Fraction number	Wt%	Cum. wt%[a]	Molecular weights by various methods			
			VPO[b]	FIMS M_n	FIMS M_W	M_W/M_n
1	1.4	99.3	1570[c]	1054	1183	1.16
2	10.6	93.3	1544	1062	1182	1.12
3	14.3	80.8	1488	1050	1170	1.11
4	16.7	65.3	1239	962	1084	1.11
5	17.8	48.0	1033	982	1112	1.13
6	15.6	31.3	952	992	1130	1.13
7	9.1	19.0	633	860	992	1.14
8	5.1	11.9	553	753	883	1.15
9	9.3	4.7	357	666	822	1.17

[a]See Table 7.
[b]In benzene at 0°C.
[c]Corrected (by KHA) for 7 wt% butylated hydroxy toluene left over from the THF used for the SEC separation.
Source: Data adapted from Ref. 1.

and VPO, but augmented them by FIMS and viscosity measurements. The M_n values calculated from FIMS closely match those measured by VPO for the lower-MW fractions. In the upper range, the FIMS MWs are lower than those from VPO, probably because of incomplete sample vaporization. Aggregation effects seem to be a somewhat less likely cause in view of the viscosity results obtained at three temperatures (between 10 and 50°C). The molecular-weight distribution of this sample is extraordinarily steep, possibly due to insufficient separation during the SEC separation.

Note that all the molecular-weight distributions discussed here were derived from whole crude oils or from residues without removal of insolubles (asphaltenes). The topic of asphaltenes will be taken up later in Chapter 10.

III. OTHER PHYSICAL PROPERTIES OF HEAVY PETROLEUM FRACTIONS

A. Density, API Gravity

1. Primary Density Relations

Density, d, is one of the most important properties of petroleum samples for several reasons: (1) it is a good indicator of crude oil quality; (2) it can be easily and precisely measured; and (3) it correlates with aromaticity, naphthenicity, and paraffinity (as will be discussed in Chapter 5). Densities are measured at a standard temperature, mostly 15 or 20°C (59 or 68°F). They are often written in the form of d_4^{20} to indicate that they were measured at 20°C and in pyknometers calibrated with water at 4°C, i.e., with a medium of $d = 1.0000$.

Specific gravity, SpGr, is the ratio of sample density to the density of water at specified temperatures. It is very easy to measure, namely, with sinkers. For petroleum samples, both sample and water densities are measured at 60°F (156°C). The specific gravity values of petroleum samples are close to those of density and range from about 0.80 for light paraffinic crude oils to about 1.00 for heavy asphaltic ones.

To expand the narrow range in specific gravity values, the API gravity scale was introduced:

$$°\text{API} = \frac{141.5}{\text{SpGr (60°F)}} - 131.5 \tag{4.8}$$

This unfortunate construct is a modified inverse specific gravity with values ranging from about 40 for very light paraffinic crudes to about 10 for highly asphaltic crudes. Samples with specific gravities >1, for instance, some nondistillable residues and most pentane insolubles, have negative API values. API gravity is commonly used by petroleum chemists and engineers to characterize crude oils and refinery streams.

We believe a better scale might be what we call "incremental density," which is defined (38) by

$$D_i = (d - 0.5)(200)$$

Here crude oils range from about 60 to 100, which is a similar span (40 units) as that on the API gravity scale. But the incremental density has three advantages over the API gravity: (1) it is a simpler expression, easy to use even without calculator; (2) it increases with increasing density rather than going the opposite way; and (3) it has only positive values, starting with 25.2 for *n*-pentane and going up to about 100 for heavy oils. The incremental density serves the same purpose as the API gravity but without its disadvantages.

Tables 4.12 and 4.13 show the specific and API gravities as well as incremental densities of various crude oils and selected heavy oil fractions. The corresponding values for model compounds are shown in Table 4.14. The densities for coronene and indole could not be found in the literature and were estimated by Hirsch's method (see below).

An interesting aspect of density is its change with molecular weight. In Fig. 4.8, the density at 20°C of *n*-paraffins and of several *n*-alkyl ring compounds are plotted versus 1/MW. All data fall on linear curves, one curve for each compound type. All the curves converge on the same density (0.8513) at infinite MW. This is to be expected because the compositional differences of the diverse compound types disappear at infinitely high alkyl content. The densities of most isocompounds are slightly greater than those

Table 4.12 Densities and API Gravities of Selected Crude Oils

Sample	Characterization	Specific gravity[a]	API gravity[a]	Incremental density[b]
Altamont	Paraffinic	0.8238	40.3	64.8
Arabian Heavy	Intermediate	0.8888	27.7	77.6
Alaskan North Slope	Intermediate	0.8894	27.6	77.8
Offshore Calif.	Intermediate	0.919	22.5	83.8
Maya	Intermediate	0.921	22.2	84.2
Kern River	Naphthenic	0.9698	14.4	94.0
Boscan	Asphaltic	0.999	10.1	99.8

[a]Data from Chevron Research and Technology Company.

[b]These numbers may be slightly too high because they were calculated from the specific gravities rather than from density.

Table 4.13 Densities and API Gravities of Selected Crude Oil Fractions

Crude oil		AEBP range (°F)	Specific gravity	API gravity	Incremental density
Altamont		Whole crude	0.8238	40.3	64.8
		<300	0.7263	63.3	45.2
		300–400	0.7663	53.1	53.2
		400–500	0.7919	47.2	58.4
		500–650	0.8080	43.6	61.6
		650–865	0.8319	38.6	66.4
		865–1000	0.8496	35.1	70.0
	Residue	>650	0.8559	33.8	71.2
	Residue	>1000	0.8844	28.5	76.8
Arabian Heavy		Whole crude	0.8888	27.7	77.8
		<300	0.6892	73.8	37.8
		300–400	0.7775	50.5	55.6
		400–500	0.8132	42.5	62.6
		500–650	0.8550	34.0	71.0
		650–810	0.9071	24.5	81.4
		810–1000	0.9390	19.2	87.8
	Residue	>650	0.9833	12.4	96.6
	Residue	>1000	1.0351	5.2	107.0
Kern River		Whole crude	0.9698	14.4	94.0
		<366	0.7923	47.1	58.4
		366–423	0.8338	38.2	66.8
		423–513	0.8660	31.9	73.2
		513–650	0.9053	24.8	81.0
		650–1250	0.9799	12.9	96.0
		650–832	0.9580	16.2	91.6
		832–1050	0.9861	12.0	97.2
		1050–1250	1.0151	7.9	103.0
	Residue	>650	0.9938	10.6	98.8
	Residue	>1050	1.0400	4.6	108.0
	Residue	>1250	1.0536	2.8	110.8

Source: Chevron Research and Technology Company.

of the normal hydrocarbons, but some are smaller. Ring compounds and heterocompounds have significantly higher densities than the paraffins.

In Chapter 2, we had talked about the misleading aspect of the term "heavy" as in petroleum heavy ends. Heavy in this context is used instead of the bulkier term "high-boiling." In reality, of course, it means "of higher density." For most crude oils, especially the heavy (!) asphaltic ones, this

Table 4.14 Densities and API Gravities of Some Model Compounds

Sample	Molecular weight	Density[a] 20°C	Specific gravity	API gravity[b]	Incremental density[c]
n-Pentane	72	0.6261	0.630	92.4	25.2
n-Hexane	86	0.65925	0.6623	82.1	31.8
n-Hexadecane	226	0.773	0.777	50.5	54.6
Cyclohexane	84	0.779	0.781	49.7	55.8
Perhydropyrene	218	0.983	0.985	12.2	96.6
Perhydrocoronene	324	1.204[d]	—	6.3[e]	105.3
Benzene	78	0.8790	0.884	28.6	75.8
Toluene	92	0.8669	0.872	30.8	73.4
Ethylbenzene	106	0.8670	0.871	31.0	73.4
Dodecylbenzene	246	0.8551	0.858	33.4	71.0
Pyrene	202	1.271	1.27	−20.2[e]	154.2
Coronene	300	1.4049[f]	—	−30.8[e]	181.0
Ovalene	416	1.5268[f]	—	−38.8[e]	205.3

Decalin, *cis*	138	0.895	—	26.6[e]	79.0
Decalin, *trans*	138	0.872	—	30.8[e]	74.4
Tetralin	132	0.970	—	14.4[e]	94.0
Naphthalene	128	1.15	—	−8.54	130.0
Indene	116	0.991[g]	0.9974	10.4	98.2
Benzothiophene	134	1.1484[h]	1.1684	−10.4	129.6
Indole	117	—	1.2200	−15.5	(144.0)

[a]From TRC Tables of Thermodynamic Data.
[b] °API = (141.5/SpGr) − 131.5.
[c]D_i = 200(d − 0.5).
[d]Boduszynski, 1990. Unpublished result.
[e]These API gravity values are somewhat too high because they were calculated from densities measured at 20°C rather than from specific gravities measured at 60°F.
[f]Calculated by Hirsch's method (39).
[g]20/20°C.
[h]13/4°C

	16°C	20°C
d(water)	0.99913	0.99823

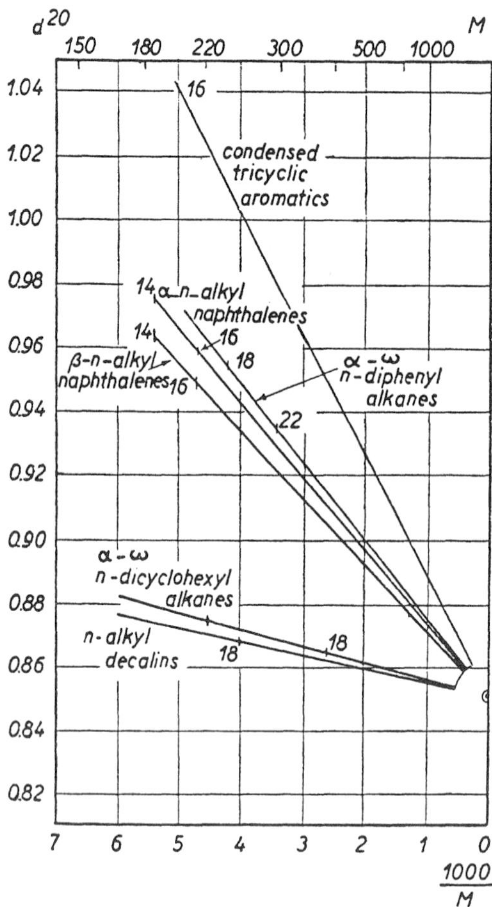

Figure 4.8 Smittenberg's plot of density versus reciprocal MW for *n*-paraffins and several *n*-alkyl ring compounds. (From Ref. 40. Reproduced with permission of the publisher.)

substitution of terms is acceptable. Crude oils with gravities between 10 and 20° API have been traditionally considered heavy. These are usually rich in high-density naphthenes, aromatics, and polar compounds. Boscan crude is a classic example with its API gravity of 10°. Light crudes, which have low densities (high API gravities), are rich in paraffins. Altamont crude oil (42°API) is an extreme example of those. For the residues of these light paraffinic crudes, the terms heavy and high boiling are no longer interchangable.

Figure 4.9 shows the API gravities of six crude oils and their fractions. Although, in general, the API gravity distinctly decreases with increasing boiling point, the atmospheric and vacuum residues of the paraffinic Altamont crude have remarkably high gravities (low densities). Even its vacuum residue is much less dense than the other entire crude oils. Its nondistillable residue (>1300°F, >700°C) has a density of only 0.888 (27.5° API), much less than the ("lighter") atmospheric residues (>650°F, >345°C) of the other five crudes which have densities ranging from 0.982 to 1.027 (12.6–6.3° API). In fact, even this *nondistillable* residue is lighter than four of the other whole crude oils. Another exception to the heavy–high-

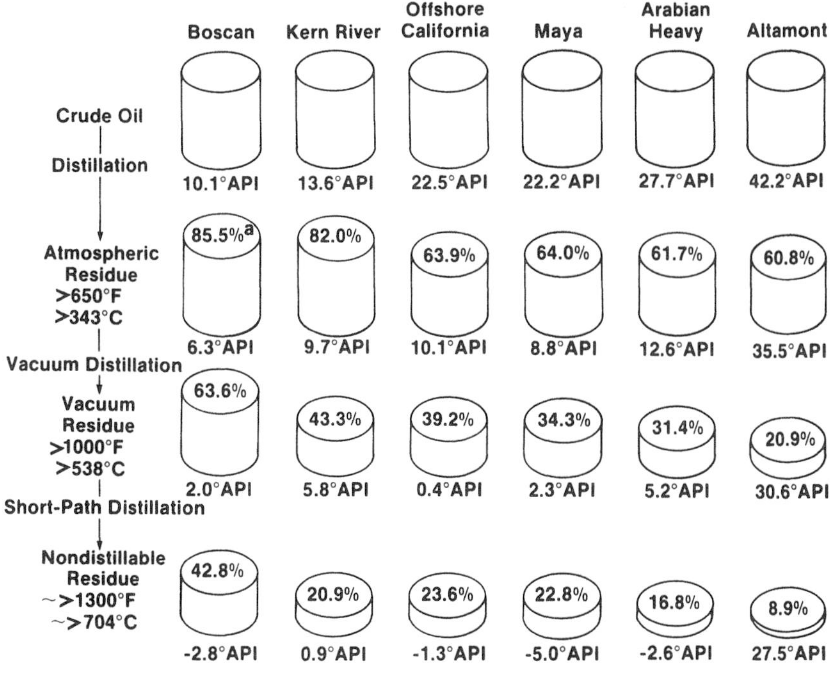

Figure 4.9 Effect of distillation on API gravity of various crude oils. (From Ref. 2. Reproduced with permission of the publisher.)

boiling interchangability is the nondistillable residue of our classic Boscan crude which is lighter (−2.8° API) than the corresponding residue of the otherwise lighter Mayan crude oil (−5.0° API).

(This pair of API gravities is one more example for the problems of this scale. The sample with −2.8° API is lighter than that with −5.0° API. On the other hand, a sample with 2.8° API is heavier than one with 5.0° API. One must step back and think to get it right. The corresponding values in incremental density would be 119.9 and 123.7 for the first pair, and 110.7 and 107.3 for the second pair. No room for confusion here. The greater number always stands for the higher density.)

An important feature of the inverse of density, the specific volume, is its additivity from atomic contributions. Actually, it is the molar volume, MV = *d*/MW, which is the sum of the atomic volumes. This additivity played an important part in the use of density for average structure determinations. In reality, other features in addition to atomic volumes are needed for the calculation of molar volumes. In alkyl benzenes, and even more so in condensed saturated as well as aromatic-ring systems, the so-called juncture carbon atoms differ from the regular carbons in their volume contribution. Montgomery and Boyd (41) exploited this feature and derived an equation for MV as the sum of paraffinic, regular (peripheral) naphthenic, regular (peripheral) aromatic, and internal naphthenic and internal aromatic carbon atoms.

Hirsch (39) refined this concept, distinguishing eight types of hydrocarbons and calculating correction terms for each class, "overlap volumes," in addition to the diverse structural contributions. Using graphical and regression methods, he evaluated 55 data points from various sources* and calculated the reduced molecular volumes for the 8 hydrocarbon types. Hirsch's equations serve two purposes. One is the possibility to obtain structural information from density measurements. The other is the calculation of molar volume or density of compounds whose density is unknown. The accuracy of his approach was quite good: about 1.5% average error for ring compounds without side chains and about 0.5% for compounds with side chains.

Satou et al. (42) arrived at almost equally good values for the molar volume of alkylnaphthenes and alkylbenzenes up to two rings per molecule as Hirsch did but by a much simpler method. Plotting the molar volumes of 160 hydrocarbon compounds, they obtained linear curves with the same slope but different intercepts. The intercept was smaller than that of par-

*Ref. 66 (Class 1); Ref. 47 (Classes 1–7); Ref. 67 (Class 3); Ref. 68 (Classes 4, 5, 6, 8); Ref. 69 (Classes 4, 5, 6, 8); Ref. 47 (Class 7).

affins by 20 ml/mol per aliphatic ring and by 40 ml/mol per aromatic ring. Thus, the molar volumes of such simple hydrocarbons can be calculated from those of the normal paraffin of equal carbon number by applying these simple corrections. The deviations for pyrene and other pericondensed aromatic hydrocarbons, however, are prohibitive. Further corrections may make it possible to include such structures, too, but we are not aware of such attempts.

2. Composite Density Relations

As we mentioned earlier, density in its various forms by itself or in combination with the boiling point has been used for the classification of crude oils and petroleum fractions. Some examples of such composite parameters are shown in Table 4.15. The Watson *K* factor, K_W, also called the UOP *K* factor, was selected by Watson and Nelson (43), two research chemists at UOP, to estimate the paraffinicity or naphthenicity of crude oils. It was widely employed for several decades and is still in use. However, according to Riazi and Daubert (44), K_W is not as good an indicator as the viscosity gravity constant (VGC) and the refractivity intercept, R_i. In Table 4.16, these parameters are compared to each other in their effectiveness to separate model hydrocarbon compounds into paraffins, naphthenes, and aromatics.

The viscosity gravity constant was introduced by Hill and Coats (45) as a simple and sensitive measure of paraffinicity and naphthenicity of crude oils and, thus, as a means for their characterization. They defined the VGC for viscosity measurements at either of two temperatures, 110 or 210°F:

Table 4.15 Use of Density (as API Gravity) in Different Criteria for the Characterization of Petroleum Samples

Characterization	°API range	*K* range[a]	CI range[b]
Paraffinic	>30	>12.2	<29.8
Intermediate	20–30	11.45–12.2	29.8—57.0
Naphthenic	<20	<11.5	>57.0

[a] *K* = UOP characterization factor (from Ref. 43):

$$K = \frac{\sqrt[3]{(50\%\ \text{AEBP} + 460)}}{\text{SpGr}} \quad \left(\text{originally } \frac{\sqrt[3]{\text{av. bp} + 460}}{\text{SpGr}}\right)$$

[b] CI = correlation index (from Ref. 70):

$$\text{CI} = 87552/(50\%\ \text{AEBP} + 460) + 473.7\text{SpGr} - 456.8$$

Table 4.16 Comparison of Several Parameters and Their Effectiveness in Separating Compounds of Different Hydrocarbon Types

Hydrocarbon type	Value range			
	MW	R_i	VGC	K_W
Paraffin	337–535	1.048–1.05	0.74–0.75	13.1–13.5
Naphthene	248–429	1.03–1.046	0.89–0.95	10.5–13.2
Aromatic	180–395	1.07–1.105	0.95–1.13	9.5–12.53

Source: Data from Ref. 44 (reproduced with permission of the publisher).

$$\mathrm{VGC} = \frac{10\mathrm{SpGr} - 1.0752\log(V_{100} - 38)}{10 - \log(V_{100} - 38)} \tag{4.9a}$$

and

$$\mathrm{VGC} = \frac{\mathrm{SpGr} - 0.24 - 0.022\log(V_{210} - 35.5)}{0.755} \tag{4.9b}$$

SpGr is the specific gravity of the sample at 60°F (15.6°C) and V_{100} and V_{210} are the Saybold universal viscosities at 100 and 210°F (37.8 and 100°C). For those who prefer metric quantities, Kurtz et al. (46) provided the equation

$$\mathrm{VGC} = \frac{d - 0.1384\log(V_{20} - 20)}{0.1526[7.14 - \log(V_{20} - 20)]} + 0.579 \tag{4.9c}$$

where d is the density and V_{20} the kinematic viscosity in centistokes, both at 20°C. All three equations give the same VGC value for a given sample. Riazi and Daubert (44) tested the VGC formulas with data from API Research Project 42 (47) and found that they cleanly separated hydrocarbons of the three types (paraffins, naphthenes, aromatics) into distinctly different groups without overlap; see Table 4.16.

The refractivity intercept, R_i, was proposed by Kurtz and Ward (48,49). It is defined by the equation $R_i = n - d/2$, where n is the refractive index, measured with light of the sodium d-line at 20°C, and d is the density, also measured at 20°C. As seen in Table 4.16, R_i too discriminates well between the different types of hydrocarbons.

In Chapter 5, Section II, we will discuss the use of density in combination with other properties to determine the distribution of paraffinic, naphthenic, and aromatic carbon in petroleum fractions.

B. Viscosity

In this book we treat the subject of viscosity only as far as it contributes to the compositional analysis of heavy petroleum samples. Thus, we barely touch on such popular subjects as blending formulas or temperature dependence. We restrict ourselves to three aspects:

1. Viscosity as a function of AEBP
2. Viscosity–MW and viscosity–structure relations
3. Intrinsic viscosity as a means to determine molecular size, i.e., molecular weight and configuration

1. General Principles

The viscosity of a fluid (or gas) is defined by Newton's Law of Viscous Flow:

$$f = \nu S \frac{dv}{dr} \tag{4.10}$$

which says the frictional force, f, which a fluid experiences flowing past a stationary plane surface, is proportional to the surface area, S, of this plane and to the velocity gradient, dv/dr, where dv is the difference in velocity between two adjacent layers and dr is the distance between them. The proportionality constant, ν, is the "coefficient of viscosity," or more simply, the "viscosity." Newton's law only holds for laminar flow, not for turbulent flow. The viscosity of a liquid, in contrast to that of a gas, increases with pressure and decreases exponentially with temperature. We may write

$$\nu = Ae^{\Delta E_{vis}/RT} \tag{4.11}$$

or

$$\ln \nu = A' + \Delta E_{vis}/RT \tag{4.12}$$

The quantity ΔE_{vis} is a measure of the energy barrier that must be overcome before flow can begin. According to Eyring (50), the activation energy of flow, ΔE_{vis}, can be visualized as the energy required to create a hole in the liquid large enough for a molecule to change place with a neighbor. Interestingly, in many cases it is one-fourth to one-third that of the vaporization energy, i.e., of the energy required to create a hole in the liquid large enough for a molecule to move into or out of. The term $e^{\Delta E_{vis}/RT}$ can be seen as a Boltzman factor determining the fraction of molecules in the liquid with the minimum energy to overcome the barrier. The larger the molecules and the greater the interactions between them, the greater the required energy and the higher the viscosity. Qualitatively, the activation energy of flow, ΔE_{vis}, and, thus, ν can be deduced from the vaporization energy. Quantitatively, large differences are observed with hydrogen-bonding materials and others prone to association.

The common unit of viscosity is the centipoise (cP). Water at 20°C (68°F) has a viscosity of 1.005 cP. The viscosity of several small hydrocarbon molecules are given in Table 4.17. The differences in these simple molecules between *n*-alkanes, alkylcyclohexanes, and alkylbenzenes are quite small and determined only by their bulkiness. The same holds true for the viscosities of low-boiling oil fractions, but not for the polar, high-boiling ones where interactions come into play. Here, viscosities may rise very sharply with AEBP as shown in Table 4.18.

Most viscosities are measured with capillary viscometers. Fluids of low and intermediate viscosity are run in viscometers under the driving force of gravity rather than by externally applied pressure. In those cases where the liquid flows by gravity, the *kinematic* viscosity, η, is measured. The kinematic viscosity is the absolute viscosity (in cP) divided by the density, and its unit is the centistoke (cS).

We have already mentioned that the viscosity decreases with increasing temperature. The reason is that at higher temperature more fluid molecules have the energy required to create a hole in the liquid, allowing them to move in or out. For lubricants and other petroleum fractions, the ASTM viscosity–temperature chart is the standard method of plotting the viscosity–temperature data for interpolation and extrapolation. It has no basis in theory, but it is quite general and holds over a large temperature range, providing straight lines. The underlying double logarithmic relationship was developed by Wright (51) and has the form

$$\ln \ln Z = A + B \ln T \qquad (4.13)$$

where

Table 4.17 Viscosity (in cP) of Some Pure Hydrocarbons at 50°C (122°F)

C-no.	*n*-Alkanes	*n*-Alkylcyclohexanes	*n*-Alkylbenzenes
6	0.33	0.604	0.435
10	0.617	0.828	0.682
15	1.586	2.19	1.740
20	3.306	4.58	3.60

Source: TRC Tables of Thermodynamic Data.

Table 4.18 Viscosity at 50°C (122°F) of Crude Oil Distillates. Values under 100 are in Centistoke, those above 100, in Centipoise

Crude oil	°API		Mid-AEBP						
		°F	300	350	450	575	730	910	1150
		°C	150	175	230	300	390	490	620
Altamont[a]	42.2		—	0.87	1.54	3.10	9.1	23	—
Arabian Light[b]	33.3		0.67	—	1.39	2.99	13.2	—	—
Arabian Heavy[a]	27.7		—	—	1.49	3.38	15	83	—
Alaskan NS[a]	27.6		—	0.84	1.49	3.48	16.6	230	23,000
Maya[a]	22.2		—	—	1.53	3.72	18.9	122	—
Kern River[a]	13.6		—	—	—	—	47.7	2600	2×10^6

Crude oil		Mid-AEBP					
	°F	540	647	720	783	852	—
	°C	285	340	280	217	455	—
Midcontinent[c]	3.1	7.05	12.1	27.2	60	—	

Sample		Mid-AEBP					
	°F	—	600	670	785	—	—
	°C	—	315	355	420	—	—
Cracking residue from midcontinent[c]		—	6.0	15	102	—	—

[a]Chevron Research and Technology Company.
[b]Reference 71.
[c]Reference 43.

$$Z = \eta + 0.7 + \exp[F(Z)] \tag{4.14}$$

$F(Z)$ becomes negligible at viscosities >2 cS, i.e., for heavy petroleum fractions. The constants A and B for a given sample can be determined from viscosity data at two temperatures. Then the viscosity at any temperature can be calculated from T.

2. Viscosity as a Function of AEBP

Figure 4.10 shows the viscosities of distillation cuts from several crude oils plotted versus their mid-boiling-points. The corresponding data are listed in Table 4.18. We see the same general type of curves as in the MW–AEBP plot, except that the viscosity curves of different crudes diverge

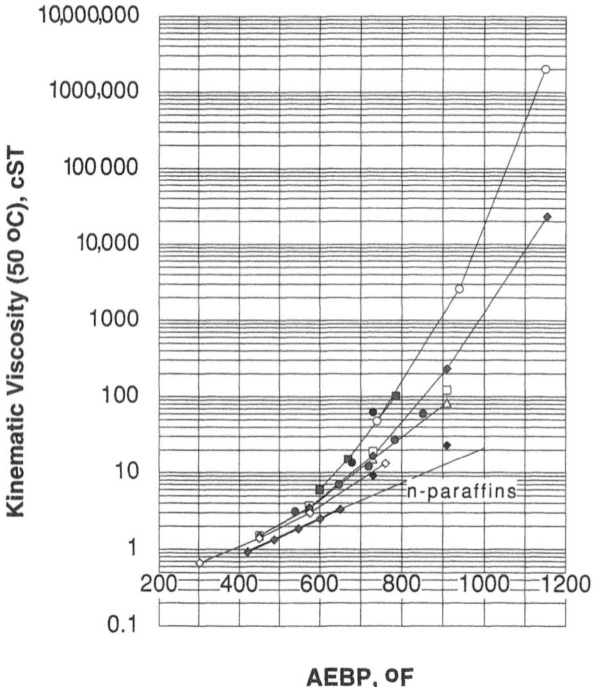

Figure 4.10 (Kinematic) viscosity of selected crude oils versus their AEBP. Values <100 are in centistoke; those above are in centipoise. ○—Kern River (Ref. 4); ■—distillate from cracking resid of Midcontinent (Ref. 43); ●—Midcontinent (Ref. 43); ◇—Alaskan North Slope (Ref. 4); □—Maya (Ref. 4); △—Arabian Heavy (Ref. 4); ◈—Arabian Light (Ref. 71); ◆—Altamont (Ref. 4).

considerably, indicating a fairly strong compositional effect. Furthermore, the curve of the normal paraffins is the one closest to the bottom right. The paraffinic Altamont crude oil data are only slightly higher, whereas the biodegraded, naphthenic Kern River crude and the cracking residue of Midcontinent (43) give the steepest curves. Obviously, interactions caused by aromaticity and polarity have a much greater effect on viscosity than on boiling point.

Twu (52) correlated, with rather high accuracy, the kinematic viscosities, η, of *n*-alkanes in the range C_1-C_{100} and 210°F with their boiling points, using API 42 data (47). These are his equations:

$$\ln(\eta_{210} + 1.5) = 4.73227 - 27.0975\alpha + 49.4491\alpha^2 - 50.4706\sigma^4 \quad (4.15)$$

and

$$\ln(\eta_{100}) = 0.801621 + 1.37179 \ln(\eta_{210}) \quad (4.16)$$

with

$$\alpha = 1 - T_b/T_c \quad (4.17)$$

and

$$T_c = T_b(0.533272 + 0.191017 \times 10^{-3}\, T_b + 0.779681 \times 10^{-3}\, T_b^2 + 0.284376 \times 10^{10}\, T^3 + 0.959468 \times 10^{28}/T^3) \quad (4.18)$$

T_b is the normal boiling point and T_c the critical temperature, both in °Rankine.

Twu needed these equations as reference data for a bigger project, namely, the calculation of kinematic viscosity of oil samples from specific gravity and boiling point by means of a perturbation expansion method. Because SpGr and T_b are routinely determined for most oils, such a method could save the additional effort of measuring viscosity. For this project, Twu regressed a total of 563 data points, assembled from 6 sources, and arrived at a correlation which he shows in the form of 20 equations. A test of these equations with the data from Table 4.19 showed most predicted viscosities to be within about ±100% except for those higher than 200 cS; see Table 4.19. This accuracy is quite good, considering the large range of viscosities covered (from 1 to about 1,000,000).

Twu presented his correlation also in the form of two figures with η at 100 and 210°F plotted versus API gravity using the Watson characterization factor $K = T_b(°R)^{1/3}/\text{SpGr}$ as the parameter for the second independent variable. When checking these graphs with experimental data of Table 4.19, we found them to be less accurate than the equations. In our eval-

Table 4.19 Kinematic Viscosities of Heavy Distillation Cuts Estimated from Boiling Point and Specific Gravity by Two Methods

Crude oil	Rounded Mid-AEBP[a] (°F)	SpGr[a]	Measured Vis(122) (cS)	Calculated Vis(122) Altgelt	Vis(122) Twu 1	Vis(100) Twu 2
Kern River	740	0.9580	47.7	67	21	18
	940	0.9861	2,600	716	160	1,500
	1150	1.0151	2×16^6	2.1×10^6	4,440	—
Arabian Heavy	450	0.8132	1.5	0.9	4.5	—
	575	0.8550	3.4	2.4	6.8	—
	730	0.9071	15.0	12.7	16.5	3
	910	0.9390	83	87	76	60
Altamont	450	0.7919	1.5	1.6	4.5	—
	575	0.8080	3.1	2.6	6.5	—
	730	0.8319	9.1	5.5	12.6	2
	910	0.8496	23	13.6	34	4

Maya	450	0.8270	1.5	1.6	4.5	0.7
	575	0.8702	3.7	4.0	6.9	0.9
	730	0.9170	18.9	19.2	17.1	4
	910	0.9516	122	157	80	70
Alaskan	450	0.8388	1.5	1.6	4.5	0.5
North Slope	575	0.8735	3.5	3.8	6.9	0.9
	730	0.9403	16.6	58	18.3	9
	910	0.9497	230	203	83	70
	1155	0.9870	23,045	5,200	2,418	>1,000
n-Alkanes						
C_{10}	345	0.7301	0.9	1.0	3.8	1.0[b]
C_{20}	651	0.7924	3.3	3.2	4.2	5.5[b]
C_{30}	841	0.8133	10[c]	7.0	18.3	16.8[b]
C_{40}	972	0.8241	20[c]	12.7	35.5	37.3[b]

[a]AEBP data of crude oil fractions from Chevron Research and Technology Company.
[b]Extrapolated from log(liquid viscosities) versus AEBP of lower *n*-alkanes (Fig. 4.9).
[c]Calculated by Twu's equations for *n*-alkanes.

uation, Twu's equations can be used for most oils with good success, but not his figures as published.

Altgelt (53) derived another, purely empirical, correlation for the same purpose:

$$\eta(50°\mathrm{C}) = f(\mathrm{MBP}f\{f(\mathrm{SpGr})\}(\mathrm{SpGr})^8(\mathrm{SpGr}_{\mathrm{ref}})^4 \qquad (4.19)$$

with MBP = mid-boiling-point, $f(\mathrm{MBP}) = [(1600/\mathrm{SpGr}^2 - \mathrm{MBP})/1000]^5$, $f(\mathrm{SpGr}) = (1.0153 - \mathrm{SpGr})^{1.09}$, ref = reference SpGr, i.e., SpGr of the corresponding fraction with a mid-boiling-point around 730°F, and $f\{f(\mathrm{SpGr})\} = f(\mathrm{SpGr})$ for $f(\mathrm{SpGr}) < 0.022$; whereas, for $f(\mathrm{SpGr}) > 0.022$,

$$f\{f(\mathrm{SpGr})\} = (0.017 + f\{f(\mathrm{SpGr})\} - \mathrm{SpGr})^{\mathrm{Exp}}$$

with

$$\mathrm{Exp} = \frac{1}{\mathrm{SpGr}^{1.2}} - \frac{0.003}{f\{f(\mathrm{SpGr})\}} \frac{d\{f(\mathrm{SpGr})\}}{0.04} \mathrm{SpGr}^3$$

His predictions are also listed in Table 4.19, together with that of Twu. Altgelt's correlation seems to give somewhat better predictions than Twu's, especially for the very high boiling cuts. However, both methods yield only approximate results. Actually, it is surprising that the estimated values are as close as they are, considering the diverse nature of the crude oils and their fractions. The viscosities of some samples, e.g., the high-boiling cuts of Kern River and Maya, are clearly determined by strong interactions due to their polar nature in terms of both heteroatom content and aromatic-ring condensation. On the other hand, the paraffinic Altamont fractions have unusually low interactions even in the high-boiling cuts and, consequently, have comparatively low viscosities.

3. Intrinsic Viscosity

The intrinsic viscosity, $[\eta]$, is a concept widely used in polymer chemistry. It is defined as the extrapolation of the reduced specific viscosity, $\langle\eta\rangle$, to infinite dilution, where $\langle\eta\rangle$ is the difference of sample solution and solvent viscosities divided by the solvent viscosity and concentration:

$$\langle\eta\rangle = \frac{\eta_{\mathrm{solut}} - \eta_{\mathrm{solv}}}{\eta_{\mathrm{solv}} c} \qquad (4.20)$$

and

$$[\eta] = \lim_{c \to \infty} \langle\eta\rangle \qquad (4.21)$$

[η] is a measure of molecular size. From [η] and MW, we can draw conclusions about the molecular configuration. This is what Kyriacou et al. (1) did for a vacuum residue (VR) sample. They separated their VR by size exclusion chromatography into nine fractions and measured the MWs and [η]. The [η] values, measured at three temperatures and in two solvents, are listed in Table 4.20. The values obtained in 1-methylnaphthalene at 50°C (122°F) are practically the same as those measured at 10 and 20°C (50 and 68°F) in THF, at least they are not lower. This agreement is no proof of nonaggregation because 1-methylnaphthalene is not a very polar solvent, and 50°C is still a moderate temperature. On the other hand, the molecular weights determined by VPO under even milder conditions—in benzene at 40°C (104°F)—are rather low, which also suggests little, if any, aggregation.

In Fig. 4.11, the intrinsic viscosities measured at 20°C in THF by Kyriacou et al. (see Table 4.20) are plotted versus MW together with similar data by Altgelt (54). The latter include intrinsic viscosities of various SEC fractions obtained from the pentane solubles (maltenes) and an aromatics fraction of a Boscan asphalt. All data points fall fairly close to a straight line with a slope considerably steeper than that of a randomly coiled polymer (polyisobutene), which is also shown for comparison. In view of the scatter of the points, we feel that, without further verification, a detailed

Table 4.20 Intrinsic Viscosities of the Fractions of a Vacuum Residue Measured Under a Variety of Conditions

		[η] ($cm^3\ g^{-1}$)		
Fraction	MW VPO	THF 10°C	THF 20°C	1-Methylnaphthalene 50°C
1	1570[a]	8.45	8.20	9.12
2	1544	7.74	7.56	8.01
3	1488	5.56	5.31	7.69
4	1239	5.00	4.49	6.36
5	1033	3.60	3.56	5.37
6	952	3.21	3.05	5.27
7	633	2.75	2.70	3.35
8	553	2.66	2.63	2.22
9	357	1.18	1.15	1.38

[a]Corrected (by KHA) for 7 wt% butylated hydroxy toluene left over from the THF used for the SEC separation.

Source: Data adapted from Ref. 1.

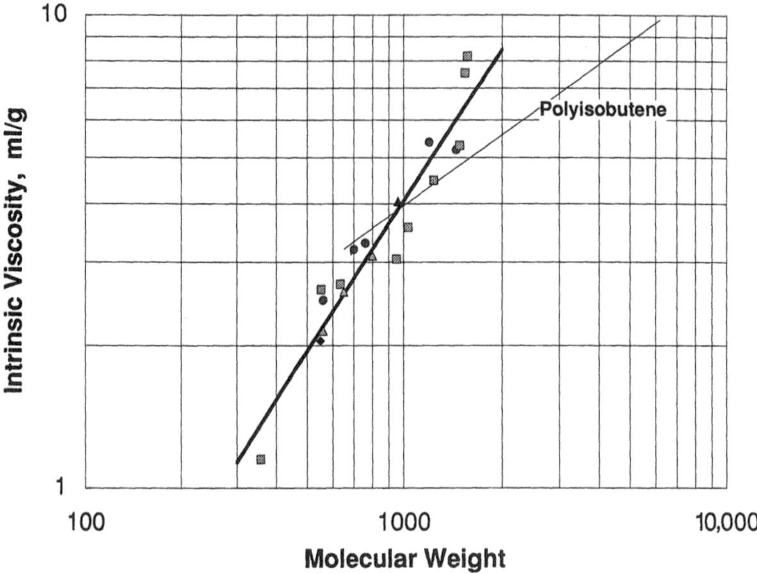

Figure 4.11 Intrinsic viscosity of SEC fractions obtained from a vacuum residue and from the pentane maltenes of a Boscan asphalt versus MW. ●—SEC fractions of n-C_5 maltenes from Boscan asphalt (1); ♦—saturates from a low-MW SEC fraction of Boscan asphalt (1); △—polars from low and medium MW SEC fraction of Boscan asphalt (1); ▲—low MW SEC fraction of n-C_5 asphaltenes from Boscan asphalt (1); ▣—SEC fractions of cold lake bitumen VR (2); (—)—polyisobutene curve (3). Note: 1–Ref. 54; 2–Ref. 1; 3–Ref. 54.

model calculation of the kind done by Kyriacou et al. (1) can only be tentative.

The fact that the curve of the petroleum fractions crosses the polymer curve deserves consideration. In simple terms, this means that below a MW of about 1000, petroleum molecules are more compact than the polymer molecules, and above this limit they are either less compact or less spherical than coiled polymer molecules, e.g., rodlike or, more plausably, like platelets. In view of the content of aromatic- and naphthenic-ring systems, the platelet hypothesis appears reasonable. Furthermore, there is some experimental evidence for this shape from small-angle neutron scattering measurements by Ravey et al. (6).

The picture is quite different for asphaltene fractions, i.e., for fractions obtained from *previously precipitated* alkane insolubles. In Fig. 4.12 we

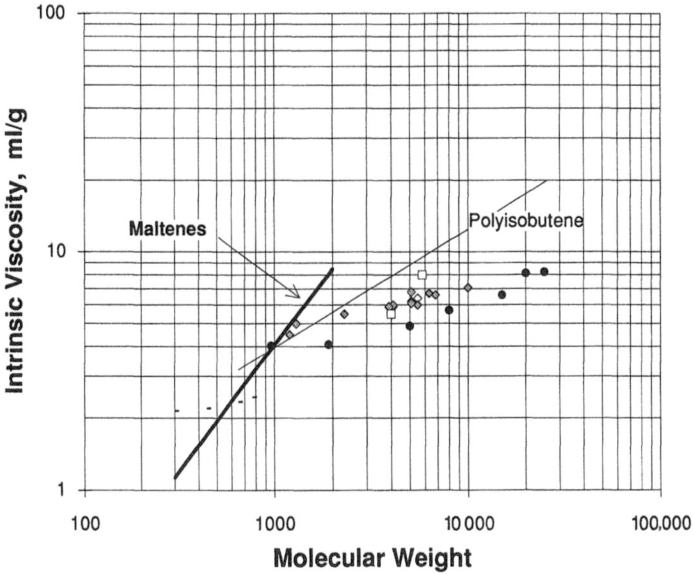

Figure 4.12a Intrinsic viscosity of asphaltenes and asphaltene fractions versus MW. (Data adapted from Refs. 12, 21, 57, and 58.) Altgelt (12): ●—Boscan *n*-pentane asphaltene SEC fractions. Reerink (21): ◈—Kuwait *n*-pentane asphaltene fractions by *n*-heptane reprecipitates; □—*n*-heptane asphaltenes from two crude oils, unfractionated. Sakai (57,58): (–) Petroleum pitch asphaltene fractions (benzene–*n*-hexane).

plotted the intrinsic viscosity of whole asphaltene samples and their fractions (from Refs. 12, 21, 57, and 58) against MW. Here all the asphaltene points lie below the polymer curve. Altgelt and Harle (56) suggested stacked aggregation as an explanation. In their view, asphaltene molecules look like platelets which aggregate in stacks, if not perfectly, yet forming relatively compact particles. The fractions then differ primarily by their particle rather than their molecular size. Moreover, their particle weight increases linearly with the degree of aggregation, but, because of their increasing compactness, the viscosity rises at a much smaller rate. The observed aggregation into such compact shapes may be a consequence of prior asphaltene precipitation.

The data by Sakai et al. (58) are very different from Altgelt's (12) and Reerink's (21). Their asphaltenes were derived from petroleum pitch, i.e., from thermally cracked material, and are very different from regular pe-

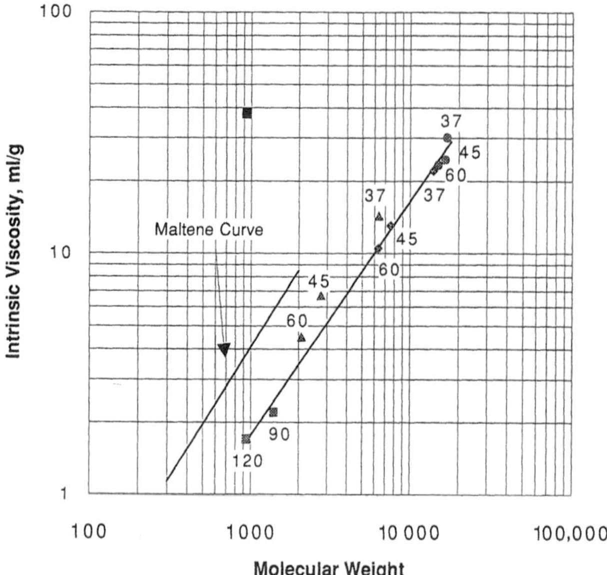

Figure 4.12b Intrinsic viscosity of asphaltene fractions versus MW. Whole *n*-hexane asphaltenes from northern Iraq crude oil in different solvents and at various temperatures. (Data from Ref. 5.) ◎—benzene; △—chlorobenzene; ◇—THF; ▨—nitrobenzene; ■—nonaggregated asphaltenes at 25°C (our extrapolation). The numbers next to the symbols indicate the temperature in °C.

troleum asphaltenes, having an even greater tendency to aggregate than those.

Al-Jarrah and Al-Dujaili's data (5) do not fit this graph at all. If we take their MW and [η] in benzene at 37°C, their lowest temperature, the point would be higher than the other asphaltenes by a factor of 5, and if we accept their lowest MW, 935, as correct and pair it with the extrapolated [η] at 25°C, the point would be exactly 10 times as high as the corresponding viscosity on the oil and maltene curve for this MW. Possibly, all their intrinsic viscosities were misprinted and are high by a factor of 10.

We will say more on the subject of asphaltene aggregation and viscosity results in Chapter 10. Here we can summarize: We think measurements of [η] can complement MW measurements and point to gross structural features or indicate aggregation effects as demonstrated by the different behavior of asphaltenes, pentane soluble petroleum components, and linear polymers.

IV. CHEMICAL PROPERTIES OF HEAVY PETROLEUM FRACTIONS

A. C and H Concentrations, Hydrogen Deficiency

It is customary to distinguish among paraffinic, intermediate, naphthenic, and, sometimes, asphaltic crude oils even though there are now better ways for their categorization (59,60). The distinction is primarily based on differences in density (see, e.g., Tables 4.12 and 4.15); the overall carbon concentration and the atomic H/C ratio vary only within narrow limits (83–87 and 1.4–1.9, respectively). Greater differences in these latter properties show up between different fractions of the same crude oil and also between the equivalent high-boiling fractions of different crude oils; see Table 4.21. The lowest-boiling members of a crude oil—propane and butanes—are obviously purely paraffinic, whereas the high-boiling cuts, and especially the residues, can contain 40–50% aromatic carbon and up to 7% heteroatoms.

The H/C ratio is a popular measure of the nonparaffinicity and widely used for the assessment of crude oils and refinery streams as well as for that of other fossil fuel samples. A related, very descriptive quantity is the hydrogen deficiency value, Z, which was introduced by mass spectroscopists and is defined by the formula $C_nH_{2n+Z}X$. X stands for one or several heteroatoms. For paraffins, $Z = 2$, giving the familiar formula C_nH_{2n+2}. It decreases by 2 for every double bond and for every ring closure. Thus, cyclohexane (one ring) has a Z value of 0, and benzene (one ring and three double bonds) has one of -6.

We have listed the Z-numbers and H/C ratios of several types of saturated and aromatic-ring systems in Table 4.22. Here we see both the strength and the weakness of this concept. On the one hand, the Z-numbers are fairly specific and vary from $+2$ to -52 for the compounds listed in Table 2.2. For petroleum compounds with very high AEBPs, Z can be lower than -200, as shown in Figure 4.13. On the other hand, the same Z-number can represent different compound types. For instance, alkyl benzenes and tetracycloalkanes both have a Z-number of -6.

The same confusion arises from the H/C ratios. In certain ways, the two numbers (Z and H/C) complement each other. Both depend on the number of rings and double bonds. In addition, at least for the lower members, the H/C ratio changes with the length of the alkyl groups. H/C varies continuously from 4.0 for methane to zero for fullerenes. Other examples of compound types with different Z-numbers, including some heterocompounds, were given in Chapter 2. Figure 4.14 gives an impression of the H/C ratios of petroleum fractions and their change with AEBP.

Table 4.21 Elemental Composition of Selected Crude Oils

Crude oil AEBP range °F		C (%)	H (%)	H/C (%)	S (%)	Total N (ppm)	Basic N (ppm)
Altamont							
Whole crude		85.9	14.1	1.97	0.04	—[a]	—
>300	Lt fractions	85.3	14.7	2.07	0.005	—	—
300–400	Hvy naphtha	85.3	14.7	2.07	0.001	—	—
400–500	Kerosene	85.5	14.5	2.04	0.003	0.95	—
500–650	AGO	84.6	14.4	2.04	0.02	15	5
650–865	LightVGO	85.6	14.4	2.02	0.03	60	20
865–1000	Heavy VGO	86.2	14.0	1.95	0.04	135	45
>650	Atm residue	86.2	14.1	1.96	0.05	235	—
>1000	Vac residue	85.8	13.2	1.85	0.09	580	150
Arabian Heavy							
Whole crude		84.5	12.5	1.78	2.8	—	—
>300	Lt fractions	85.0	14.9	2.10	0.01	—	—
300–400	Hvy naphtha	84.5	14.4	2.04	0.11	0.3	<1
400–500	Kerosene	84.4	14.1	2.00	0.46	1.0	<2
500–650	AGO	84.4	13.1	1.86	1.5	30	15
650–810	Light VGO	85.4	11.7	1.64	2.9	475	89
810–1000	Heavy VGO	84.6	12.2	1.73	3.1	940	208
>1000	Vac residue	83.5	10.9	1.57	5.5	4600	661

1025[b]		84.11	11.15	1.58	4.04	1530	330
1090[b]		83.25	10.66	1.53	4.25	2040	370
1145[b]		83.61	10.83	1.54	4.09	1950	380
1205[b]		83.51	10.71	1.53	4.64	1960	400
1305[b]		83.47	10.70	1.53	4.79	2700	520
(1560)[c]	SEF-1[d]	83.89	10.22	1.46	5.44	4000	—
(2310)[c]	SEF-2[d]	81.68	7.78	1.14	7.57	8900	—
(2570)[c]	SEF-3[d]	82.60	7.53	1.09	8.04	10000	—
Alaskan North Slope							
Whole crude		—	12.3	1.72	1.47	1980	—
>300	Lt fractions	85.3	14.7	2.07	0.002	—	—
300–400	Hvy naphtha	86.3	13.7	1.90	0.03	0.2	—
400–500	Kerosene	86.5	13.2	1.83	0.11	1.5	1
500–650	AGO	86.0	12.8	1.79	0.55	88	40
650–840	Light VGO	86.5	12.2	1.69	1.10	1020	—
840–1060	Heavy VGO	85.6	11.7	1.64	1.45	2500	—
1060–1250	Molec. dist	88.6	11.4	1.54	1.96	4400	—
>650	Atm residue	88.1	11.4	1.55	1.82	3450	—
>1060	Vac residue	85.6	10.5	1.47	2.77	6800	—
>1250	Nondist resid	86.5	9.9	1.37	3.25	7600	—

[a]Dash stands for "not determined in this table."

[b]Mid-AEBPs of short-path distillates.

[c]Mid-AEBPs of solubility fractions calculated from VPO MWs.

[d]Solubility fractions of nondistillable residue.

Source: Chevron Research and Technology Company.

Table 4.22 *Z*-Numbers and H/C Ratios of Various Hydrocarbon Molecules. $Z = -2[(R + DB) - 1]$

Hydrocarbon	Formula	No. of rings	No. of double bonds	*Z*-number	H/C ratio
Saturates					
n-octane	C_8H_{18}	0	0	2	2.250
Iso-octane	C_8H_{18}	0	0	2	2.250
Hexane	C_6H_{14}	0	0	2	2.333
Hexene	C_6H_{12}	0	1	0	2.000
Cyclohexane	C_6H_{12}	1	0	0	2.000
Cyclohexene	C_6H_{10}	1	1	−2	1.667
Decalin	$C_{10}H_{18}$	2	0	−2	1.800
Perhydrophenanthrene	$C_{14}H_{24}$	3	0	−4	1.714
Perhydropyrene	$C_{16}H_{26}$	4	0	−6	1.625
Perhydropicene	$C_{22}H_{36}$	5	0	−8	1.636
Perhydroperylene	$C_{20}H_{32}$	5	0	−8	1.600
Aromatics					
Benzene	C_6H_6	1	3	−6	1.000
Toluene	C_7H_8	1	3	−6	1.143
Hexyl benzene	$C_{12}H_{18}$	1	3	−6	1.500
Dodecyl benzene	$C_{18}H_{30}$	1	3	−6	1.667
Tetralin	$C_{10}H_{14}$	2	3	−8	1.004
Naphthalene	$C_{10}H_8$	2	5	−12	0.800
Fluorene	$C_{13}H_{10}$	3	6	−16	0.769
Phenanthrene	$C_{14}H_{10}$	3	7	−18	0.556
Chrysene	$C_{18}H_{12}$	4	9	−24	0.667
Pyrene	$C_{16}H_{10}$	4	8	−22	0.625
Perylene	$C_{20}H_{12}$	5	10	−28	0.600
Benzo[ghi]perylene	$C_{22}H_{12}$	6	11	−32	0.583
Coronene	$C_{24}H_{12}$	7	12	−36	0.500
Ovalene	$C_{32}H_{14}$	10	17	−52	0.438

In addition, note the H/C ratios of some major (very broad) fractions of the paraffinic Altamont crude in Table 4.21. As to be expected, they start out at the same level (2.07) as the light fractions of the other, more aromatic–naphthenic crude oils. But the value of the Altamont vacuum residue (1.85) is about the same as those of much lower-boiling fractions from heavier crude oils, e.g., the atmospheric gas oil from Arabian Heavy and the kerosine fraction from Alaskan North Slope.

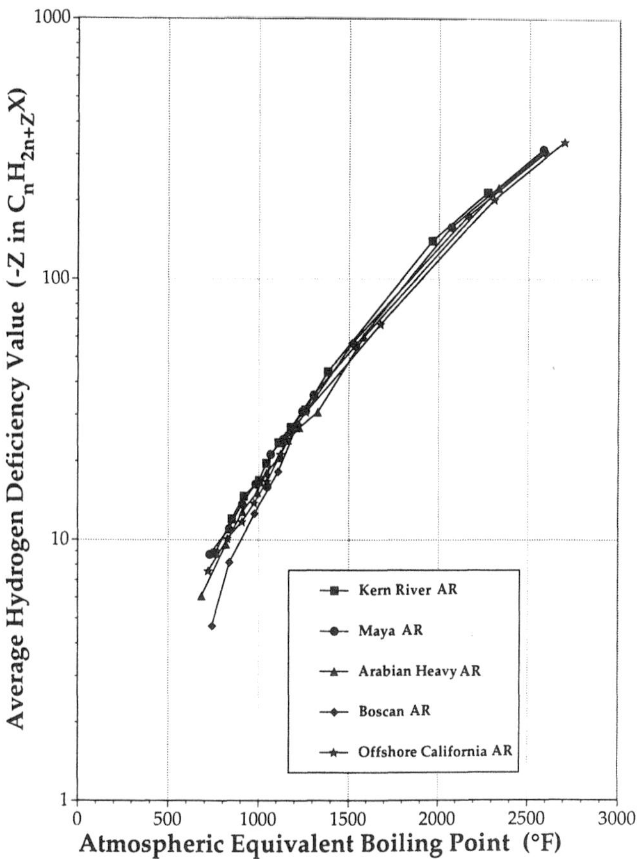

Figure 4.13 Hydrogen deficiency as a function of AEBP for five heavy crude oils. (From Ref. 61. Reproduced with permission of the publisher.)

B. Heteroatom Concentrations

Sulfur is the most abundant heteroatom in crude oils, followed by nitrogen, oxygen, and metals. Most crude oils contain between 0.01 and 1 wt% S. A much smaller group may have between 1 and 10% S (59). The nitrogen content is usually much lower, with an average of about 0.1 wt%. Basic nitrogen accounts for 12–35% of total N, depending on boiling point and crude oil origin. General oxygen data are scarce. Mostly, oxygen contents of crude oils are low, also around 0.1 wt%. Organometallic compounds in

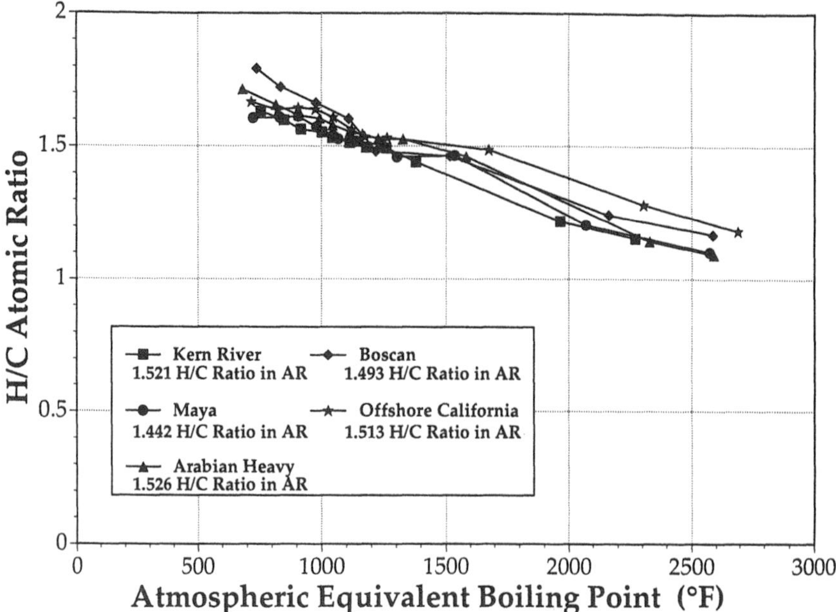

Figure 4.14 H/C ratio as a function of AEBP for five heavy crude oils.

petroleum contain predominantly vanadium, and nickel and, to a lesser extent, iron. Other metals have also been reported (62), but systematic data are available only for V and Ni. These may be present in greatly variable amounts, ranging from <1 ppm to 1200 ppm V and 150 ppm Ni, depending on the crude oil (59).

Going back to Table 4.21, we find the elemental composition of three crude oils: Altamont, a very paraffinic light crude, and Arabian Heavy and Alaskan North Slope, two representative heavy crude oils. Even this limited selection gives an impression of the great changes in composition with increasing AEBP. The low-boiling cuts consist almost entirely of hydrocarbons with some sulfur (up to 1 wt% in some cases) and a few ppm N. Sulfur levels rise from <0.01 wt% for light distillates to >3 wt% in nondistillables. SEF-3 fractions were found to contain as much as 8% sulfur (Fig. 4.16). Nitrogen and oxygen start at much lower levels, <0.2 ppm ($<2 \times 10^{-5}$ wt%) for light and even middle distillates. They may rise to between 1 and 2.5 wt% for nitrogen (Fig. 4.18) and to between 0.3 and 1.5 wt% for oxygen in SEF-3 fractions.

In Table 4.23, we take a closer look at the S and N distribution of three crude oils as a function of boiling point. The atmospheric residues of these crudes had been separated by short-path distillation into 5–10 cuts, and the nondistillable residues were further fractionated by solubility into three more fractions. Thus, the moderate increase in S content with boiling point

Table 4.23 Sulfur and Nitrogen Contents of Various Crude Oils as a Function of AEBP

Atmospheric residue	Fraction	Mid-AEBP (°F)	S (%)	N (%)
Arabian Heavy	Cut 1	662	2.62	0.03
	Cut 2A	795	3.14	0.07
	Cut 2B	890	3.36	0.10
	Cut 3A	972	3.75	0.12
	Cut 3B	1025	4.01	0.15
	Cut 4A	1090	4.27	0.20
	Cut 4B	1145	4.22	0.20
	Cut 5A	1205	4.62	0.19
	Cut 5B	1305	4.84	0.27
	SEF-1	1560	5.45	0.39
	SEF-2	2310	7.54	0.88
	SEF-3	2570	8.02	0.98
Kern River	Cut 1	720	0.99	0.48
	Cut 2A	825	1.02	0.59
	Cut 2B	885	1.11	0.67
	Cut 3A	955	1.19	0.81
	Cut 3B	1025	1.34	0.96
	Cut 4A	1090	1.39	1.07
	Cut 4B	1160	1.34	1.20
	Cut 5	1270	1.30	1.29
	SEF-1	1345	1.33	1.41
	SEF-2	1925	1.45	2.21
	SEF-3	2230	1.43	2.43
Boscan	Cut 1	720	4.47	0.16
	Cut 2	815	4.57	0.22
	Cut 3	955	4.68	0.32
	Cut 4	1090	5.56	0.48
	Cut 5	1195	5.67	0.63
	SEF-1	1495	6.14	0.63
	SEF-2	2140	7.07	1.53
	SEF-3	2570	7.11	1.79

is demonstrated in great detail as is the totally different behavior of the N content with its initially slow but finally steep increase. Figures 4.15 and 4.16 show these trends in graphic form for five crude oils.

Table 4.24 gives more examples of the sulfur and nitrogen contents in crude oils. Also listed are S and N contents in the respective atmospheric and vacuum residues and the sums of S and N.

More detail about the distribution of these quantities is presented in Tables 4.25–4.27, which give the average molecular parameters for several molecular distillates and SEF fractions of three crude oils. Table 4.25, for instance, shows the data for nine molecular distillates and three SEF fractions of Arabian Heavy crude. The first two molecular distillation cuts of these, 1 and 2A, together would make up the light VGO of this crude oil; cuts 2B and 3A represent the heavy VGO, and all the fractions beyond that constitute the vacuum residue (>1000°F, >540°C). The remaining molecular distillation cuts and the first SEF fraction together are equivalent to the so-called maltenes (the pentane-soluble portion of a vacuum residue or asphalt), and the SEF fractions 2–4 may be viewed as subfractions of what would traditionally be called asphaltenes. These latter two equivalencies are only approximate as discussed in Chapter 10. More about SEF fractions is found in Chapters 3 and 10.

Tables 4.25–4.27 present the *average molecular parameters* rather than just the percentages of the various elements. This has the advantage of giving a clearer impression of a sample's chemical composition. It is more instructive, for instance, to learn that in Cut 5 of Kern River crude every molecule contains 0.75 nitrogen atoms on average than to see an abstract number of 0.85 weight % N. A number of 0.75 N per molecule can be immediately translated into 75 N atoms per 100 molecules, whereas a weight percent number remains abstract. The average molecular parameters, of course, depend critically on the molecular weight. This is no problem with the relatively well-behaved distillates. However, as we have seen earlier, the MWs of the polar fractions of truly nondistillable residues are not unequivocally accessible. Even so, this concept is so powerful in conveying a graphic picture of the composition of a crude oil and its increasing complexity with increasing AEBP that some uncertainty in the MW of the heaviest fractions can be tolerated in exchange for clarity.

The change in composition with increasing boiling point is dramatic. Although light distillates (naphtha) have hardly any heteroatoms other than some sulfur, the heavier ones contain steadily increasing amounts of S and especially of N and O as their boiling point goes up. Some examples of this increase are shown in Figs. 4.17 and 4.18. The same trend is observed with H/C ratio (except that it goes down) and with the hydrogen deficiency, Z (which becomes more negative); see Fig. 4.13. Compare the information

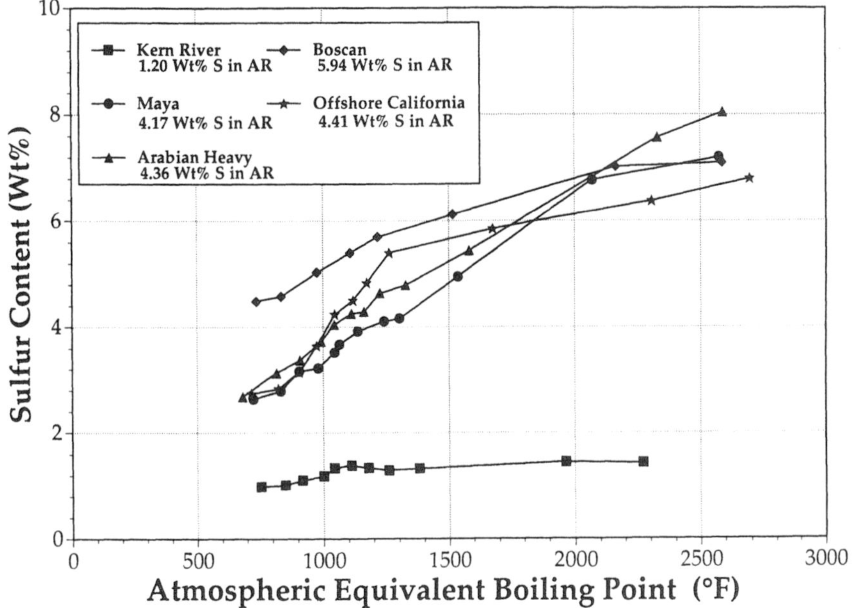

Figure 4.15 Sulfur content (wt%) of five heavy crude oils as a function of AEBP.

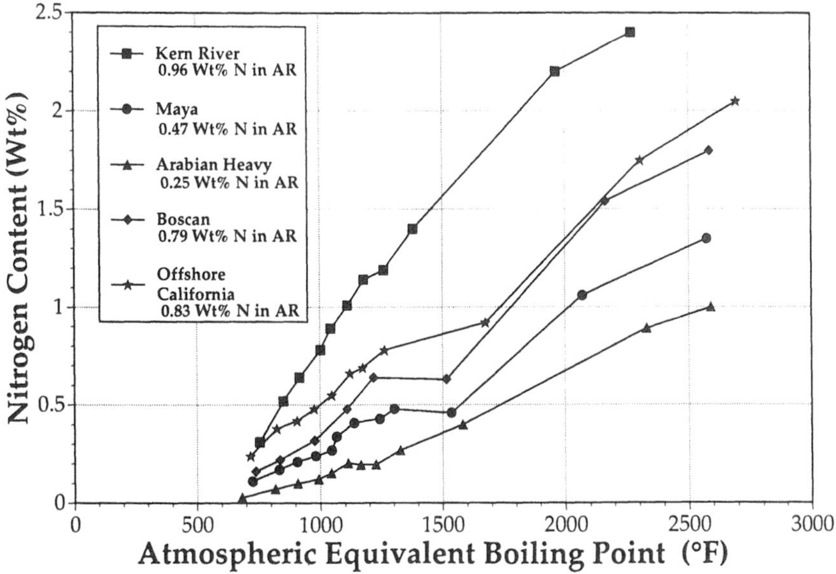

Figure 4.16 Nitrogen content (wt%) of five heavy crude oils as a function of AEBP.

Table 4.24 Sulfur Contents and Nitrogen of Various Crude Oils and Their Atmospheric and Vacuum Residues

		Sulfur (wt%)			Nitrogen (ppm)			Sulfur + nitrogen (wt%)		
Crude	Source	Whole crude	AR $>650^{+}$	VR $>1000^{+}$	Whole crude	AR $>650^{+}$	VR $>1000^{+}$	Whole crude	AR $>650^{+}$	VR $>1000^{+}$
A	Nigeria	0.07	0.18	0.34	250	1020	5540	0.10	0.28	0.89
B	Indonesia	0.18	0.21	0.23	500	1200	3860	0.23	0.33	0.62
C	Louisiana	0.27	0.49	0.82	700	1800	4200	0.34	0.45	1.44
D	California	1.02	1.21	1.43	1050	2230	4130	1.17	1.43	1.84
E	Alaska	1.10	1.73	2.37	1550	2960	4620	1.26	2.03	2.83
F	Texas	1.65	2.89	3.80	1600	3170	6550	1.81	3.29	4.46
G	Kuwait	2.52	4.21	5.57	2700	3700	5380	2.79	4.58	6.11
H	Saudi Arabia	2.80	4.37	5.40	3250	5400	7870	3.13	4.91	6.19
I	Mexico	3.43	4.75	5.68	5200	8200	11460	3.95	5.57	6.83
J	Venezuela	4.10	4.31	4.72	5360	7040	14070	4.64	5.01	6.13
K	California	5.10	6.37	7.59	5400	6060	8150	5.64	6.98	8.41
L	Venezuela	5.40	6.02	6.23	7000	8250	9950	6.10	6.85	7.23

Table 4.25 Average Molecular Parameters of Kern River AR

Fraction	Mid-AEBP (°F)	Cum. wt% from AR	Mol. wt.[a]	Z^b	Average number of atoms/molecule								
					C_t	C_{ar}	H	S	N	O	V	Ni	Fe
Distillate data													
Dist 1	<400	—	190	−1.8	13.7	—	25.6	0.020	0.001	—	—	—	—
Dist 2	~570	—	260	−6.0	19.0	—	32.0	0.047	0.012	0.037	—	—	—
Atmospheric residue data													
Cut 1	722	21.6	335	−8.9	24.1	6.4	39.3	0.104	0.075	0.088	—	—	—
Cut 2A	825	32.8	420	−12.0	30.0	8.0	48.0	0.134	0.156	0.118	—	—	—
Cut 2B	885	42.8	470	−14.7	33.7	9.0	52.7	0.163	0.215	0.177	—	—	—
Cut 3A	955	51.4	525	−16.9	37.7	10.0	58.5	0.196	0.293	0.198	—	—	—
Cut 3B	1025	58.3	595	−19.7	42.3	12.4	64.9	0.249	0.378	0.264	—	—	—
Cut 4A	1090	64.3	680	−23.6	48.6	14.0	73.6	0.296	0.492	0.349	0.001	0.001	—
Cut 4B	1160	68.9	760	−27.0	53.8	16.1	80.6	0.316	0.615	0.415	0.001	0.002	—
Cut 5	1240	74.3	880	−31.5	62.4	17.7	93.3	0.356	0.745	0.515	0.001	0.001	—
SEF-1	1365[c]	90.3	1110[d]	−44.1	78.9	29.3	113.7	0.462	1.112	0.737	0.001	0.002	0.001
SEF-2	1945[c]	96.0	2520[d]	−140.3	179.3	80.0	218.3	1.141	3.957	1.888	0.009	0.019	0.013
SEF-3	2250[c]	99.1	3620[d]	−215.4	254.7	137.8	294.0	1.617	6.204	2.986	0.019	0.025	0.027
SEF-4	—	99.6	—	—	—	—	—	—	—	—	—	—	—

[a]Number average molecular weight determined by VPO in toluene except as noted.
[b]Z in the general formula $C_nH_{2n+Z}X$.
[c]Mid-AEBP calculated from MW.
[d]Number average molecular weight determined by VPO in pyridine.
Source: Data adapted from Refs. 4 and 63.

Table 4.26 Average Molecular Parameters of Arabian Heavy AR

Fraction	Mid-AEBP (°F)	Cum. wt% from AR	Mol. wt.[a]	Z[b]	Average number of atoms/molecule								
					C_t	C_{ar}	H	S	N	O	V	Ni	Fe
Cut 1	660	13.9	305	−6.1	21.5	4.6	36.9	0.255	0.006	—	—	—	—
Cut 2A	795	32.5	400	−9.6	28.2	6.3	47.8	0.392	0.021	—	—	—	—
Cut 2B	890	41.1	480	−12.8	33.6	7.2	54.3	0.506	0.034	—	—	—	—
Cut 3A	972	48.1	550	−15.2	38.3	8.7	61.4	0.642	0.048	—	—	—	—
Cut 3B	1025	54.6	620	−18.1	43.3	10.4	68.5	0.783	0.068	—	—	—	—
Cut 4A	1090	59.5	675	−20.7	46.6	10.9	72.5	0.894	0.098	—	0.0005	0.0001	—
Cut 4B	1145	62.7	760	−24.1	52.9	13.5	81.6	1.019	0.106	—	0.001	0.0001	—
Cut 5A	1205	69.5	825	−26.9	57.2	14.6	87.4	1.193	0.115	—	0.001	0.0001	—
Cut 5B	1305	72.3	935	−30.8	65.1	18.2	99.4	1.401	0.180	—	0.001	0.0001	—
SEF-1	1560[c]	87.7	1615[d]	−60.2	112.0	36.3	163.7	2.745	0.461	—	0.002	0.0004	—
SEF-2	2310[c]	95.9	3870[d]	−224.1	261.4	133.8	298.7	9.155	2.460	—	0.034	0.009	—
SEF-3	2570[c]	99.0	5060[d]	−313.0	345.7	192.2	378.3	12.723	3.617	—	0.051	0.016	—
SEF-4	—	99.5	—	—	—	—	—	—	—	—	—	—	—

[a]Number average molecular weight determined by VPO in toluene except as noted.
[b]Z in the general formula $C_nH_{2n+Z}X$.
[c]Calculated from MW.
[d]Number average molecular weight determined by VPO in pyridine.
Source: Ref. 30 with permission of the publisher.

Table 4.27 Average Molecular Parameters of Boscan AR

Fraction	Mid-AEBP (°F)	Cum. wt% from AR	Mol. wt.[a]	Z^b	Average number of atoms/molecule								
					C_t	C_{ar}	H	S	N	O	V	Ni	Fe
Cut 1	720	7.7	335	−4.7	23.0	5.7	41.3	0.470	0.038	0.046	—	—	—
Cut 2	815	24.2	430	−8.2	29.6	7.9	51.0	0.614	0.067	0.064	—	—	—
Cut 3	955	34.1	540	−12.6	37.5	9.0	62.4	0.852	0.124	0.163	0.0001	—	—
Cut 4	1090	43.4	675	−18.3	46.2	12.5	74.0	1.136	0.231	0.214	0.011	0.0006	—
Cut 5	1195	49.9	875	−27.8	60.5	16.3	93.2	1.562	0.401	0.307	0.022	0.001	—
SEF-1	1495[c]	74.2	1420[d]	−56.9	96.4	29.5	135.9	2.718	0.639	—	0.019	0.002	—
SEF-2	2140[c]	87.3	3460[d]	−175.0	231.0	103.2	287.0	7.590	3.806	—	0.223	0.018	—
SEF-3	2570[c]	98.0	5510[d]	−307.2	369.6	192.2	432.0	12.230	7.087	—	0.479	0.037	—
SEF-4	—	100.0	—	—	—	—	—	—	—	—	—	—	—

[a]Number average molecular weight determined by VPO in toluene except as noted.
[b]Z in the general formula $C_nH_{2n+Z}X$.
[c]Calculated from MW.
[d]Number average molecular weight determined by VPO in pyridine.
Source: Ref. 30 with permission of the publisher.

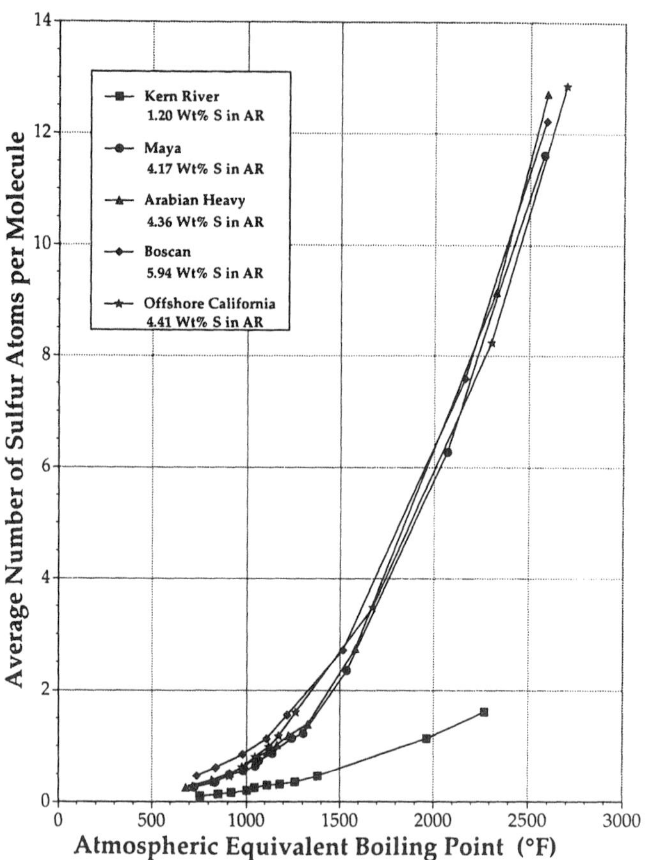

Figure 4.17 Average number of S atoms per molecule as a function of AEBP for five heavy crude oils. (From Ref. 61. Reproduced with permission of the publisher.)

content of Figs. 4.17 and 4.18, where the average numbers of heteroatoms per molecule are plotted, with that in Figs. 4.15 and 4.16, where the weight percent heteroatoms is plotted.

A look at the numbers in Tables 4.23–4.25 is also quite revealing. The first fraction in Table 4.25, Cut 1, has a mid-AEBP of 722°F (383°C) and contains about 24 C atoms on average. The average sulfur content is slightly greater than 0.1 S atom per molecule. This means that more than 10% of all the molecules in this cut would have one sulfur atom on average if each

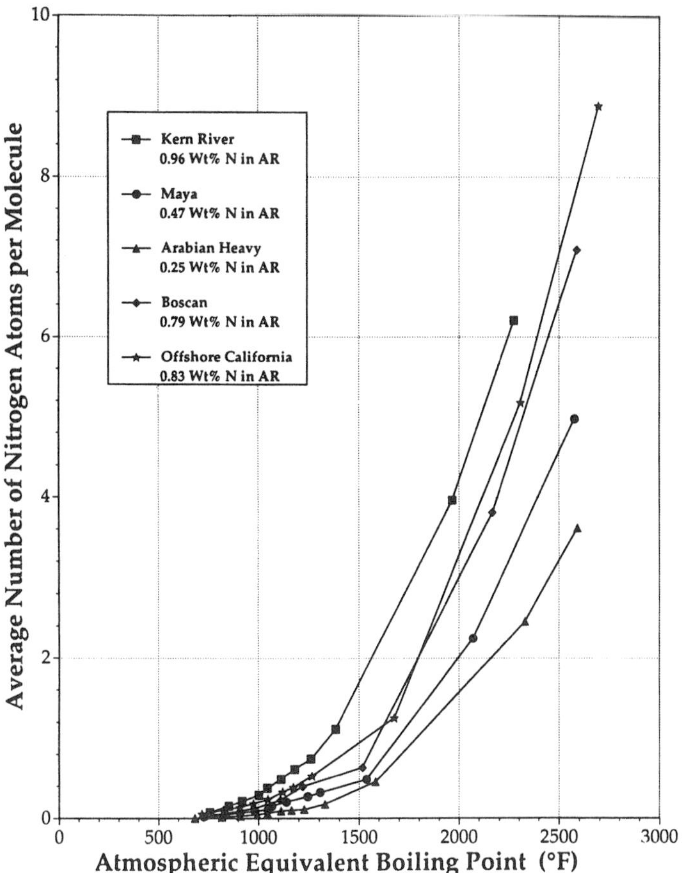

Figure 4.18 Average number of N atoms per molecule as a function of AEBP for five heavy crude oils. (From Ref. 61. Reproduced with permission of the publisher.)

of these had only one. In reality, fewer than 10% of the molecules in this sample may contain sulfur atoms since some of them may carry two. Counting also the N and O atoms, each molecule in this cut has 0.27 heteroatoms on average, or 27% of the molecules have one heteroatom on average, which is more than one might expect from a light VGO fraction.

In the heavy VGO fraction, Cut 3A, the average number of heteroatoms per molecule is about 0.7 (0.2 S, 0.3 N, and 0.2 O). If there were no more than one heteroatom in *any* molecule, two out of three molecules would be nonhydrocarbons. Or, looking at it differently, if only every other molecule were a heterocompound, about half of these would have to contain two heteroatoms. This in a VGO fraction, not even in a vacuum residue fraction!

The average hydrogen deficiency, Z, in the molecular distillation Cut 3A is -17, its number average MW is 525 daltons, and on average it has 38 C atoms. In reality, many molecules in this cut will be smaller and have lower hydrogen deficiencies, but many others will be larger and may have considerably higher Z numbers.

Boscan crude, Table 4.27, is even richer in heteroatoms than Kern River crude. Its first cut, corresponding to a light VGO, has an average MW of 335 daltons. On average, each molecule has 23 carbon atoms, 5.7 of these being aromatic, and 0.47 S, 0.038 N, and 0.046 O atoms. Assuming each molecule had only one heteroatom, 47 of 100 molecules would contain one S, 3.8 one N, and 4.6 one O atom. The heaviest distillate is Cut 5, which has a mid-AEBP of almost 1200°F (650°C). The molecules in this cut contain about 60 C atoms on average, including 16.3 aromatic ones, and almost 2.4 heteroatoms: 1.56 S, 0.40 N, and 0.31 O. Some of the N compounds are likely to comprise porphyrin structures (with four nitrogens). Therefore, the number of molecules having any nitrogen could be considerably lower than 40 in 100. Similarly, the number of molecules containing any oxygen would be smaller than the average of 31 per 100 because some of the oxygen would be in the form of carboxylic acids. Even so, almost every molecule of this cut would contain at least one, but more likely, close to two heteroatoms.

The numbers really become staggering with the nondistillable residues and their fractions. The SEF-1 fraction of Boscan crude (mid-AEBP about 1520°F or 830°C, *n*-pentane soluble) has an average molecular weight of about 1420, an average number of 96 C atoms, 30 of these aromatic, an average Z-number of -57, and an average number of >3.7 heteroatoms per molecule. Assuming that some molecules are free of heteroatoms or contain only 1, the remainder must then have that many more, say 5 on average and maybe up to 10 in some of the compounds. Now imagine the structure of the molecules in fraction SEF-3, which has an average molecular weight of about 5500, an average Z-number of -307, and, on average, more than 19 heteroatoms per molecule! Such mixtures are beyond meaningful further fractionation.

In view of these numbers it is clearly impossible to analyze the nondistillables on the molecular level, e.g., trying to identify individual compounds by MS, no matter how carefully they may be fractionated. This, of course, does not mean compositional analysis would be useless here.

On the contrary. It only means we cannot expect to get the detailed information we are used to from the lighter fractions.

A few examples of the compositional information available in this AEBP range are given in Figs. 4.19a and 4.19b, which show the distributions of vanadium, nickel, and iron in three crude oils (61). Both metals exhibit distinct humps in fractions around AEBPs of 1200°F (650°C), minima around 1600°F (870°C), followed by linear increases with further rising AEBPs. This behavior is of interest to the process engineer and chemist. It was only unraveled by careful fractionation, elemental analysis of the fractions, and then plotting the metals concentration versus AEBP.

Or look at the percent sulfur in Fig. 4.15. In Kern River AR (atmospheric residue), the sulfur concentration remains almost constant throughout the entire AEBP range, whereas, in contrast, that in Arabian Heavy AR increases from a low of 2.5 wt% to a high of 8.5 wt%. The sulfur curves of Arabian Heavy AR cross over Boscan AR and those of Maya AR. Note that the last three points of each of these curves belong to SEF fractions. Without the concept of AEBP, much of the individual variation of sulfur content with increasing boiling point would have been hidden.

The distribution of percent nitrogen in the same residues is presented in Fig. 4.16. Only one slight crossover here, but still interesting and steep increases with AEBP. The hydrogen contents do not reveal any detail, showing little variation among oils and only moderate change with AEBP (Fig. 4.20). In contrast, the hydrogen deficiencies range from $Z = -8$ to over -300 for the some of the SEF-3 fractions (Fig. 4.13). The aromatic carbon content again rises less dramatically, from around 20 to about 60% for Kern River and California offshore atmospheric residues as shown in Fig. 4.21.

Figures 4.13 to 4.21 demonstrate the unifying and simplifying value of the concept of AEBP. By being able to survey the elemental as well as the chemical group composition of an entire petroleum, including the nondistillable portion, one gets the whole picture at a glance. This makes the analytical information much easier to understand and use.

C. Carbon Residue

The coking propensity of petroleum fractions can be measured by three ASTM test methods: the microcarbon residue (MCR, ASTM D4530), the older and equivalent Conradson carbon residue (CCR, ASTM D189), or the Ramsbottom carbon residue (RCR, ASTM D524) test method. Not only do distillation residues form coke, but so do components of the higher-boiling distillates. Figure 4.22 portrays the increase of MCR with AEBP. The carbon residues are additive (64), i.e., if a sample is fractionated into

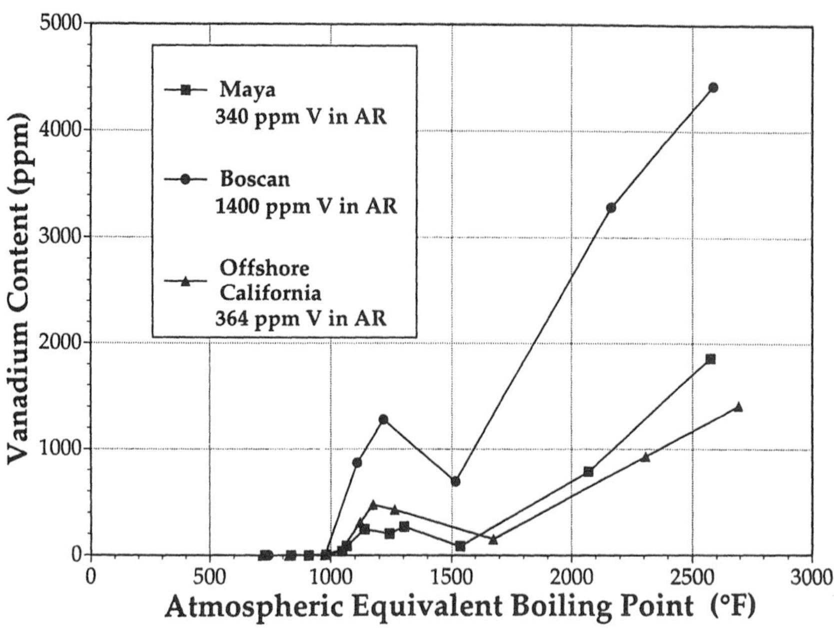

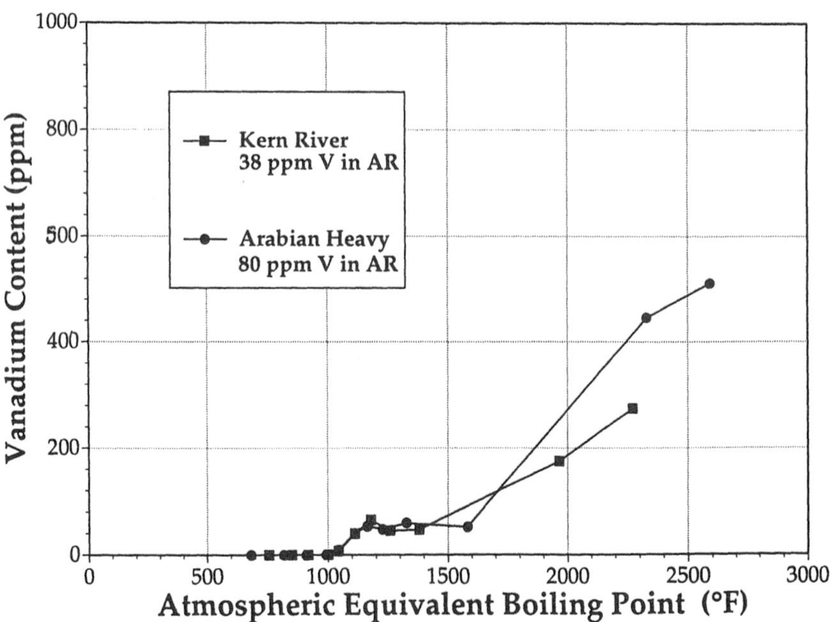

Figure 4.19a (top and bottom) Vanadium contents (ppm) of five heavy crude oils as a function of AEBP. (From Ref. 61. Reproduced with permission of the publisher.)

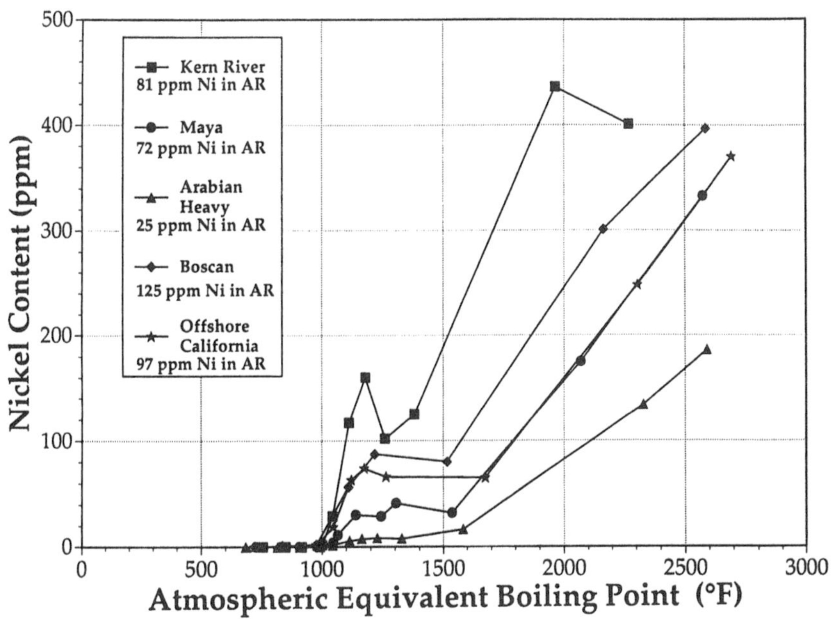

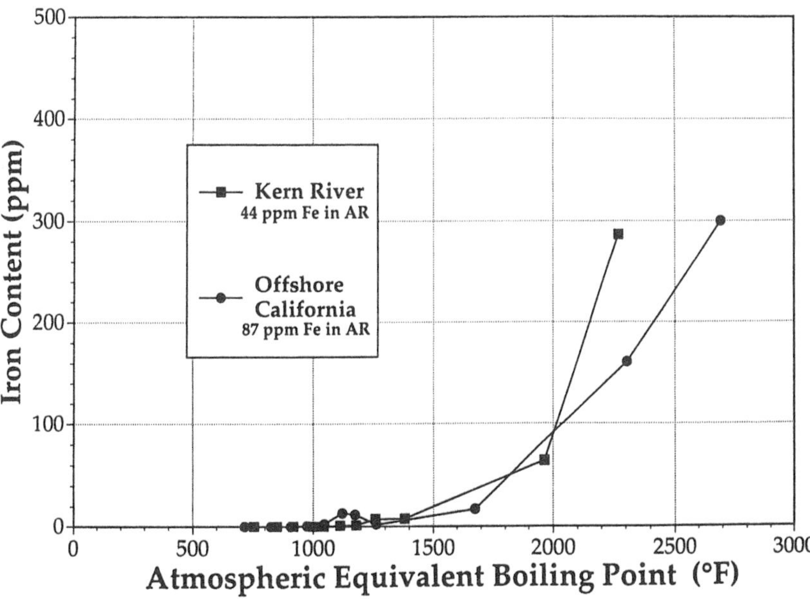

Figure 4.19b (top and bottom) Nickel and iron contents (ppm) of five heavy crude oils as a function of AEBP. (From Ref. 61. Reproduced with permission of the publisher.)

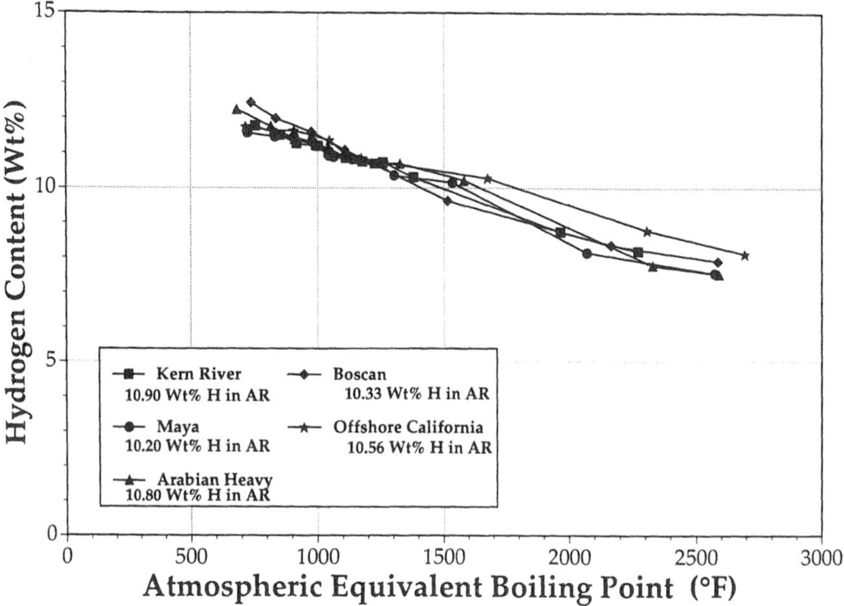

Figure 4.20 Hydrogen content (wt%) of five heavy crude oils as a function of AEBP.

several cuts, the weighted sum of the carbon residue (CR) numbers of the fractions will be equal to the CR number of the original sample. Two examples, shown in Table 4.28, illustrate this fact.

Roberts (64) also found a fairly universal linear correlation between CCR and the H /C ratio: H/C = 171 − 0.0115CCR. This equation holds within two limits; at H/C values ≥1.71, the CCR = 0 (no coke formation); at H/C ratios ≤0.5, it is 100 (all the material converts to coke under test conditions). The correlation is based on 114 data points derived from 15 different crude oils and has a correlation coefficient of 0.988. Roberts found a similar relation between CCR and N content. The fact that these correlations exist can only be explained in terms of the distribution of chemical species in these fractions. In each fraction, there are some molecules with larger aromatic-ring systems and more heteroatoms than average which, because of these characteristics, have low H/C ratios and a high propensity for coking. The proportion of these coke precursers, of course, is higher in the higher-boiling fractions than in the lower-boiling ones.

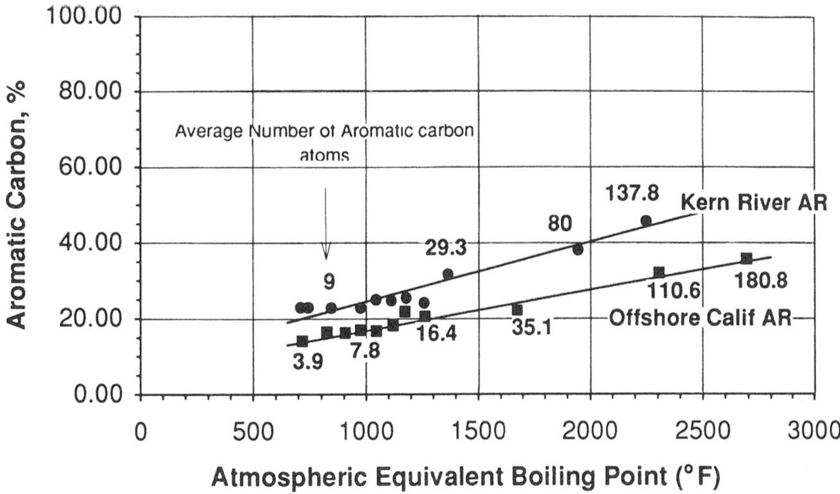

Figure 4.21 Aromatic carbon content (wt%) of five heavy crude oils as a function of AEBP.

Recently, Fixari et al. (65) described a rapid micromethod, which gives results that correlate quite well with the CCR. Its main advantage is the additional information it generates in the process: the amounts of C, H, S, and N in four fractions of the sample, and, by implication, the amount of material distilled over in atmospheric and vacuum distillation. The latter information is deduced from the amounts of carbon lost by evaporation equivalent to distillation up to the respective endpoints.

The apparatus is shown in Fig. 4.23. The sample is placed in the heat programmed oven where it is exposed to a set of four temperatures and two types of atmospheres. During the first three stages, a controlled flow of inert gas sweeps over it, carrying the volatiles into the combustion ovens. In the last stage, a mixture of oxygen and inert gas is used.

First, the sample is "distilled" at two temperatures, 220 and 360°C (428 and 680°F), i.e., the volatile portions of the sample are swept away by the nitrogen stream. The oven temperature of 220°C, combined with the gas flow, corresponds to an upper AEBP of 370°C (700°F) in a regular still, i.e., according to the authors' definition, to the endpoint of atmospheric distillation. Ordinarily though, 345°C (650°F) is considered the endpoint of atmospheric distillation, and the oven temperature would have to be

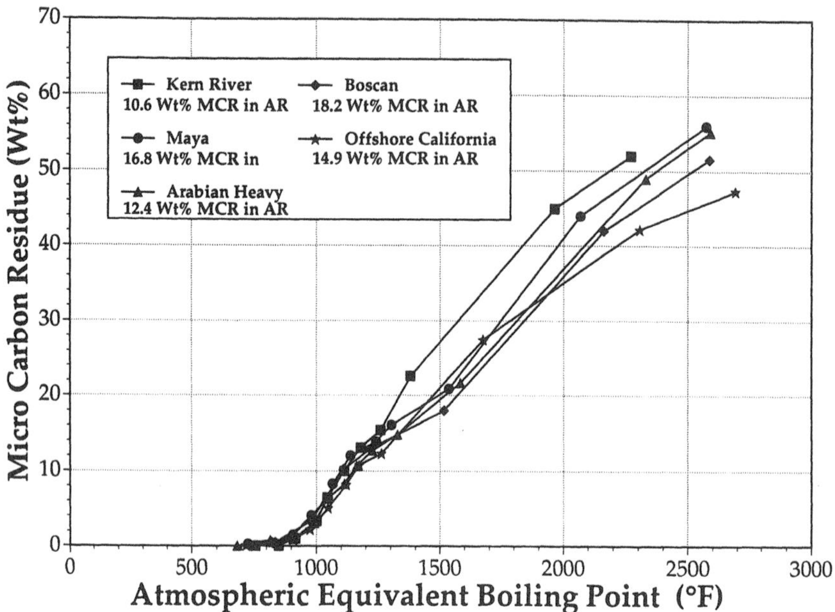

Figure 4.22 Micro carbon residue (wt%) of five heavy crude oils as a function of AEBP. (From Ref. 61. Reproduced with permission of the publisher.)

adjusted correspondingly lower. The oven temperature of 360°C (680°F) corresponds to an upper boiling point of 500°C (930°F) in a regular still, the endpoint of vacuum distillation, again according to the authors' definition (whereas most others consider 600°C, or 1100°F, as the endpoint). In these two stages, the distillables obtainable with atmospheric and vacuum distillation are determined.

Then the sample is cracked at 500°C oven temperature, and the fragments are swept into the combustion zone, giving the amount of crackable material. All the volatiles from the first oven go immediately to the combustion ovens where they are burned to CO_2, H_2O, SO_2, and NO. In the last stage, the remaining coke is burned in place with air at 800°C.

The combustion gases from the four stages are collected and analyzed separately. Thus, the amount and elemental composition of the two distillates, the cracking products, and the coke-forming part of the sample are measured in one operation over a period of about 80 min. The results are displayed in terms of %C, %H, %S, %N, and the ratios H/C, S/C, and N/C for each of the four fractions.

Table 4.28 Conradson Carbon Residue (CCR) Data for Distillates and Residues of Two Crude Oils

	Fourties		Arabian light	
AEBP (°C)	Yield (wt%)	CCR (wt%)	Yield (wt%)	CCR (wt%)
370–400 Distillate	13.0	<0.01	15.9	<0.01
400–435 Distillate	13.0	0.01	11.5	0.05
435–470 Distillate	16.2	0.07	10.1	0.34
470–510 Distillate	14.5	0.82	8.6	1.3
510–550 Distillate	8.9	2.7	12.9	4.0
550–510 Distillate	9.8	5.4	7.5	6.4
600–650 Distillate	7.5	7.6	9.2	11.2
650–700 Distillate	4.5	12.1	4.0	16.6
00 >700 Residue	12.6	30.8	20.3	35.3
>369 Residue, experimental	100	5.4	100	10.6
>369 Residue, calculated	100	5.9	100	10.0

Source: Data adapted from Ref. 64.

The %C in fraction 4 is related to the Conradson carbon residue and can be converted to it by calibration. Fixari et al. defined "crackability indices" for C, H, S, and N calculated from the contents of these elements (X) in fractions 3 and 4 by the equation

$$\text{Cra} = 100 \frac{\%X_{\text{F3}}}{\%X_{\text{F3}} + \%X_{\text{F4}}} \tag{4.22}$$

Figure 4.24 shows the cracking index of various fractions and products of a Safaniya vacuum residue plotted versus the H/C ratio. This pyrolysis method gives fairly detailed information which can be very valuable, especially when applied to feed and product of a processing step or to different fractions of a heavy oil. It depends on the calibration of the oven temperatures in terms of distillation temperatures and on precise standardization.

V. SUMMARY

In this chapter, we discussed the physical and chemical properties of heavy petroleum fractions. For compositional analysis, the MW has an importance equivalent to that of boiling point. The principal methods for its determination in our field are VPO, SEC, and MS. All of these have certain limitations in the upper ranges. MS methods lose sensitivity beyond 2000

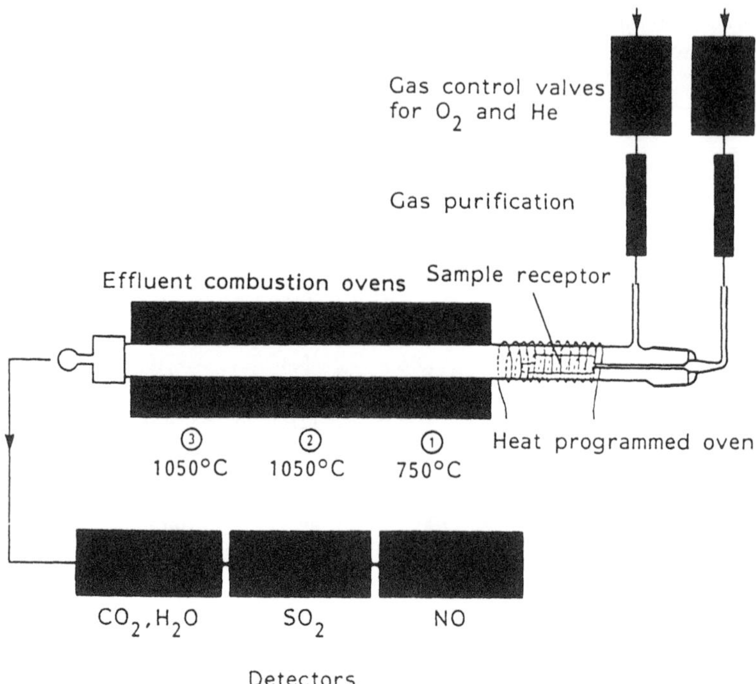

Figure 4.23 Oxidative pyroanalytical apparatus. (From Ref. 61. Reproduced with permission of the publisher.)

MW, and possibly lower, because such large molecules are not sufficiently volatile. VPO and SEC become unreliable around 2000 MW because the polar petroleum components form aggregates even in "good" solvents such as toluene and, thus, inflate MWs. Other potential problems of VPO, caused by its colligative nature, e.g., low-MW contaminants, may actually lower MWs.

Combination of VPO and intrinsic viscosity measurements of fractions allows one to estimate aggregation effects and molecular or particle shapes. Density in its various forms (specific gravity, API gravity) is extensively used in oil analysis, often together with the boiling point in composite values such as the UOP *K* factor. Density increases steadily with AEBP for paraffins and petroleum fractions. It depends strongly on composition, increasing with aromaticity and polarity. High-boiling fractions of a par-

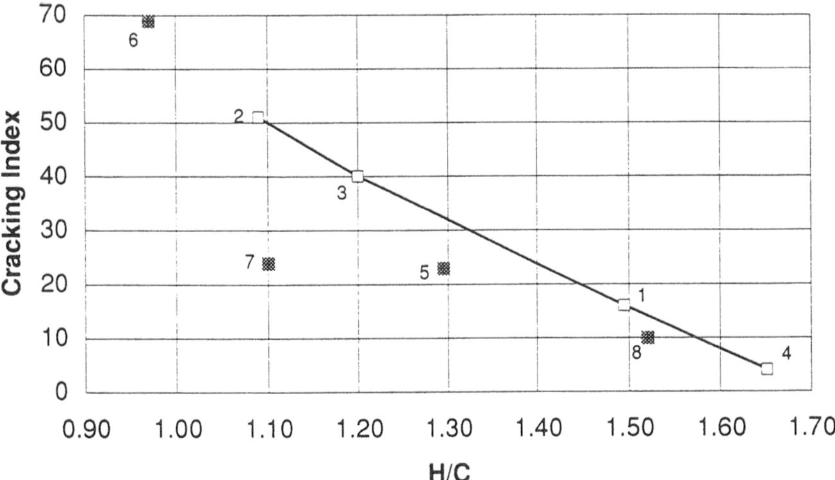

Figure 4.24 Cracking indices of fractions and products obtained from a safaniya vacuum residue (Data redrawn from Ref. 65.). 1 = original vacuum residue; 2 = pentane asphaltenes; 3 = heptane asphaltenes; 4 = pentane maltenes; 5 = oils from visbreaking; 6 = precipitate from visbreaking; 7 = oils from coking; 8 = hydrotreated atmospheric residue.

affinic crude may be lighter than lower-boiling fractions of aromatic or naphthenic crudes.

Viscosity was discussed only insofar as it is part of compositional analysis. It increases very strongly with increasing boiling point, much more steeply than MW. The viscosity increase is most pronounced with polar samples and least with paraffinic ones, contrary to the behavior of MW. The use of intrinsic viscosity has already been mentioned.

The main chemical properties of heavy petroleum fractions and their change with AEBP were described. Carbon, hydrogen, and heteroatom contents ordinarily vary more with boiling point than with the origin of crude oils. Those of C and H remain in relatively narrow limits (83–87 and 7.5–14%, respectively), whereas those of the heteroatoms change by several orders of magnitude from a few parts per million to several percent. Nondistillable petroleum fractions (>1300°F or 700°C) may contain as much as 40–50% aromatic C, 8% S, 2.5% N, and 2% O.

The continuous changes in chemical and physical properties with AEBP were pointed out. The transition from gas oils to nondistillables, having only hypothetical "equivalent" boiling points, is completely smooth. The steady increase of average MW, sulfur and nitrogen content, aromaticity, and carbon residue with AEBPs up to about 2600°F (1430°C) not only justifies the proposed AEBP concept, but actually demonstrates its virtues. By allowing extrapolations, the AEBP concept helps us understand the composition of the least tractable, most polar, insoluble fractions (SEF fractions, asphaltenes) and provides a means to predict their properties to some extent.

REFERENCES

1. Kyriacou, K. C., Baltus, R. E., and Rahimi, P. 1988. Characterization of Oil Residual Fractions Using Intrinsic Viscosity Measurements. *Fuel*, **67**:109–113.
2. Boduszynski, M. M. 1987. Composition of Heavy Petroleums. 1. Molecular Weight, Hydrogen Deficiency, and Heteroatom Concentration as a Function of Atmospheric Equivalent Boiling Point up to 100°F (760°C). *Energy Fuels*, **1**:2–11.
3. Moschopedis, S. E., Fryer, J. F., and Speight, J. G. 1976. Investigation of Asphaltene Molecular Weights. *Fuel*, **55**:227–232.
4. Boduszynski, M. M. 1988. Composition of Heavy Petroleums. 2. Molecular Characterization. *Energy Fuels*, **2**:597–613.
5. Al-Jarrah, M. M., and Al-Dujaili, A. H. 1989. Characterization of Some Iraqi Asphalts. II. New Findings on the Physical Nature of Asphalts. *Fuel Sci. Tech. Int.*, **7**(1): 69–88.
6. Ravey, J. C., Ducouret, G., and Espinat, D. 1988. Asphaltene Macrostructure by Small Angle Neutron Scattering. *Fuel*, **67**:1560–1566.
7. Altgelt, K. H. 1969. Unpublished results.
8. Chung, K. E., Anderson, L. L., and Wiser, W. H. 1978. New Procedure of Molecular Weight Determination by Vapor-Phase-Osmometry. *Fuel*, **58**:847–852.
9. Selucky, M. L., Ruo, T. C. S., Skinner, F., and Kim, S. S. 1978. Presented at Confab, Saratoga, Wyoming, July 26, 1978. Cited by Selucky et al. (10) and witnessed by one of the authors (KHA).
10. Brulé, B. 1979. Characterization of Bituminous Compounds by Gel Permeation Chromatography (GPC). *J. Liq. Chromatogr.*, **2**:165–192.
11. Selucky, M. L., Kim, S. S., Skinner, F., and Strausz, O. P. 1981. Structure-Related Properties of Athbasca Asphaltenes and Resins as Indicated by Chromatographic Separation. *The Chemistry of Asphaltenes*, edited by J. W. Bunger and N. C. Li, Advances in Chemistry Series 195 (6), American Chemical Society, Washington, DC, pp. 83–118.
12. Altgelt, K. H. 1968. Asphaltene Molecular Weights by Vapor Pressure Osmometry. *Amer. Chem. Soc. Div. Petr. Chem.*, **13**(3): 37–44.

13. Yau, W. W., Kirkland, J. J., and Bly, D. D. 1979. *Modern Size Exclusion Liquid Chromatography. Practice of Gel Permeation and Gel Filtration Chromatography*. John Wiley and Sons, New York.
14. Hagel, L., and Janson, J.-C. 1992. Size-Exclusion Chromatography. *Chromatography, 5th Edition, Fundamentals and Applications of Chromatography and related migration methods*, edited by E. Heftmann, Elsevier, Amsterdam, Chap. 6, pp. A267–A307.
15. Altgelt, K. H. 1979. Gel Permeation Chromatography (GPC). *Chromatography in Petroleum Analysis*, edited by K. H. Altgelt and T. H. Gouw, Marcel Dekker, New York and Basel, pp. 287–312.
16. Biggs, W. R., Fetzer, J. C., Brown, R. J., and Reynolds, J. G. 1985. Characterization of Vanadium Compounds in Selected Crudes. I. Porphyrin and Non-Porphyrins Separation. *Liq. Fuels Tech.*, **3**(4): 397–421.
17. Buchanan, D. H., Warfel, L.C., Baley, S., and Lucas, D. 1988. Size Exclusion Chromatography of Coal Extracts Using Pyridine as Mobile Phase. *Energy Fuels*, **2**:32–36.
18. Del Paggio, A. A., Kamla, G. C., Forster, A. R., and Shephard, M. A. 1990. Characterization of Metal-Containing Molecules in Various Hydrotreated Atmospheric Vacuum Bottoms. *Amer. Chem. Soc. Div. Petr. Chem.*, **35**:606–613.
19. Bartle, K. D., Mulligan, M. J., Taylor, N., Martin, T. G., and Snape, C. E. 1984. Molecular Mass Calibration in Size-Exclusion Chromatography of Coal Derivatives. *Fuel*, **63**:1556–1560.
20. Acevedo, S., Escobar, G., Gutierrez, L. B., and D'Aquino, J. 1992. Synthesis and Isolation of Asphaltene Standards for Calibration of G.P.C. Columns and Determination of Asphaltene Molecular Weights. *Fuel*, **71**:1077–1079.
21. Reerink, H., and Lijzenga, J. 1975. Gel-Permeation Chromatography Calibration Curve for Asphaltenes and Bituminous Resins. *Anal. Chem.*, **47**:2160–2167.
22. Rodgers, P. A., Creagh, A. L., Prange, M. M., and Prausnitz, J. M. 1987. Molecular Weight Distributions for Heavy Fossil Fuels from Gel Permeation Chromatography and Characterization Data. AICHE1987 Spring National Meeting Houston, March 3–April 2, Prepr. No. 20B.
23. Daubert, T. E., and Danner, R. P. 1989. *Physical and Thermodynamic Properties of Pure Chemicals*. Hemisphere Publishing Corporation, New York.
24. Bartle, K. D., Mills, D. G., Mulligan, M. J., Amaechina, I. O., and Taylor, N. 1986. Errors in the Determination of Molecular Mass Distribution of Coal Derivatives by Size-Exclusion Chromatography. *Anal. Chem.*, **58**:2403–2408.
25. Aczel, T., Colgrove, S. G., and Reynolds, S. D. 1985. High Resolution Mass Spectrometric Analysis of Coal Liquids. *Amer. Chem. Soc. Div. Fuel Chem.*, **30**:209–220.
26. Ali, L. H. 1971. A Method for the Calculation of Molecular Weights of Aromatic Compounds, and Its Application to Petroleum Fractions. *Fuel*, **50**:298–307.
27. Leon, V. 1987. Average Molecular Weight of Oil Fractions by Nuclear Magnetic Resonance. *Fuel*, **66**:1445–1446.

28. Rousseau, B., and Fuchs, A. H. 1989. Determination of Average Molecular Weights of High-Boiling Aromatic Oil Fractions by ^{13}C and ^{1}H nuclear magnetic resonance. *Fuel*, **68**:1158–1161.
29. Kryashchev, A. N., Popov, O. G., Posadov, I. A., Rozental, D. A., Kushnarev, D. F., and Rokhin, A. V. 1991. Evaluation of Molecular Mass of Petroleum Asphaltenes. *J. Appl. Chem. USSR*, **64**(3): 644–649. (Translation by Plenum Publishing, New York 1991, pp. 576–581.)
30. Boduszynski, M. M. 1987. Composition of Heavy Petroleums. Chevron Report.
31. Altgelt, K. H., and Boduszynski, M. M. 1992. Composition of Heavy Petroleums. 3. An Improved Boiling Point–Molecular Weight Relation. *Energy Fuels*, **6**:68–72.
32. Acevedo, S., Leon, O., Rivas, H., Marquez, H., Escobar, G., and Gutierrez, L. 1987. Neutral Fraction Compounds: A Representative Sample of Petroleum Asphaltenes? *Amer. Chem. Soc. Div. Petr. Chem.*, **32**(2): 426–431.
33. Shaw, J. E. 1989. Molecular Weight Reduction of Petroleum Asphaltenes by Reaction with Methyl Iodide–Sodium Iodide. *Fuel*, **68**:1218–1220.
34. Ettinger, M, Nardin, R., Mahasay, S. R., and Stock, L. M. 1986. An Investigation of the O- and C-Alkylation of Coal. *J. Org. Chem.*, **51**:2840–2842.
35. Rose, K. D., and Francisco, M. A. 1987. Characterization of Acidic Heteroatoms in Heavy Petroleum Fractions by Phase-Transfer Methylation and NMR Spectroscopy. *Energy Fuels*, **1**:233–239.
36. Champagne, P. J., Manolakis, E., and Ternan, M. 1985. Molecular Weight Distribution of Athabasca Bitumen. *Fuel*, **64**:423–425.
37. Khan, Z. H., and Hussain, K. 1989. Non-Destructive Analysis of Crude Oil by Gel Permeation Chromatography. *Fuel*, **68**:1198–1202.
38. Altgelt, K. H. 1991. Unpublished work.
39. Hirsch, E. 1970. Relations Between Molecular Volume and Structure of Hydrocarbons at 20°C. *Anal. Chem.*, **42**:1326–1329.
40. Van Nes, K., and van Westen, H. A. 1951. *Aspects of the Constitution of Mineral Oils*, Elsevier Publishing Co., New York.
41. Montgomery, D. S., and Boyd, M. L. 1959. New Method of Hydrocarbon Structural Group Analysis. *Anal. Chem.*, **31**:1290–1298.
42. Satou, M., Nemoto, H., Yokoyama, S., Sanada, Y. 1991. Calculation of Molar Volume of Hydrocarbons in Coal-Derived Liquids by a Group Contribution Method. *Energy Fuels*, **5**:638–642.
43. Watson, K. M., Nelson, E. F., Murphy, G. B. 1935. Characterization of Petroleum Fractions. *Ind. Eng. Chem.*, **27**:1460–1464.
44. Riazi, M. R. and Daubert, T. E. 1980. Prediction of the Composition of Petroleum Fractions. *Ind. Eng. Chem. Process Des. Dev.*, **19**:289.
45. Hill, J. B. and Coats, H. B. 1928. The Viscosity–Gravity Constant of Petroleum Lubricating Oils. *Ind. Eng. Chem.*, **20**:641.
46. Kurtz, S. S., King, R. W., Stout, W. J., Partikian, D. G., and Skrabek, E. A. 1956. Relationship Between Carbon-Type Composition, Viscosity-Gravity

Constant, and Refractivity Intercept of Viscous Fractions of Petroleum. *Anal. Chem.*, **28**:1928–1936.

47. API Research Project 42. 1962. *Tables of Physical Properties*.
48. Kurtz, S. S., Jr. and Ward, A. L. 1936. The Refractivity Intercept and the Specific Refraction Equation of Newton. I. *J. Franklin Inst.*, **222**:563.
49. Kurtz, S. S., Jr., and Ward, A. L. 1937. The Refractivity Intercept and the Specific Refraction Equation of Newton. II. The Electronic Interpretation of the Refractivity Index and of the Specific Refraction Equations of Newton, Eykman, and Lorentz–Lorenz. *J. Franklin Inst.*, **222**:563.
50. Eyring, H. 1936. Viscosity, Plasticity, and Diffusion as Examples of Absolute Reaction Rates. *J. Chem. Phys.*, **4**:283–291.
51. Wright, W. A. 1969. An Improved Viscosity–Temperature Chart. *J. Mater.*, **4**(1): 19–27.
52. Twu, C. H. 1985. Internally Consistent Correlation for Predicting Liquid Viscosities of Petroleum Fractions. *Ind. Eng. Chem. Process Res. Dev.*, **24**:1287–1293.
53. Altgelt, K. H. 1992. Unpublished result.
54. Altgelt, K. H. 1970. Gel Permeation Chromatography (GPC) in the Structural Analysis of Asphalt. *Bitumen, Teere, Asphalte, Peche*, **21**: 75–86 (in German).
55. Fox, T. G., and Flory, P. J. 1949. Intrinsic Viscosity–Molecular Weight Relationships for Polyisobutylene. *J. Phys. Chem.*, **53**:197–212.
56. Altgelt, K. H., and Harle, O. H. 1975. The Effect of Asphaltenes on Viscosity. *Ind. Eng. Chem. Prod. Res. Dev.*, **14**:240–246.
57. Sakai, M., Sasaki, K., and Inagaki, M. 1981. Determination of Intrinsic Viscosity of Fractionated Pitches and Discussion on the Shape and Size of Pitch Molecules. *Carbon*, **19**:83–87.
58. Sakai, M., Sasaki, K., and Inagaki, M. 1983. Hydrodynamic Studies of Dilute Pitch Solutions: The Shape and Size of Pitch Molecules. *Carbon*, **21**:593–596.
59. Tissot, B. P., and Welte, D. H. 1984. *Petroleum, Formation and Occurrence*. Springer-Verlag, Heidelberg, pp. 415–423.
60. Speight, J. G. 1980. *The Chemistry and Technology of Petroleum*. Marcel Dekker, New York and Basel, pp. 40–47.
61. Boduszynski, M. M., and Altgelt, K. H. 1992. Composition of Heavy Petroleums. 4. Significance of the Extended Atmospheric Equivalent Boiling Point (AEBP) Scale. *Energy Fuels*, **6**:72–76.
62. Yen, T. F. 1975. *The Role of Trace Metals in Petroleum*. Ann Arbor.
63. Boduszynski, M. M. 1985. Characterization of "Heavy" Crude Components. *Amer. Chem. Soc. Div. Petr. Chem.*, **30**(4): 626–635.
64. Roberts, I. 1989. The Chemical Significance of Carbon Residue Data. *Amer. Chem. Soc. Div. Petr. Chem.*, **34**(2): 251–255.
65. Fixari, B., Le Perchec, P., and Bigois, M. 1991. Fractions Distribution and Elemental Composition of Heavy Oils by Pyroanalysis. *Fuel Sci. Tech. Int.*, **9**(3):321–335.
66. Yen, T. F., Erdman, J. G., and Hanson, W. E. 1961. Reinvestigation of Densimetric Methods of Ring Analysis. *J. Chem. Data*, **6**:443.

67. Kurtz, S. S., King, R. W., Stout, W. J., and Peterkin, M. E. 1958. Carbon-Type Composition of Viscous Fractions of Petroleum. Density-Refractivity Intercept Method. *Anal. Chem.*, **30**:1224–1236.
68. Ferris, S. W. 1955. *Handbook of Hydrocarbons.* Academic Press, Inc., Publishers, New York.
69. Sergienko, S. R. 1965. *High Molecular Compounds in Petroleum.* Israel Program for Scientific Translation (from Russian).
70. Smith, H. M. 1940. Correlation Index to Aid in Interpreting Crude-Oil Analyses. *BuMines Tech. Paper* **610**.
71. Beg, S. A., Al-Mutawa, A. H., and Amin, M. B. 1989. Arab Light Crude Study on Kinematic Viscosity, *Fuel Sci. Tech. Int.*, **7**:187–205.

5

Structural Group Characterization of Heavy Petroleum Fractions

I. WHAT IS STRUCTURAL GROUP CHARACTERIZATION?

The last chapter dealt with some bulk properties of petroleum. Here we take the next step in the characterization of heavy petroleum fractions, that is, determination of paraffinic, naphthenic, and aromatic carbon on the one hand and of carbon in certain structural groups on the other hand. Examples of the latter category are methyl, methylene, methine groups, aliphatic groups next to aromatic rings or farther away, long alkyl chains, and chain branch points. Not only can the presence or absence of these groups be ascertained, but also their abundance in the sample.

In their book *Aspects of the Constitution of Mineral Oils*, Van Nes and van Westen (1, p. 242) write:

> The separation of heavier oil fractions into individual components is an altogether hopeless enterprise. Even the preparation and identification of uniform fractions—containing exclusively molecules of the same size and type—is extremely complicated and time-consuming.

We can only add that this conclusion still holds today. Van Nes and van Westen continue:

> The . . . failure of molecular type analysis made it necessary to devise methods for estimating the chemical composition of hydrocarbon mixtures, without complete separation, preferably even without any chemical or physical pretreatment. When chemical methods failed, recourse was made to methods for *structural group* analysis. . . . Two principles underlie practical methods for structural group analysis: a) Instead of molecules certain structural groups are considered as components of the mixture. . . . b) Information about chemical composition as defined under a) is obtained by determining physical constants which can be measured rapidly.

The physical constants Van Nes and van Westen referred to are bulk properties of molecules in the mixture, mainly refractive index, density, and molecular weight. Their structural group analysis, known as the *n-d-M* method, uses these simple measurements for the calculation of percent paraffinic, naphthenic, and aromatic carbon, and the average number of aromatic and naphthenic rings in the mixture. Nuclear magnetic resonance (NMR) spectroscopy greatly expanded the number of structural group accessible to analysis. Now, such specific structural groups as CH_3, CH_2, aromatic as well as aliphatic CH and C, and C in various specific groupings can be directly determined by NMR techniques.

It is important to bear in mind that the **structural group analysis methods** describe a complex mixture in terms of average structural features of the molecules, that is, **groups of carbon and/or hydrogen types**, in contrast to the **compound group-type methods** which provide information in terms of compound types such as paraffins, mononaphthenes, dinaphthenes, trinaphthenes, and so on, that is, **groups of molecular species** in the mixture. Compound group-type methods are based on mass spectrometry (MS) and will be discussed in Chapter 7.

Aromatics are defined as a group (fraction) of aromatic molecules obtained by suitable chromatographic methods. Besides aromatic carbon atoms these aromatic molecules also contain numerous paraffinic and naphthenic carbon atoms. Therefore, the percentage of aromatics (molecules) in an oil fraction, measured by hydrocarbon-type methods, is always larger than the percent aromatic carbon (atoms) measured by the *n-d-M* method or by NMR.*

*Assume, for instance, a fraction containing 50 mole% saturates of C_{30} and 50 mole% aromatics of C_{30}, where the aromatic molecules consist of 15 aromatic and 15 aliphatic carbon atoms (all of these being average numbers). Then the fraction would contain 25 mole% aromatic and 75 mole% aliphatic carbon. The aromatic carbon would be reported as 25%, but the aromatics as close to 50 (wt)% (somewhat less because the aromatics contain 15 + 2 fewer H atoms per molecule).

The *n-d-M* and related methods are discussed here in greater detail than justified by their present use. However, they represent an ingenious approach, relating specific structural elements with easily measured bulk properties which is still an approach of interest. They also entail various structural relations which are of general value beyond their original scope. Some of these have been used, for instance, in the development of the extended Brown Ladner method presented in Chapter 8.

NMR techniques give insight into the skeletal details of molecules. In such complex mixtures as heavy petroleum fractions, much of the detail may be lost, but some is still visible, for example, aromatic and aliphatic carbon; aliphatic CH_3, CH_2, CH, and Cq (quaternary C), and aromatic CH and Cq; C next to aromatic rings and farther away; and long paraffinic chains and, to some extent, naphthenic carbon. These results can be enhanced and extended by combination with additional information, for instance, elemental composition, as in the Brown Ladner method. More advanced methods of this kind are the extended Brown Ladner methods and Allen's functional group analysis. These are discussed in Chapter 8.

II. THE *n-d-M* AND RELATED METHODS

A. The *n-d-M* Method

1. Basics

The *n-d-M* method is a means of calculating the distribution of carbon among paraffinic, aromatic, and naphthenic groups in a sample from such simple measurements as refractive index (n), density (d), and molecular weight (M). It is a collection of several empirical relations established by Tadema (1, pp. 317, 318) between the carbon atom distribution among aromatic, naphthenic, and paraffinic groups and such physical properties as refractive index, density, and molecular weight in petroleum fractions.

The first experimental basis for the *n-d-M* method was a set of 34 fractions and their hydrogenation products derived from 5 fairly diverse crude oils (1, pp. 264–267). The oils had been topped and then distilled to give between five and eight fractions boiling between about 400 and 900°F (about 200 and 500°C), that is, in the atmospheric and light vacuum gas oil range. Parts of these fractions were analyzed by the fairly accurate but tedious "Direct Method"* to establish the compositional reference base. The Direct Method includes a multistage by hydrogenation procedure, molecular-weight determination by boiling-point elevation, the meas-

*See discussion in Section II.A.3.

urement of hydrogen content by two methods, and the measurement of sulfur and nitrogen contents (1, pp. 265–276).

The paraffinic carbon of the fractions ranged from a low of 33% for the most aromatic one to about 70% for the most paraffinic one. Aromatic carbon ranged from 8 to 33%, and naphthenic carbon from about 20 to 36%.

Smittenberg and Mulder (2) had collected the most accurate density and refractive index data available at the time and plotted them versus the reciprocal carbon number (plus a parameter related to the hydrogen deficiency, Z); see Fig. 5.1. The result was a set of straight lines for the homologous series of normal alkanes and α-alkenes, *n*-alkylcyclopentanes, *n*-alkylcyclohexanes, and *n*-alkylbenzenes. Within their experimental error of ±0.2%, Smittenberg's data fell precisely on these lines with the exception of the first two or three members. In both graphs—for density and refractive index—all lines meet exactly in one point at the infinite carbon number. These points represent the "limiting paraffin," that is, the hypothetical *n*-paraffin with infinite chain length in the liquid state. Because the curves are linear and have a common point at infinite molecular weight, data read off these graphs are more precise than any single measurements.

Tadema combined Smittenberg and Mulder's (2) collection of densities and refractive indices with his group's compositional data obtained on the 34 fractions mentioned above and established a set of equations of the type

$$\%\mathrm{C} = \frac{a}{M} + b\Delta d + c\Delta n \tag{5.1}$$

Here, %C represents the percent carbon atoms in one of the three structures, paraffinic, aromatic, and naphthenic. It can also stand for %C in ring structures (naphthenic, aromatic, or all rings, naphthenic plus aromatic ones). The coefficients a, b, and c are constants found by regression analysis. M is the molecular weight and Δd and Δn are the differences between the measured values of d and n of the fractions and those of the "limiting paraffin" (extrapolated values in Smittenberg's plots), respectively.

Another set of equations has the form

$$R = a' + b'M\Delta d + c'M\Delta n \tag{5.2}$$

where R is the number of all, or aromatic, or naphthenic rings in the hypothetical mean molecule; and a', b' and c' are constants (different from a, b, and c). The coefficients a, b, c, a', b', and c' are functions of blending coefficients for aromatic, naphthenic, and paraffinic components derived from Smittenberg's equations (1, p. 319).

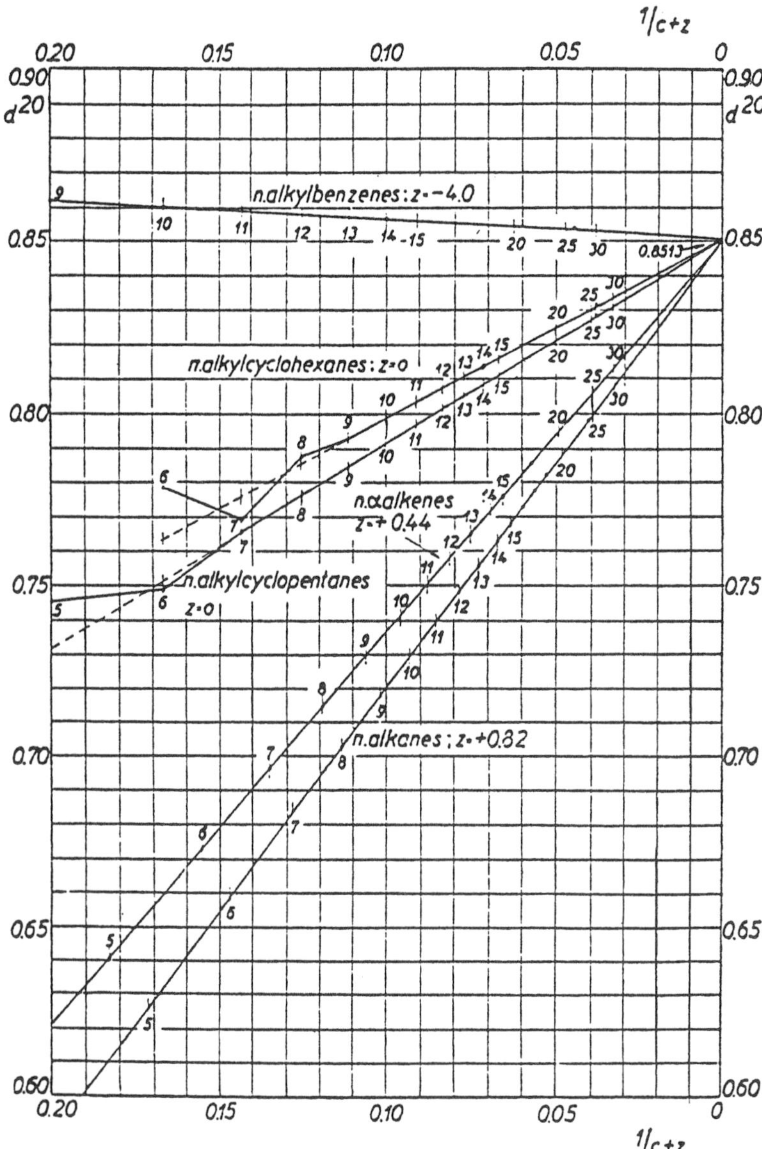

Figure 5.1 Density of various hydrocarbons according to Smittenberg and Mulder. (From Ref. 2. Reproduced with permission of the publisher.)

Tadema determined the values of these six coefficients by various methods and found the ratios c/b and c'/b' to be the most important and also the most accurate constants. Substituting k and k' for these and rearranging gave the following equations for the percent carbon and for the ring number:

$$\%C = a/M + b(\Delta d + k\Delta n) \tag{5.3}$$

$$R = a' + b'(\Delta d + k\Delta n) \tag{5.4}$$

The distillation range had to be divided into two parts with different sets of coefficients for satisfactory results from the equations. Table 5.1 shows the final results of his regression analysis, the formulas for %C in aromatic rings and in all rings, and the number of aromatic and all rings per molecule. The percent naphthenic C and the number of naphthenic rings were more accurately found by difference:

$$\%C_N = \%C_R - \%C_A \tag{5.5}$$

and

$$R_N = R_T - R_A \tag{5.6}$$

Table 5.1 Tadema's Formulas for the *n-d-M* Method

Measurements at 20°C	Measurements at 70°C
High	High
$\%C_A = 3660/M + 430(2.51\ \Delta n - \Delta d)$	$\%C_A = 3660/M + 410(2.42\ \Delta n - \Delta d)$
Low	Low
$\%C_A = 3660/M + 670(2.51\ \Delta n - \Delta d)$	$\%C_A = 3660/M + 720(2.42\ \Delta n - \Delta d)$
High	High
$\%C_R = 10{,}000/M + 820(\Delta d - 1.11\Delta n)$	$\%C_R = 11{,}500/M + 775(\Delta d - 1.11\Delta n)$
Low	Low
$\%C_R = 10{,}600/M + 1440(\Delta d - 1.11\Delta n)$	$\%C_R = 12{,}100/M + 1400(\Delta d - 1.11\Delta n)$
High	High
$R_A = 0.44 + 0.055M(2.51\Delta n - \Delta d)$	$R_A = 0.41 + 0.055M(2.42\Delta n - \Delta d)$
Low	Low
$R_A = 0.44 + 0.080M(2.51\Delta n - \Delta d)$	$R_A = 0.41 + 0.080M(2.42\Delta n - \Delta d)$
High	High
$R = 1.33 + 0.146M(\Delta d - 1.11\Delta n)$	$R = 1.55 + 0.146M(\Delta d - 1.11\Delta n)$
Low	Low
$R = 1.33 + 0.180M(\Delta d - 1.11\Delta n)$	$R = 1.55 + 0.180M(\Delta d - 1.11\Delta n)$

Source: Ref. 1, p. 323.

than directly by regression. Similarly,

$$\%C_P = 100 - \%C_A - \%C_N \tag{5.7}$$

The group at the Koninklijke/Shell Laboratory established the standard errors of the *n-d-M* method by applying both the *n-d-M* method and the Direct Method (as reference) to 166 new oil fractions and comparing the results (1, p. 346). They also included some pure hydrocarbons in this study.

2. Scope and Application

As originally proposed, the *n-d-M* method was meant to be applied exclusively to petroleum distillates boiling between about 400 and 900°F (about 200 and 500°C), that is, to atmospheric and light vacuum gas oils. Lower-boiling fractions do not follow the underlying relations between physical properties and chemical structure with sufficient regularity, and in higher-boiling fractions these relations become too insensitive. The sample should also contain no more than 75% carbon in ring structures (aromatic + naphthenic) with the additional restriction that the percentage of carbon in aromatic rings should not exceed 1.5 times that in naphthenic rings. Only catacondensed and six-membered rings were included in the basic calculations of the method. Later studies (3) showed that the presence of five-membered rings in the sample, even at high concentration, makes little difference.

Further restrictions called for no more than four rings altogether per molecule and no more than two aromatic ones. Also, heteroatom contents of more than 2% sulfur, 0.5% nitrogen, and 0.5% oxygen are not allowed. For samples exceeding these limits, there may be large differences between actual and calculated values (1, p. 339); nor is the method meant for individual hydrocarbons, even though its predictions may be quite good in some exceptional cases (3). It works poorly for most single aromatic compounds, though surprisingly well for mixtures of alkylated aromatics.

Even oils of the proper range, after treatment by severe hydrogenation, cracking, or other processing steps, may be sufficiently changed from the average composition of virgin petroleum fractions; they are out of range for the *n-d-M* method. In these cases, the method can still be applied (with caution) to monitor the progression of such processing steps, but then it will show only trends rather than quantitative compositional changes. For measuring the conversion of small or trace amounts of aromatic rings, the *n-d-M* method is clearly not sensitive enough.

The *n-d-M* method is extenion of the earlier Direct Method and the Waterman Ring Analysis developed by Waterman and his associates (4, 5). Several improvements over the following 20 years by various researchers in the Koninklijke/Shell Laboratory, notably by Tadema, led to the present form, which was described in great detail by Van Nes and van Westen (1, pp. 242–399).

3. Procedure of the *n-d-M* Method

Refractive index (n), density (d), and molecular weight (M) of the oil sample are determined and heteroatom contents are measured or estimated. Refractive index and density must be measured at the same temperature, preferably at 20°C. For subsequent calculations, it is convenient to introduce two new variables v and w:

$$v = 2.51(n - 1.4750) - (d - 0.8510) \tag{5.8}$$

$$w = (d - 0.8510) - 1.11(n - 1.4750) \tag{5.9}$$

The values 1.4750 and 0.8510 are the extrapolated data points at 20°C for the limiting paraffin in Smittenberg's plots of refractive index and density. The other two constants, 2.51 and 1.11, are empirical coefficients.

Now the percentages of aromatic carbon (C_A), total ring carbon (C_R), the mean number of aromatic rings (R_A), and total rings (R_T) per molecule are calculated from v and M by the, also empirical, equations shown in Table 5.2. (The symbol S there stands for sulfur.)

For measurements of n, d, and M at 70°C, the numerical parameters in these equations change to some extent. All the equations for measurements at 20 and 70°C are presented in Table 5.2 for easy reference. If data measured at other temperatures must be used, the appropriate parameters can be calculated with sufficient accuracy by linear interpolation from those at 20 and 70°C.

If refractive index and density are measured with a precision of ± 0.0001 and the molecular weight with a precision of $\pm 2\%$, duplicate results by the same chemist should differ by no more than the amounts given in the second column of Table 5.3. Van Nes and van Westen (1, p. 346) applied the *n-d-M* method to 166 oil fractions which they had also analyzed by accurate ultimate analysis before and after hydrogenation (Direct Method). The root mean squares of the differences between the values calculated by the *n-d-M* method and those found by the Direct Method at 20 and 70°C are shown in the third and fourth columns of Table 5.3. The precision of $\pm 1.7 - 2\%$ for C_N is remarkably good.

One of the equations of the *n-d-M* method relates the number of rings in saturated hydrocarbons to hydrogen content and molecular weight. Vlugter, Waterman, and Van Westen (4) suggested the possibility of replacing %H by the specific refraction r_{sp}. The latter is defined by the relation

$$r_{sp} = \frac{n^2 - 1}{n^2 + 2} \frac{1}{d} \tag{5.10}$$

Table 5.2 Collection of All the Formulas of the *n-d-M* Method

Measurements at 20°C	Measurements at 70°C
Calculate: $v = 2.51(n - 1.4750) - (d - 0.8510)$ $w = (d - 0.8510) - 1.11(n - 1.4750)$	Calculate: $x = 2.42(n - 1.4600) - (d - 0.8280)$ $y = (d - 0.8280) - 1.11(n - 1.4600)$
$\% \text{ C}_A \begin{cases} \text{if } v \text{ is positive:} \\ \quad \% \text{ C}_A = 430v + 3660/M \\ \text{if } v \text{ is negative:} \\ \quad \% \text{ C}_A = 670v + 3660/M \end{cases}$	$\% \text{ C}_A \begin{cases} \text{if } x \text{ is positive:} \\ \quad \% \text{ C}_A = 410x + 3660/M \\ \text{if } x \text{ is negative:} \\ \quad \% \text{ C}_A = 720x + 3660/M \end{cases}$
$\% \text{ C}_R \begin{cases} \text{if } w \text{ is positive:} \\ \quad \% \text{ C}_R = 820w - 3S + 10{,}000/M \\ \text{if } w \text{ is negative:} \\ \quad \% \text{ C}_R = 1440w - 3S + 10{,}600/M \end{cases}$	$\% \text{ C}_R \begin{cases} \text{if } y \text{ is positive:} \\ \quad \% \text{ C}_R = 775y - 3S + 11{,}500/M \\ \text{if } y \text{ is negative:} \\ \quad \% \text{ C}_R = 1400y - 3S + 12{,}100/M \end{cases}$
$\% \text{ C}_N \{\% \text{ C}_N = \% \text{ C}_R - \% \text{ C}_A$ $\% \text{ C}_P \{\% \text{ C}_P = 100 - \% \text{ C}_R$	$\% \text{ C}_N \{\% \text{ C}_N = \% \text{ C}_R - \% \text{ C}_A$ $\% \text{ C}_R \{\% \text{ C}_P = 100 - \% \text{ C}_R$
$R_A \begin{cases} \text{if } v \text{ is positive:} \\ \quad R_A = 0.44 + 0.055Mv \\ \text{if } v \text{ is negative:} \\ \quad R_A = 0.44 + 0.080Mv \end{cases}$	$R_A \begin{cases} \text{if } x \text{ is positive:} \\ \quad R_A = 0.41 + 0.055Mx \\ \text{if } x \text{ is negative:} \\ \quad R_A = 0.41 + 0.080Mx \end{cases}$
$R_T \begin{cases} \text{if } w \text{ is positive:} \\ \quad R_T = 1.33 + 0.146M(w - 0.005S) \\ \text{if } w \text{ is negative:} \\ \quad R_T = 1.33 + 0.180M(w - 0.005S) \end{cases}$	$R_T \begin{cases} \text{if } y \text{ is positive:} \\ \quad R_T = 1.55 + 0.146M(y - 0.005S) \\ \text{if } y \text{ is negative:} \\ \quad R_T = 1.55 + 0.180M(y - 0.005S) \end{cases}$
$R_N \{R_N = R_T - R_A$	$R_N \{R_N = R_T - R_A$

Source: Ref. 1, p. 344.

where n is the refractive index, generally measured at 20°C for the sodium d-line, and d is the density, also measured at 20°C. Making use of the fact that the molar refraction, $r_{\text{mol}} = r_{\text{sp}}M$, is an additive function of the atomic refractions, Geelen et al. (in Ref. 6) arrived at an equation which relates the specific refraction to the number of (naphthenic) rings and MW:

$$r_{\text{sp}} = 0.3310 + \frac{1.419 - 1.654R_N}{M} \tag{5.11}$$

Figure 5.2 shows a plot of r_{sp} versus 1000/MW. The lines for the different R_N values all converge in one point at infinite MW, which is the point of infinite paraffinic C content. Because of this fact, the equation is useful

Table 5.3 Repeatability of Results Obtained by the *n-d-M* Method and Standard Deviations in Relation to the Direct Method

Variable	Repeatability	Standard deviation	
		20°C	70°C
% C_A	0.8	±1.0%	±1.2%
% C_N	1.8	±1.7%	±2.0%
% C_P and % C_N	1.6	±1.3%	±1.7%
R_A	0.06	±0.04	±0.05
R_N	0.14	±0.08	±0.12
R_T	0.12	±0.08	±0.11

Source: Data adapted from Ref. 1.

only up to a molecular weight of about 2000. At higher molecular weights, it becomes too imprecise. Surprisingly, it can be applied to mineral oils that contain aromatic structures even though it was derived only for saturated hydrocarbons (6).

B. The Direct Method

The Direct Method, developed by Waterman and associates (5), is quite accurate despite certain limitations. Its main feature is the comparison of the percent carbon in ring structures before and after careful exhaustive hydrogenation of aromatic to naphthenic rings. The hydrogenation is performed in such a way as to ensure the absence, or at least a minimum, of side reactions (1, pp. 258–263). Under ideal conditions, the increase in hydrogen content is an exact measure of the total amount of aromatic carbon atoms present in the sample because each aromatic C atom takes up one hydrogen atom.

Another aspect of the Direct Method is the use of several rigorous relations between structural parameters which allowed the calculation of the number of rings in aromatic and naphthenic molecules from hydrogen content, carbon content in ring structures, and molecular weight.

The equation for naphthenic molecules (no aromatic rings),

$$R_N = 1 + \frac{(8.326 - 0.5793\ \%\text{H})M}{100} \tag{5.12}$$

is based on the fact that a molecule loses two H atoms with each ring closure. For hydroaromatic molecules (naphthenic and aromatic rings), two cases had to be distinguished: mixtures of monocyclic and noncyclic hydrocarbons ($R_T \leq 1$) and mixtures of monocyclic and polycyclic hydro-

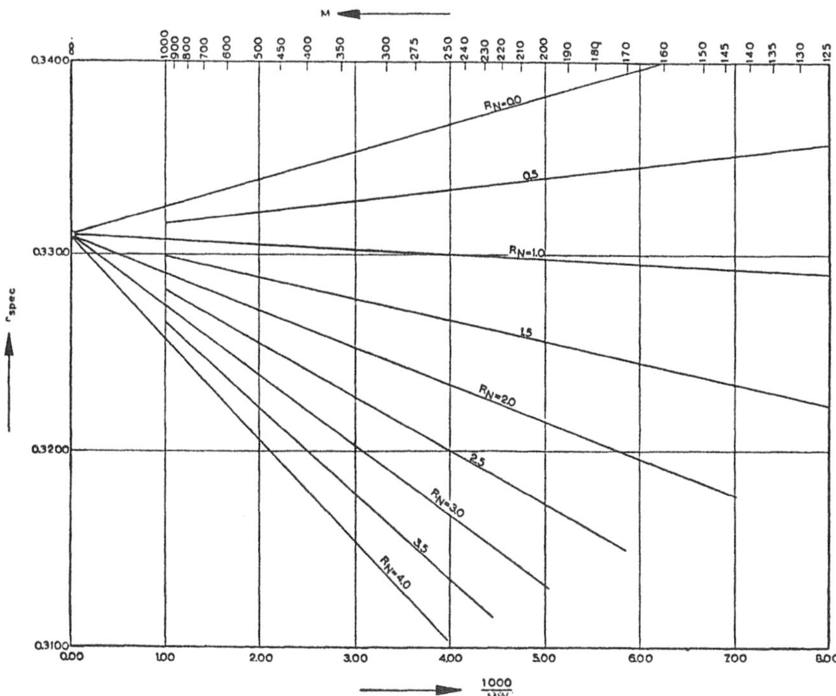

Figure 5.2 Plot of specific refraction versus 1000/MW of saturated hydrocarbons with ring number as parameter. (From Ref. 6. Reproduced with permission of the publisher.)

carbons ($R_T \geq 1$). For the former,

$$R_T = \frac{\%CR_TM(100 - \%H)}{720600} \tag{5.13}$$

and for the latter,

$$R_T = [\%CR_TM(100 - \%H)/480400] - \tfrac{1}{2} \tag{5.14}$$

(from Ref. 1, pp. 257–258).

These relations are rigorous, but they involve several assumptions:

1. The calculation of aromatic carbon content assumes hydrogenation of only aromatic double bonds with no other reactions occurring, including no breaking of any bonds. Van Nes and van Westen (1, p. 262)

point out the need for better ways to eliminate cracking and ring opening during hydrogenation.

2. No olefinic double bonds are allowed in the sample, a fairly safe assumption for most petroleum fractions.
3. Only catacondensed ring systems are considered. Noncondensed and pericondensed ring systems are assumed either to be absent or to be present in such amounts that they would balance each other, that is, together they would make the same contribution to the hydrogen content as the catacondensed ring systems (1, p. 254). Again, for the range of ring numbers permissible for the *n-d-M* method, this assumption should not cause great errors.
4. Only six-membered rings are assumed to occur in the fractions even though it is known that five-membered rings are usually present in fair amounts in petroleum samples (1, p. 253). It turned out that this assumption was not as serious as it might appear (3).

C. The Density–Refractivity Intercept Method

Kurtz and co-workers (7, and references cited therein) combined density measurements with those of refractive index in their density–refractivity intercept (DRI) method. As the *n-d-M* method, the DRI method was intended for the determination of carbon-type composition in the boiling range of gas oils and lubricating oils (400–900°F, 200–500°C). It is based on the observation that the molar volume is the sum of segmental increments (atomic volumes) and a constant, 3.12 cm^3, which they named "kinetic impact-free volume":

$$\text{Mol. vol} = 16.28n_1 + 13.15n_2 + 9.7n_3 + 6.2n_4 + 31.2 \qquad (5.15)$$

where n_1 is the number of carbon atoms in open-chain structures, n_2 is the number of carbon atoms in ring structures, except ring junction carbons, n_3 is the number of carbon atoms at ring junctures, n_4 is the number of double bonds, and 31.2 is the kinetic impact-free volume at 20°C and 1 atm.

The density is readily calculated by dividing the molar volume by molecular weight. The values obtained with this equation are within 0.15% of experimental data, even for compounds with five naphthenic, aromatic, or mixed rings. The structural effects on the density caused by the presence of aromatic and even of naphthenic rings are at least 10 times as large as the experimental error.

By applying this equation to a set of hydrocarbons, Kurtz et al. constructed a triangular diagram as shown in Fig. 5.3. This plot was arbitrarily chosen to represent only samples containing 30 carbon atoms and equiv-

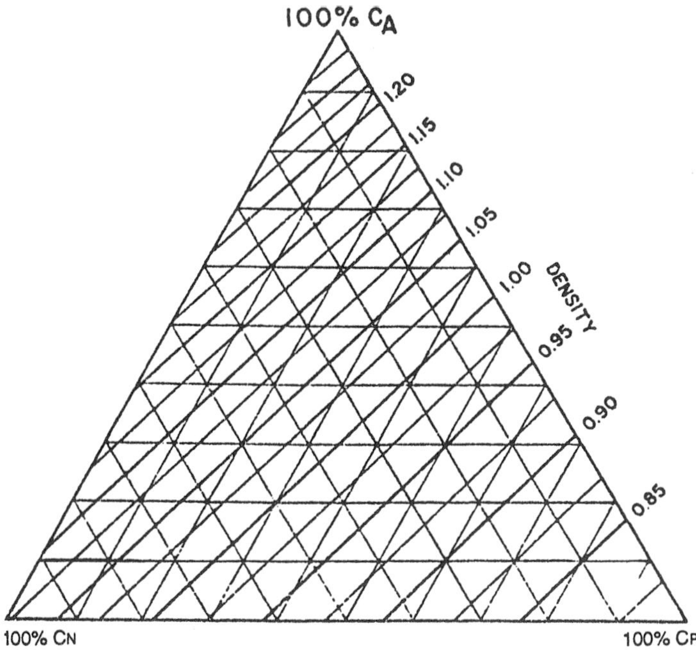

Figure 5.3 Triangular plot of density versus aromaticity, naphthenicity, and paraffinicity for hydrocarbons with 30 carbon atoms. (From Ref. 7. Reproduced with permission of the publisher.)

alent densities from samples with more or fewer carbons. Kurtz et al. obtained these equivalent densities from a plot of density versus carbon number where the data from all samples with a given hydrocarbon-type composition fall on one line; see Fig. 5.4. The equivalent density (of a hypothetical sample) with 30 carbon atoms is taken from the intercept of this line with that of the 30 carbon atom line. Thus, Fig. 5.3 actually represents samples of carbon numbers between 15 and 40 per molecule, but all reduced to a nominal 30. In this plot, each of the lines of constant density connects the points belonging to real or hypothetical compounds (of nominal carbon number 30) having this particular density but different composition.

To find the composition of a specific sample, that is, a specific point on a density curve in this plot, an additional piece of information was needed. Kurtz et al. chose the refractive intercept, $RI = n - d/2$. When the RI for hydrocarbons of 30 carbon atoms of different hydrocarbon-type com-

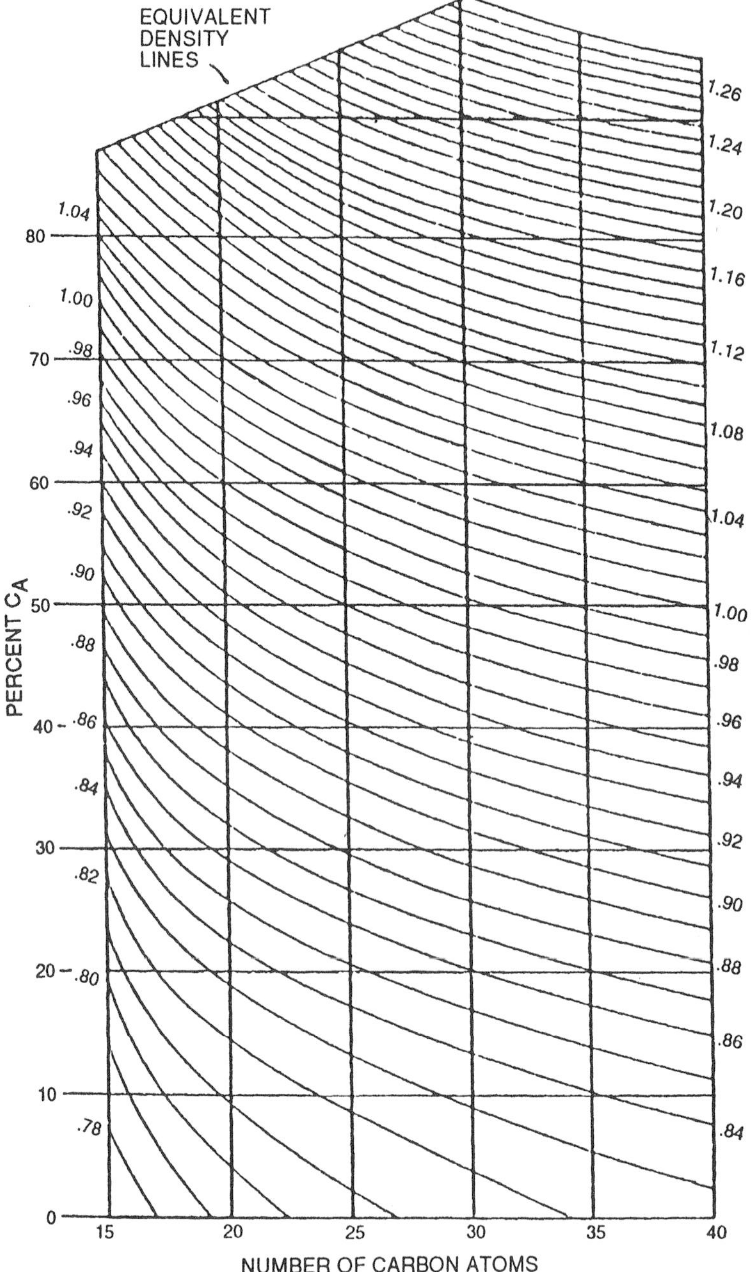

Figure 5.4 Graph for obtaining equivalent density at 30 carbon atoms. (From Ref. 7. Reproduced with permission of the publisher.)

position is plotted in this triangular diagram, the curves of constant RI cross those of constant density. The resulting graph is shown in Fig. 5.5. The intercept of a density curve with a RI curve fixes a compound on the plot and allows the determination of its composition in terms of its aromaticity, naphthenicity, and paraffinicity. The final diagram had a finer grid and allowed Kurtz et al. to read off the composition with sufficient precision. Today, computer programs would have replaced the density–MW and RI–MW plots as well as the triangular diagram.

Because fairly large errors arose from the application of Fig. 5.3 to petroleum samples, Kurtz et al. recalculated the densities of the model compounds used for establishing the triangular diagram, making two different assumptions. First, they assumed that only 50% (rather than 100%) of the naphthenic rings were condensed, the remainder being noncondensed; and second, that with increasing aromatic carbon content the naphthenic

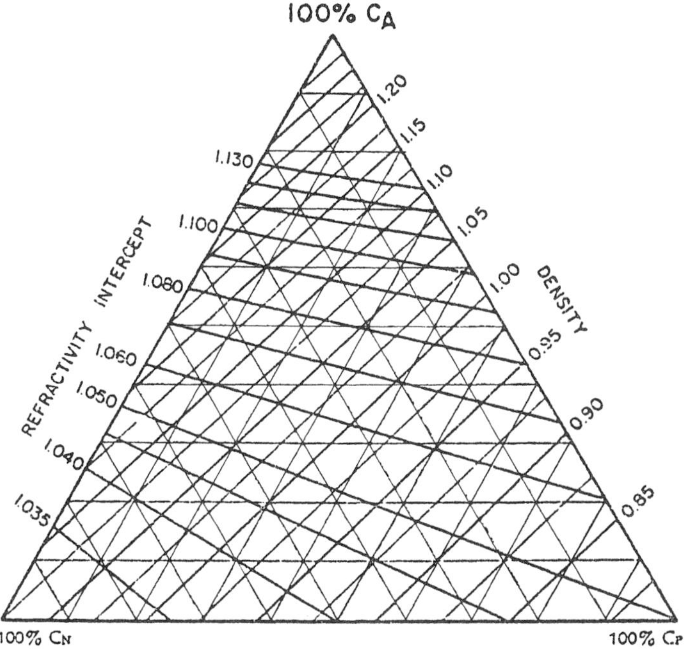

Figure 5.5 Triangular plot of density and refractivity intercept versus aromaticity, naphthenicity, and paraffinicity for hydrocarbons with 30 carbon atoms assuming 100% naphthenic ring condensation. (From Ref. 7. Reproduced with permission of the publisher.)

ring condensation increased, or rather that its noncondensation decreased. Thus, 50% isolated naphthenic rings were assumed for molecules with no aromatic rings, 25% isolated naphthenic rings for those with one aromatic ring, 12.5% for those with two aromatic rings, and 6.25% for those with three aromatic rings. No correction was applied for more than three aromatic rings. Interestingly, recent findings confirm the validity of these assumptions (see Chapter 10).

The new calculation, applied to 369 oils, including 83 oils studied by API Research Project 6, gave standard deviations from experimental data of 1.4% for C_A, 2.6% for C_N, and 2.5% for C_P. One carbon atom equals 3.3% in an oil with 30 carbon atoms. Kurtz et al. (7) found little systematic deviation in their calculations. The effect of sulfur on the results is small. For high sulfur oils, the C_N and C_P values can be improved somewhat by the corrections

$$\%C_{N_{\text{corr}}} = \%C_{N_{\text{orig}}} - \frac{\text{wt\% S}}{0.416} \tag{5.16}$$

and

$$\%C_{P_{\text{corr}}} = \%C_{P_{\text{orig}}} + \frac{\text{wt\% S}}{0.478} \tag{5.17}$$

An interesting sidelight of Kurtz et al.'s paper was their suggestion to calculate the wt% carbon of an oil from MW and the viscosity gravity constant, VGC, in those cases where it is not readily available otherwise, by the equation

$$\text{wt\%C} = 0.0735\text{MW} + 7.0\text{VGC} \tag{5.18}$$

The VGC was proposed by Hill and Coats (8) as a means for the determination of the paraffinic or naphthenic character of crude oils from simple physical measurements and is still frequently used for this purpose (see Chapter 4).

III. NUCLEAR MAGNETIC RESONANCE SPECTROSCOPY

A. Overview

Nuclear magnetic resonance (NMR) methods have gained a prominent place in the compositional analysis of petroleum fractions. In its basic applications, NMR is fast and relatively inexpensive. Because of its convenience, speed, and greater wealth of detailed information, particularly from ^{13}C-NMR, it has displaced the *n-d-M* and related methods in most laboratories. NMR measures directly aromatic and aliphatic carbon as well as hydrogen distributions. Beyond these results, both C and H in various structural groupings in a molecule can be determined.

^{1}H and ^{13}C nuclei are the most common ones used in NMR spectroscopy; ^{15}N, ^{14}N, and ^{33}S have been employed on occasion with petroleum fractions for special applications. Our discussion here is restricted to the use of regular NMR techniques for the structural group-type analysis of heavy petroleum fractions. Advanced NMR methods, such as spectral editing, 2D techniques, and computer-assisted methods are discussed in Chapter 8. For an easy basic text on the principles of NMR spectroscopy and more on the instrumentation involved, we refer to a book by Kemp (9), Chapter 3.

B. The Tools

1. ^{1}H-NMR Spectroscopy

Proton NMR has been widely employed in the structural analysis of organic compounds and mixtures, including petroleum fractions. With this fast and relatively inexpensive technique, we can measure the hydrogen distribution between aromatic and aliphatic groups and some subgroups. In particular, it allows us to distinguish between hydrogen attached to aliphatic carbons next to an aromatic ring (α position) and those farther away; hydrogen atoms in monoring and multiring aromatics; and olefinic hydrogen.

Very little of the sample is required, usually about 10 mg. A solution in deutero chloroform, contained in a glass tube of 5 mm diameter, is placed in a highly homogeneous magnetic field where it is surrounded by one or more coils. The coils serve to subject the sample to a weak field of radio-frequency (rf) perpendicular to the direction of the main magnetic field. The hydrogen nuclei of the sample can be visualized as spinning magnets with their spin axis precessing about the direction of the strong magnetic field. When the radio frequency is equal to the precession frequency, resonance occurs between the two, and the spin axis is tilted. This resonance is detected by a receiver coil (which often is the same as the rf coil). Resonance conditions for protons in different chemical groups differ slightly (by 1–10 ppm) because of the electronic environment in the molecule. The variation of rf power in the receiver coil with frequency at constant magnetic field strength is measured and comprises the nuclear resonance spectrum. The position of a sample resonance peak is compared with that of tetramethylsilane (TMS) as a reference material, and the difference is reported as "chemical shift," δ, a dimensionless and field-independent number expressed in terms of ppm difference in frequency from the reference.

Sample concentrations are chosen high enough to give a good resonance signal, but low enough to keep the solution viscosity moderate and to allow rapid tumbling of the molecules. Rapid tumbling contributes decisively to the generation of highly resolved spectra by averaging out magnetic dipole interactions between hydrogen nuclei on the same molecule. Concentrations lower than 2% stabilize the chemical shifts of the aromatic protons, which are slightly concentration-dependent above this value (10).

Table 5.4 shows the currently accepted band assignments for ^{1}H-NMR spectra. The basic assignments go back to Bartle and Smith (11). Those presented in our table were taken from a more recent review article by Snape, Ladner, and Bartle (12) with some added refinements by Bartle et al. (13), Cookson and Smith (10), and Netzel (14). The refinements, marked by a slight offset in the table, require more resolution than is usually

Table 5.4 Assignment of Bands in ^{1}H-NMR Spectra

Hydrogen type	Symbol[a]	Chemical shift (ppm)	Source
Aromatic H	HA	6.0–9.0	12
Triaromatic H	HA3	7.8–8.3	15
Diaromatic H	HA2	7.2–7.8	15
Monoaromatic H	HA1	6.6–7.25	10,15
Phenolic H	HAoh	5.0–9.0	12
H in CH_2 α to 2 aromatic rings	HL2$\alpha\alpha$	3.4–5.0	12
H in CH_3 α to an aromatic ring	HLα	1.9–3.4	12
H in α-CH_2	HL2α	2.4–2.8	16
H in α-CH_2 and CH	HL1,2α	2.3–4.0	15
H in α-CH_3	HL3α	1.9–2.3	15
		2.1–2.4	16
H in β and β^+ CH_2 and CH. H in β-CH_3 and in CH_2 and CH β and farther from an aromatic ring. Also CH and CH_2 in paraffins.	HLβ	1.0–1.9	12
		1.1–2.1	16
H in β-CH and β-CH_2	HL1,2β	1.6–2.0	13
H in CH_3 γ or farther from an aromatic ring. Also CH_3 in paraffins.	HL3g	0.5–1.0	12
		0.5–1.1	16

[a]For a discussion of these symbols see Chapter 8 and Table 5.5.
Source: Data primarily taken from Ref. 12.

achieved with heavy petroleum fractions. Additional information about the presence or absence of interfering groups may also help with the evaluation.

Quantitative accuracy of proton NMR for aromatic and aliphatic hydrogen is about 1% for distillates and 2–3% for residues; that for the distinction of aliphatic hydrogen atoms α to an aromatic ring from those β and farther away is somewhat lower. The other differentiations shown in Table 5.5 are much less certain. CH_3, CH_2, and CH hydrogen can ordinarily not be distinguished except for methyl hydrogen γ or farther away from aromatic rings, which is represented by the band at $\delta = 1$ ppm. Even this γ-CH_3 peak is sometimes difficult to quantify because of interference by naphthenic CH and CH_2 (10). In some spectra of narrow compound-class fractions, regions belonging to CH_3, CH_2, and CH in α position to aromatic rings are separated by valleys and can be used for the estimation of these groups.

Protons attached to single-ring and multiring aromatics can usually be distinguished with reasonable accuracy, especially when the sample concentration is 2% or less. At such low concentrations, the dividing line between these protons is at 7.25 ppm (10).

Figure 5.6A shows an example of a ^{1}H-NMR spectrum of a relatively narrow heavy petroleum fraction, namely, the aromatics fraction of a light vacuum gas oil (VGO), that is, a compound group fraction obtained from a narrow distillation cut of low mid-AEBP (825°F, 440°C). Dividing lines between the peak regions for the various structural group types were drawn to illustrate the potential and the limits of the method.

In spectra of samples with greater complexity and higher boiling point, much of the detail is lost and differentiation between group types becomes more difficult. The spectrum of the aromatics fraction of a super heavy VGO (1240°F, 670°C mid-AEBP) in Fig. 5.6B illustrates the effect of boiling point. Here only the distribution between aromatic and aliphatic H can be determined. All other H species are not sufficiently separated from each other. The spectrum of the unfractionated whole super heavy VGO would have been even more blurred. Still worse are the spectra of such complex fractions as asphaltenes. On the other hand, the spectra of compound-class fractions (e.g., mono-, di-, and triaromatics) from narrow distillation cuts are much better than those of the parent compound groups (aromatics).

2. ^{13}C-NMR Spectroscopy

The basic instrumentation for ^{13}C-NMR is the same as that for ^{1}H-NMR except that there are now two radio-frequency fields orthogonal (at right angles) to the main magnetic field, one for observing the ^{13}C nuclei and

Table 5.5 Assignment of Bands in ^{13}C-NMR Spectra

Carbon type	Symbol[a]	Chemical shift (ppm) B[b]	M[c]	A[d]
Carbonyl	C=O	170–210		
Aromatic C–OH	CAPS–O	148–168	150–157.3	
Aromatic C bonded to heterocyclic N	—		150–157.3	
Aromatic quaternary C	CAq		124–148[e]	
C, CH, or CH_2 substituted peripheral aromatic C	CAPS1,2	138–148	138–150	140–155
Peripheral aromatic C fused to a naphthenic C	CANF		135.3–138[e]	
Methyl substituted peripheral aromatic C	CAPS3		133.1–135.3[f]	131–139
Iternal aromatic C	CAI		127.1–133.1[f]	
Methyl substituted peripheral aromatic C and internal aromatic C	CAPS3, CAI		124–135	
Internal internal aromatic C	CAII		123–125[b]	
Mainly aromatic C–H (unsubstituted peripheral aromatic C)	CAPH	115–129.5	119–129	120–131
Aromatic C para to C–OH	—		116.9–122.0	
Aromatic C meta to heterocyclic N	—		116.9–122.0	
Aromatic C–H ortho to C–OH	—	100–115	106.9–116.9	
Naphthenic CH and CH_2	CN1,2	25–60		
	Unresolved in petroleum samples (Ref. 25)			
Internal (bridgehead) naphthenic C or CH	CNI	37–60		
CH in alkyl side chains (not isopropyl or isobutyl)	CL1	37–60		
CH_2 in alkyl side chains adjacent to CH	CL2ad1	37–60		
Ring-joining methylene	CL2αα	32–43		
Naphthenic CH	CN1	25–50		

Naphthenic CH_2	CN2	22.5–37		
Paraffinic CH_2	CL2ch	22.5–37		
CH_3	CL3	10–24		
CH_2 for γ^+-carbon in alkyl chains $\geq C_5$	CLγ			29–30
CH_2 for δ,ε-carbon in alkyl chains $\geq C_8$	CLδ	29.7		
CH_2 α to two aromatic rings	CL2αα	32.1–54.2		
C, CH, CH_2 α to an aromatic ring	CLqα, CL1α, CL2α		32.1–54.2	
CH_2 in butyl group β to an aromatic ring	CL2βbut		32.1–54.2	
CH_2β to a terminal CH_3 in chains ≥5	CL2		31.5–32.6	
CH_2 in propyl group α to an aromatic ring	CL2αprop		30.8–32.1	
CH_2 in ethyl group α to an aromatic ring	CL2αeth		26.9–30.8	
Shielded CH_2 α to an aromatic ring	CL2αshld	24–27.5		
β CH_2 in propyl side chains	CL2prop	24–27.5		
β CH_3 in isopropyl side chains	CL2iprop	24–27.5		
CH_3 on naphthenic and hydroarom. groups	CL3αN	18–24		
CH_2 next to terminal CH_3 in chains ≥4	CL2n3t	22.5–24		22.5–23.5
CH_3 next to terminal CH in chains ≥6	CL2n3t	22.5–24		22.5–22.7
α-CH_3 not shielded by adjacent groups	CL3α	20.5–22.5	20.4–21.9	
α-CH_3 shielded by one adjacent ring or group	CL3αshld1	18–22.5		
CH_3 β to CH in chains ≤6	CL3β			18–20
CH_3 β to an aromatic ring	CL3β	15–18	14.3–17.2	
CH_3 γ^+ to an arom. ring and CH	CL3γ	11–15	8.5–14.3	13–15
CH_3 γ^+ to an arom. ring but β to CH	CL3γ			10.5–11.4

[a]The letters are arranged in sequence of importance: first C or H for carbon or hydrogen. Next A, L, or N for aromatic, aliphatic, or naphthenic. Then P or I for peripheral or internal, or 1 to 3 for CH, CH_2, or Ch_3, and so on. For a more complete list of symbol nomenclature see Table 8.8.
[b]From Ref. 13. Established for coal liquids, but not restricted to those.
[c]From Ref. 20. Established for coal liquids, but not restricted to those.
[d]From Ref. 16. Established for lower-boiling hydrocarbons, but not restricted to those.
[e]See also Ref. 24.
[f]From Ref. 20, Table 3.

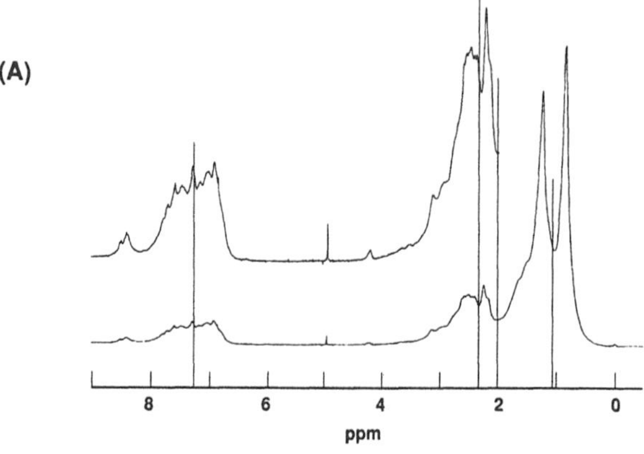

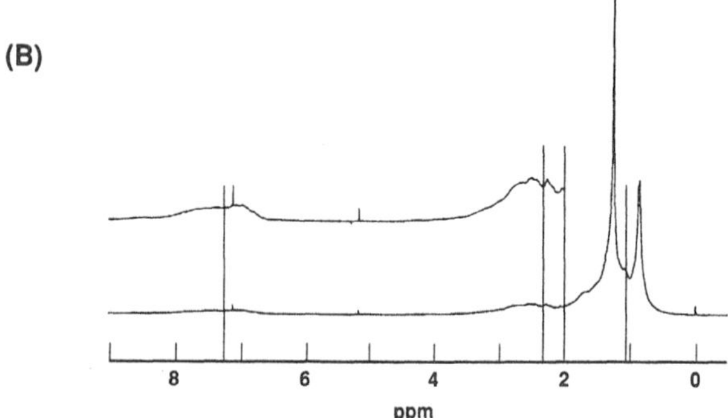

Figure 5.6 ^{1}H-NMR spectra of two Kern River crude fractions: (A) the aromatics fractions from Cut 2A (825°F, 440°C mid-AEBP); (B) the aromatics fractions from Cut 5 (1240°F, 670°C Mid-AEBP). (From Ref. 17.)

the other for decoupling ^{1}H nuclei. The low abundance of ^{13}C among the C isotopes (1.1%) and the lower gyromagnetic ratio of ^{13}C makes the signal weaker by more than two orders of magnitude; and, moreover, the carbon nuclei have longer relaxation times. The effect is that, even with Fourier transform (FT) data aquisition, ^{13}C-NMR measurements can take several hours to perform.

FT spectroscopy is much faster than the old magnetic field scanning techniques because here the different ^{13}C nuclei are all excited simultaneously, each time producing the entire spectrum for co-addition. Each pulse lasts only a few microseconds, but data acquisition, including a waiting period for complete relaxation of the nuclei between measurements, requires about 10 s. Thousands of spectra are usually taken, and the signals are accumulated by computer for averaging. The signal-to-noise ratio increases with the square root of the number of measurements. A spectrum taken in 16 h is twice as good as one taken in 4 h and 4 times as good as one taken in 1 h (if everything else is equal). If only 10 mg or less of a high-boiling sample is available, in typical circumstances the spectrometer must be run for at least 10–20 h to produce a spectrum with a reasonable signal-to-noise ratio.

Nuclear coupling would add complex splitting patterns and, thus, many additional peaks to an NMR spectrum. Fortunately, $^{13}C-^{13}C$ coupling between neighboring nuclei is statistically unlikely because of the low isotopic concentration of ^{13}C and can be ignored. C–H coupling is usually suppressed by means of a decoupler, that is, by broad-band rf irradiation of the protons. As a consequence of the natural low C–C coupling and the instrumental C–H decoupling, distinct high-resolution spectra are obtained free of splitting patterns.

In contrast to ^1H-NMR spectroscopy, the peak areas arising from the ^{13}C nuclei in different molecular positions ordinarily are not proportional to their concentration, and quantitative measurements require certain precautions. Two effects must be overcome: (1) the different relaxation times of ^{13}C nuclei in different chemical groups, especially of aliphatic versus aromatic C; (2) the nuclear Overhauser enhancement (NOE). The latter refers to the rise in signal intensity when C–H coupled protons are saturated by the decoupling field. The signal strength can increase up to threefold, which would be desirable if the increase were always the same. Instead, because of its variability, the NOE must be avoided or suppressed for quantitative measurements.

One way to do this is to add a small amount of a paramagnetic relaxation reagent, such as trisacetylacetonatochromium(III), $Cr(AcAc)_3$, which changes the dominant relaxation mechanism into one involving the interaction between unpaired electrons and ^{13}C nuclei. It also reduces the long relaxation times of some carbons, thus taking care of the first one of the two detrimental effects. Other ways will be mentioned in Chapter 8.

In its simplest form, ^{13}C-NMR can distinguish between aliphatic and aromatic carbon. But much more detail is accessible. Bartle et al. (13) put together an extensive list of carbon types discernable by ^{13}C-NMR based on spectra published in the Aldrich Library of NMR Spectra (18) and on a survey made by Snape, Ladner, and Bartle (19). Their list is shown with slight modifications in Table 5.5. Included in Table 5.5 are assignments by two other groups, Maekawa et al. (20) and Abu-Dagga and Rüegger (16).

Maekawa et al. had based theirs on their evaluation of model compound data published by others (21–23). Abu-Dagga and Rüegger refer to a book of spectroscopic tables by Pretsch et al. (26), apparently with their own confirmation and possibly sharpening of the assignments by spectral editing and 2D NMR techniques applied to low-boiling crude oil fractions. In general, the assignments made by these three groups are in agreement, although there are some specific ones chosen only by one group or another.

Very striking is the overlap between some of the assignment ranges. The overlaps are so wide that they would severely diminish the value of this compilation were it not for further developments, summarized by the term "spectral editing" and "two-dimensional" (2D) NMR. Both are briefly described in Chapter 8. Despite the rather broad ranges for some of the parameters, the actual spectra usually display a number of sharp, distinct peaks which often can be identified even without these advanced techniques, especially if the spectra were obtained on compound-class fractions rather than on broad cuts.

In the aliphatic region of petroleum ^{13}C-NMR spectra, several sharp peaks stand out and are used for quantitative evaluation. As an example, Fig. 5.7 shows the NMR spectrum of an Arabian Heavy VGO. We have selected 12 of the peaks for identification and listed the shift data and assignments in Table 5.6. For the assignments, we chose those by Netzel et al. (27), Thiel and Gray (28), and Cookson and Smith. (29). That for the range of CH_3 groups next to naphthenic rings came from Bartle et al. (13). Some of the CH_n positions giving rise to these peaks are illustrated in Fig. 5.8. The most prominent peak is usually that at 29.7 ppm. It is attributed to CH_2 in long alkyl chains, positioned four or more carbons away from an aromatic ring ($\delta^{(+)}$ CH_2) and from terminal CH_3. Most of Thiel and Gray's peak assignments are identical with those used by Cookson and Smith. (29) to determine the distribution of paraffinic CH_n groups in their diesel fuel samples. Netzel et al. (27) discuss the appearance of some peaks outside the normally assigned positions observed in some spectra in terms of minor shifts due to chain branching within four C atoms from the group responsible for the peak.

Normally, the absorption of 29.5–30.3 ppm allows us to estimate the amount of carbon in long aklyl chains ($\geq C_5$). Because this band represents CH_2 groups two or more carbons away from an aromatic ring and a terminal group, there must be four more carbons per chain than indicated by the area under these peaks. The *number of long chains*, n_{LCh}, can be estimated from the peaks at 14.2 (ω CH_3) and 28.1 ppm (CH_2 next to a terminal branch point, i.e., a CH group). On the one hand, the peak at 14.2 ppm gives results too high for this purpose because it also indicates CH_3 groups from chain branches. Subtracting half of the 37.6-ppm peak area (CH_2 next to CH groups inside a chain) corrects for this feature. On the other hand, the 14.2 peak does not cover twin CH_3 groups (as in an isopropyl group). This is why the 28.1 peak (CH next to two terminal CH_3 groups)

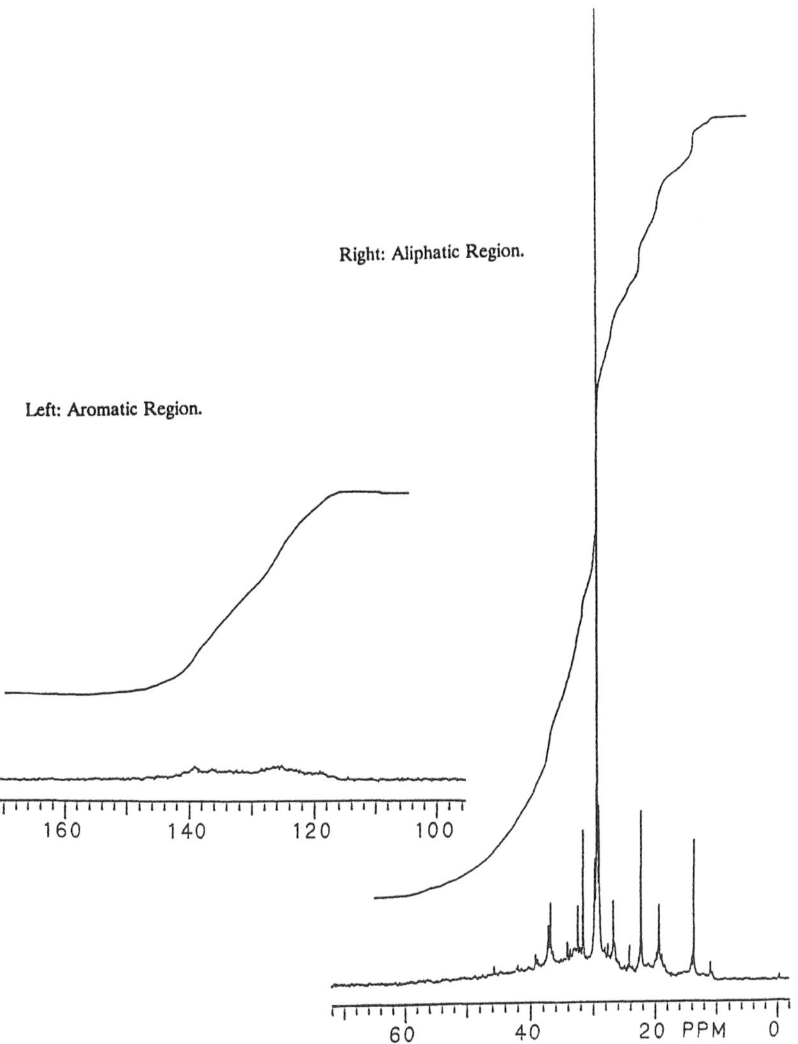

Figure 5.7 Aliphatic ^{13}C-NMR spectrum of an Arabian Heavy VGO. (Cut 2A, 795°F Mid-AEBP, see Table 4.25 of Chapter 4.) (From Ref. 17.)

Table 5.6 Carbon-Type Assignments for Peaks and Peak Areas in Figs. 5.7 and 5.9

Peak no.	Chemical shift (ppm)	Reference	Assignment
1	14.1–14.4	a, b, c	ω CH_3
2	19.3–20.4	a, b	γ,δ CH_3 α to CH
3	22.6–22.8	a, b, c	ε^+ CH_2 next to CH_3
		a, b	ε CH_3 next to CH at end of chain
	18–24	d	CH_3 on naphthenic groups
4	24.4–24.65	a, b, c	ε^+ CH_2 β to two CH groups
5	24.8–25.1	a, b, c	γ CH_2 β to two CH groups
6	28.08–28.14	a, b, c	CH α to two terminal CH_3 groups
7	29.4–30.3	a, b, c	γ^+ CH_2 in long chains away from branch points
a	29.5	a	γ CH_2 in long chains away from branch points
b	30.0	a	δ CH_2 in long chains away from branch points
8	31.8–32.5	a, b, c	ε^+ CH_2 2 C removed from CH_3
9	32.4–33.2	a, b	CH in chain next to CH_3 (methyl branch)
10	37.2–37.3	a, b	δ^+ CH next to CH_3 (isolated methyl branch)
11	37.5–37.7	a, b, c	ε^+ CH_2 next to CH in chain far from end
12	39.3–39.6	a, b, c	ε^+ CH_2 next to CH in chain near end

[a]Data from model compounds by Netzel et al. (27).
[b]Data for bitumen by Thiel and Gray (28).
[c]Data for saturates from diesel fuel by Cookson and Smith (29).
[d]From Ref. 13.

is needed. The final equation for n_{LCh} is then (30)

$$n_{\mathrm{LCh}} = A(14.2) - \tfrac{1}{2}A(37.6) + \tfrac{1}{2}A(28.1) \tag{5.19}$$

The number of CH_2 groups in long chains is the sum of those represented by the peak at 29.7 ppm, the number of CH_2 groups at the two ends of each long chain, that is, six times the number of long chains (three times for each of the two ends; see Fig. 5.8), and the number of CH_2 groups close to branch points inside the chain, namely, those next to them and one C removed, on both sides of the CH group. Thus, the total number is

$$n_{\mathrm{CH_2,LCh}} = \mathrm{C}(29.7) + 6n_{\mathrm{LCh}} + 4\mathrm{C}(37.6) \tag{5.20}$$

For molecules with long chains connecting two rings or ring clusters, the calculation is more involved (30).

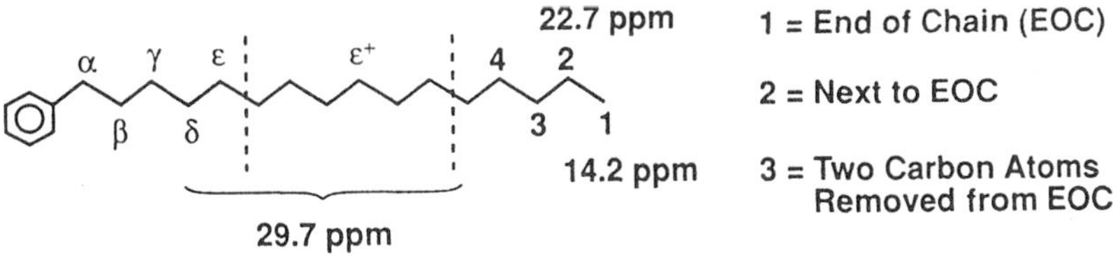

22.7 ppm
α γ ε ε⁺ 4 2
β δ 3 1
14.2 ppm
29.7 ppm

1 = End of Chain (EOC)

2 = Next to EOC

3 = Two Carbon Atoms Removed from EOC

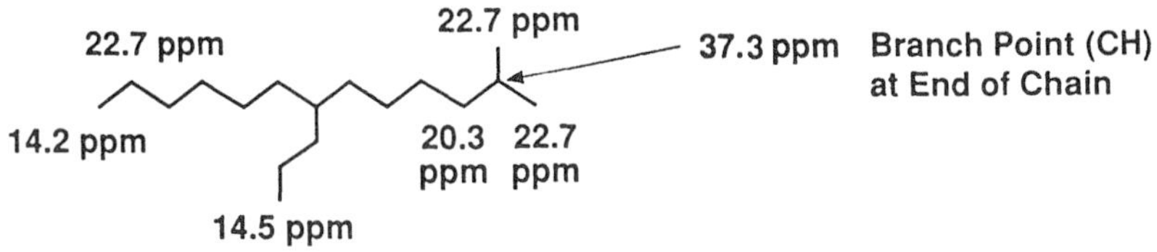

22.7 ppm
22.7 ppm
37.3 ppm Branch Point (CH) at End of Chain
14.2 ppm
20.3 ppm 22.7 ppm
14.5 ppm

Structure	Marked group CH_3 ○	CH_3 ●	CH □
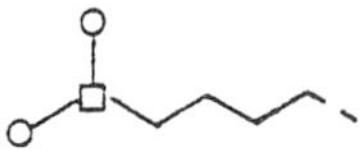	22.7 22.8	-	28.1
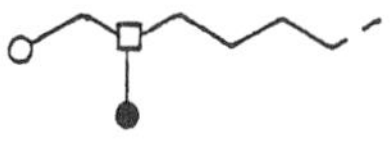	11.5	19.3	34.6
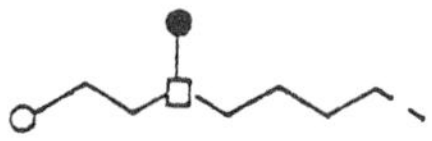	14.5	19.8	32.7
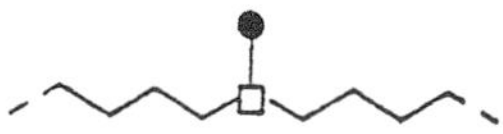	-	19.8	33.0
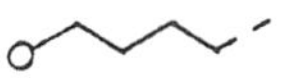	14.2	-	-

Figure 5.8 (top and bottom) Illustration of CH_n groups in hydrocarbon molecules and their respective chemical shifts in the ^{13}C-NMR spectrum.

The number of CH groups in long chains is found from the absorbances at 37.6 and 39.5 ppm. Because the 37.6 peak represents CH_2 next to CH carbons rather than CH directly, we must divide its intensity by 2. Thus,

$$n_{CH,LCh} = \frac{1}{2} C_{(37.6)} + C_{(39.5)}$$

The CH_3 groups give rise to at least four peaks in the spectra, namely, at 14.2, 22.7, 11.5, and one around 19.5 ppm; see Fig. 5.8. The 22.7-ppm peak signifies twin CH_3 groups as in isopropyl, and it also has a contribution from CH_2 next to a CH_3 group ending a long-chain segment. The latter, therefore, does not need to be represented again by its 14.2 peak. The number of all the CH_3 groups is therefore

$$n_{CH_3,LCh} = C(22.7) + C(11.5) + C(19.5)$$

The final result for C in long chains is then the sum of the three CH_n types. These equations are a first approximation only. Other factors need be considered for a better estimate.

The average length of long chains derives from dividing the number of carbon atoms in long chains, C_{LCh}, by the number of long chains, n_{LCh}.

The CH_3 group gives rise to several peaks depending on its position in the molecule. A peak at 14.2 ppm signals such a group positioned at the end of an unbranched chain segment of at least two or three CH_2 groups. A CH_3 group next to a branch point (CH group) at the end of a chain with at least two CH_2 groups produces a peak at 22.7 ppm. Farther away from the end of the chain, such a group gives a peak at or near 19.8 ppm. The CH_3 group at the end is also affected by these branch points. Some examples are presented in the bottom half of Fig. 5.8.

Numerous peak positions of CH_n groups in paraffinic model compounds with isoprenoid and other branched structures were published by Netzel et al. (27), Ward and Burnham (31), and Cookson et al. (32). Even more detailed model compound data on paraffinic, naphthenic, aromatic, and heterocompound molecules are found in the book by Breitmaier and Voelter (33).

Naphthenic carbon in petroleum samples, especially in heavy ones, usually occupies so many slightly different positions that the peaks are unresolved and form a broad hump in the range 25–60 ppm, under the generally well-resolved paraffinic peaks.

This hump has been exploited for the determination of naphthenic carbon (25,34). Yamashita et al. (35) compared two computer-based methods for the quantitative measurement of the hump with the old cut-and-weigh method. The evaluation of the hump is the only direct method for the determination of naphthenic CH_n groups. However, its measurement may not always be reliable. In very high-boiling petroleum samples, many paraffinic resonances are also only partly resolved. Although this leads mainly to broader peaks, the overlap may in some cases add to the hump and, thus, cause erroneously high results for naphthenic carbon. On the other hand, in the spectra of such heavily biodegraded oils rich in naphthenic biomarkers as, for example, Kern River crude fractions, some of the naphthenic groups stand out from the rest as sharp peaks and can be mistaken for paraffinic groups. The hump would then give low results for the naphthenic carbon.

Because the naphthenic hump is so common in spectra of heavy petroleum fractions, the popular evaluation of such spectra via the integral curve can lead to serious errors. Instead, an integration method taking into account an individualized segmented baseline; one for each peak or group of peaks must be employed. The spectrum shown in Fig. 5.9 may illustrate the case. Without the segmented baseline, the peak areas would be overestimated by 50–200%.

The aromatic region of the ^{13}C-NMR spectrum can be evaluated by conventional integration. Figure 5.10 gives an example with vertical lines indicating the boundaries of the main aromatic group types as defined in Tables 5.5 and 5.7.

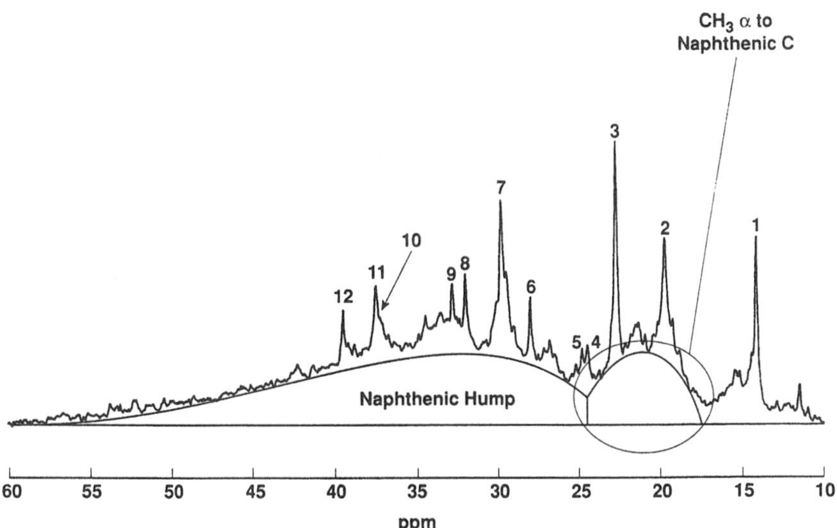

Figure 5.9 Partial ^{13}C-NMR spectrum (Aliphatic region) of the aromatics fraction of Kern River Cut 2A (825°F, °C Mid-AEBP). (From Ref. 17.)

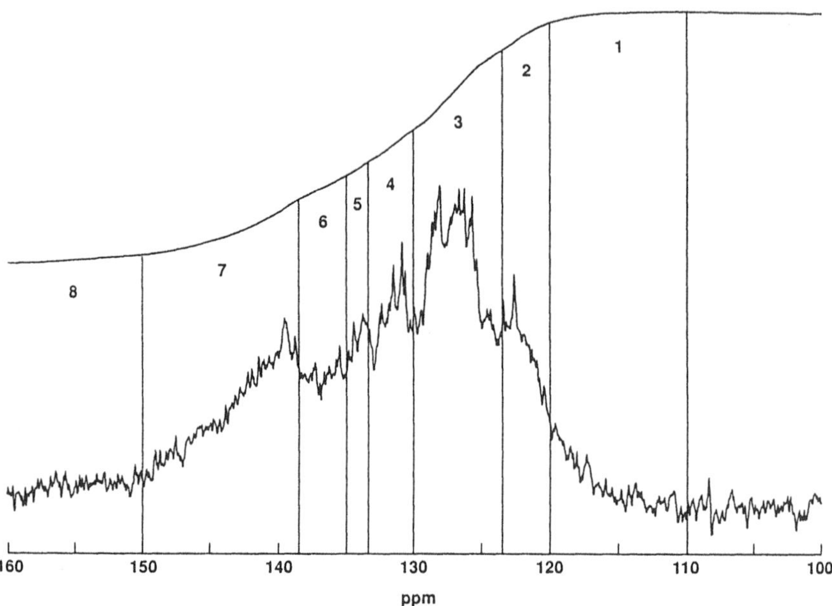

Figure 5.10 Partial ^{13}C-NMR spectrum (Aromatic region) of the aromatics fraction of Kern River Cut 2A (825°F, °C Mid-AEBP). (From Ref. 17.)

So far we have only dealt with those atomic groups which are directly accessible from the band assignments. More information can be extracted from NMR data when they are combined with carbon and hydrogen contents from elemental analysis. Brown and Ladner (36) first used this principle to calculate the aromaticity, f_a = aromatic C/total C, the degree of aromatic ring substitution, and the ratio of peripheral to total aromatic C from ^{1}H-NMR. Since the advent of ^{13}C-NMR, it has become possible to calculate more structural parameters in addition to those directly read off the spectra. For instance, Abu-Dagga and Rüegger (16) calculate the H/C ratio for the aliphatic part in the mixture from the aromatic to aliphatic hydrogen and carbon ratios:

$$\frac{H_L}{C_L} = \frac{H}{C} \frac{1 + C_A/C_L}{1 + H_A/H_L}$$

This ratio depends on both chain length and naphthenic character of the sample. It can be used to calculate the CH_2 and CH contents from the

Table 5.7 Evaluation of the Aromatic Region of the ^{13}C-NMR Spectra of the Kern River Crude Fractions Shown in Fig. 5.11

	Fraction Cut 2A	Mid-AEBP (°F)	(°C)	MW	f_a
(A)	Cut 2A	825	440	420	0.27
(B)	Cut 3A	955	510	525	0.27
(C)	Cut 5	1240	670	880	0.28
(D)	SEF-1	1345	730	1110	0.37
(E)	SEF-2	1925	1050	2520	0.45
(F)	SEF-3	2230	1220	3620	0.54

Absorption band (ppm)	Structural parameter[a]	% of total aromatic C (A)	(B)	(C)	(D)	(E)	(F)
110–120	Heterocompounds	4	9	5	10	20	20
120–124	CH[a]	10	14	11	14	15	14
124–130	CH + CAI[b]	27	28	16	29	27	28
130–133	CAI[b]	11	9	8	8	10	8
133–135	CAPS-3[c]	5	6	7	6	6	7
135–138	CANF[d]	9	9	9	10	9	9
138–150	CAPS-1,2[e]	25	21	28	19	12	14
>150	Heterocompounds	8	2	7	3	0	1
	$f_a = C_{arom}/C_{total}$	0.27	0.27	0.28	0.37	0.45	0.54

[a]Unsubstituted peripheral aromatic carbon.
[b]Internal aromatic carbon.
[c]CH_3 substituted peripheral aromatic carbon.
[d]Aromatic carbon fused to naphthenic rings.
[e]CH and CH_2 substituted peripheral aromatic carbon.

equations

$$CH_2 = \{H_L(1 - C_L/H_L) - \tfrac{1}{3}\,CH_3\}$$

and

$$CH = H_A - 2CH_2 - 3CH_3$$

where CH_3 is obtained from the ^{13}C-NMR spectrum. These two groups cannot otherwise be distinguished by regular NMR procedures, and they are sometimes difficult to evaluate quantitatively even with spectral editing (see Chapter 8).

Clutter et al. (37) distinguish three approaches to such additional information:

- The structural parameter approach
- The average structure approach
- The functional analysis approach

The first and third are further discussed in Chapter 8. The second has little promise because in most cases the average structure of such complex mixtures as petroleum fractions is quite different from the actual structures of most molecules present. This means the average structure usually is not the same as a representative structure. Clutter et al. recognized this fact and recommended the functional analysis approach.

C. Applications

Numerous applications of NMR spectroscopy to fossil fuel fractions have been described in the literature, but relatively few on petroleum fractions. References can be found in the biannual reviews of *Analytical Chemistry*, for example, by Sutton (38,39), Yonko (40), and Cooper (41–43). Most of the published applications involved measurements on rather broad fractions such as atmospheric or vacuum residues and asphaltenes. These measurements give only a general impression of a sample's composition. More detailed information can be gained by applying NMR to narrow compound-class fractions obtained from distillation cuts by chromatography. In particular, the ^{1}H-NMR spectra of broad fractions are largely unresolved, whereas those of compound-class fractions reveal a surprising amount of spectroscopic structure.

In the following pages we will explore the effect of fractionation on the NMR spectra of heavy petroleum fractions. Three levels of fractionation will be investigated,

1. By AEBP, that is, by distillation and by solubility fractionation
2. By chromatographic separation into *compound groups*
3. By subfractionation of one of these (the aromatics) into *compound classes* (mono-, di-, and triaromatics)

Let us first see the effect of AEBP on the ^{13}C-NMR spectrum of a petroleum sample. In Fig. 5.11 we assembled the spectra of three high-boiling distillates and three nondistillable solubility fractions of Kern River crude. The AEBPs and f_a values (aromatic to total carbon ratio) are listed in the figure. More compositional data of these fractions were presented in Chapter 4, Table 4.24. The most striking observation with these spectra is their amazing similarity. At first glance, there seems to be hardly any

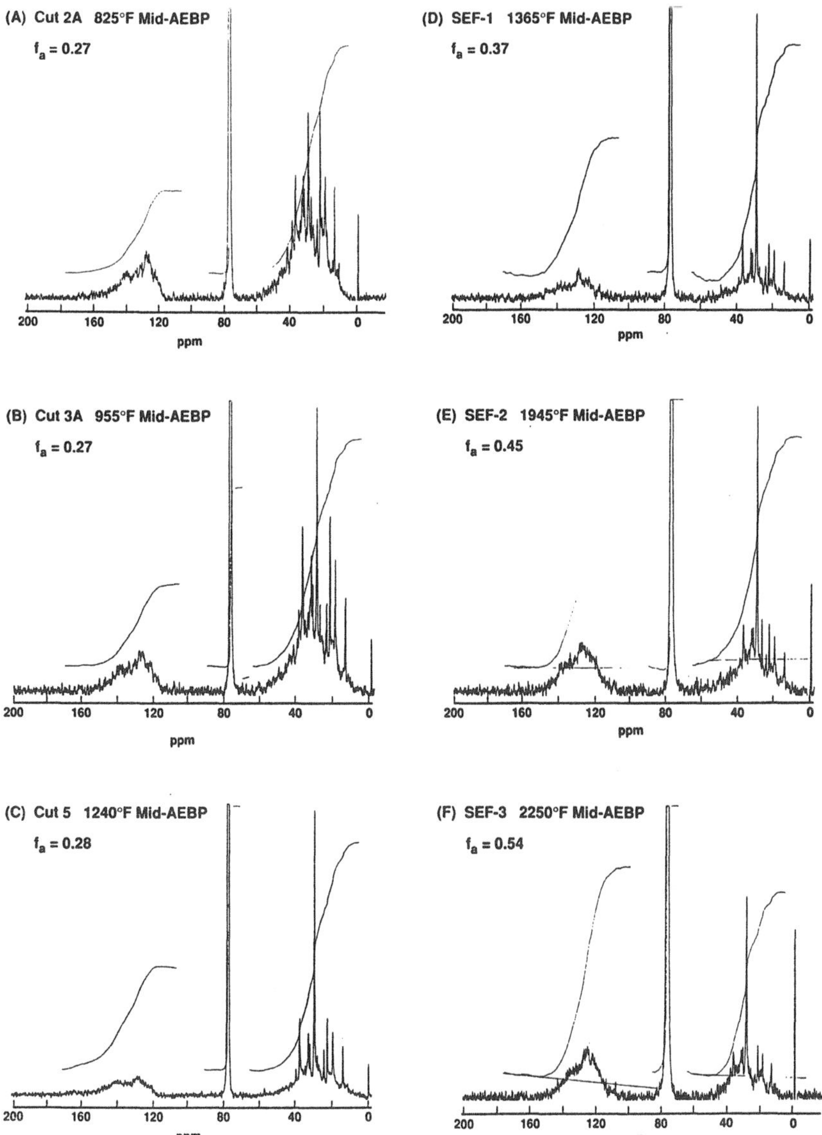

Figure 5.11 (A–F) ^{13}C-NMR spectra of six Kern River crude fractions of increasing AEBP. (From Ref. 17.)

change in either the aromatic or the aliphatic region. It takes close inspection and numerical evaluation to reveal some differences.

Remarkably, at least eight of the peaks from Table 5.6 stand out clearly even in the heaviest samples, SEF-2 and SEF-3. Thus, ^{13}C-NMR spectra are much less sensitive to the complexity of the sample than ^{1}H-NMR spectra.

The integral curves of the aromatic regions of the spectra in Fig. 5.11 were evaluated for the bands of interest, and the data were compiled in Table 5.7. Only the f_a values reveal a significant change with AEBP. The 30-ppm group of peaks changes in relation to both the naphthenic hump and the CH_3 peaks, though not in a continuous steady trend. Although, in general, the aromaticity and the number of carbon atoms in intermediate- and long-chain CH_2 groups increase with AEBP, the aromaticity increases the most in the nondistillable residue fractions, and the long-chain carbon increases most in the higher distillates and somewhat less among the solubility (SEF) fractions of the nondistillable residue. None of the other features of the spectra or the data in Table 5.7 would draw attention to the dramatically increased complexity of the higher AEBP fractions revealed by the rapidly increasing MW and heteroatom content.

In contrast, the difference between different crude oils is quite obvious. Compare, for example, the spectra of Fig. 5.11 with those of Fig. 5.7. Note the much larger size of the prominent peak at 29.7 ppm in the Arabian Heavy VGO cut. The small size of this peak in the Kern River VGO spectrum indicates a lack of long straight alkyl chains as would be expected from a biodegraded oil.

Greater differentiation than in the spectra of Fig. 5.11 occurs when we compare the ^{13}C-NMR spectra of compound groups. Figure 5.12 shows those of the saturates, aromatics, and polars of Kern River Cut 2A. Here, the aliphatic regions display obvious differences, for example, in the ratio of the CH_3 peaks to those caused by paraffinic CH_2 and CH groups. The differences in the aromatic regions are even greater. Aside from the absence of aromatics in the saturates fraction, the types of aromatics in the other two fractions (aromatics and polars) are obviously quite different. For easier reference, we have drawn two vertical lines at 124 and 133 ppm (see Table 5.7 for their significance). The results of the integration are listed in Table 5.8. They are obviously not very precise, but we recognize, for instance, that the polars have only about half as many substituted aromatic carbon atoms as the aromatics. The differentiation between internal aromatic carbon (CAI) and unsubstituted aromatic carbon from the ^{13}C-NMR spectra is impossible without additional information, for example, from ^{1}H-NMR spectra. We will come back to this subject in Chapter 8.

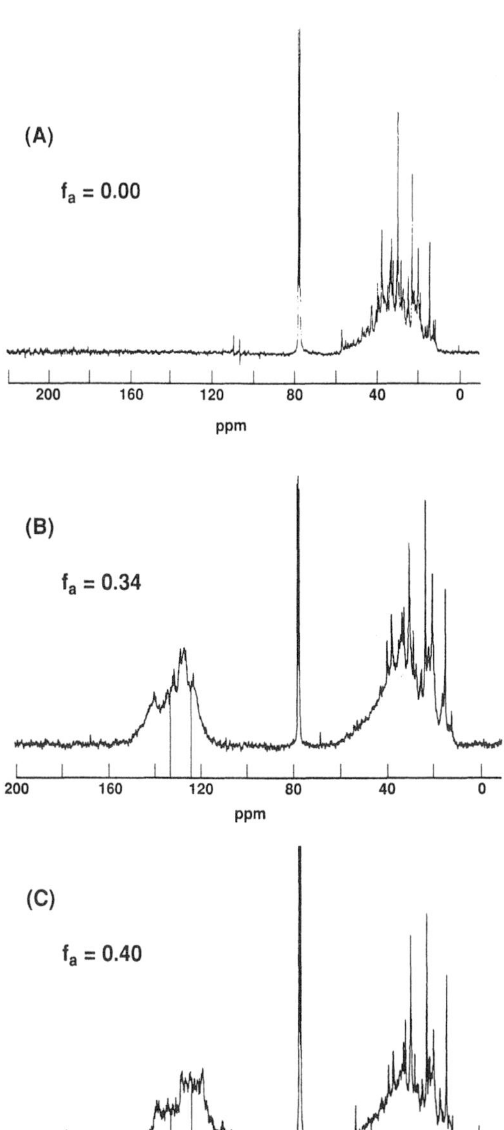

Figure 5.12 ^{13}C-NMR spectra of (A) saturates (36.5 wt%), (B) aromatics (46.9 wt%), and (C) polars (15.7 wt%) fractions from Kern River crude VGO, Cut 2A. (From Ref. 17.)

Table 5.8 Evaluation of the Aromatic Region of the ^{13}C-NMR Spectra of the Kern River VGO Fractions Shown Figs. 5.12 and 5.13

	Fraction Cut 2A	f_a
(A)	Saturates	0.00
(B)	Aromatics	0.34
(C)	Monoaromatics	0.18
(D)	Diaromatics	0.27
(E)	Triaromatics	0.55
(F)	Polars	0.40

Absorption band (ppm)	Structural parameter[a]	% of total aromatic C (A)	(B)	(C)	(D)	(E)	(F)
110–120	Heterocompounds	0	6	3	9	5	26
120–124	CH[a]	0	15	9	15	29	18
124–130	CH + CAI[b]	0	32	28	28	10	26
130–133	CAI[b]	0	12	11	12	6	9
133–135	CAPS-3[c]	0	7	5	7	9	5
135–138	CANF[d]	0	8	9	7	22	7
138–150	CAPS-1,2[e]	0	20	33	19	8	9
>150	Heterocompounds	0	1	2	2	0	0
	$f_a = C_{arom}/C_{total}$	0	0.34	0.18	0.27	0.55	0.40

[a]Unsubstituted peripheral aromatic carbon.
[b]Internal aromatic carbon.
[c]CH_3 substituted peripheral aromatic carbon.
[d]Aromatic carbon fused to naphthenic rings.
[e]CH and CH_2 substituted peripheral aromatic carbon.

Going into greater detail yet, we look at compound classes in Fig. 5.13. This figure shows the ^{13}C-NMR spectra of the mono-, di-, and triaromatics fractions of cut 2A.* Here the differences are even more pronounced. The aromatic carbon content increases from 12% in the monoaromatics to 54% in the triaromatics. The distinct decrease in paraffinic compared to naphthenic carbon, the increase in aromatic–naphthenic ring fusion, and the shift in CH_3 to CH_2 groups with increasing aromatic ring number is new information. The aromatic–naphthenic ring fusion is indicated by the size of the CANF region (135–138 ppm) and the change in the size of the paraffinic CH_2 and CH substituted aromatic C (138–150 ppm). Note the

*The method of their preparation will be described in Chapter 6.

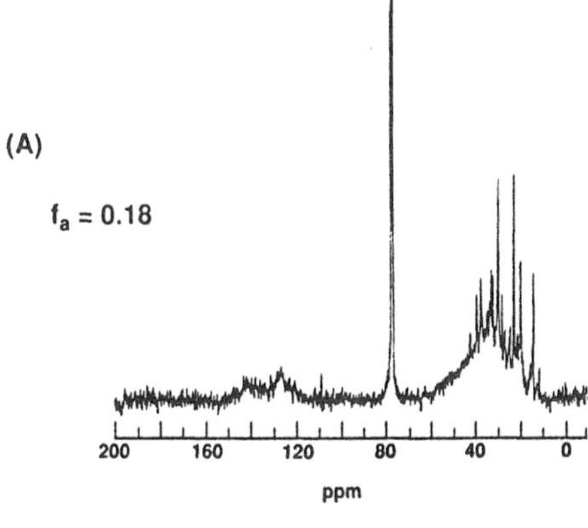

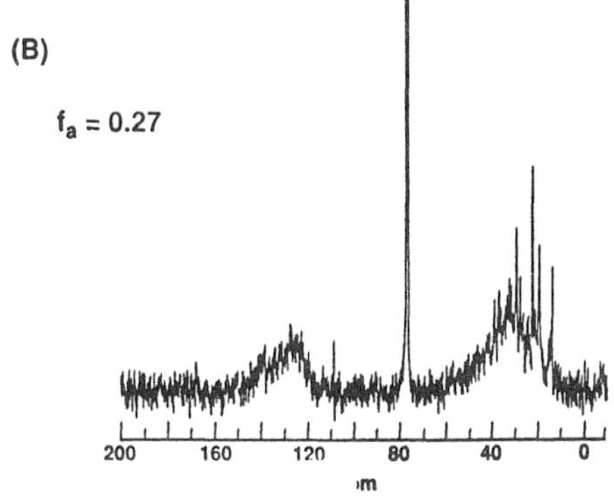

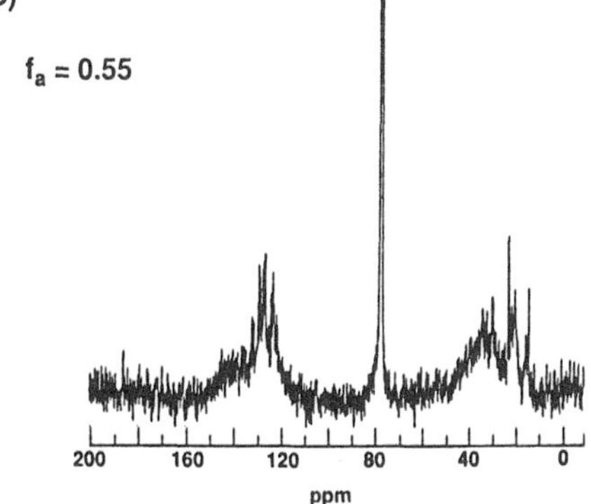

Figure 5.13 ^{13}C-NMR Spectra of Kern River crude (A) monoaromatics (37.7%), (B) diaromatics (29.0%), and (C) triaromatics (20.0%) fractions of Kern River Cut 2A. (From Ref. 17.)

increase of the former and the decrease of the latter with increasing aromatic ring number.

Figure 5.14 presents the ^{1}H-NMR spectra of the same monoaromatics and triaromatics shown in Fig. 5.13 and in addition the spectrum of the parent aromatics fraction. The ^{1}H-NMR spectra of the two aromatic compound class fractions display much greater detail than that of the whole aromatics fraction. The α-H sector has between two and four peaks allowing the estimation of the α-CH_3, CH_2, and CH groups. The aromatic region, too, has distinct sectors. In the triaromatics spectrum, the characteristic phenanthrene peaks stand out. This spectrum has also significant absorption in the monoaromatic region indicating molecules with up to three separate single, interconnected aromatic rings in addition to those with three fused aromatic rings. From MS measurements, we know that this fraction contains hydroaromatic molecules with three aromatic rings separated by naphthenic ones or fused together as in phenanthrenes, and some with heteroatoms.

The numerical evaluation of the integral curves in Figs. 5.6 (aromatic region) and 14 gave the results listed in Table 5.9. Most striking here is the strong increase in aromatic H with increasing aromatic ring number and its decrease with increasing boiling point. The H of α CH and CH_2 increases as the aromatic H does, whereas the β H goes the opposite way. The H from γ CH_3 decreases with increasing aromatic ring size and somewhat with increasing boiling point.

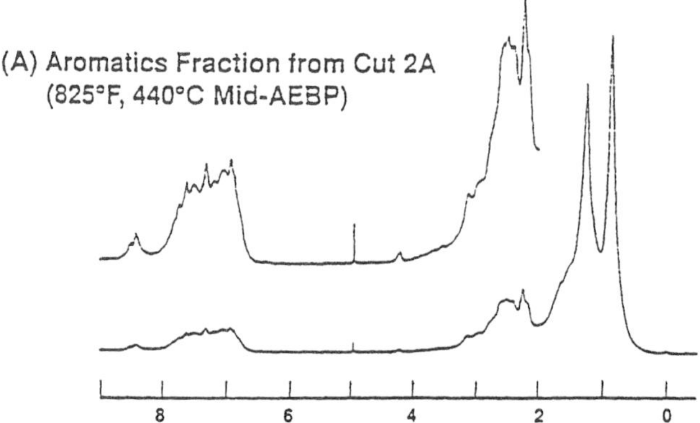

Figure 5.14 ^{1}H-NMR spectra of the (A) total aromatics fraction of Kern River Cut 2A, (B) their monoaromatics, and (C) triaromatics subfractions. (From Ref. 17.)

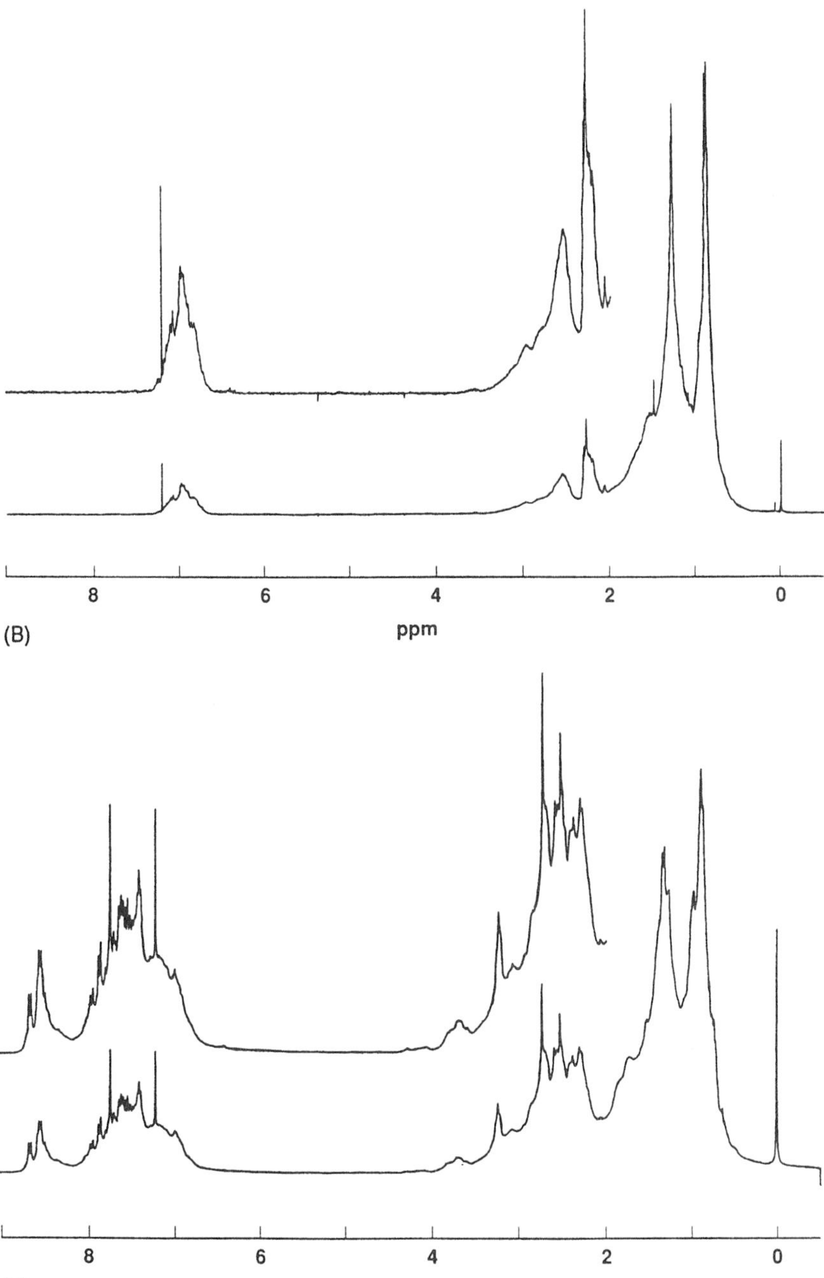
8
6
4
2
0
(B)
ppm
8
6
4
2
0
(C)
ppm

Table 5.9 Evaluation of the ^{1}H-NMR Spectra Shown in Figs. 5.6 and 5.14

	Fraction	Mid-AEBP (°F)	Mid-AEBP (°C)	MW
(A)	Aromatics, Cut 2A	825	440	420
(B)	Monoaromatics, Cut 2A	825	440	
(C)	Diaromatics, Cut 2A	825	440	
(D)	Triaromatics, Cut 2A	825	440	
(E)	Aromatics, Cut 3A	955	510	525
(F)	Aromatics, Cut 5	1240	670	880

Absorption band (ppm)	Structural parameter	% of total aromatic C (A)	(B)	(C)	(D)	(E)	(F)
0.0–4.5	Aliphatic H	87.9	95.7	91.8	83.5	92.4	94.3
0.0–1.0	γ CH_3	26.7	27.9	25.4	19.1	23.4	23.9
1.0–1.9	β CH_2 + β CH + β CH_3	40.6	51.8	44.2	34.6	48.4	55.1
1.0–1.6	β CH_2 + β CH_3	34.3	43.9	35.6	27	39.6	48.3
1.6–1.9	β CH_2 + β CH	6.3	8	8.6	7.6	8.9	6.8
1.9–3.4	α CH_3 + α CH_2 + α CH	21.3	13.9	22.1	29.3	20.3	15.6
1.9–2.25	α CH_3	7	8	4.9	5.2	4.7	3.7
2.2–2.8	α CH_2	10.2	5.6	11.5	15.3	9.4	6.8
2.8–3.4	α CH_2 + α CH	4.1	0.3	4.9	6.7	4.9	4.3
3.4–4.5	CH_2 + CH α to 2 aromatic rings	0	0	0.8	2	1.5	1
6.5–10	Aromatic H	12	4.5	8	16.5	7.5	5.5
6.5–7.3	Monoaromatic H	—	4.5	3.5	4.5	—	—
7.3–7.8	Diaromatic H	—	0	4.5	7.5	—	—
7.8–10	Triaromatic H	—	—	—	4.5	—	—

IV. SUMMARY

Structural group-type characterization methods make it possible to group the molecular features of complex mixtures for easier data handling and comprehension. Instead of describing a petroleum fraction in molecular detail, they allow its description in relatively few relevant parameters. The *n-d-M* and related methods, for instance, provide the distribution of aromatic, naphthenic, and paraffinic carbon in a sample. If applied to the permissible range of boiling point and chemical composition, the *n-d-M* method is quite accurate, better than $\pm 2\%$ even for the naphthenic carbon content.

NMR, especially the combination of ^{1}H- and ^{13}C-NMR and elemental analysis, allows the determination of numerous average structural groups in a petroleum fraction, such as all the aromatic and aliphatic carbon; to some extent, naphthenic and paraffinic C; the number and length of long paraffinic chains; the degree of chain branching; the number of aromatic methyl substituents; the aromatic ring size; the degree of aromatic ring substitution and, to some extent, the type of substituent carbon (CH, CH_2, CH_3); and many more. Again, these NMR results are all average numbers, namely, the concentrations of these groups in the mixture.

NMR works very well with petroleum distillates, especially with narrow fractions. Chromatographic separation of distillation cuts into compound-class fractions further enhances NMR data. ^{13}C-NMR is acceptable even for nondistillable pentane-insoluble petroleum fractions. ^{1}H-NMR loses detail with these fractions, but it still permits the determination of aliphatic and aromatic H.

REFERENCES

1. Van Nes, K., and van Westen, H. A. 1951. *Aspects of the Constitution of Mineral Oils*, Elsevier Publishing Company, New York.
2. Smittenberg, J., and Mulder, D. 1948. Relations between refraction, density, and structure of series of homologous hydrocarbons. I. Empirical formula for refraction and density at 20° of *n*-alkanes and *n*-alpha-alkanes. *Rec. Trav. Chim.*, **67**:813–825, 826–838.
3. McClellan, A. 1979. Carbon structure typing. Some comments on n-d-M and other methods. Hydrocarbon structures. Chevron report, unpublished.
4. Vlugter, J. C., Waterman, H. I., and van Westen, H. A. 1932. Improved methods of examining mineral oils, especially the high boiling components of nonaromatic character. *J. Inst. Petroleum Technol.*, **18**:735–750.
5. Vlugter, J. C., Waterman, H. I., and van Westen, H. A. 1935. Improved methods of examining mineral oils, especially the high boiling components. *J. Inst. Petroleum Technol.*, **21**:661–676.

6. Waterman, H. I., Boelhower, C., and Cornelissen, J. 1958. *Correlation Between Physical Constants and Chemical Structure*. Elsevier Publishing Co., New York, p. 24.
7. Kurtz, S. S., Jr., King, R. W., Stout, W. J., and Peterkin, M. E. 1958. Carbon-type composition of viscous fractions of petroleum. *Anal. Chem.*, **30**:1224–1236.
8. Hill, J. B., and Coats, H. B. 1928. The viscosity–gravity constant of petroleum lubricating oils. *Ind. Eng. Chem.*, **20**:641.
   Huang, P. K. 1977. PhD Thesis, Dept. of Chem. Engineering, The Pennsylvania State University, University Park, PA.
9. Kemp, W. 1991. *Organic Spectroscopy*, 3rd ed. W. H. Freeman and Company, New York, Chapter 2, pp. 19–99.
10. Cookson, D. J., and Smith, B. E. 1987. An Investigation of the Utility of ^{1}H and ^{13}C NMR Chemical Shift Data When Applied to Fossil Fuel Products. *Coal Science and Chemistry*, edited by A. Volborth. Elsevier Science Publishers, Amsterdam, pp. 31–60.
11. Bartle, K. D., and Smith, J. A. 1967. A high-resolution proton magnetic resonance study of refined tars II—high molecular weight fractions (of coal tars). *Fuel*, **46**:29–46.
   Boduszynski, M. M. 1985. Unpublished work.
   Bouquet, M., and Bailleul, A. 1981. Various Applications of High-Field Nuclear Magnetic Resonance on Petroleum Products. *PETROANALYSIS '81*, edited by G. B. Crump. Wiley, Chichester, Chapter 35, pp. 394–408.
12. Snape, C. E., Ladner, W. R., and Bartle, K. D. 1983. Structural Characterization of Coal Extracts by NMR. *Coal Liquifaction Products. Vol. 1. NMR Spectroscopic Characterization and Production Processes*, edited by H. D. Schultz. Wiley, New York, Chapter 4, pp. 69–84.
13. Bartle, K. D., Ladner, W. R., Martin, T. G., Snape, C. E., and Williams, D. F. 1979. Structural analysis of supercritical-gas extracts of coals. *Fuel*, **58**:413–422.
14. Netzel, D. A. 1987. Quantitation of carbon types using DEPT/QUAT NMR puls sequences: Application to fossil-fuel-derived oils. *Anal. Chem.*, **59**:1775–1779.
15. Cookson, D. J., and Smith, B. E. 1987. One- and two-dimensional NMR methods for elucidating structural characteristics of aromatic fractions from petroleum and synthetic fuels. *Energy Fuels*, **1**:111–120.
16. Abu-Dagga, F., and Rüegger, H. 1988. Evaluation of low boiling crude oil fractions by n.m.r. spectroscopy. Average structural parameters and identification of aromatic components by 2D n.m.r. spectroscopy. *Fuel*, **67**:1255–1262.
17. Boduszynski, M. M., and Wilson, D. M. 1985. Unpublished.
18. Aldrich Library of NMR Spectra (1974).
19. Snape, C. E., Ladner, W. R., and Bartle, K. D. 1979. Survey of carbon-13 chemical shifts in aromatic hydrocarbons and its application to coal-derived materials. *Anal. Chem.*, **51**:2189–2198.

20. Maekawa, Y., Yoshida, T., and Yoshida, Y. 1979. Quantitative ^{13}C n.m.r. spectroscopy of a coal-derived oil and the assignment of chemical shifts. *Fuel*, **58**:864–872.
21. Johnson, L. F., and Jankowski, W. C. 1972. *Carbon-13 NMR Spectra*. Wiley, New York.
22. JEOL-JAPAN, Ltd. Tokyo-Company brochures: *Carbon-13 NMR Spectra*. Nos. 1–3.
23. Wilson, N. K., and Stothers, J. B. 1974. ^{13}C and ^{1}H spectra of some methylnaphthalenes. *J. Magn. Resonance*, **15**:31–39.
24. Fisher, P., Stadelhofer, J. W., and Zander, M. 1978. Structural investigation of coal-tar pitches and coal extracts by ^{13}C n.m.r. spectroscopy. *Fuel*, **57**:345–352.
25. Young, D. C., and Galya, L. G. 1984. Determination of paraffinic, naphthenic, and aromatic carbon in petroleum derived materials by carbon-13 NMR spectrometry. *Liq. Fuels Technol.*, **2**(3):307–326.
26. Pretsch, E., Clerc, T., Seibl, J., and Simon, W. 1976. *Tabellen zur Strukturaufklärung organischer Verbindungen mit spectroskopischen Methoden*. Springer-Verlag, Berlin.
27. Netzel, D. A., McKay, D. R., Heppner, R. A., Guffey, F. D., Cooke, S. D., Varie, D. L., and Lynn, D. E. 1981. ^{1}H- and ^{13}C-n.m.r. studies on naphtha and light distillate saturate hydrocarbon fractions obtained from in-situ shale oil. *Fuel*, **60**:307–320.
28. Thiel, J., and Gray, M. R. 1988. NMR spectroscopic characteristics of Alberta bitumens. *AOSTRA J. Res.*, **4**:63–73.
    Thiel, J., and Wachowska, H. 1989. ^{1}H and ^{13}C n.m.r. characteristics of aliphatic component of coal extracts. *Fuel*, **68**:758–762.
29. Cookson, D. J., and Smith, B. E. 1985. Determination of structural characteristics of saturates from diesel and kerosene fuels by carbon-13 nuclear magnetic resonance spectroscopy. *Anal. Chem.*, **57**:864–871.
30. Altgelt, K. H. 1991. To be published.
31. Ward, R. L., and Burnham, A. K. 1984. Identification by ^{13}C n.m.r. of carbon types in shale oil and their relation to pyrolysis conditions. *Fuel*, **63**:909–914.
32. Cookson, D. J., Rolls, C. L., and Smith, B. E. 1989. Structural charcteristics of branched plus cyclic saturates from petroleum and coal derived diesel fuels. *Fuel*, **68**:788–762.
33. Breitmaier, E., and Voelter, W. 1987. *Carbon-13 NMR Spectroscopy*, 3rd ed. VCH Verlagsgesellschaft mbH, Weinheim, Germany.
34. Galya, L. G., and Young, D. C. 1983. A new carbon-13 NMR method for determining aromatic, naphthenic, and paraffinic carbon. *Amer. Chem. Soc. Petrol. Chem. Prepr.*, **28**(5):1316–1318.
35. Yamashita, G. T., Saetre, R., and Somogyvari, A. 1989. Evaluation of integration procedures for PNA analysis by ^{13}C-NMR *Amer. Chem. Soc. Petrol. Chem. Prepr.*, **34**(2):301–305.
36. Brown, J. K., and Ladner, W. R. 1960. A study of the hydrogen distribution in coal-like materials by high-resolution nuclear magnetic resonance spec-

troscopy II—A comparison with infra-red measurement and the conversion to carbon structure. *Fuel*, **39**:87–96.
37. Clutter, D. R., Petrakis, L., Stenger, Jr., R. I., and Jensen, R. K. 1972. Nuclear magnetic resonance spectrometry of petroleum fractions. *Anal. Chem.*, **44**:1395–1405.
38. Sutton, D. L. 1989. Crude oils. *Anal. Chem.*, **61**:165R–167R.
39. Sutton, D. L. 1991. Crude oil, shale oil, and coal oil. *Anal. Chem.*, **63**:50R–52R.
40. Yonko, T. 1987. Crude oils. *Anal. Chem.*, **59**:252R–254R.
41. Cooper, J. R. 1987. Asphalts, bitumens, tars, and pitches. *Anal. Chem.*, **61**:259R–261R.
42. Cooper, J. R. 1989. Asphalts, bitumens, tars, and pitches. *Anal. Chem.*, **61**:171R–173R.
43. Cooper, J. R. 1991. Asphalts, bitumens, tars, and pitches. *Anal. Chem.*, **61**:56R–58R.

6

Chromatographic Separation of Heavy Petroleum Fractions

I. GENERAL THOUGHTS ON SEPARATIONS

The structural group techniques discussed in the previous chapter may be viewed as the first level in the characterization of fractions. The next level brings us to their separation into compound groups and compound classes.

Well-planned separations serve two purposes. First, they are an analytical tool in themselves in that they yield information about the composition of a sample in terms of its main compound groups and compound classes (see Chapter 2 for definitions). The concentration of saturates, aromatics, and polars in a sample may be all the information required. For more detailed analysis, each of these may be further divided; for example, the aromatics by aromatic ring number into the compound classes of mono-, di-, tri-, tetra-, and pentaaromatics. Having done this, we already know a great deal about the sample without any additional measurements.

At least equally important as providing this information is the fact that the separation of a petroleum fraction into compound classes aids subsequent measurements performed for molecular characterization. As will be discussed further in Chapter 7, interpretation of nonfragmenting MS, although relatively easy with well-defined compound-class fractions, is almost impossible with entire distillation cuts. A molecular ion peak can represent several compound types or even various compound classes rather than just one compound type, which makes its assignment difficult or impossible. Only mass peaks from well-defined fractions can be assigned unequivocally. Careful separation provides such well-defined fractions for good mass spectra and the additional analytical information necessary for their interpretation. More will be said on this topic in Chapter 7. Chromatographic separation into well-defined compound-class fractions also facilitates the interpretation of other spectra, for example, from NMR and IR spectroscopy.

Another advantage of good separations is the mitigation of peak "crowding." Crowded mass spectra are hard to read for two reasons: (1) because of peak overlap resulting from several different compound type molecules with similar masses; and (2) because of dilution which prevents some compounds from being seen and distinguished from the background noise. The dilution by other components in broad fractions is especially severe for molecules of high MW. Although the latter may have reasonable concentrations in terms of weight percent, their mole percent, which determines the peak height in mass spectra, may be prohibitively low. Both advantages of good separations, the gain of complementary information and the avoidance of the dilution effect, are not restricted to MS but aid the measurements by other techniques as well.

Numerous methods have been employed for the separation of petroleum fractions into the main compound groups such as saturates, aromatics, and polars. Some of these have been reviewed, for example, by Altgelt and Gouw (1). If performed only on one column, for example, a silica or alumina column, separation is poor, frequently leaving saturates contaminated with substantial amounts of aromatics, aromatics with polar compounds, and polars with aromatic hydrocarbons.

Separating the polar fractions first on a low-polarity adsorbant greatly helps subsequent separations of the remaining "neutrals," primarily by saving the columns used in the latter process from being deactivated and destroyed by irreversible adsorption and, thus, justifying the expense of high-efficiency columns. The "purified" neutrals can then more easily and cleanly be separated into saturates, aromatics, and neutral heterocompounds. This sequence, removal and speciation of all polars first, followed by the separation of the neutrals on one or several separate columns, was an important innovation by the authors of the API Research Project 60 (2; see also Refs. 3 and 4). They set a trend of combining different columns for these separations, each designed to optimize a specific task. With this specialization, good separations into specific, well-defined fractions have become commonplace.

The API separation scheme involves more than just the *removal* of polar compounds to clean up the sample. It subdivides them on ion exchange resins and on $FeCl_3$ (deposited on clay) into at least three fractions, acids, bases, and neutral nitrogen compounds. However, despite various improvements, these separations are still fraught with problems. Ion exchange resins are tedious to prepare to proper standards and may even cause artifacts. An alternative is the separation of the polars on two types of active alumina—first basic, then acidic. Minor amounts of very acidic and basic compounds may be lost in this process by irreversible adsorption. In most routine applications, the advantages, ease of operation (commercial columns), and a high degree of standardization, outweigh the disadvantage of this loss.

For the highest degree of sorting, that is, for separations leading to the simplest and best defined fractions, at least two fractionation steps must be chosen in such a way that they separate the molecules by different ("orthogonal") principles, for example, one by molecular size and the other by chemical structure. Altgelt and co-workers (1,5) recommended separating by size first, using size exclusion chromatography (SEC),* and then separating each size fraction by chemical nature, using normal phase liquid chromatography (LC). With LC on deactivated alumina, the SEC cuts were subdivided into compound groups such as saturates, aromatics, and various polar groups. Because of the reduced molecular size ranges, the subsequent LC separations were easier to perform on the SEC fractions and led to less overlap between fractions than without prior treatment. Unfortunately, SEC has several problems in this application. One is the large column size required for SEC if it is chosen as the first step in a preparative multistep separation scheme. Second, with certain highly polar petroleum fractions, the packing materials tend to adsorb part of the sample. Small polar compounds added to the solvent as modifiers, for example, methanol, reduce the adsorption of sample components. The most effective compounds, pyridine and other amines, however, are hard to remove and interfere with further separations and measurements in this application.

A more practical choice for the first fractionation of a heavy petroleum is distillation. Not only can it handle large samples without the need for solvent removal (and without adsorption problems), but it provides us with customary, repeatable, well-defined, and extensively researched fractions in the distillates range. Distillation of residues into a number of reasonably

*We prefer the more descriptive term size exclusion chromatography (SEC) to the older term gel permeation chromatography (GPC).

narrow cuts decisively aids subsequent chromatographic separations into compound groups and compound classes.

Indeed, in most cases it is indispensable for good separations because the chromatographic methods become increasingly difficult with increasing AEBP of the petroleum fractions. Methods which work well for vacuum gas oils (650–1000°F, 350–540°C AEBP) may give problems with super heavy vacuum gas oils (VGOs) (1000–1300°F, 540–700°C AEBP) and fail for nondistillables (>1300°F, 700°C AEBP). For example, the aromatic moiety in a monoaromatic molecule with 40 carbon atoms represents only 15% of the mass. It stands to reason that for fractions of C_{40} (heavy VGOs) the separation of saturates and aromatics is by far not as sharp as that of samples with C_{25} (light VGOs). Obviously, the effect gets worse with larger molecules.

The separation of aromatics by ring number is similarly affected by molecular size, that is, by aliphatic rings and alkyl substituents. The problem is even greater in the separation of aromatics from heterocompounds, especially from neutral ones. Deep distillation helps the analysis of heavy petroleum fractions by removing the most refractory nondistillable molecules and producing sufficiently narrow cuts for good chromatographic separations.

The cuts below 850°F (450°C) are easy to analyze; those above this AEBP pose increasing problems. These changes in ease of separation and analysis with AEBP make it imperative to check out any separation method for its applicability to truly high-boiling fractions. Most methods become strictly operational rather than well defined when applied to nondistillable residues (>1300°F, >700°C AEBP).

Because distillation was the subject of Chapter 3, it will not be treated here again. This leaves mainly the chromatographic procedures for the remainder of this chapter. The theory and practice of chromatography have been treated, for example, by Snyder and Kirkland (6) and in many other books and reviews. In the most recent book on this subject, *Chromatography, 5th Edition*, edited by Heftmann (7), a number of experts discuss in fair detail the theory of the chromatographic process in general, eight types of chromatography (and two more other separation methods), column technology, stationary and mobile phases, and instrumentation. Several applications, including to fossil fuels, are also reviewed there, though somewhat briefly. The impressive progress in chromatographic methodology during the last decade has also been reviewed by another panel of practitioners in this field (8,9). There is no need for us to cover this material again. In the following pages of our chapter we will describe only those aspects of chromatography which are directly related to the separation and analysis of heavy petroleum fractions.

II. THE TOOLS

A. Gas Chromatography

Gas chromatography (GC) is the method of choice for light and middle distillates because of its high resolution. However, for high-boiling petroleum distillates, it has much more limited use because of its propensity to separate primarily by carbon number and because of the immense number of compound types and homologues in these materials, let alone their low volatility. The two main applications here are simulated distillation (see Chapter 3) and the separation and identification of compound types and their homologues in well-defined LC fractions. Trusell (10) has reviewed the field of GC applied to heavy petroleum fractions.

Whereas GC of low-boiling materials is often performed with packed columns, that of heavy petroleum fractions is done with open tubular column GC. Lee et al. (11) described the theory and practice of this latter technique. They cover general concepts such as zone spreading, peak resolution, separation efficiency, and so on; type of stationary phases and their temperature stability; column design, injectors, detectors, and other instrumentation. Among the numerous applications described in their book, there are also some of interest in our context.

A recent review by Poole and Poole (12) of the principles of GC covered general principles and practical aspects. Among the latter were discussions of packed as well as open tubular columns, selection of stationary phases and carrier gas, detectors, pumps, and other instrumentation.

Despite successful applications to certain high-boiling petroleum fractions, GC by itself is generally inadequate for the primary separation of such heavy distillates. Although its temperature range may well be high enough, its resolution is not sufficient for the large number of compound types and homologues in these complex materials. Resolution is further impaired by the peak broadening caused by large molecules (13, p. 71). For these heavy materials, liquid chromatography (LC) and supercritical fluid chromatography (SFC) are clearly superior. The wide choice of mobile phases for these methods permits fine-tuning to the compositional subtleties of the analytes in a way unachievable by GC. SFC has the additional advantage of pressure programming.

The prime use of GC in the analysis of heavy petroleum fractions is in simulated distillation (SIMDIS) with flame ionization detection (FID) (86–88), in GC with atomic emission detection (20–23,89) and other element specific detection (26,90), and in combination with mass spectrometry as GC-MS (see Chapter 8).

Simulated distillation has been discussed in Chapter 3. (See also more references there.) The various detectors, though mainly for LC, will be described briefly in a separate section of this chapter. Here it may suffice to point out that modern atomic emission detectors (20,87,89), used with GC, have high sensitivities, comparable to those of the flame ionization detector. They can monitor C, H, S, N, O, and metals, that is, all the important elements for petroleum fractions, and several of these simultaneously.

Fourier transform IR (FTIR) detectors used with GC for the analysis of heavy petroleum fractions allow us to monitor the chromatogram by specific functionalities, for example, OH and NH, and to distinguish certain polar compound types. Special flow cells and other interfaces for subnanogram identification by GC-FTIR are available and have been reviewed by Norton et al. (19).

GC-MS combines the ease of operation and the separation ability of GC with the characterization power of MS (see Chapter 8). GC-MS has the advantage over LC-MS that the gas chromatograph is easier to interface with MS than the liquid chromatograph. The GC carrier gas is more easily removed than liquid solvents used with LC because of its smaller amount (in the gaseous state) and its smaller molecular size, that is, its greater size difference to the analytes. The interfacing of liquid chromatographic columns with MS will be discussed further in Chapter 8.

Wadsworth and Villalanti (25) combined GC with a special chemical ionization MS technique (see Chapter 7) for the rapid analysis of middle distillates and VGOs up to 950°F (510°C) AEBP. They took advantage of the complementary strengths of the two methods, separation by boiling point with GC and distinction between hydrocarbon and sulfur compound types by chemical ionization MS.

Ordinarily, the upper temperature range of GC is around 1000°F (540°C). High temperature GC (25,86,88,91–93) extends this range to about 1400°F (700°C), that is, to about C_{130} for paraffins. High temperature GC requires special techniques, for example, the use of short, wide capillaries coated with a thin film of thermally stable phase and carrier gas flow programming, often in addition to temperature programming. Nondistillable residues (>1300°F, ~700°C AEBP) and their fractions cannot be analyzed by this method because of their insufficient volatility and temperature stability.

A very important application of GC to our field is in combination with LC and SFC methods, that is, in "multidimentional chromatography." After initial separation of a distillate by LC into hydrocarbon compound classes, GC can often be used to identify the compound types and carbon number members in these narrow fractions (see, e.g., Refs. 14–16). Calibration of the GC column with known *n*-alkanes or other model compounds allows the determination of certain indices for the identification of unknown compounds. The best known of these are the Kovats indices (17; see also Ref. 90, p. 217; and Ref. 85, pp. A33–A35). For PAH samples,

Lee et al. (18) introduced a retention index based on a series of PAH standards.

Coulombe and Sawatzky (24) exploited GC's propensity to separate by boiling point when they developed a method for measuring the number average MWs of aromatic and other compound fractions. FID served as their detector. They obtained distinct and separate calibration curves for paraffinic, naphthenic, and aromatic model compounds. Polyaromatics with three to six rings fell on one and the same curve with moderate scatter. An internal standard helped to avoid dilution errors.

B. Liquid Chromatography

Snyder and Saunders (27) reviewed the principles and practice of LC and its application to petroleum fractions. An excellent general guide for the LC practitioner with a wealth of useful advice was written by Snyder and Kirkland (6). Poppe (28) wrote a very recent, though briefer review of LC. Despite these and other treatments of this subject, it seemed worthwhile to repeat here some basic aspects of direct interest to our special field.

1. Normal and Reversed Phase Chromatography

In normal phase chromatography, the mobile phase is nonpolar and the column packing is polar. Separation occurs because sample molecules dissolved in the mobile phase interact with the packing. The molecules are temporarily adsorbed and, after some time, returned to the mobile phase. This process of adsorption–desorption occurs many times for a molecule on its way through the column. The greater the compound's polarity, the stronger it interacts with the sorbent surface, the more it is retained, and the later it elutes. Various types of interaction occur, from a weak induced dipole–induced dipole interaction (London forces) to hydrogen bonding to strong acid–base attraction.

Normal phase partitioning chromatography is a variant of normal phase chromatography. Here the column packing is coated with a layer of polar liquid which is immiscible with the nonpolar mobile phase and remains on the packing as the stationary phase. Sample molecules move back and forth between the two phases and are partitioned between these according to their respective solubilities in them. Polar sample molecules ordinarily have a greater solubility ratio in favor of the stationary phase than less polar ones and spend more time there. Consequently, they elute later than nonpolar molecules. During the last two decades, so-called bonded phases have replaced the liquid stationary phases in partitioning chromatography.

Bonded phase packings are usually silica particles with chemically modified surfaces. The OH groups on the silica are either silanized or esterified and, in effect, replaced by organic groups, mostly paraffinic hydrocarbons of eight or more carbon atoms. These hydrocarbons are nonpolar and offer no advantage in normal phase chromatography. However, they are very useful in reversed phase chromatography where the sample is eluted with a very polar solvent such as acetonitrile and even mixtures of acetonitrile or similar solvents with methanol or water. Here the sample molecules can partition again between a polar phase and a nonpolar one, except that the phases are reversed from those in normal phase chromatography. Here, the *nonpolar* molecules are more soluble in the stationary phase than in the polar mobile phase and are, therefore, retained more strongly than polar molecules.

The advantage of reversed phase chromatography with its combination of an inert packing surface and a polar mobile phase is that, theoretically at least, it is a very mild method and prevents sample loss due to irreversible adsorption of highly polar molecules, which can be a serious problem in normal phase chromatography of high-boiling petroleum fractions. Unfortunately, it is difficult to completely cover the silica surface with the organic phase. Often a small percentage of the silica OH groups remains unreacted and is available for interaction with polar sample molecules which may then be strongly adsorbed.

In the last 10–15 years, other bonded phases with moderately polar groups became available, for example, with amino, cyano, nitro, and phenyl endgroups. These can be used in either way, normal or reversed phase chromatography, depending on the polarity of the mobile phase. They may have special applications as, for instance, separation of hydrocarbons by aromatic ring number; or they are useful in combination with other column packings (in column trains) for the fractionation of samples with widely diverse polarity. In the following sections of this chapter we will encounter several such applications.

2. Chromatography on Ion Exchange Resins

This subject has been treated recently by Walton (29). In contrast to most workers interested in this method, we are concerned here only with *nonaqueous* chromatography on ion exchange resins. Munday and Eaves (30) were the first to apply the latter version to petroleum fractions. It found widespread use when the scientists of the API-60 project made it a part of their separation scheme, allowing them to isolate acidic and basic components from heavy petroleum fractions (see, e.g., Refs. 2–4 and 31–33). The ion exchange resins are commercially available and consist of macroreticular styrene–divinylbenzene copolymers bonded with strong base or acid groups. For high activity and low bleeding, they need to be prepared by a lengthy sequence of solvent washes* as shown in Table 6.1. The conditioned resins can be stored for several weeks as a slurry in cyclohexane under nitrogen in dark bottles with airtight caps.

Before a run, a column needs to be flushed with four void volumes of mobile phase to mitigate bleeding during a run. The sample, dissolved at about 20 wt% concentration in the appropriate solvent, is introduced into a pair of columns connected in series, usually with the anion column first and the cation column second. The "neutral" fraction is eluted at 40°C (104°F) with 2.5 times the void volume of solvent. Green et al. suggest different solvents depending on the sample's boiling range: cyclopentane for distillates boiling below 500°C (930°F), benzene–cyclohexane 1:3 for short-path distillates boiling between 500 and 700°C (930–1290°F), and ethanol–THF–benzene 1:4.5:4.5 for nondistillable residues; see Fig. 6.1. The eluate obtained with the latter mixture contains weak acids and bases besides the neutrals. These can be separated from the neutrals by running them again on the dual columns, this time with the benzene–cyclohexane mixture as eluent.

After elution, the resins are transferred from the columns to a Soxhlet apparatus. The acids are extracted from the anion exchange resins with a

Table 6.1 Sequence of Solvent Washes for Preparing Anion Exchange Resins

Order	Reagent	Amount (dm^3/1.5 kg resin)
1	1–2 N HCl	25
2	Water	4
3	1–2 N NaOH	25
4	Water	12
5	1:1 Propanol–water	25
6	Propanol	25
7	Ethyl ether	Soxhlet, 24 h (N_2)
8	*n*-Pentane	Soxhlet, 24 h (N_2)

Note: Steps 1–6 are performed while the resin is in a large Buchner funnel. For 3–8, the resin is transferred to a large extraction apparatus. For steps 5 and 6, 1- or 2-propanol may be used when activating the anion resins. Anion resins should be protected from CO_2 once they are in the hydroxide form (steps 3–8). For activation of cation resins, 1 and 3 are reversed and 35 dm^3 (liters) each of HCl and NaOH are used. 1-Propanol should be used for step 6 when activating cation resins.
Source: From Ref. 33. Reproduced with permission of the publisher.

*The following description relies to a great extent on a paper by Green et al. (33).

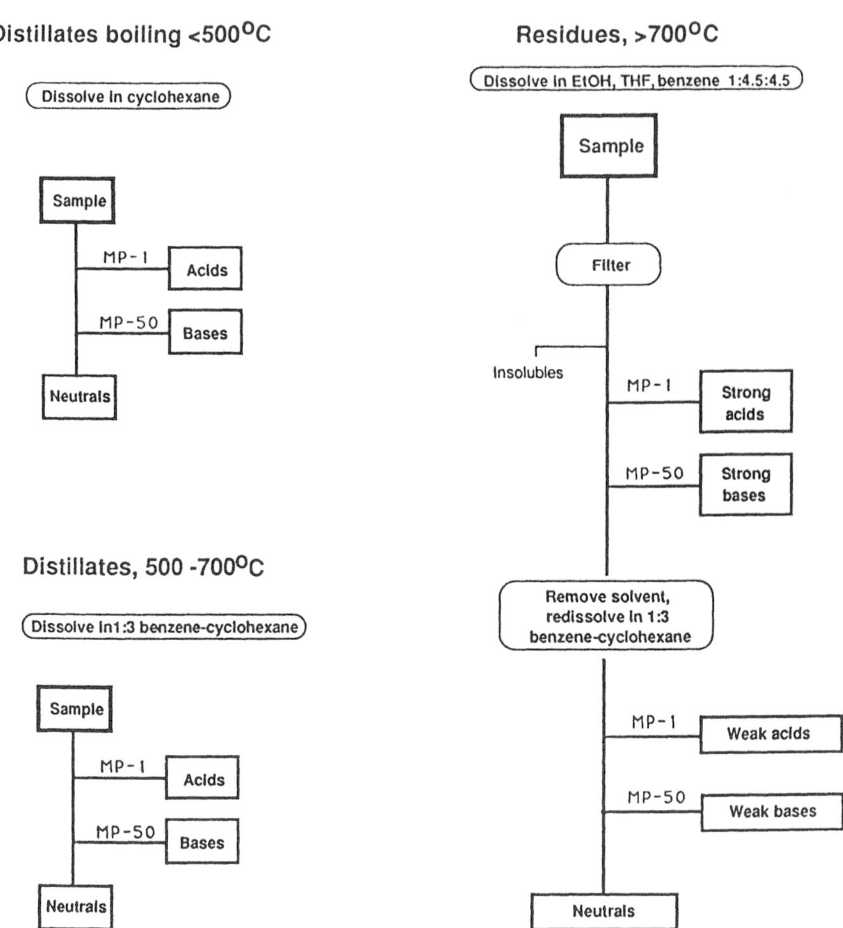

Figure 6.1 Schemes for the separation of petroleum samples into acid, base, and neutral fractions. (From Ref. 33. Reproduced with permission of the publisher.)

benzene–formic acid azeotrope, and the bases from the cation exchange resins, with benzene plus propylamine. The amine must be added slowly, in contrast to the formic acid. Modifications of resin preparation and elution and extraction conditions allow the separation and isolation of neutrals, weak and strong acids, and weak and strong bases (33).

The technique is complex and tedious. The preparation and conditioning of the resins is critical for their stability as well as for their performance.

For instance, bleeding, that is, the release of decomposition products from the ion exchange resins, can be a problem (34). One must also keep in mind the amphoteric character of certain compounds which causes them to be classified either as acids or as bases, depending on the sequence of the ion exchange resins the sample has to pass. With anion exchange resin first, they are collected as acids; but with cation exchange resins first, they are counted as bases. By performing the analysis both ways, the amount of amphoteric material can be measured.

3. Size Exclusion Chromatography

Size exclusion chromatography (SEC), or gel permeation chromatography (GPC), is the separation method which comes closest to differentiating by MW only, almost unaffected by chemical composition. Actually, it separates by molecular size. SEC was discussed in Chapter 4 as one of the ways to measure molecular weights and will not be elaborated here any further except for recent applications with element-specific detection. Hagel and Johnson (35) recently reviewed fundamental and practical aspects of this field, though more from the biochemist's point of view. A review by Altgelt (95) covered the field with special regard for the petroleum chemist.

SEC, usually practiced with refractive index detection, yields a mass profile (concentration vs. time or elution volume) which can be converted to a mass vs. molecular weight plot by means of a calibration curve. The combination of SEC with element specific detection has widened this concept to provide the distribution of heterocompounds in the sample as a function of elution volume and MW. Fish and co-workers (36,37) pioneered the use of reverse phase HPLC and SEC with a graphite furnace atomic absorption (GFAA) detector for measuring the distribution of vanadyl and nickel in heavy petroleum samples and asphaltenes, and for similar applications. Using variants of this technique, inductively coupled and direct current plasma atomic emission spectroscopy (ICP and DCP), Biggs et al. (22,23) extended and improved the former SEC/GFAA method. Now they could continuously monitor the separation, whereas the GFAA approach had allowed only one data point about every 25 s. Biggs et al. included iron and sulfur traces and established detection limits and the absence of concentration related artifacts. As mobile phase they employed a mixture of xylene, chosen for its low volatility, with 20 vol% pyridine, to suppress adsorption of polar compounds on the column packing, and 0.5 vol% *o*-cresol, to mitigate aggregation of solute molecules due to hydrogen bonding. They measured the elemental profiles of V, Ni, Fe, and S of a number of heavy crude oils and their residues. Figure 6.2 gives an impression of the high quality of elution profiles achieved with this method.

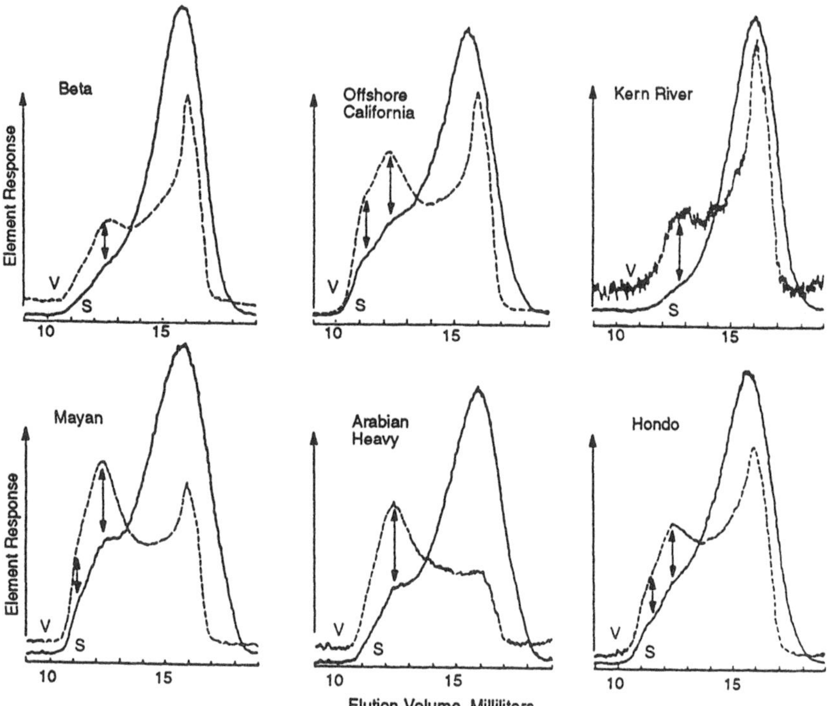

Figure 6.2 Vanadium and sulfur SEC–ICP profiles of six atmospheric residues. (From Ref. 23. Reproduced with permission of the publisher.)

4. Ligand Exchange Chromatography

Ligand exchange chromatography (LEC) has been used as a way to separate sulfur compounds from hydrocarbons. It exploits the specific interactions of sulfur compounds—or other heteroatoms—with certain metal ions. Silver, mercury, copper, and zinc ions, for example, form coordination complexes with (aliphatic) sulfides, whereas palladium ions interact strongly with (aromatic) thiophenic compounds. These metals, deposited as chlorides or other salts on silica, can, therefore, selectively slow down the elution of sulfur compounds relative to hydrocarbons on a chromatographic column. Nishioka et al. (38–41) reported the separation of polycyclic aromatic sulfur heterocycles (PASHs) as a group from PAHs on a palladium chloride. With chloroform/*n*-hexane (1 : 1) as mobile phase, they achieved essentially complete separation as demonstrated by GC and GC-MS.

C. Supercritical Fluid Methods

1. Supercritical Fluid Chromatography

Supercritical fluid chromatography (SFC) combines the advantages of gas and liquid chromatography in that its mobile phase has the low viscosity of a gas and the variable solvent strength of liquids. Under the simplest conditions (constant pressure, only one mobile phase), SFC has nearly the resolution of GC, but its performance can be further enhanced by the variation of solvent and by pressure programming (in addition to temperature programming) during a separation. Schoemakers and Uunk (42) described the fundamentals as well as some practical aspects of the method.

At the high pressure applied with SFC, *n*-butane and *n*-pentane are good solvents even for polar compounds. CO_2 is only a poor to fair solvent for most heavy petroleum fractions, but it has the additional virtue of being easily removed from the fractions. It permits the use of a flame ionization detector (FID) for SFC. (The FID provides a uniform response over a wide range of sample compounds and comes closest to being a true mass detector; see below.) CO_2 is also the solvent of choice for the combination of SFC and MS. Commercial equipment for analytical as well as preparative SFC separations is available.

Despite these advantages, SFC has not yet been used much for heavy petroleum fractions. One reason is the tendency of the restrictor to plug. The restrictor is needed to maintain the required high pressure throughout the entire column and consists either of a short, very fine capillary or a plate with a tiny hole positioned at the outlet. Even if complete plugging of the restrictors can be prevented, molecular clusters may still form in it and then cause spikes in the FID output. These annoyances become worse with increasing sample molecular weight and polarity and, therefore, with increasing sample AEBP.

By placing a linear restrictor (a capillary of 130 mm length and 25 μm diameter) very close to the FID flame, Fuhr et al. (43) solved the problem for their system. Except for an asphaltene sample, all their analytes were successfully run. These workers characterized a heavy crude oil and a bitumen, employing capillary SFC with CO_2 as mobile phase. They found an excellent correlation between the boiling point of aliphatic compounds and their capacity factors* on the SFC column. Even the capacity factors for all hydrocarbons, that is, both aliphatic and aromatic compounds together, correlated quite well with boiling point.

Nishioka et al. (41) employed SFC for separating a vacuum residue fraction of polyaromatic sulfur-containing compounds into subfractions with

*The capacity factor, k, is the ratio of retention time of a compound, t_r, minus that of an unretained compound, t_0, to t_0: $k = (t_r - t_0)/t_0$.

three, four, and five aromatic rings. Their preparative supercritical fluid extraction/fractionation system with UV detection was described by Campbell and Lee (44). By GC and MS they identified numerous S compounds in these subfractions and estimated their abundance. Without prior separation by other means, they obtained clearly defined fractions differing by the number of aromatic rings.

Campbell and Lee (44) noted a difference in the SFC separation of coal and petroleum oils. Their coal oils were separated exactly by ring number, whereas a crude oil was separated mainly by boiling point. The main difference is the degree of alkyl substitution which is quite small in coal liquids and can be very large in petroleum fractions. Although in the coal oils the effect of aromatic ring number was predominant, in the petroleum molecules it may have been overshadowed by that of the number of C atoms in the aliphatic part of the molecules.

SFC with a smectic liquid crystalline stationary phase is very effective in separating aliphatic from aromatic hydrocarbons and also in separating different types of PAHs and other PACs (45). For instance, a mixture of 11 PAHs, each with 5 catacondensed but differently arranged aromatic rings, gave a chromatogram of clearly separated peaks. Similarly, good chromatograms were obtained for two mixtures of model compounds, one containing five catacondensed naphthylcarbazolylethane isomers, and the other, nine catacondensed six-ring thiophene derivatives.

Wright et al. (46) discussed the merits of SFC and its easy interfacing with MS. They also described some applications to polar, high-MW compounds from petroleum fractions and from carbon black extracts containing (presumably) up to 12 pericondensed aromatic rings. Complex, reasonably well-resolved chromatograms resulted from these runs, but, aside from pyrene and coronene, no compounds were identified.

Campbell and Lee (44) pointed out an additional advantage of SFC: Because of density programming, the columns can be packed with larger, less expensive particles than required by GC and still maintain the same selectivity. The low-cost column packings can be replaced more often, which eliminates the need for tedious sample cleanup before an SFC run.

2. Supercritical Fluid Extraction

Supercritical fluid extraction (SFE) is a nonchromatographic method in which a sample is extracted by a solvent under supercritical conditions. Its advantages over regular extraction methods include speed, low temperature, the use of CO_2 (nonflammable and easily removed) or other very low-boiling solvents, and selectivity. By judicious gradual variation of pressure, the solvent power can be adjusted, and a number of successive fractions may be obtained.

Taguchi et al. (47) described a modern semimicro scale SFE apparatus (6.5 ml) which they tested for efficiency and optimum conditions with model compounds and by micro column SFC. Extraction recoveries depended on a number of experimental conditions. The authors concluded that "from the trace analytical point of view, investigation of the quantitative extraction conditions including matrix effects is indispensible." Four other instruments were tested and compared by Lopez-Avila et al. (48). Restrictor plugging and leaks were the main problems encountered.

Singh et al. (49) described a macro scale SFE unit (500 ml) and its application to cutting deeper into a vacuum residue. They collected only one extract and compared its composition with that of several short-path distillation cuts from the same residue.

D. Detectors

Our discussion here will cover some general aspects of detectors, used with LC (and some with GC and SFC) of heavy petroleum fractions, which appear important to us. We also describe the virtues of several specific detectors, namely, the flame ionization (FID), the photodiode array (PDAD), the inductively coupled plasma (ICPD), and the atomic emission detector (AED). For a survey of all the detectors used in our field, the reader may consult Snyder and Kirkland's book (6) or one of the later reviews.

The most common ones for LC are the refractive index detector (RID), the single- and variable-wavelength UV absorbance detectors, and the relatively new but now widely employed diode array detector. For special applications, primarily for heterocompounds, infrared detectors are frequently used. The powerful detection method of MS in on-line chromatographic arrangements will be discussed in Chapter 7.

Calibration is needed for the evaluation of signal intensity in terms of concentration. The response factors of different compound classes, and even those of sufficiently different homologues, usually vary and should theoretically all be known for quantitative measurements. However, for many applications, general (average) response factors for an entire fraction are sufficient. Other times, separate average response factors for each of the main compound classes in the fraction are required. The choice of varying response factors is most important for UV, IR, and visible light detectors, but should not be neglected even for refractive index (RI) detectors. Simple chromatograms with distinct peaks for different compound types can be reasonably well evaluated, provided, of course, the peaks have been properly assigned. For complex chromatograms with broad humps, it is more difficult to assign the appropriate response factors.

For detection by RI, the response factor is proportional to the difference in RI between the sample molecules of a given fraction and the solvent. Some examples of RIs are given in Table 6.2. This table also shows the difference in RI between various model compounds and THF, the most common solvent used with SEC for which RI is a very popular detector. Note the increase in RI of the *n*-alkanes with increasing carbon number in contrast to the decrease in RI of the alkylbenzenes with increasing carbon

Table 6.2 Refractive Indices of THF and of Various Model Compounds

Compound	MW	RI	$RI - RI_{THF}$
THF	72	1.4070	—
Alkanes			
n-Decane	142	1.4102	0.0032
n-Pentadecane	212	1.4315	0.0245
n-Eicosane	282	1.4425	0.0355
n-Pentacosane	352	1.4491	0.0421
Cyclohexane	84	1.4262	0.0192
Decylcyclohexane	224	1.4534	0.0464
Aromatics			
Benzene	78	1.5011	0.0941
Toluene	92	1.4961	0.0891
Decylbenzene	218	1.4832	0.0762
Eicosylbenzene	358	1.4805	0.0735
Naphthalene	128	1.7 ± 0.2[a]	0.3
Methylnaphthalene	142	1.6176	0.2106
Decylnaphthalene	268	1.5434	0.1364
Phenanthrene	178	1.5943	0.1873
Pyrene	202	1.7700[b]	0.3630
Heterocompounds			
Pyrrole	67	1.5085(78)[b]	0.1015(08)
Indole	117	1.6300	0.2330
Pyridine	79	1.5075[b]	0.1005
Quinoline	131	1.6091[b]	0.2021
		1.6235[a]	0.2165
Thiophene	84	1.5289	0.1219
2-Propylthiophene	126	1.5049	0.0979
Benzothiophene	134	1.6374	0.2304

[a]*CRC Handbook of Physical and Chemical Properties*, 1990.
[b]Reference 57.
Note: Data from TRC Thermodynamics Tables except where indicated.

number of the substituents. The first effect reflects the bearing of molecular weight on RI; the second one is a consequence of the dilution of the aromatic nature by the alkyl groups. In samples containing alkanes and aromatics in roughly equal amounts, the RI effect between MW fractions tends to cancel out in SEC, and one RI calibration for the entire sample is often sufficient (50). However, for predominantly paraffinic fractions, a single calibration leads to excessive responses for the aromatic components.

The UV and visible spectra of different compound classes may vary by orders of magnitude for a given wavelength. This feature can be turned into an advantage by the use of full spectrum detectors. Photodiode array detectors are designed as on-line detectors, typically measuring the spectrum in the UV and visible range at several hundred wavelengths in about 10 ms. They can be run either in the absorbance or fluorescence mode. Data handling requires computer programs which are ordinarily provided by the instrument manufacturer. Gluckman et al. (51,52) reported the development of a PDAD for use with fluorescence spectroscopy and its application to fossil fuel fractions (e.g., Refs. 53 and 54). In a trade brochure, Owen (55) described the basic instrumentation of the PDAD. The principles of the calculations were discussed by Vandeginste et al. (56).

Photodiode array absorbance detectors permit the mapping of a sample by compound type in a single HPLC run. An example of an HPLC-UV map of the aromatics fraction in a vacuum gas oil is shown in Fig. 6.3. Because the spectra of many aromatic compounds, including many with condensed ring systems, are known, the ring types can be recognized in the map. HPLC-UV maps are quickly established and can graphically show the differences in composition brought about by hydrocracking and other processing steps.

An almost truly universal detector is the flame ionization detector (FID) in that it gives the same response for all hydrocarbons. In an FID, the eluate is mixed with hydrogen and burned in a controlled flame. The resulting ionized CO_2 is monitored by measuring its electric charge. Heteroatoms give close to zero responses, possibly even negative ones. Their presence, therefore, distorts the results and should, ideally, be corrected for. FIDs are very sensitive and have excellent baseline stability. Their linear response covers a concentration range of seven orders of magnitude. But they are easily overloaded and may, therefore, require stream splitters. Their advantages, limitations, and some other characteristics have been described, for example, by Lee et al. (11, pp. 128–133).

FIDs find their widest use in GC and SFC. In LC they require special interfaces such as moving wires (or chains or ribbons) and rotating disks, which are designed to remove the solvent before the sample is burned in

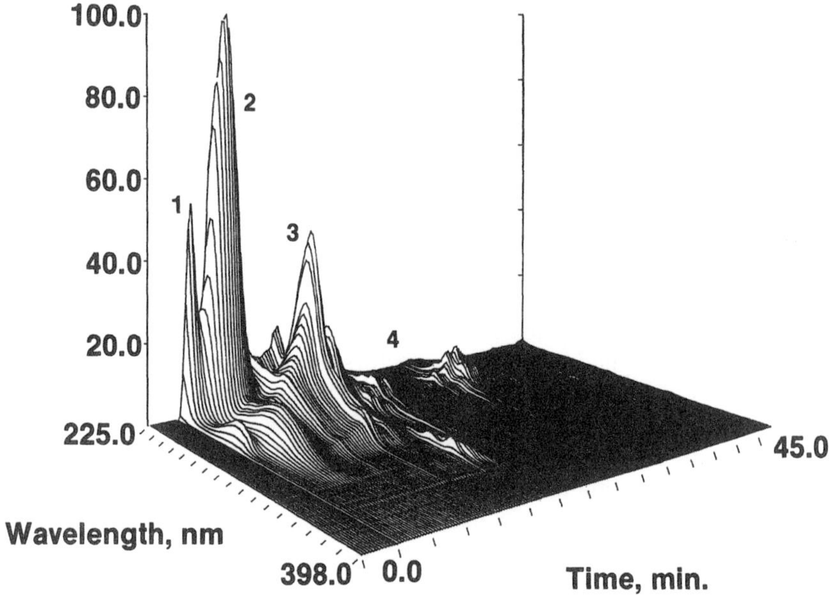

Figure 6.3 HPLC-UV map of an aromatics fraction obtained with a photodiode array absorbance detector. Monoaromatics (1), diaromatics (2), triaromatics (3), and tetraaromatics (4) can be distinguished. Among the latter, two tetraaromatic types are recognized: (a) pyrenes and (b) chrysenes. (From Ref. 58a. Reproduced with permission of the publisher.)

the flame. Until recently, these interfaces were hard to run consistently. Therefore, FIDs are rarely used in LC. However, as discussed in the previous subsection, FID can be an excellent choice for SFC with CO_2 as the mobile phase, which does not require such an interface.

The inductively coupled plasma (ICP) and the atomic emission (AE) detectors are element-specific and allow the simultaneous detection of metals and other heteroatoms in chromatographic separations of petroleum fractions (e.g., Refs. 3, 4, 20, and 21). The two detectors are related in that both use a plasma at 4000–6000°C to atomize the sample molecules and excite the resulting atoms which they emit photons characteristic of the various elements. The spectra exhibit sharp peaks which are well suited for multielement detection and quantitation. Most of the modern AEDs utilize a microwave-induced helium plasma in contrast to the inductively coupled plasma of the ICP detector.

The Hewlett-Packard AED employs a highly sensitive photodiode array sensor which increases selectivity over earlier versions by allowing real-time multipoint background correction (21). Its dynamic range is in the order of 10^4 for C, H, S, N, and O (20). This makes it very useful for the analysis of petroleum fractions. Kosman demonstrated its utility and superiority over the FPD in combination with GC for several applications in the petroleum industry. Figure 6.4 shows his C, H, S, and N selective chromatograms obtained by GC-AED of a petroleum middle distillate as an example of the capabilities of this detector. Note the almost superimposable patterns for C and H and the very different distributions of S and N.

For many applications, the use of two or more detectors is recommended because each may give incomplete information. UV detectors would be useless for aliphatic hydrocarbons, RI detectors are insensitive for analyte components of similar RI as the solvent, and FIDs must be used in LC with interfaces which may have their own limitations.

Coulombe (58) tested four types of detectors for SEC: UV, RI, LC-FID, and an evaporative detector. They concluded that neither should be the only one in a chromatographic system. The LC-FID showed only part

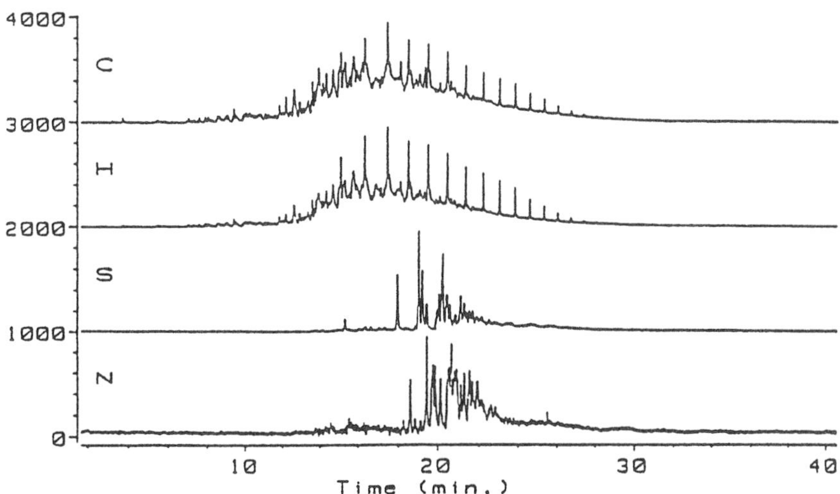

Figure 6.4 C, H, S, and N selective chromatogram obtained by GC-AED of a petroleum middle distillate. (From Ref. 20. Reproduced with permission of the publisher.)

of the chromatogram and was judged unsatisfactory. The UV and RI detectors were easy to use and had excellent linearity over several orders of magnitude. But, as discussed earlier, their response depends strongly on the structure of the analyte and, in case of the RI detector, also on the solvent. The RI detector, therefore, does not work for gradient elution or other modes of changing solvent.

The evaporative detector had to be carefully optimized, especially with respect to temperature. It loses volatile components and should not be employed for samples having hydrocarbons of <300 MW. Once it is stabilized, it "is solid and relatively easy to use," according to Coulombe. Our experience with the evaporative detector was rather discouraging in that its widely different response to the components in our samples made it difficult to calibrate, even with reasonably well-defined compound-class fractions.

III. EARLY APPLICATIONS

A. The "Oils, Resins, Asphaltenes" Approach

The earliest attempt to separate petroleum residues and asphalts into simpler fractions was by precipitation with petroleum ether into an insoluble part, the "asphaltenes," and a soluble portion, the "maltenes." The asphaltenes turned out to contain very high concentrations of the detrimental heteroelements (metals, S, N), and the maltenes were much easier to handle in analytical work. This separation by precipitation is still in vogue, although nowadays it is performed with single low-boiling alkanes, usually *n*-pentane or *n*-heptane. Other precipitants have occasionally been used in search for greater specificity, for example, acetone and even acids. After precipitation of the asphaltenes, the remaining material is said to have been "deasphalted," or better, "deasphaltened," and is called "maltenes" or "deasphalted oil" (DAO). (The term deasphalted oil is usually reserved for a refinery product obtained from vacuum residue by propane or butane extraction.)

The maltenes have been further separated into resins and oils, originally by extraction, later by chromatography on columns packed with Fuller's earth, Attapulgus clay, anhydrous or partially deactivated alumina, and similar adsorbents. Numerous methods were employed with variations in adsorbent, solvent type, and number of fractions collected. Generally, the hydrocarbon used for asphaltene precipitation is also the solvent for the subsequent separation of the maltenes. The fraction which comes off the adsorbent with this solvent is called "oil," and the retained fraction is called "resins." The resins can be removed from the adsorbent all at once with a powerful solvent or in steps with solvents of increasing strength to yield several fractions of increasing polarity. The resulting oil and resin fractions, as the asphaltenes, are strictly operationally defined and have little chemical significance.

The problems with the "oils, resins, asphaltenes" approach range from poor separation efficiency to poor definition. Some of the shortcomings will be discussed in Chapter 10. Suffice it to state here that separations are poor and the resulting fractions inadequate for molecular characterization. Examples for poor definition of terms are the differences in the asphaltenes' yield and composition, depending on the type of residue and the precipitant used. The yield and composition of the other two fractions vary in corresponding fashion.

Modern separation methods render operationally and chemically much better defined fractions: *compound groups* (saturates, aromatics, acids, bases, etc.) and *compound classes* (e.g., mono-, di-, tri-, and tetraaromatics, sulfides, or pyrrolic N compounds, etc.). We believe the terms oils, resins, and asphaltenes are archaic and are being superseded by these or similar chemically more distinct designations.

B. Early Attempts at Chemical Distinctions

1. Corbett's Approach

Corbett (59,60) already tried to introduce chemical terms for his fractions. After dividing his asphalt samples by *n*-heptane precipitation into asphaltenes and maltenes, he chromatographed the latter on active alumina into three fractions with four solvents of increasing strength: heptane, benzene, and a 50/50 mixture of benzene and methanol, followed by trichlorobenzene. He called the heptane eluate "saturates"; the benzene eluate, "naphthene-aromatics"; and the last fraction, "polar-aromatics." However, these separations, too, were not good enough yet to yield chemically well-defined fractions.

What makes Corbett's work still interesting are two issues. First, he already pointed out the need to carry distillation to the limit and to continue the separation by solubility methods to produce fractions of increasing MW and viscosity. (Viscosity was his prime criterion because he was studying the effect of fractionation on asphalt properties.) Second, despite his rather inefficient chromatographic separations, he could demonstrate that the composition of his asphalt fractions (distillable and nondistillable) changed considerably with their boiling point and beyond.

Corbetts' procedure started with subjecting a vacuum residue to distillation in a short-column high-vacuum laboratory still (60) and generating several cuts up to an AEBP of 565°C (1050°F). Treatment of the high-

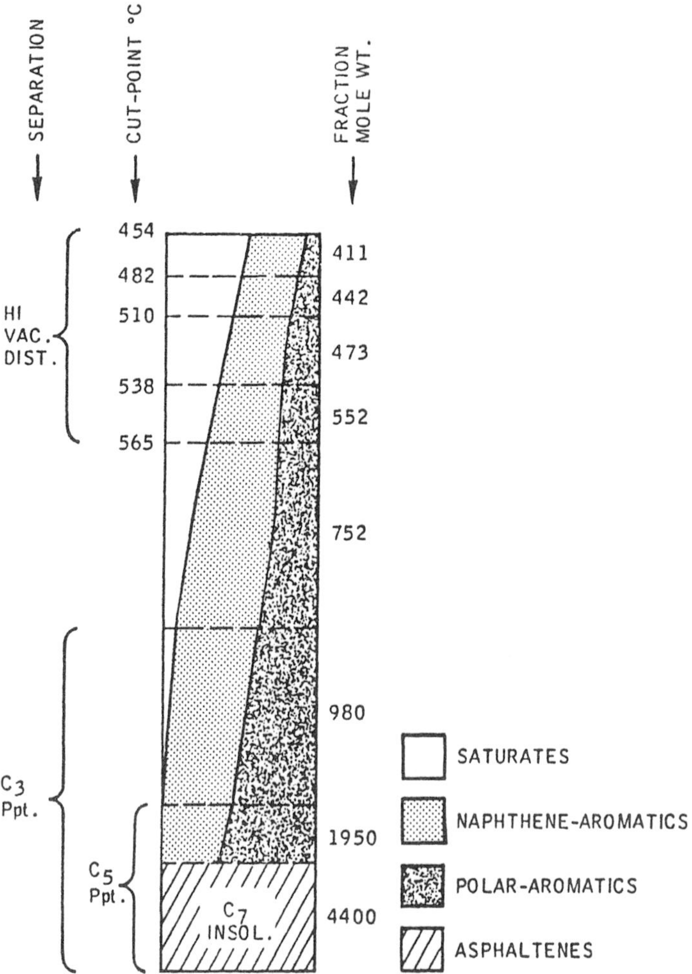

Figure 6.5 Corbett's compositional profile of a vacuum residue. (From Ref. 60. Reproduced with permission of the publisher.)

vacuum residue, first with propane, followed by *n*-pentane and *n*-heptane, yielded another two "equivalent distillation" fractions (our term) besides the insoluble heptane asphaltenes. All of these, except for the asphaltene fraction, were extracted with *n*-heptane and subjected to his chromatographic separation. This way Corbett and Petrossi (60) obtained the compositional profile of a vacuum residue shown in Fig. 6.5.

The top cut, of 454°C (850°F) AEBP, contained about 50 wt% saturates, 40 wt% naphthene-aromatics, and 10 wt% polar-aromatics. With increasing AEBP, the amount of saturates decreased to zero in the last heptane soluble fraction, whereas that of the polar aromatics increased to about 60 wt%. Corbett did not yet draw the conclusion that this compositional trend might continue into the domain of the asphaltenes.

Another way to present these data is shown in Fig. 6.6. In this bar graph, the first column illustrates the composition of the original vacuum residue.

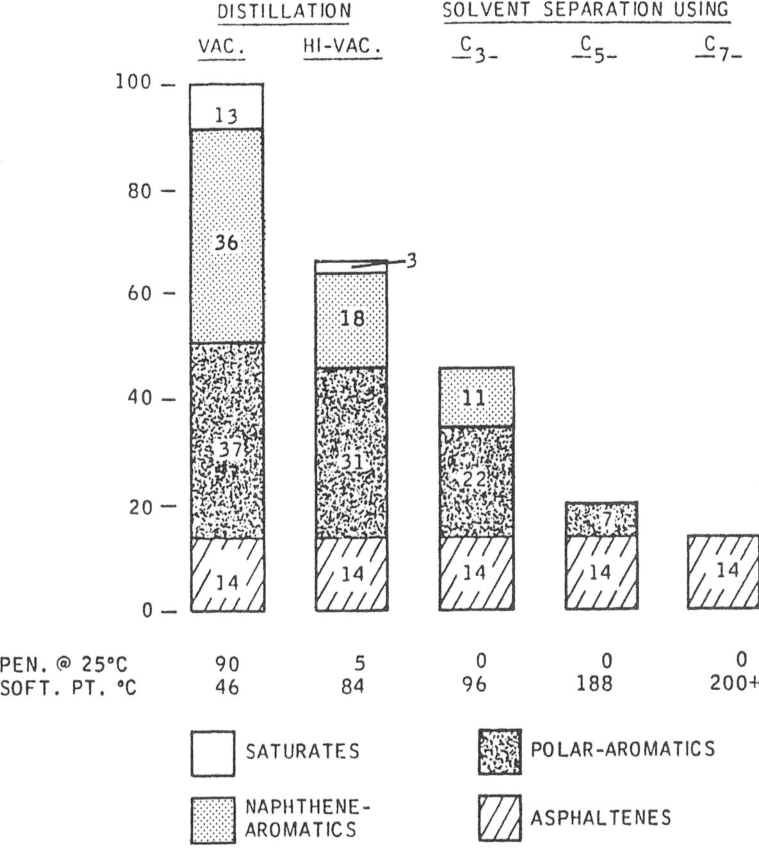

Figure 6.6 Effect of distillation and solvent extraction on the composition of a vacuum residue. (From Ref. 60. Reproduced with permission of the publisher.)

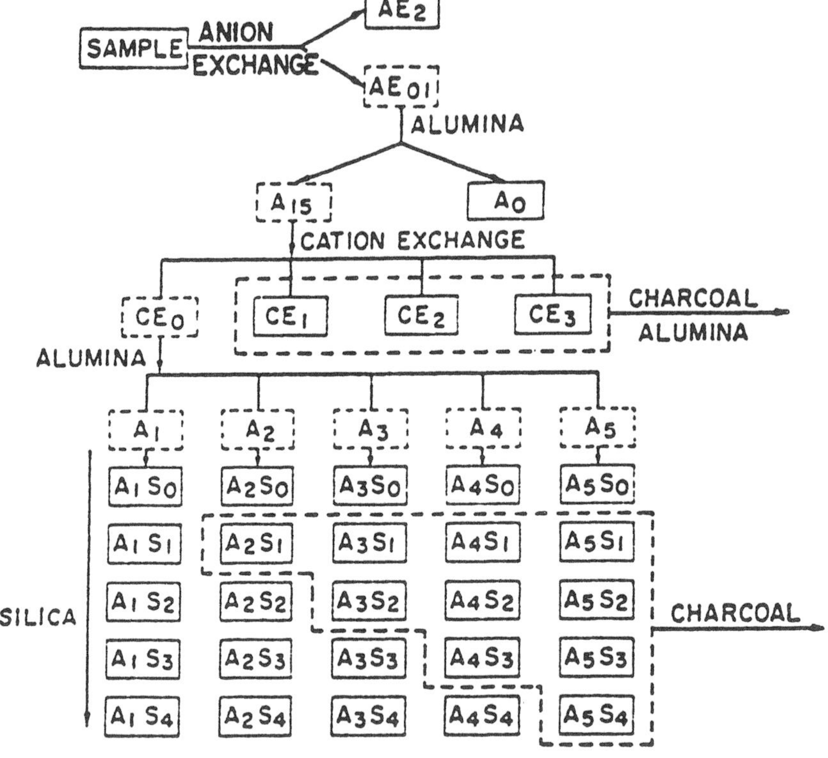

Figure 6.7 Snyder and Buell's fractionation scheme. (From Ref. 61. Reproduced with permission of the publisher.)

The following bars show the composition of the residues of high-vacuum distillation and of three solvent extractions. High-vacuum distillation removes primarily saturates and naphthene-aromatics; propane and pentane extractions take out both naphthene-aromatics and polar-aromatics. Heptane finally removes the remainder of the latter. Corbett and Petrossi (60) already point out the equivalence of solvent extraction with distillation in its ability to "cut deeper into the residuum in order to recover high yields of extract and to concentrate asphaltene-containing fractions."

2. Snyder's Early Approach

Snyder and co-workers (61–64) pioneered some of the techniques which we now call modern. They, too, recognized the importance of distillation to make the chromatographic separations more manageable. They also introduced nonaqueous chromatography on ion exchange resins for the isolation of carboxylic acids and of N compounds early in their scheme (61). They even pointed out the possibility of omitting the anion exchange resin step for greater convenience and speed in those cases where the determination of the carboxylic acids was immaterial. These workers were the first to identify numerous N and O compounds in petroleum distillates and to measure their amounts.

Their fractionation schemes were very complicated with 30–40 fractions. Their first one, shown in Fig. 6.7, started out with an anion exchange column for separating carboxylic acids from the rest. Chromatography on deactivated alumina then separated the neutrals from the polars. The latter were subdivided into four fractions of increasing basicity. The first one of these contained neutral to very weakly basic compounds which were further separated on deactivated alumina into five fractions, each of which was subdivided into five fractions on deactivated silica. One more step, chromatography on charcoal, was applied to several of these fractions to distinguish between the aliphatic and the aromatic compounds.

This general scheme was developed for a distillate boiling between 700 and 850°F (370 and 455°C). It was varied slightly for the other two distillates with their boiling ranges of 400–700°F (205–370°C) and 850–1000°F (455–540°C). For these, the anion exchange column was omitted and the first alumina column was replaced by a silica column. The neutrals were divided into saturates, mono-, di-, and polyaromatics, and heterocompounds. The sulfides were separated from the hydrocarbons, essentially by LEC, on cation exchange resin in the mercuric ion form. Thus, many of our present separation techniques were already employed by Snyder and co-workers, even if only at the less sophisticated stage possible at that time.

This is true for the characterization methods, too. These entailed low-voltage high-resolution MS; NMR, IR, UV, x-ray fluorescence spectroscopy; and chemical methods such as nonaqueous titration and elemental analysis. Altogether, this was a pioneering effort aimed at the complete characterization of the atmospheric and vacuum gas oil fractions of a crude oil. They did not, at that time, extend the distillation beyond 1000°F (540°C).

IV. MODERN APPLICATIONS

A. Separation of Acids, Bases, and Neutrals

1. The API-60 Method and Newer ABN Versions

The authors of the API-60 separation and characterization scheme introduced two innovations: First, they extended the distillation of their petroleum or other fossil fuel samples up to about 700°C (1290°F) by employing short-path distillation (see Chapter 3). They also abandoned the precipitation of pentane insolubles (asphaltenes) as a first step in the separation of the nondistillables. Their second major innovation was the use of ion exchange resins for the separation and isolation of nitrogen compounds from crude oil fractions. This gave them chemically much better defined fractions, namely, one or several concentrates of acids and one or several concentrates of bases. It also eliminated or mitigated losses which were common on the harsher silica or alumina columns employed before. Neutral heterocompounds passed the ion exchange resins and were removed from the remainder of the neutrals by charge-transfer chromatography on Attapulgus clay coated with $FeCl_3$. The hydrocarbons were finally subdivided into saturates and aromatics in a separate column.

The original API-60 separation scheme used open glass columns. The first and main one was packed with macroreticular anion exchange resin, the second one with macroporous cation exchange resin. A third column was packed with large layer of $FeCl_3$-coated clay, followed by a layer of macroporous anion exchange resin. The latter had the task to destroy the complexes formed with the Fe ions so that the original compounds could be collected after the separation. The last column, in which the hydrocarbons were separated, contained either fully activated silica gel or fully activated basic alumina. For a more detailed description, see, for example, Ref. 65.

The ion exchange resins, as well as the $FeCl_3$-coated clay, required lengthy, careful preparation and conditioning. Even so, they gave several kinds of problems, the most important one of which was the slow deterioration of the ion exchange resins and the release of the products into the mobile phase (34). The decomposition products would then interfere with the later spectroscopic characterization of the fractions. Not necessarily a problem, but a fact to remember, is the amphoteric character of certain compounds which causes them to be classified as bases or acids depending on the sequence of the ion exchange resin in the column.

Green and co-workers (33) made a careful study of the acid–base–neutrals (ABN) separation. Their improvements, which were discussed in Section III.B, made the method much more reliable, but it is still laborious.

Desbene et al. (66) described an HPLC modification of the API-60 method. In only 6 h it produces seven fractions, three acid, three base, and a neutrals fraction with results equivalent to those of the API-60 method.

2. Simpler Methods

If no subdivision of the polars into acid and base fractions is desired, much simpler methods can be used for the separation of petroleum fractions into saturates, aromatics, and polars. A problem with most of these methods is that they were not tested for their applicability to fractions of increasing AEBP up to nondistillable fractions. When they are applied to whole atmospheric or vacuum residues, they may perform well for the lower-boiling components of these samples but not for the higher-boiling ones. With this proviso, we still find it useful to mention some of the methods described in the recent literature.

Lundanes and Greibrokk (67) employed a dual-column system with column switching and backflushing for the separation of an atmospheric petroleum residue into saturates, aromatics, and polars. They, too, removed the polars first, using a cyano column in normal phase operation with hexane as mobile phase. The retained polars were recovered from this column by backflushing with hexane. The unretained saturates and aromatics were separated on another column packed with silica, which let the saturates pass through and, on backflushing, released the aromatics in a narrow peak. The separation seems to be quick and quite satisfactory. The larger aromatic ring systems are retained along with the polars. The authors co-injected chrysene, a four-ring PAH, as a standard to indicate the time when the columns had to be switched for separating the aromatics from the polars. Chrysene is the largest aromatic ring compound to pass as an aromatic rather than eluting together with the polars.

Lundanes and Greibrokk (68) also developed an SFC method for the same purpose. They chose CO_2 as their mobile phase so they could use the universal flame ionization detector. CO_2 is a relatively poor solvent, too poor in combination with the strongly polar silica. But with a (less polar) cyano column it worked well, nicely separating the polars from the neutrals. The neutrals, in turn, were separated into saturates and aromatics on silica coated with silver nitrate. The final results of this SFC method compare very well with those from a different, preparative separation as shown in Table 6.3.

Pearson and Garfeh (69) described the separation of saturates, aromatics, and polars from an atmospheric petroleum residue by normal phase LC on a single cyano column and an amino/cyano column pair using two solvents and FID. But they applied this method to hexane soluble samples, free from asphaltenes. The sample solution is first pumped through the single cyano column, which retains the polars, and then through the amino/cyano column combination for separating the saturates from the aromatics. In this system, the saturates are allowed to pass through all the columns;

Table 6.3 Results of Lundanes and Greibrokk's Group Separation of Saturates, Aromatics, and Polars by SFC Using CO_2 as Mobile Phase

	Oil A		Oil B	
Fraction	By weight[a] (%)	SFC[b] (%)	By weight[a] (%)	SFC[c] (%)
Saturates	44.5	45.6 ± 0.6	57.6	59.6 ± 0.6
Aromatics	35.4	35.3 ± 1.1	28.7	27.1 ± 0.6
Polars	20.0	19.1 ± 1.2	13.6	13.2 ± 1.1

[a]Gravimetric determination.
[b]Based on 13 measurements.
[c]Based on 8 measurements.
Source: Ref. 68.

the aromatics are backflushed with hexane from the dual column, bypassing the first one; and, finally, the polars are released from the cyano column by backflushing with methyl tert-butyl ether (MTBE). MTBE has a low boiling point (56°C) and performed equally well as methylene chloride but without its disadvantages (occasional contamination with traces of HCl, and problems with the FID).

Pearson and Garfeh's detector was a commercial rotating disk with quartz fiber belt and dual-jet FID. For universal response throughout the entire sample range, they had to lower the block temperature of this apparatus from 150 to 69°C (302 to 156°F). This mitigated the evaporative loss of the *n*-paraffins above C_{23} and presumably that of other hydrocarbons in this range. They tested the response factors of the three compound classes and found those of the saturates and aromatics to be equal. Those of the polars were 20% lower with very little variation (± 1%) between six different sample sources.

Carbognani and Izquierdo (70) presented a preparative HPLC method for the separation of deasphaltened vacuum residues into the same three compound groups. Their system separates up to 250 mg sample per run into four fractions: saturates, aromatics, weak polars, and strong polars. It can be programmed for multiple runs to fractionate up to 3 g overnight. Cyclopentane and methylene chloride serve as their solvents. Their first adsorbent is again a cyano-bonded silica for retention of the strongly polar components. The remainder of the sample is pumped through two silica columns which allow the saturates to pass through. The aromatics are collected by backflushing with cyclopentane. The weak polars are retained on the silica columns until, at the end of a batch of runs, they are recovered all together by backflushing with methylene chloride. The strong polars are backflushed from the cyano column with this solvent after each run. Two detectors in series, UV and RI (refractive index), monitor the separation.

Boduszynski's (58a) preparative HPLC method for the separation of heavy petroleum fractions (see Section VI.C.4) takes about equally long and delivers nine fractions. It is performed in two dual-column systems and requires somewhat more labor because of the evaporation of so many fractions. It has the advantage of speciation of the polars into three chemically well-defined fractions and nearly complete removal of nitrogen compounds. Furthermore, its performance was evaluated over a wide AEBP range.

B. Subfractionation of the Main Compound Groups

1. Separation of Aromatics by Ring Number

Fractionation of aromatics by ring number is best done on column packings with bonded phases, for example, amino-, cyano-, phenyl-, or nitro-derivatized silicas (13; more references there and in the following paragraphs). Separation on these packings proceeds only nominally by ring number; actually it discriminates by the number of pi electrons or double bonds in the ring system. For catacondensed aromatics, this is equivalent to ring number separation; however, pericondensed molecules have fewer double bonds than catacondensed ones with the same ring number. For instance, pyrene (four rings, eight double bonds) elutes after phenanthrene (three rings, seven double bonds) and before chrysene (four rings, nine double bonds).

Molecules with nonplanar aromatic ring systems, usually sterically hindered, elute before planar ones (71,72a), and long, stretched-out PAHs elute later than more compact ones (13,71). Generally, molecules with a larger (aromatic) surface area can interact with more polar groups on the packing surface and will, therefore, be more strongly retained. Figure 6.8 shows the elution profile of several aromatic prototype model compounds on an amino column. The plot of log(capacity factor) versus the number of double bonds for these compounds is quite linear with little scatter; see Fig. 6.9.

Paraffinic substituents have a shielding effect and tend to decrease retention because of reduced interaction, causing slightly earlier elution. Naphthenic groups attached to aromatic rings, in contrast, add slightly to the interaction with the packing and, therefore, tend to increase retention. When both types of substituents are present, the effects often cancel out. The effects of alkyl substitution on separation efficiency become more pronounced with increasing AEBP. This is illustrated in Fig. 6.10, which shows the HPLC-UV maps of the aromatics from six fractions of increasing

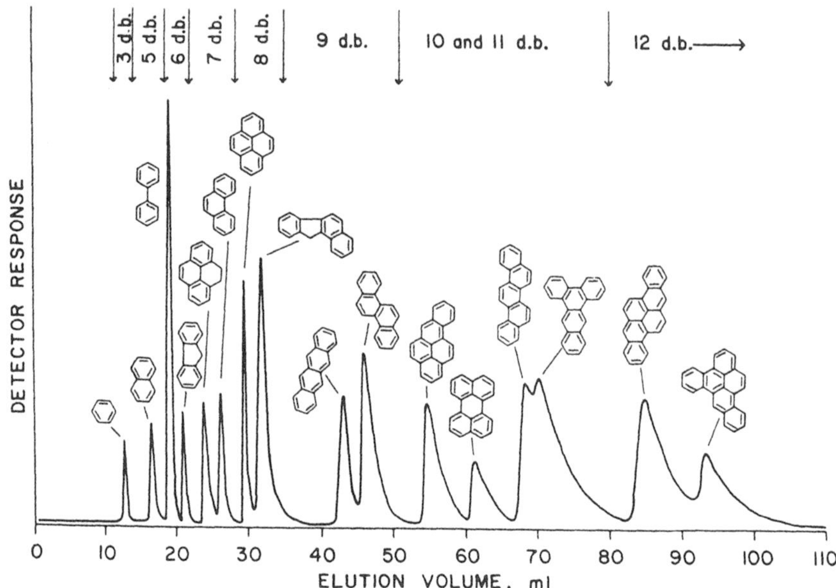

Figure 6.8 Elution profiles of model compound PAHs on an amino column with *n*-heptane as mobile phase. UV detector set at 254 nm; d.b. = double bonds. (From Ref. 72. Reproduced with permission of the publisher.)

AEBP obtained from a Kern River residue by short-path distillation and solubility fractionation (SEF). The progressive overlapping between adjacent ring-type fractions is caused by the increasing sample complexity due largely to increasing alkyl substitution on the aromatic rings. (Another contributing factor may be the increased sulfur compound content.) At AEBPs below 1000°F (540°C), the overlap is less than 20%. Above this limit, it becomes considerably larger and is difficult to quantify with present analytical methods.

When we (72) began work in this area, we chose amino columns because these had been found suitable for this purpose (73–76), and they were commercially available. We calibrated our amino columns with numerous model compounds for our work with coal liquids and petroleum fractions, investing a fair amount of time and effort. These packings worked well, and, with some care, they lasted up to 2 years with constant use. Thus, we had no reason to switch to other types. However, cyano and other bonded phases may work equally well or, possibly, even better.

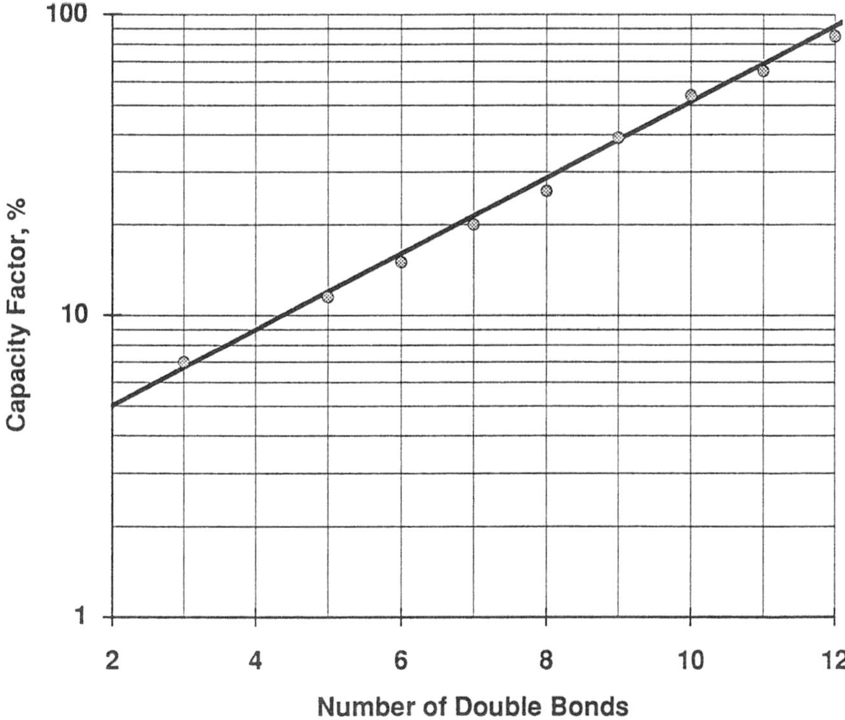

Figure 6.9 Plot of capacity factors of model compounds from Fig. 6.8 versus number of double bonds.

Grizzle and Thomson (77), for instance, reported better separation of a shale oil sample on a self-prepared dinitroanilinopropyl-bonded silica column than on a commercial diamine column. They also tested an alumina column which gave good separation, but apparently retained substantial amounts of tetraaromatic and especially of penta$^+$ aromatic hydrocarbons. Figure 6.8 shows an early example of the separation of model compounds achieved on a semipreparative amino column, and Fig. 6.11 displays the elution profiles of a 460–800°F (240–430°C) coal oil distillate and a coal oil residue (>800°F, 427°C), on the same column.

Unless the sulfur compounds are preseparated from the aromatics, many of them elute together with the corresponding aromatic ring number fractions, that is, benzothiophenes with diaromatics, dibenzothiophenes with triaromatics, and so on. The last eluates of aromatic ring number sepa-

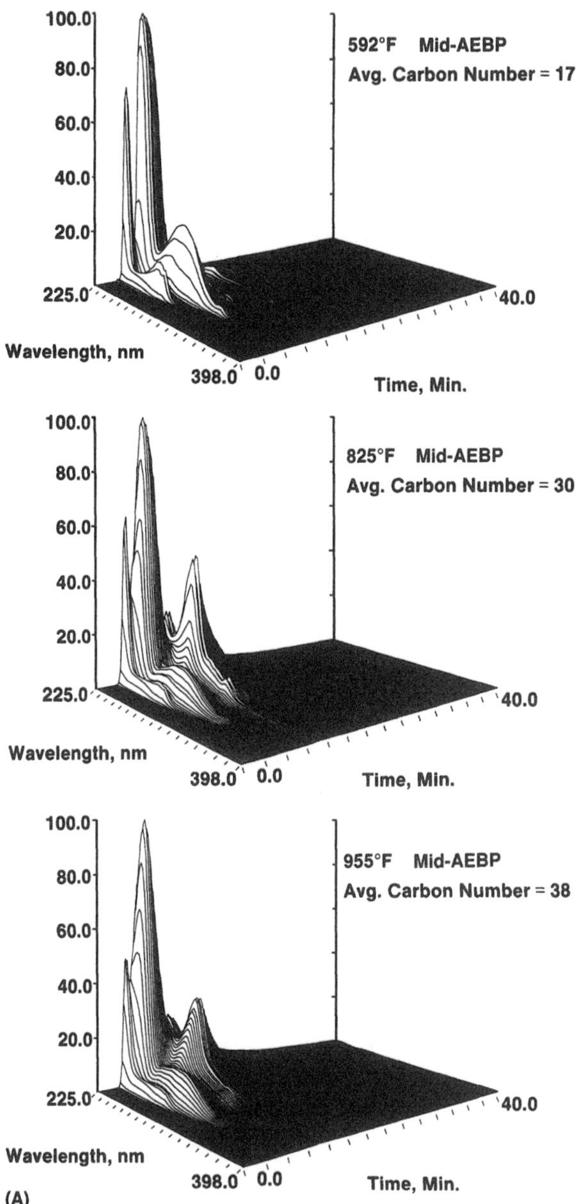

Figure 6.10 HPLC-UV maps of the aromatics fractions from five short-path distillation cuts and a SEF-1 (solubility) fraction of a Kern River residue. (A) Mid-AEBPs below 1000°F (540°C); (B) mid-AEBPs above 1000°F (540°C). (From Ref. 58a. Reproduced with permission of the publisher.)

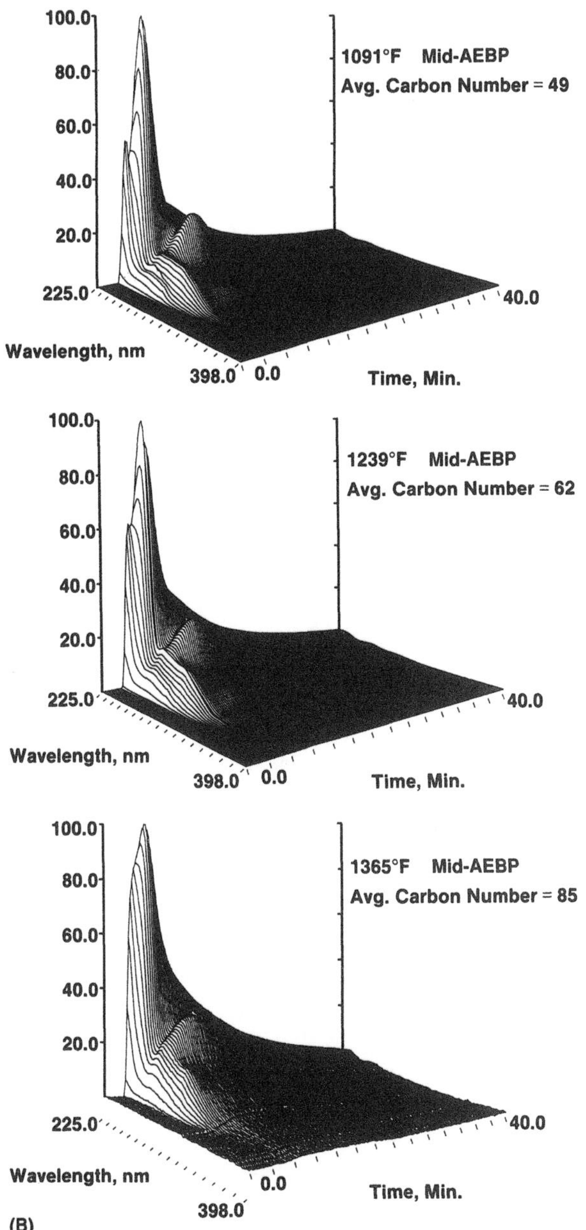
100.0
80.0
60.0
40.0
20.0
1091°F Mid-AEBP
Avg. Carbon Number = 49
225.0
40.0
Wavelength, nm
398.0
0.0
Time, Min.
100.0
80.0
60.0
40.0
20.0
1239°F Mid-AEBP
Avg. Carbon Number = 62
225.0
40.0
Wavelength, nm
398.0
0.0
Time, Min.
100.0
80.0
60.0
40.0
20.0
1365°F Mid-AEBP
Avg. Carbon Number = 85
225.0
40.0
Wavelength, nm
0.0
Time, Min.
398.0
(B)

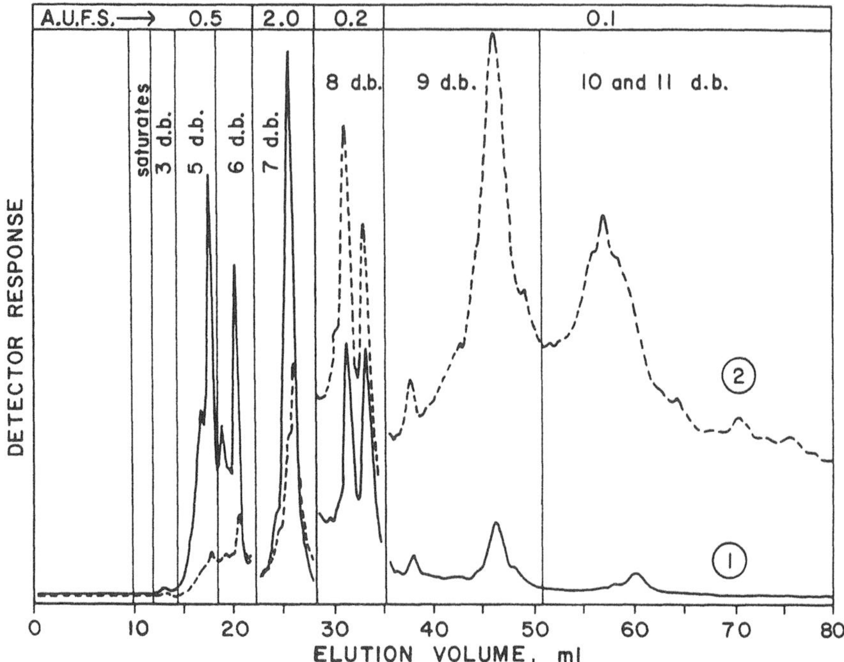

Figure 6.11 Elution profiles of coal liquid hydrocarbons from (1) a 460–800°F (240–430°C) distillate and (2) a <800°F (<430°C) residue on an amino column with *n*-heptane as mobile phase. UV detector set at 254 nm; d.b. = double bonds. (From Ref. 72. Reproduced with permission of the publisher.)

rations, the pentaaromatics and, to some extent, even the tetraaromatics, are not only contaminated with sulfur compounds but often also with other neutral heterocompounds containing N or O. These have generally one aromatic ring less than the hydrocarbons in that fraction. The degree of contamination increases sharply with increasing AEBP.

In most separation schemes, the pentaaromatics fraction contains all the leftover material not collected in any of the other fractions. These molecules were passed through as neutrals by the first column(s) and were retained by the last one (amino or similar). Thus, the pentaaromatics contain all the compound types which did not fit the appropriate categories of the scheme including many heterocompounds. The aromatics fractions of the higher AEBP cuts (>1000°F, 540°C) often contain more heterocompounds than hydrocarbons. In the lower aromatic ring number fractions, these are mainly sulfur compounds.

A rapid method for the analysis of hydrocarbon fractions by their aromatic ring number was presented by Yamamoto (78). It uses small thin-layer rods consisting of silica with aminopropyl or cyanopropyl bonded phases and very low sample loads (50 μg). By development with *n*-hexane under specified conditions, a hydrocarbon fraction is separated on the rod into four bands. The saturates, mono-, di-, and triaromatics making up these bands are measured in a commercial apparatus by FID directly off the rod. Then the remaining material is further developed with toluene into bands of polyaromatics and polars, and these are measured again by FID. Thus, six fractions are detected and measured altogether. Because two development and three drying peaks are required by this process, one determination takes roughly 30 minutes. However, by running several measurements in a staggered fashion, 10 samples can be handled in about 90 minutes.

2. Separation of Polars into Basic and Acidic Compound Group Fractions

This is usually the first step in a separation scheme for petroleum samples as discussed above (Section IV.A.1). It is also part of our scheme (Section IV.C.3) and will be described further there. The SARA and ABN methods use ion exchange resins for the separation, whereas we and others (e.g., 93,94) perform it on acidic and basic alumina. Because these are the main techniques for this application, and both are covered in the two subsections, we can move on to the next subject.

3. Subfractionation of Basic and Acidic Compound Group Fractions

Green et al. (81) described the subfractionation of both acid and base concentrates from petroleum distillates. Their fractionation method for the acids had been worked out by Green (82). It consists of normal phase chromatography on silica with a mixed solvent of three components in varying ratios. The important feature of the method is that one of these components is a tetraalkylammonium hydroxide (TAAH), usually tetramethylammonium hydroxide (TMAH). The TAAH is adsorbed by the silica and converts its surface from a mildly acidic to a mildly basic one. Acids are retained on this surface according to their acidity, weak ones weakly and strong ones strongly. For high efficiency, the amount of TAAH is varied during the course of the separation; see Table 6.4.

For those heavy petroleum residues which are not completely soluble in the MTBE–methanol mixture called for by Table 6.4, a chloroform

Table 6.4 Gradient Program for the Subfractionation of Acids on Silica with a Mobile Phase Containing TMAH

Time (minute)	Volume %	
	A	B
0	98	2
2	98	2
11	94	6
35	50	50
49	0	100
50	0	100
53	98	2
70	98	2

Note: A = methyl *tert*-butyl ether (MTBE); B = 60% (v/v) MTBE–40% methanol + $1.5 \times 10^{-3} M$ TMAH.
Source: From Ref. 82. Reproduced with permission of the publisher.

gradient can be used. Figure 6.12 compares the chromatograms of a strong petroleum acid concentrate in these two solvents. Green measured the retention indices of numerous acids and other compounds in four solvent–silica systems, including the two of Fig. 6.12. These retention indices allow the clear distinction of some major acid groups, such as pyrrolic compounds, amides and lactams, hydroxy-aromatics, carboxylic acids, and multifunctionals. On this basis, five to eight acid fractions may be collected from real samples.

For the subfractionation of base concentrates, Green and co-workers (81,83,84) used the same principle as applied to the acid concentrates. They separated the different basic compound types by normal phase chromatography on silica, using a mobile phase containing propanoic or a similar acid. Typical subfractions of petroleum-derived acid and base concentrates are shown in Fig. 6.13.

C. Complete Separation Schemes

1. General Considerations

The purpose of a separation scheme is the efficient separation of a complex mixture into the optimum number of fractions required for their characterization. For heavy petroleum, this usually means distillation into several cuts (fractions) up to the limit of about 1300°F (700°C). These cuts are then separated further by chromatography. First, the polars are removed and possibly subdivided in the process into several acid and base fractions.

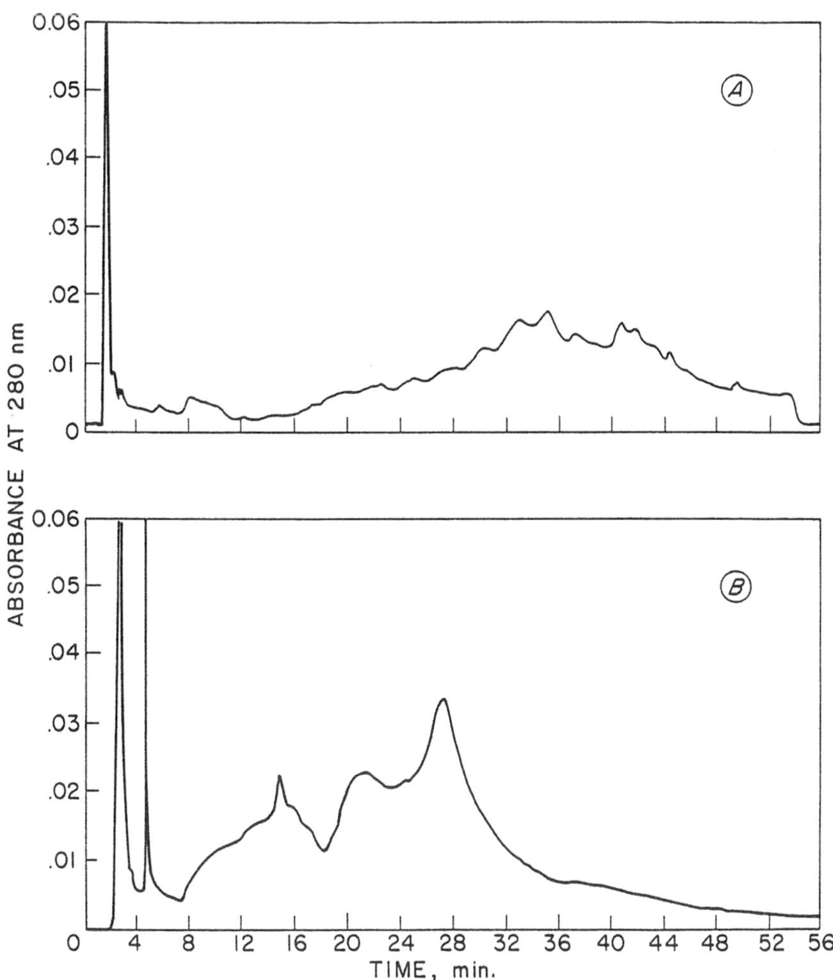

Figure 6.12 HPLC chromatograms of Wilmington <1250°F (<675°C) strong acid concentrate in (A) MTBE-TMAH (Table 6.4) and (B) chloroform-TMAH gradients. Separation A is solubility-limited; separation B looks promising. MTBE = methyl *tert*-butyl ether; TMAH = tetramethylammonium hydroxide. (From Ref. 82. Reproduced with permission of the publisher.)

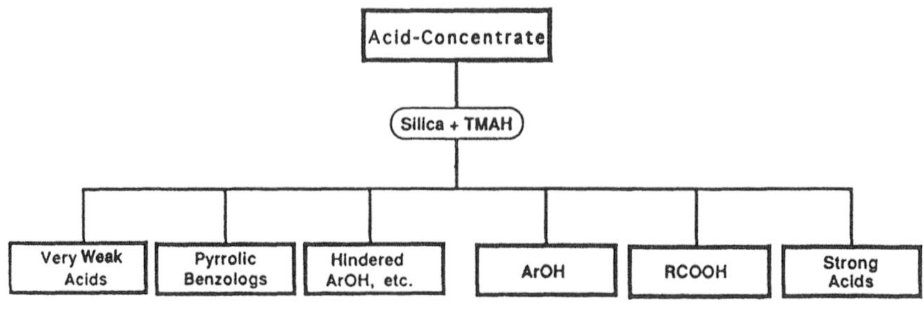

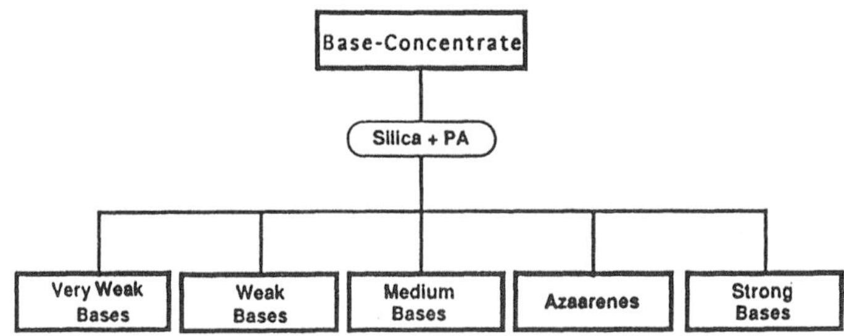

Figure 6.13 Typical subfractions of petroleum-derived acid and base concentrates obtained by Green's methods. TMAH = tetramethylammonium hydroxide; PA = propanoic acid. (From Ref. 81. Reproduced with permission of the publisher.)

Then the neutrals are separated into saturates and three to five aromatic compound classes. In most cases, this is sufficient for characterization by MS and some auxiliary measurements.

Nondistillables (<1300°F AEBP) can be subdivided by chromatographic or extraction methods (SEF) into fractions of increasing complexity and MW. Both SEF (sequential elution fractionation) and SFE (supercritical fluid extraction) seem particularly suitable since their mode of separation is similar to that of distillation (see Section III.C and Chapter 2). They can be seen as continuing the latter into the nondistillable range ("equivalent distillation").

Numerous fractionation schemes have been tested and employed which do not fit the terms of our definition, in that they may not be efficient, yielding more fractions than necessary for characterization. Or they may not generate sufficiently clean fractions for characterization. This may be the case because the MW range is too wide as, for example, in atmospheric or vacuum residues, or because the chemical complexity is too large as, for example, in aromatics which have not been subdivided by aromatic ring number. Many of these less efficient schemes had been developed along the way and served their purpose well until better ones came along. Rather than discussing the various stations along the way, we will restrict our description to two examples which, in our biased view, fit our criteria for an efficient separation scheme and may serve as a guide to other schemes which may be better suited for other applications.

2. The NIPER Scheme

A very extensive, detailed, and well-defined separation scheme is that of the National Institute for Petroleum and Energy Research (NIPER) at Bartlesville, Oklahoma (81). It was designed to test simpler methods and to find heterocompound types overlooked by those. Otherwise, it would not have been necessary to go to such great length in their separations.

Various parts of this scheme have already been discussed in the foregoing subsections. Here they will be put in context. The NIPER scheme separates a heavy petroleum sample first by molecular distillation into several distillation cuts and a nondistillable residue. All or some of these cuts are then divided by a modified API-60 method into three to five main fractions: one or two acid concentrates, one or two base concentrates, and a neutrals fraction (Fig. 6.14). All of these are then further subdivided, the acids and bases as illustrated in Fig. 6.13, and the neutrals as shown in Fig. 6.14.

The neutrals consist primarily of hydrocarbons and sulfur compounds although considerable amounts of nitrogen and oxygen compounds may also be present, depending on the crude oil. Normal phase chromatography of this fraction on silica removes some leftover polars. Then the sulfides are taken out by ligand exchange chromatography (LEC). Charge-transfer chromatography on a dinitroanilinopropyl-bonded silica column divides the remainder into four fractions: saturates plus mono-, di-, tri-, and tetraaromatics, plus higher aromatics. The first of these is then subdivided on silica into saturates and monoaromatics. Thus, the neutrals are divided into seven fractions: saturates, five aromatic hydrocarbon fractions (with $1–5^{+}$ aromatic rings), and a sulfide fraction (96). In an earlier version (81), the LEC separation was performed later in the scheme, namely, on all the aromatic ring number fractions, thus producing three or four thiophene fractions rather than just one. Either way, the total number of fractions *per distillate* achieved by this scheme is greater than 30.

3. Boduszynski's Scheme

Boduszynski (58a), too, first separates a crude oil sample into several distillation cuts (eventually by high-vacuum short-path distillation) and a

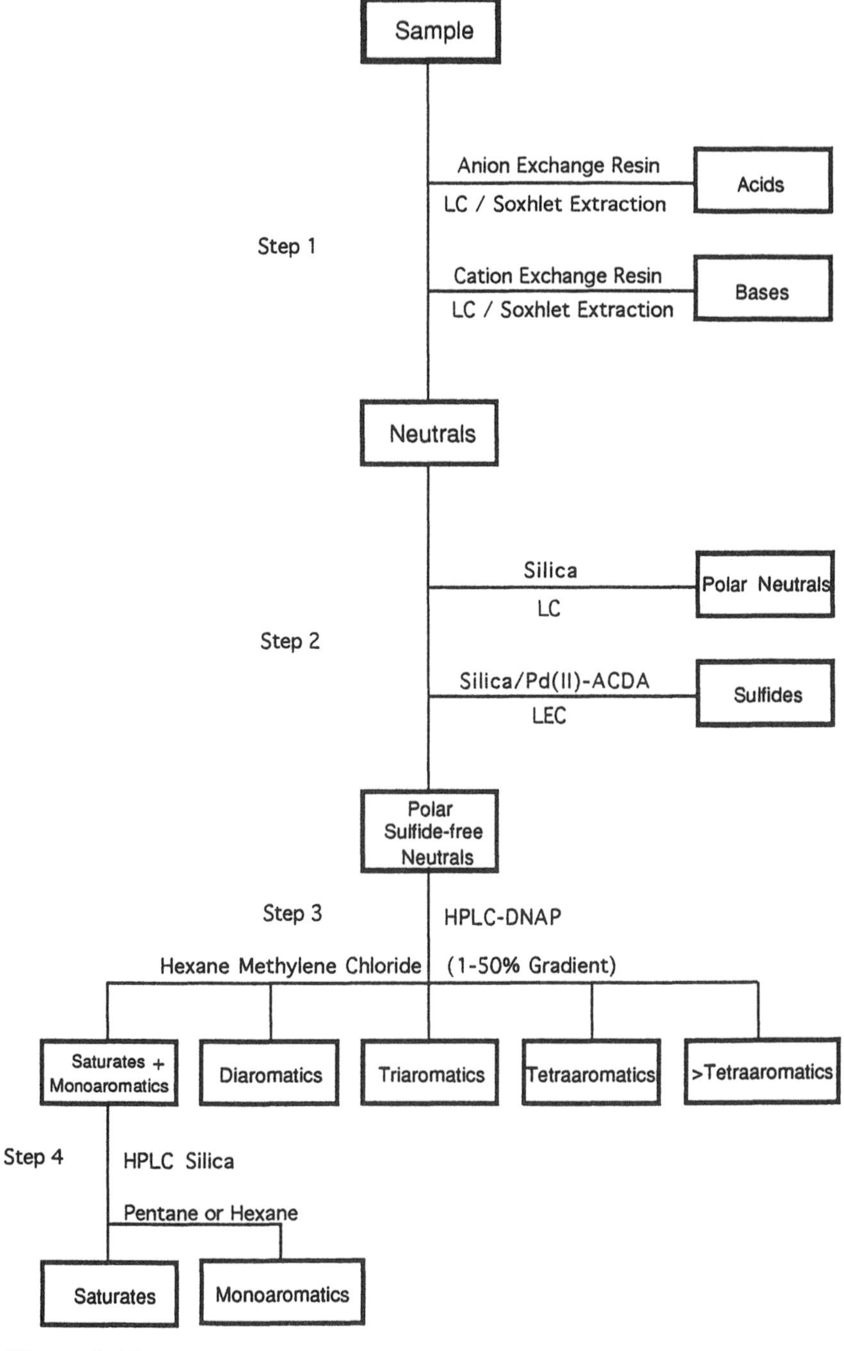

Figure 6.14 The present NIPER fractionation scheme. (From Ref. 96.)

nondistillable residue. The latter is then submitted to "equivalent distillation," that is, to solubility fractionation by SEF into four fractions as described in Chapter 3.

The (true) distillation cuts are fractionated by preparative HPLC in two separate operations as shown in Fig. 6.15. In the first one, the sample is separated on two alumina columns, a basic and an acidic one. We call this part the HPLC-BA/AA method. Four to six reasonably distinct fractions are obtained: "acidic" compounds, "pyrrolic nitrogen" compounds, "basic" compounds, and three "neutral compounds" fractions which often are combined. The second method, HPLC-NH_2/SIL, separates the neutral compounds in another dual-column system into saturates, monoaromatics, diaromatics, triaromatics, tetraaromatics, and pentaaromatics plus neutral heterocompounds.

The separation on the dual basic alumina/acidic alumina column (HPLC-BA/AA) is conceptually similar to the first step of the API-60 scheme in that it first separates the sample on the basis of acidity and basicity. The major difference is the use of basic and acidic alumina instead of anion and cation exchange resins. Although the latter cause less irreversible adsorption than alumina or none, and are believed to be more selective, they are tedious to prepare and tend to deteriorate. Therefore, we prefer the dual alumina columns for routine separations.

The HPLC-BA/AA method makes use of two columns (25 cm × 9 mm i.d.), four solvents, and several valves for column switching and backflushing. One of the columns is packed with basic alumina (BA), the other with acidic alumina (AA). The solvents used are cyclohexane, toluene, methylene chloride, and a 4:1 (v/v) mixture of methylene chloride and methanol.

In step 1, the sample solution, 200–400 mg dissolved in cyclohexane, is injected first into the BA column where the "pyrrolic" and "acidic" fractions are retained. The remaining sample goes on to the AA column and "neutrals" are eluted there with cyclohexane as Fraction 1 (F1) and then, in step 2, with toluene as Fraction 2 (F2). A third fraction of "neutrals" (F3) is recovered, from the AA column only, with methylene chloride in step 3. The "basic" fraction (F4) is retained on the AA column and recovered from it in step 4 by backflushing with the methylene chloride–methanol mixture. In step 5, the "pyrrolic N compounds" (F5) are eluted from the BA column with methylene chloride. The "acidic" compounds (F6), finally, are obtained by backflushing the BA column with the methylene chloride–methanol mixture. The procedure is fully automated which makes it simpler than it appears from the figure and the description.

Fractions from duplicate runs are usually combined for further work. Reproducibility between duplicates is about 5% or better for the "neutral"

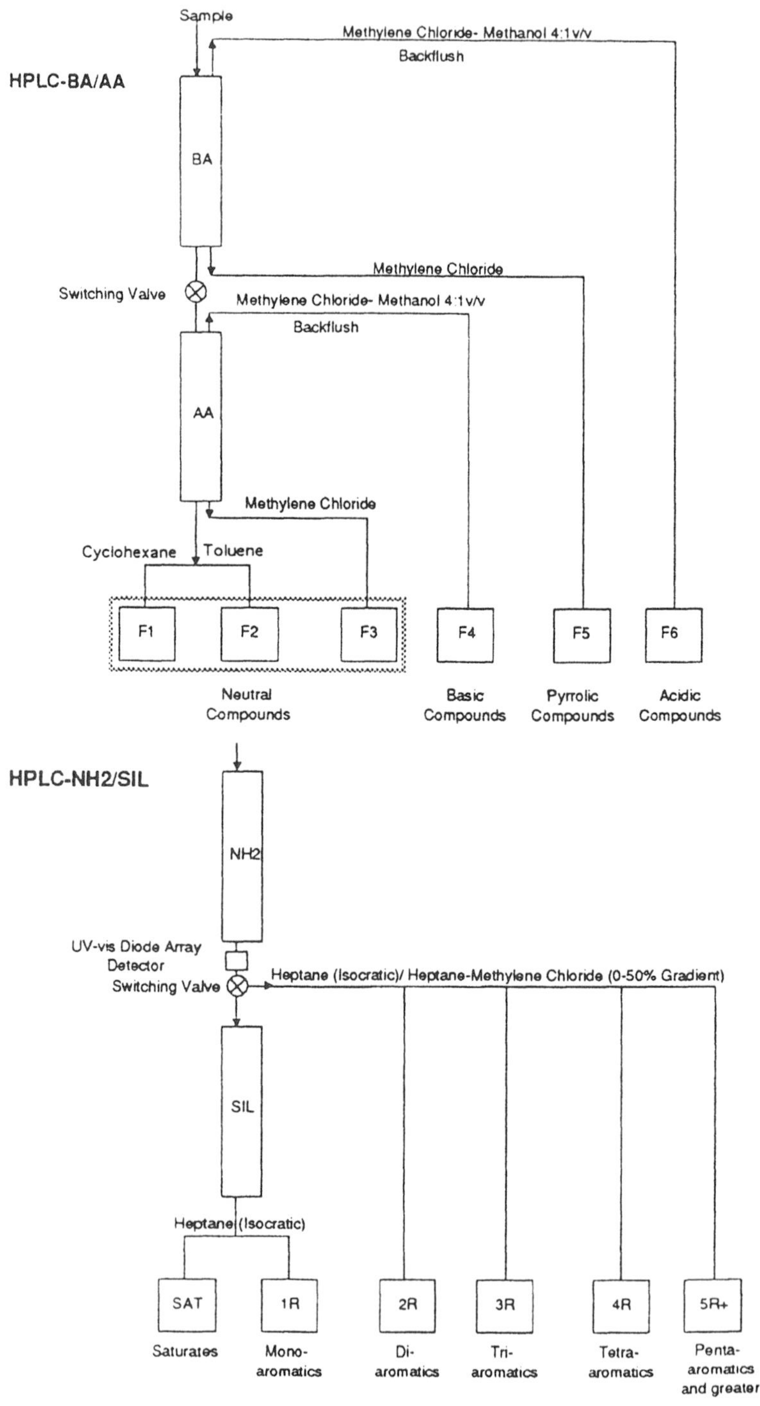
Sample
Methylene Chloride- Methanol 4:1 v/v
Backflush
HPLC-BA/AA
BA
Methylene Chloride
Switching Valve
Methylene Chloride- Methanol 4:1 v/v
Backflush
AA
Methylene Chloride
Cyclohexane
Toluene
F1
F2
F3
F4
F5
F6
Neutral Compounds
Basic Compounds
Pyrrolic Compounds
Acidic Compounds
HPLC-NH2/SIL
NH2
UV-vis Diode Array Detector
Switching Valve
Heptane (Isocratic)/ Heptane-Methylene Chloride (0-50% Gradient)
SIL
Heptane (Isocratic)
SAT
1R
2R
3R
4R
5R+
Saturates
Mono-aromatics
Di-aromatics
Tri-aromatics
Tetra-aromatics
Penta-aromatics and greater

fractions and may vary between 5 and 15% for the others. The greater error of some of the fractions is caused by their small amounts which make weighing less accurate.

The neutral fractions 1–3 are usually combined and stripped of their solvents. They contain all the hydrocarbons and most of the S compounds. Sometimes, a qualitative HPLC-UV survey on an analytical amino column gives sufficient information about the composition of the combined "neutrals" or one or two of the subfractions. Otherwise, the combined "neutrals" are redissolved in dry *n*-heptane and separated on a preparative ZORBAX-NH_2/SIL columns into five fractions.

The first eluate (saturates + monoaromatics) from the NH_2 column goes directly into the silica (SIL) column for separation of the saturates from the monoaromatics. The fact that only dry heptane enters this column enables us to use it over and over again without regeneration. Two detectors in series, UV and RI, monitor the separation for the correct cut point. Reproducibility is better than 5%. Cut points between the aromatic ring-type fractions are determined by a UV/vis diode-array spectrophotometer. Field ionization mass spectrometry confirms the high degree of separation achieved. Yields (in wt%) of the fractions are obtained gravimetrically. Reproducibility again is about 5–15%. The total number of final fractions from the entire scheme is 9 per distillate sample.

Our entire scheme is built on the requirement of routine operations. After each run, the packing of our first dual column is replaced by fresh, "as received" basic and acidic alumina. The other two columns need not be repacked since they are protected by the respective preceding ones, the NH_2 column by the alumina columns, and the silica column in turn by the NH_2 column. Both of these can be used hundreds of times.

Our system may seem quite complex, but it is fully automated and fast. Besides, it has several other advantages:

1. High flexibility and good control for handling very diverse samples spanning a wide AEBP range
2. Effective removal of nitrogen compounds from neutrals and their separation into basic and pyrrolic species
3. Greater differentiation of the aromatic compound classes and the polar compound fractions than most other schemes

Figure 6.15 Boduszynski's HPLC separation scheme for high-boiling (650–1300°F AEBP) petroleum fractions.

Table 6.5 Comparison of Several Modern Chromatographic Schemes for the Separation of Heavy Petroleum Fractions

Method Sequence	Compound type separated	Column type	Eluent[a]	Comments
API-60, preparative LC. Short-path distillates in cyclohexane				Well-defined, broad fractions
1	Acids	Anion exch. res	Benzene/methanol/CO_2	
2	Bases	Cation exch. res	Benzene/methanol/2-propylamine	
3	Neutral N compds-1	$FeCl_3$	Dichloroethane	
	Neutral N compds-2	$FeCl_3$	Benzene/ethanol/water	
4	Hydrocarbons +		Cyclohexane *n*-Pentane	
5	Aromatics-1	Silica	Pentane/benzene	
	Aromatics-2		Benzene/methanol	
6	Saturates	Silica	*n*-Pentane	
NIPER, preparative LC and Soxhlet extraction. Short-path distillates in cyclohexane/benzene For distillates 300–700°C (570–1300°F)				Well-defined, good fractions
1	Acids	Anion exch. res	Benzene/formic acid	
2	Bases	Cation exch. res	Benzene/1-propylamine	
3	Neutrals		3:1 cyclohexane/benzene	
4	Polar-neutrals	Silica	4.5:4.5:1 THF/$MeCl_2$/ethanol backflush	
5	Sulfides	Pd(II)-ACDA[c]	1% acetonitrile in $MeCl_2$, backflush	
6	Sulfide-free neutrals	Silica/Pd(II)-ACDA[c]	5% $MeCl_2$ in hexane	
HPLC				
6–8	Aromatics 2–4$^+$	DNAP[b] silica	Hexane/methylene chloride (1–50% gradient)	
9	Monoaromatics	Silica	Pentane or hexane	
10	Saturates	Silica	Pentane or hexane	

Boduszynski (58a)				Well-defined fractions
Stage 1. 650–1300°F AEBP fractions in cyclohexane				
1	"Acidic" compounds (F6)	Basic alumina[d]	Methylene chloride/methanol backflush	
2	"Pyrrolic" compounds (F5)	Basic alumina	Methylene chloride	
3	"Basic" compounds (F4)	Acidic alumina[e]	Methylene chloride/methanol backflush	
4	"Neutral" compounds (F3)	Acidic alumina	Cyclohexane/toluene	
5	"Neutral" compounds (F2)	Basic/acidic alumina	Toluene	
6	"Neutral" compounds (F1)	Basic/acidic alumina	Cyclohexane	
Stage 2. The combined neutrals fractions in *n*-heptane				Fractions 1–5 are sharp and well defined, fraction 6 is a mixture of pentaaromatics and higher aromatics with neutral hetero-compounds
1	Saturates	NH_2/SIL	*n*-Heptane	
2	Monoaromatics	NH_2/SIL	*n*-Heptane	
3	Diaromatics	NH_2	Heptane/$MeCl_2$ gradient	
4	Triaromatics	NH_2	Heptane/$MeCl_2$ gradient	
5	Tetraaromatics	NH_2	Heptane/$MeCl_2$ gradient	
6	Pentaaromatics and greater	NH_2	Heptane/$MeCl_2$ gradient	
Lundanes and Greibrokk (67), rapid analytical HPLC. Pentane solubles in hexane				Broad fractions
1	Polars	Cyano-bonded	Hexane backflush	
2	Aromatics	Silica	Hexane backflush	
3	Saturates	Silica	Hexane	
Pearson and Garfeh (69), rapid analytical HPLC. Pentane solubles in hexane				Broad fractions
1	Polars	Cyano-bonded	MTBE backflush	
2	Aromatics	Amino/cyano	Hexane backflush	
3	Saturates	Amino/cyano	Hexane	

[a]Solvent used for removing fraction from column.
[b]DNAP = 2,4 dinitroanilino-propyl.
[c]ACDA = 2-amino-1-cyclopentenedithiocarboxilic acid (bonded to silica via propyl group attached to amine group).
[d]Only basic alumina column by a switching technique.
[e]Only acidic alumina column by a switching technique.

4. Good separation of saturates from monoaromatics over a wide AEBP range

It is worth stressing that the chromatographic part of this scheme is applied only to distillates (up to about 1300°F, ~700°C). It was not devised for nondistillable (<1300°F AEBP) solvent-derived SEF fractions. With those, significant losses due to irreversible adsorption on the alumina columns are to be expected.

V. SUMMARY

In this chapter we discussed primarily the advances made in the LC separation methods during the last 20 years. An important innovation was the change in separation sequence by removing the polars before fractionating the neutrals. Earlier work was mentioned mainly by reference to summary papers or to reviews. Exceptions are brief discussions of the main chromatographic separation methods in our field and of some aspects of detectors and their use.

The ability of the photodiode array detector (PDAD) to differentiate between mono-, di-, tri-, tetra-, and pentaaromatics was pointed out. The PDAD can be used as an analytical tool in itself or, in preparative separations, as a means for choosing the best cut points.

Two recent examples of efficient fractionation schemes were described. (Efficiency depends, among other things, on the purpose for the separation.) The advantages and limitations of the two schemes were pointed out. Table 6.5 summarizes their main aspects and those of some other schemes for easy comparison.

The separation of acidic and basic compounds on ion exchange resins is more effective and milder (lower losses) than that on basic and acidic alumina, but it is difficult and time-consuming. It will be the method of choice in certain applications, but not in most routine analyses. In either case, the polars are best removed before separation of the neutrals. Prior distillation into several cuts up to the limit of about 1300°F (~700°C) makes the subsequent chromatographic separations simpler and cleaner.

REFERENCES

1. Altgelt, K. H., and Gouw, T. H. (Eds.). 1979. *Chromatography in Petroleum Analysis*. Marcel Dekker, New York.
2. Jewell, D. M., Weber, J. H., Bunger, J. W., Plancher, H., and Latham, D. R. 1972. Ion-exchange, coordination, and adsorption chromatography.

Chromatography of heavy-end petroleum distillates. *Anal. Chem.*, **44**:1391–1395.

3. McKay, J. F., Cogswell, T. E., Weber, J. H., and Latham D. R. 1975. Analysis of acids in high-boiling petroleum distillates. *FUEL* **54**:50–61.
4. McKay, J. F., Weber, J. H., and Latham, D. R. 1975. Characterization of nitrogen bases in high-boiling petroleum distillates. *Anal. Chem.* **48**:891–898.
5. Altgelt, K. H. and Segal, L. (Eds.). 1971. *Gel Permeation Chromatography*. Marcel Dekker, New York.
6. Snyder, L. R., and Kirkland, J. J. 1979. *Introduction to Modern Liquid Chromatography*. Wiley, New York.
7. Heftmann, E. 1992. *Chromatography, 5th Edition, Fundamentals and Applications of Chromatography and Related Migration Methods*, edited by E. Heftmann. Elsevier, Amsterdam.
8. Hupe, K. P., McNair, H. M., Kok, W. T., Bruin, G. C., Poppe, H., Poole, C., Chester, T. L., Wimalasena, R. L., Wilson, G. S., Bidlingmeyer, B., Armstrong, D. W., Scott, R. P. W., Widmer, H. M., Majors, R. E., and Ouchi, G. I. 1992. The past 10 years in chromatography. *LC-GC*, **10**:211–237.
9. Hupe, K. P., McNair, H. M., Kok, W. T., Bruin, G. C., Poppe, H., Poole, C., Chester, T. L., Wimalasena, R. L., Wilson, G. S., Bidlingmeyer, B., Armstrong, D. W., Scott, R. P., Widmer, H. M., Majors, R. E., and Ouchi, G. I. 1992. Chromatography 1992: A snapshot of the present. *LC-GC*, **10**:238–254.
10. Trusell, F. C. 1979. Gas Chromatography of Middle and Heavy Distillates. *Chromatography in Petroleum Analysis*, edited by K. H. Altgelt and T. H. Gouw. Marcel Dekker, New York, Chapter 5, pp. 91–119.
11. Lee, M. L., Yang, F. J., and Bartle, K. D. 1984. *Open Tubular Column Gas Chromatography. Theory and Practice*. Wiley, New York.
12. Poole, C. F., and Poole, S. K. 1992. *Chromatography, 5th Edition, Fundamentals and Applications of Chromatography and Related Migration Methods*, edited by E. Heftmann. Elsevier, Amsterdam, Chapter 9, pp. A393–A447.
13. Fetzer, J. C. 1989. Gas- and Liquid Chromatographic Techniques. *Chemical Analysis of Polycyclic Aromatic Compounds*, edited by Tuan Vo-Dinh. Wiley, New York.
14. Zadro, S., Haken, J. K., and Pinczewski, W. V. 1985. Analysis of Australian crude oils by high resolution gas chromatography—Mass spectrometry. *J. Chromat.*, **323**:305–322.
15. Payzant, J. D., Hogg, A. M., Montgomery, D. S., and Strausz, O. P. 1985. A field ionization mass spectrometric study of the maltene fraction of Athabasca bitumen. Part I—The saturates. *AOSTRA J. Res.*, **1**:175–182.
16. Payzant, J. D., Montgomery, D. S., and Strausz, O. P. 1988. The identification of homologous series of benzo[b]thiophenes, thiophenes, thiolanes and thianes possessing a linear carbon framework in the pyrolysis oil of Athabasca asphaltenes. *AOSTRA J. Res.*, **4**:117–131.

17. Kovats, E. 1958. Gas-chromatographische charakterisierung organischer verbindungen. Teil 1: Retentionsindices aliphatischer halogenide, alkohole, aldehyde und ketone. *Helv. Chim. Acta*, **41**:1915–1932.
18. Lee, M. L., Vassilaros, D. L., White, C. M., and Novotny, M. 1979. Retention indices for programmed-temperature capillary-column gas chromatography of polyvyclic aromatic hydrocarbons. *Anal. Chem.*, **51**:768–773.
19. Norton, K. L., Lange, A. J., and Griffiths, P. R. 1991. A unified approach to the chromatography-FTIR interface: GC-FTIR, SFC-FTIR, and HPLC-FTIR with subnanogram detection limits. *J. High Res. Chromat.*, **14**:225–229.
20. Kosman, J. 1992. Petroleum analysis using GC-AES. *Amer. Lab.*, May 1992, pp. 221–222.
21. Wylie, P. L., and Quimby, B. D. 1989. Applications of gas chromatography with an atomic emission detector. *J. High Res. Chromat.*, **12**:813–818.
22. Biggs, W. R., Fetzer, J. C., Brown, R. J., and Reynolds, J. G. 1985. Characterization of vanadium compounds in selected crudes I. Porphyrin and non-porphyrins separation. *Liq. Fuels Tech. 3* (**4**):397–421.
23. Biggs, J. C., Brown, R. J., and W. R., Fetzer 1987. Elemental profiles of hydrocarbon materials by size exclusion chromatography/inductively coupled plasma atomic emission spectrometry. *Energy & Fuels*, **1**:257–262.
24. Coulombe, S., and Sawatzky, H. 1988. H.P.L.C. separation and G.C. characterization of polynuclear aromatic fractions of bitumen, heavy oils and their synthetic crude products. *Fuel*, **65**:552–557.
25. Wadsworth, P. A., and Villalanti, D. C. 1992. Pinpoint hydrocarbons types. *Hydrocarbon Processing*, May 1992, pp. 109–112.
26. Arpino, P. J., Ignatiadis, I., and De Rycke, G. 1987. Sulphur-containing polynuclear aromatic hydrocarbons from petroleum. *J. Chromatog.*, **390**:329–348.
27. Snyder, L. R., and Saunders, D. L. 1979. Adsorption and Partition Chromatography. *Chromatography in Petroleum Analysis*, edited by K. H. Altgelt and T. H. Gouw. Marcel Dekker, New York, pp. 215–272.
28. Poppe, H. 1992. Column Liquid Chromatography. *Chromatography, 5th Edition, Fundamentals and Applications of Chromatography and Related Migration Methods*, edited by E. Heftmann. Elsevier, Amsterdam, Chapter 4, pp. A151–A225.
29. Walton, H. F. 1992. Ion-Exchange Chromatography. *Chromatography, 5th Edition, Fundamentals and Applications of Chromatography and Related Migration Methods*, edited by E. Heftmann. Elsevier, Amsterdam, Chapter 5, pp. A227–A265.
30. Munday, W. A., and Eaves, A. 1959. Analytical Applications for Ion Exchange Resins in the Petroleum Industry. *Fifth World Petrol. Cong. Proc.*, New York, Sect. V, Paper 9, p. 103.
31. Jewell, D. M. 1979. Ion-Exchange and Coordination Chromatography. *Chromatography in Petroleum Analysis*, edited by K. H. Altgelt and T. H. Gouw. Marcel Dekker, New York, Chapter 11, pp. 273–286.

32. Jewell, D. M., Albaugh, E. W., Davis, B. E., and Ruberto, R. G. 1974. Integration of chromatographic and spectroscopic techniques for the characterization of residual oils. *Ind. Eng. Chem. Fundamentals*, **13**:278–282.
33. Green, J. B., Hoff, R. J., Woodward, P. W., and Stevens, L. L. 1984. Separation of liquid fossil fuels into acid, base, and neutral concentrates. 1. An improved nonaqueous ion exchange method. *Fuel*, **63**:1290–1301.
34. Strachan, M. G., and Johns, R. B. 1987. Artifacts arising from the improper preparation and use of nonaqueous ion exchange resins. *Anal. Chem.*, **59**:636–639.
35. Hagel, L., and Janson, J.-C. 1992. Size-Exclusion Chromatography. *Chromatography, 5th Edition, Fundamentals and Applications of Chromatography and Related Migration Methods*, edited by E. Heftmann. Elsevier, Amsterdam, Chapter 6, pp. A267–A307.
36. Fish, R. H., and Komlenic, J. J. 1984. Molecular characterization and profile identifications of vanadyl compounds in heavy crude petroleums by liquid chromatography/graphite furnace atomic absorption spectrometry. *Anal. Chem.*, **56**:510.
37. Fish, R. H., Komlenic, J. J., and Wines, B. K. 1984. Characterization and comparison of vanadyl and nickel compounds in heavy crude petroleums and asphaltenes by reverse-phase and size exclusion liquid chromatography/graphite furnace atomic absorption spectrometry. *Anal. Chem.*, **56**:2452.
38. Nishioka, M., Campbell, R. M., Lee, M. L., and Castle, R. N. 1986. Isolation of sulfur heterocycles from petroleum- and coal-derived materials by ligand exchange chromatography. *Fuel*, **65**:270–273.
39. Nishioka, M., Campbell, R. M., Lee, M. L., and Castle, R. N. 1986. Sulfur heterocycles in coal-derived products. *Fuel*, **65**:390–396.
40. Nishioka, M., Lee, M. L., and Castle, R. N. 1986. Determination and structural characteristics of sulfur heterocycles in coal- and petroleum-derived materials. *Amer. Chem. Soc. Petrol. Div.*, **31**(3):827–833.
41. Nishioka, M., Whiting, D. G., Campbell, R. M., and Lee, M. L. 1986. Supercritical fluid fractionation and detailed characterization of the sulfur heterocycles in a catalytically cracked petroleum vacuum residue. *Anal. Chem.*, **58**:2251–2255.
42. Schoemakers, P. J., and Uunk, L. G. M. 1992. Supercritical-Fluid Chromatography. *Chromatography, 5th Edition, Fundamentals and Applications of Chromatography and Related Migration Methods*, edited by E. Heftmann. Elsevier, Amsterdam, Chapter 8, pp. A339–A391.
43. Fuhr, B. J., Holloway, L. R., and Reichert, C. 1989. Characterization of heavy oil by capillary supercritical fluid chromatography. *Fuel Sci. Tech. Int.* **7**(5-6):643–657.
44. Campbell, R. M., and Lee, M. L. 1986. Supercritical fluid fractionation of petroleum- and coal-derived mixtures. *Anal. Chem.*, **58**:2247–2251.
45. Chang, H-C. K., Markides, K. E., Bradshaw, J. S., and Lee, M. L. 1988. Selectivity enhancement for petroleum hydrocarbons using a smectic liquid

crystalline stationary phase in supercritical fluid chromatography. *J. Chromat. Sci.*, **26**:280–289.

46. Wright, B. W., Udseth, H. R., Chess, E. K., and Smith, R. D. 1988. Supercritical fluid chromatography/mass spectrometry for analysis of complex hydrocarbon mixtures. *J. Chromat. Sci.*, **26**:228–235.
47. Taguchi, M., Hobo, T., and Maeda, T. 1991. Evaluation and application of a supercritical fluid extraction system using capillary supercritical fluid chromatography. *J. High Res. Chromat.*, **14**:140–143.
48. Lopez-Avila, V., Dodhivala, M. S., Benedicto, J., and Beckert, W. F. 1992. Evaluation of four supercritical fluid extraction systems for extracting organics from environmental samples. *LC-GC*, **10**(10):762–769.
49. Singh, I. D., Ramaswami, V., Kothiyal, V., Brouwer, L., and Severin, D. 1992. Structural studies of short path distillates and supercritical fluid extract of petroleum short residue by NMR spectroscopy. *Fuel Sci. Tech. Int.*, **10**(2):267–280.
50. Rodgers, P. A., Creagh, A. L., Prange, M. M., and Prausnitz, J. M. 1987. Molecular weight distributions for heavy fossil fuels from gel permeation chromatography and characterization data. *AICHE1987 Spring Natl. Meet.*, Houston, 29 March–2 April 1987, Prepr. No. 20B.
51. Gluckman, J. C., Shelly, D. C., and Novotny, M. 1985. Miniature fluorimetric photodiode array detection system for capillary chromatography. *Anal. Chem.*, **57**:1546–1552.
52. Gluckman, J. C., and Novotny, M. 1985. Applications of a miniaturized fluorimetric photodiode array detector for capillary liquid chromatography. *J. High Res. Chromatog. Chromatog. Comm.*, **8**:672–677.
53. Beale, S. C., Wiesler, D., and Novotny, M. 1987. Characterization of the phenolic fraction from fossil-fuel materials by microcolumn liquid chromatography and its ancillary techniques. *J. Chromatog.*, **393**:391–406.
54. Andreolini, F., Borra, C., Wiesler, D., and Novotny, M. 1987. High-efficiency separation and characterization of high-molecular-weight heterocyclic sulfur compounds in fossil fuels by microcolumn liquid chromatography and related techniques. *J. Chromatog.*, **406**:375–388.
55. Owen, A. J. 1988. *The Diode Array Advantage in UV/Visible Spectroscopy.* Hewlett Packard, Waldbronn, FRG.
56. Vandeginste, B. G. M., Kateman, G., Strasters, J. K., Billiet, H. A. H., and de Gala, L. 1987. Data handling in HPLC–diode array UY-spectrometry. *Chromatographia*, **24**:127–134.
57. Daubert, T. E., and Danner, R. P. (1989). *Physical and Thermodynamic Properties of Pure Chemicals.* Hemisphere Publishing Corporation, New York.

58a. Boduszynski, M. M. 1988. Composition of heavy petroleums. 2. Molecular characterization. *Energy Fuels*, **2**:597.

58. Coulombe, S. 1988. Comparison of detectors for size exclusion chromatography of heavy oil related samples. *J. Chromat. Sci.*, **26**:1–6.

59. Corbett, L. W. 1969. Composition of asphalt based on generic fractionation, using solvent deasphaltening, elution–adsorption chromatography, and densimetric characterization. *Anal. Chem.*, **41**:576–579.
60. Corbett, L. W., and Petrossi, U. 1978. Differences in distillation and solvent separated asphalt residua. *I&EC Prod. Res. Develop.*, **17**:342–346.
61. Snyder, L. R., and Buell, B. E. 1968. Nitrogen and oxygen compound types in petroleum. A general separation scheme. *Anal. Chem.*, **40**:1295–1302.
62. Snyder, L. R., Buell, B. E., and Howard, H. E. 1968. Nitrogen and oxygen compound types in petroleum. Total analysis of a 700–850°F distillate from a California crude oil. *Anal. Chem.*, **40**:1303–1317.
63. Snyder, L. R. 1969. Nitrogen and oxygen compound types in petroleum. Total analysis of a 400–700°F distillate from a California crude oil. *Anal. Chem.*, **41**:315–323.
64. Snyder, L. R. 1969. Nitrogen and oxygen compound types in petroleum. Total analysis of a 850–1000°F distillate from a California crude oil. *Anal. Chem.*, **41**:1085–1094.
65. McKay, J. F., Amend, P. J., Harnsberger, P. M., Cogswell, T. E., and Latham, D. R. 1981. Composition of petroleum heavy ends. 1. Separation of petroleum >675°C residues. *Fuel*, **60**:14–16.
66. Desbene, P.-L., Lambert, D. C., Richadin, P., and Basseliere, J.-J. 1984. Preparative fractionation of petroleum heavy ends by ion exchange chromatography. *Anal. Chem.*, **56**:313–315.
67. Lundanes, E., and Greibrokk, T. 1985. Quantitation of high boiling fractions of North Sea oil after class separation and gel permeation chromatography. *J. Liq. Chromat.*, **8**(6):1035–1051.
68. Lundanes, E., and Greibrokk, T. 1985. Group separation of oil residues by supercritical fluid chromatography. *J. Chromatog.*, **349**:439–446.
69. Pearson, C. D., and Garfeh, S. G. 1986. Automated high-performance liquid chromatography—Determination of hydrocarbon types in crude oil residues using a flame ionization detector. *Anal. Chem.*, **58**:307–311.
70. Carbognani, L., and Izquierdo, A. 1990. Preparative compound class separation of heavy oil vacuum residua by high performance liquid chromatography. *Fuel Sci. Tech. Int.*, **8**(1):1–15.
71. Fetzer, J. C., and Biggs, W. R. 1985. Liquid chromatographic retention behavior of large, fused polycyclic aromatics. *J. Chromatog.*, **346**:81–92.
72. Boduszynski, M. M., Hurtubise, R. J., and Allen, T. W. 1983. Liquid chromatography/field ionization mass spectrometry in the analysis of high-boiling and nondistillable coal liquids for hydrocarbons. *Anal. Chem.*, **55**:225–231.

72a. Killops, S. D. 1986. Normal phase HPLC retention characteristics of polycyclic aromatic compounds on cyano/amino bonded silica. *J. High Resol. Chromatogr. Chromatogr. Comm.*, **9**:302–303.

73. Wise, S. A. 1977. Chemically-bonded aminosilane stationary phase for the high-performance liquid chromatographic separation of polynuclear aromatic compounds. *Anal. Chem.*, **49**:2306–2310.

74. Liphard, K. G. 1980. Hydrocarbon group type analysis of coal liquids by high-performance liquid chromatography. *Chromatographia*, **13**:603–606.
75. Chmielowiec, J., and George, A. E. 1980. Polar bonded-phase sorbents for high performance liquid chromatographic separations of polycyclic aromatic hydrocarbons. *Anal. Chem.*, **52**:1154–1157.
76. Chmielowiec, J., Beshai, J. E., and George, A. E. 1980. Separation, characterization and analysis of polynuclear aromatic hydrocarbon ring classes in Lloydminster and Medicine River oils. *Fuel*, **59**:838–844.
77. Grizzle, P. L., and Thomson, J. S. 1982. Liquid chromatographic separation of aromatic hydrocarbons with chemically bonded (2,4-dinitroanilinopropyl)-silica. *Anal. Chem.*, **54**:1071–1078.
78. Yamamoto, Y. 1986. Analysis of heavy oils by FID-TLC (Part 4). *Sekiyu Gakkaishi*, **29**(1):15–19.
79. Amat, M., Arpino, P., Orrit, J., Lattes, A., and Guichon, G. 1980. *Analusis*, **8**(5):179.
80. Schmitter, J. M., Ignatiadis, I., Arpino, P., and Guichon, G. 1983. Selective isolation of nitrogen bases from petroleum. *Anal. Chem.*, **55**:1685–1688.
81. Green, J. B., Reynolds, J. W., and Yu, S. K.-T. 1989. Liquid chromatographic separations as a basis for improving asphalt composition–physical property correlations. *Fuel Sci. Tech. Int.*, **7**(9):1327–1363.
82. Green, J. B. 1986. Liquid chromatography on silica using mobil phases containing tetrealkylammonium hydroxides. I. General separation selectivity and behaviour of typical polar compounds in fuels. *J. Chromatog.*, **358**:53–75.
83. Green, J. B. 1981. Liquid chromatography on silica using mobil phases containing aliphatic carboxylic acids. I. Effects of carboxylic acid chain length on separation efficiency and selectivity. *J. Chromatog.*, **209**:211–229.
84. Green, J. B., and Hoff, R. J. 1981. Liquid chromatography on silica using mobil phases containing aliphatic carboxylic acids. II. Applications in fossil fuel characterization. *J. Chromatog.*, **209**:231–250.
85. Snyder, L. R. 1992. Theory of chromatography. *Chromatography, 5th Edition, Fundamentals and Applications of Chromatography and Related Migration Methods*, edited by E. Heftmann. Elsevier, Amsterdam, Chapter 1, pp. AA1–A68.
86. Trestianu, S., Zilioli, G., Sironi, A., Saravalle, C., Munari, F., Galli, M., Gaspar, G., Colin, J. M., and Jovelin, J. L. 1985. Automatic simulated distillation of heavy petroleum fractions up to 800°C TBP by capillary gas chromatography. *J. High Resol. Chromatogr. Chromatogr. Comm.* **8**:771–781.
87. Quimby, B. D., and Sullivan, J. J. 1990. Evaluation of a microwave cavity, discharge tube, and gas flow system for combined gas chromatography-atomic emission detection. *Anal. Chem.* **62**:1027–1034.
88. Thomson, J. S., and Rynaski, A. F. 1992. Simulated distillation of wax samples using supercritical fluid and high temperature gas chromatography. *J. High Res. Chrom.* **15**:227–234.

89. Buteyn, J. L., and Kosman, J. J. 1990. Multielement simulated distillation by capillary GC-MED. *J. Chromat. Sci.* **28**:19–23.
90. Lee et al., 1984; see also ref. 11.
91. Buyten, J. Duvekot, J., Peene, J., and Musshe, P. 1991. A capillary column for high-temperature gas chromatography. *Amer. Laboratory*, Aug. 1991, 13–18.
92. Curvers, J., and van den Engel, P. 1989. Gas chromatography method for simulated distillation up to a boiling point of 750°C using temperature-programmed injection and high temperature fused silica wide-bore columns. *J. High Res. Chrom.* **12**:16–22.
93. Lipsky, S. R., and Duffy, M. L. 1986. High temperature gas chromatography: The development of new aluminum clad flexible fused silica capillary columns coated with thermostable nonpolar phases. Part 1. *J. High Resol. Chromatogr. Chromatogr. Comm.* **9**:376–382.
94. Lipsky, S. R., and Duffy, M. L. 1986. New advances in capillary gas chromatography. *LC-GC* **4**:898–906.
95. Altgelt, K. H. 1979. Gel Permeation Chromatography. *Chromatography in Petroleum Analysis*, edited by Altgelt, K. H., and Gouw, T. H. Marcel Dekker, New York, Chapter 12, pp. 287–312.
96. Green, J. B. 1992. Personal communication.

7

Molecular Characterization of Heavy Petroleum Fractions by Mass Spectrometry

I. OVERVIEW

After describing structural group-type characterization and fractionation, we are dealing in this chapter mainly with molecular characterization by mass spectrometry (MS), which provides the most detailed information on a sample available from any method.

MS furnishes the molecular weight and chemical formula ($C_nH_{2n+Z}X$) of compounds and their amounts. It can also provide important information about their molecular structure. The earliest and most common type of MS, electron impact (EI) MS, gives a fragmentation pattern displaying both parent ion peaks and fragment ion peaks, characteristic of each molecular type. Fragmentation is frequently employed to differentiate between isomers of pure compounds and of molecules in fairly simple mixtures. However, it is usually avoided with such complex samples as heavy petroleum fractions, which are composed of such a multitude of closely related compounds that their fragmentation patterns are nondistinctive and cannot be readily interpreted.

For heavy petroleum fractions, the use of nonfragmenting (NF)-MS methods is now preferred. These methods, also called "soft ionization" methods, produce predominantly parent ion (molecular ion) peaks and, thus, much simpler spectra than methods producing fragments. By rendering the molecular weight of each compound in a sample, and reasonably well its abundance, NF-MS gives us the molecular-weight distribution—or, as the mass spectroscopists say, the molar mass profile of a sample. Among the most important NF-MS techniques are field ionization MS (FIMS), field desorption MS (FDMS), chemical ionization MS (CIMS), and low-voltage (10–20 eV) electron impact MS (LVEI-MS). A particularly powerful version of LVEI-MS is high-resolution LVEI-MS (HR-LEVI-MS or, in its abbreviated form, LVRH-MS). All of these are "soft" ionization techniques, designed to generate "cold" ions of such low excess energy that they do not undergo fragmentation to any great extent, in contrast to the "hot" ions generated by conventional electron impact ionization (70–100 eV). With soft ionization methods, a compromise must be found between low fragmentation and sufficiently high sensitivity. Unfortunately, samples rich in aliphatic hydrocarbons are hard to keep completely from fragmenting, but most other compound types, especially aromatic compounds, are well behaved and yield clean parent ion spectra.

FIMS is among the most attractive soft ionization methods because it is capable of producing fragmentation-free spectra, even of the readily fragmenting aliphatic hydrocarbons. Nominal mass resolution FIMS has obviously limited utility because many different compound types have the same nominal (integer) molar mass. However, in many cases (with well-defined chromatographic fractions), FIMS can provide detailed compositional information in terms of compound types and carbon number distribution. For a mixture of aliphatic hydrocarbons (saturates), the molar mass series in the spectrum can be readily assigned to the so-called "apparent Z series" (Z in the general formula C_nH_{2n+Z}). The peak heights in the spectrum can be evaluated for quantitation, that is, for calculating the abundance of the various species. Thus, from one mass spectrum of the saturates fraction, we obtain the composition of a sample in terms of the apparent Z series and molecular weight (or carbon number) distribution.

Each of the apparent Z series can be represented by more than one compound type. To assign specific compound types to the respective apparent Z series, additional complementary information is needed. For our example of a saturates fraction, this information is provided by the chromatographic separation which assures that the sample consists of only aliphatic hydrocarbons. Thus, paraffins (acyclic alkanes) have the general formula C_nH_{2n+2}; mononaphthenes (monocyclic alkanes) C_nH_{2n}; dinaphthenes (dicyclic alkanes)—C_nH_{2n-2}; trinaphthenes (tricyclic alkanes)—C_nH_{2n-4}; and so on. The Z value decreases by 2 with every additional naphthenic ring. At least medium-resolution (MR) FIMS would be necessary to distinguish aliphatic hydrocarbons differing by 1 C and 12 H such as, for example, paraffins and heptanaphthenes (that have a molar mass difference of 0.0939 amu) as will be discussed later.

For other classes of compounds (e.g., aromatic hydrocarbons, sulfur compounds, nitrogen compounds, etc.), the MS analysis becomes much more complicated. The aromatic rings and each set of heterocompounds add one more element to the permutations in structure leading to the same or to very similar exact masses for different compounds. This subject will be elaborated later (Section II.A.2). Here just two examples may suffice: (1) a tetranaphthene and an alkylbenzene may both have the same exact mass of 414.42855; (2) an alkylbenzothiophene and an alkylnaphtheho-phenanthrene may have the same nominal molar mass of 414 with a difference of only 0.0034 amu. The identification of specific compound types (the exact formula) in the first case is altogether impossible. The second case requires either very high-resolution MS or chromatographic separation. However, a member of the second group can be differentiated from a member of the first group by medium-resolution FIMS.

Generally, for the identification of these compound classes in petroleum samples, distillation and highly efficient chromatographic separations must be combined with medium- or high-resolution MS. Distillation limits the molar mass range and, thus, concentrates the compounds in that range to a level where they are more readily detectable in the mass spectrum. Chromatography into well-defined compound-class fractions prevents the problems of the kind illustrated by our two examples above. Medium- or high-resolution MS extends the application of NF-MS by providing the exact chemical formulas, enabling the identification of compounds in easy cases and, in more complex fractions, the differentiation of many hetero-compounds from hydrocarbons and of other compound-type combinations. Tables have been published for the exact masses of a great number of compounds, for example, by Beynon and Williams (1).

With the high-molecular-weight compounds (of about 300 molecular weight or higher) and the greater variety of compound types found in high-boiling distillates and residual fractions, a single molar mass, *even if determined by high resolution*, may still correspond to more than one type of molecule. From the molar mass alone, the correct compound type cannot be recognized, and, if several compound types having the same molar mass are present, their distribution in the mixture cannot be determined without additional information either from chromatographic separations or from other spectroscopic methods or from both.

The MS analysis, even of well-defined petroleum fractions, becomes increasingly difficult with increasing boiling point. Two reasons account for this trend: (1) the rapidly increasing number of compound types, many with heteroatoms; and (2) their decreasing concentration. Complementary information is available from NMR, IR, and UV spectroscopy. As we saw in Chapter 5, NMR enables us to measure or to estimate certain structural elements in a sample, that is, in the entire sample rather than in its molecules or compound types. Examples of such structural elements are the percent aromatic and aliphatic carbon, the abundances of the various CH_n groups (n = 0–3), the percent aliphatic carbon in long alkyl chains and the amount of H in α position to aromatic rings, which can be taken as a rough measure of the degree of substitution. IR permits the measurement of certain heteroatom groups such as, for example, –OH, >NH, carbonyl, carboxyl, ester, and so on. UV, though much less specific, can be used to distinguish aromatic hydrocarbons of different ring number and ring type (Chapter 9). Some of the very high-boiling distillation cuts (~900–1300°F AEBP) and truly nondistillable fractions (>1300°F AEBP) are so difficult, or even impossible, to analyze by MS that we must settle for other characterization methods or even resort to the destructive methods described in Chapter 9.

So far, we have distinguished between "hot ionization" MS methods such as HVEI-MS (giving fragmentation) and "soft ionization" MS methods such as FIMS, FDMS, and LVEI-MS (giving parent ion spectra). We recognized high-resolution low-voltage MS (LVHR-MS) as a very useful extension of LV-MS. Another variant in this game is low-resolution high-voltage MS (HVLR-MS). Petrakis et al. (2) reported their use of it for generating spectra of the bare ring systems of molecules present in petroleum fractions, stripped of all alkyl groups. We have no personal experience with this approach, but we think that it may be quite difficult.

II. THE TOOLS

A. General Methodology

1. The Hardware

For simplicity, this discussion is restricted to double-focusing mass spectrometers. Other types of spectrometers are described by Roboz (3), Watson (4), and others. Functionally, a mass spectrometer consists of three main parts:

- The ion source, which generates ions from the sample
- One or more focusing sectors (which separate ions on the basis of their mass to charge ratio)
- The detector

Figure 7.1 shows the schematic of a double-focusing mass spectrometer. It includes, in addition to the three main parts, two types of sample inlet systems, the direct insertion probe and a heated gas expansion reservoir, and the data collection system (the computer and recorder).

The ion source is a vacuum chamber at about 10^{-6} Torr into which the sample is introduced by one of various methods. This may be done by

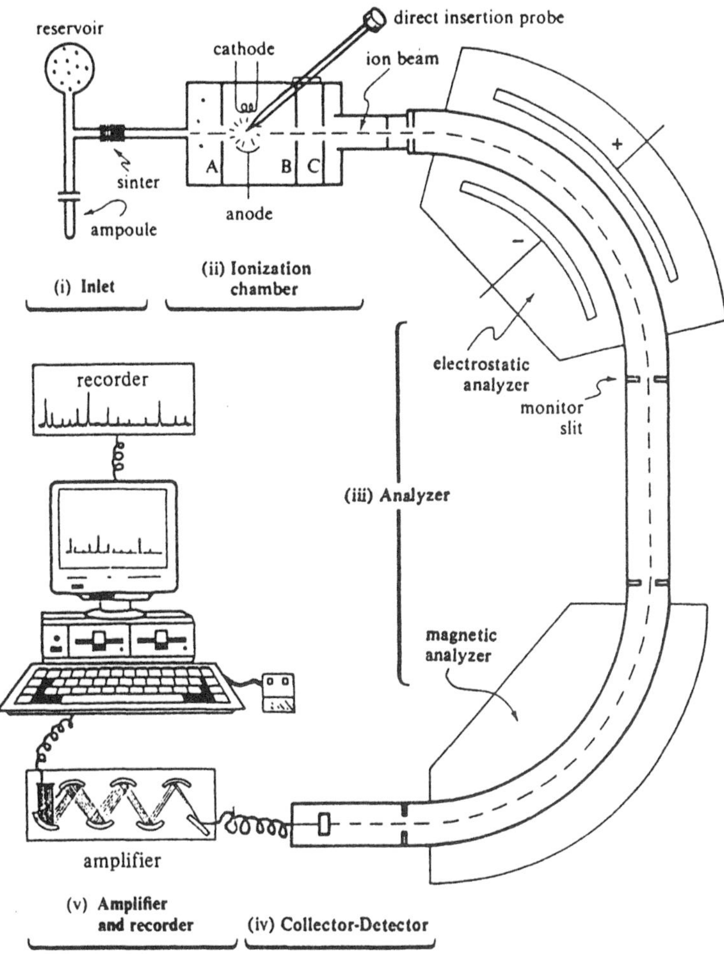

Figure 7.1 Schematic of a double-focusing mass spectrometer. (From Ref. 3. Reproduced with permission of the publisher.)

controlled evaporation from a heated surface with gently rising temperature, or by evaporation in a prechamber, with the vapors bled into the main ion source through a small hole (a "leak") slowly enough to maintain the high vacuum. Other ways are described later. In most conventional applications, the vaporized sample molecules are bombarded with a beam of 70-eV electrons. On impact, an electron transfers its energy to a sample molecule which then loses an electron—sometimes even two or more—and becomes a positively charged ion.

Not all of the transferred energy is dissipated by the loss of an electron. The ion, being left with energy in excess of that required to form an ion (about 10 eV) is in a highly excited state; it is "hot." A common way for hot ions to lose their excess energy is by breaking bonds. Mass spectra obtained by conventional electron impact (EI) have intense fragmentation patterns. By using less energetic electrons, of 10–20 eV, most of the fragmentation can be avoided, although at the price of a much weaker signal (1/10–1/100). Other ways to create "cold" ions with little or no excess energy are described in Section III.B.

The sample ions are accelerated toward the anode by an electrical potential of about 10,000 V.* In principle, if not in detail, the anode is a plate with a narrow slit leading to the next part of the mass spectrometer, the focusing sectors. The ions with the correct angular direction pass this opening; the others are deflected and lost. Single-focusing instruments have just one focusing sector (a magnetic one);† double-focusing instruments have an additional electric sector. Both are curved. In the magnetic sector, the ions are deflected in the direction perpendicular to the magnetic field according to their momentum, mv (m = mass and v = velocity). In the electric sector, the ions are deflected in the direction of the electric field according to their kinetic energy, $mv^2/2$. The combined magnetic and electric focusing mechanisms complement each other and provide high resolution more economically than can be achieved by just one sector. In single-sector machines, resolution is enhanced by reducing the slit with a concomitant loss of signal intensity.

The emerging ion beam is spread by its mass or m/e ratio. (Those few ions that carry more than one charge have proportionally lower m/e ratios: 1/2 for doubly charged, 1/3 for triply charged, and so forth.) The detector is placed at the focal point of the mass spectrometer. By sweeping the

---

*Ten kilovolts is not universal. In magnetic mass spectrometers, it ranges from 1 to about 12 kV; in quadrupole mass spectrometers, it is tens of volts.

†For the sake of simplicity, we restrict our discussion to the most rudimentary principles. Therefore, we ignore quadrupole mass spectrometers, time-of-flight mass spectrometers, and so on.

focusing sector(s), that is, by changing their magnetic (and electric) field, ions of different mass-to-charge ratio will be sequentially focused at the detector. Most mass spectrometers use electron multipliers to magnify each ion's charge by several orders of magnitude. The final (magnified) signal is stored in a computer and processed by programs provided by the manufacturer of the mass spectrometer. Additional processing programs—some commercially available, others written by the user—evaluate the raw data in a multitude of ways: some in terms of exact atomic mass units (amus) and some by applying various corrections to the data.

The mass of any peak is the sum of those of the various isotopes of each element present in the molecule. For example, for a compound with the formula $C_{30}H_{62}$, there will be molecules with all the possible combinations of carbon and hydrogen isotopes. Their distribution is based on the probability of their occurrence. That means, some molecules will have 29 ^{12}C atoms and 1 ^{13}C atom; some will have 28 ^{12}C atoms and 2 ^{13}C atoms; and so forth. Each will have a different mass and, thus, give rise (or contribute) to a different peak. For example, a paraffin molecule containing 100 carbons would have peaks at integer masses of 1402 (zero ^{13}C atoms), 1403 (one ^{13}C), 1404 (two ^{13}C), and so forth. The relative size of these peaks is determined by the statistical probability of the distribution of the C isotopes and their natural abundance (that of ^{13}C is about 1.1%). For a C_{100} hydrocarbon, the relative peak sizes are 0.869 for the peak without any ^{13}C, 1.000 for one ^{13}C, 0.570 for two, 0.214 for three, 0.060 for four, and so forth. The significance of the ^{13}C isotope containing molecules increases with increased carbon number or molecular weight. For heterocompounds, the number of peaks for any given molecule is greater yet because of the additional S, N, and O isotopes.

Mass spectra can be corrected for the distribution of the ^{13}C isotopes in such a way that only the peak without ^{13}C is shown. This peak, however, now contains the contributions from all other related peaks, that is, from those shown in the original spectrum which represented the constituents of the compound with one or several ^{13}C isotopes. This permits plots of spectra with only one peak per molecule.

The mass scale for any set of conditions is determined by using calibration standards and assigning the accurately known masses to their peaks in the spectrum. In low-resolution experiments, the calibration is performed separately. The calibration spectrum is evaluated in terms of a mathematical function of sensitivity versus *m/e*. The resultant table of mass versus the selected instrument parameters such as scan time, magnet current, or magnetic field, is applied to a subsequent run. In high-resolution experiments, a reference standard, for example, a perfluorinated kerosine sample, is usually added to the analyte. The large number of molecules in such a standard material enables the user to reference any sample peak to a nearby pair of bracketing calibration peaks. In high-resolution runs of heavy petroleum fractions, 10 to 20 sample peaks can be expected between each pair of known reference peaks.

Application computer programs for additional calculations are usually considered proprietary by their authors or by the companies where they were developed. Schmidt et al. (5) published such a program for the evaluation of low-voltage high-resolution MS (LVHR-MS) data.

2. Molar Mass and Hydrogen Deficiency

We have seen that NF-MS gives the molar mass (and abundance) of the sample molecules. *Mass and Abundance Tables for Use in Mass Spectrometry* by Beynon and Williams (1) lists a variety of chemical structures for a given mass number, based on the main isotope of each element, that is, 12 for C, 1 for H, and so on. Although this collection covers only hydrocarbons and heterocompounds containing nitrogen and oxygen, the number of choices per mass number increases rapidly with the mass number and is quite large even for molecules of moderate size (e.g., 246 amu). Table 7.1 shows a sample page of these tables for mass number 246.

Some of the various structures belonging to a given mass may differ by their hydrogen deficiency, as demonstrated by the examples in Table 7.2. Six of the compounds shown there have the same Z-number but differ in their combination of ring number and double bonds. These are only a small selection from a large number of possible isomers.

The first three are examples of compounds with equal mass but different Z-numbers. The first contains 12 more hydrogen atoms and 1 fewer carbon atom than the second. The second differs from the third the same way. The mass differences of 0.0939 amu (about 1 part in 2500) cannot be seen by nominal (integer) mass resolution mass spectrometers, only by those made for higher resolution. In samples with molar masses of about 400 or more (e.g., vacuum gas oils), hydrocarbon molecules with up to three different Z-numbers per nominal mass may be present. With heterocompounds, even more combinations may be possible.

For aliphatic hydrocarbons, Z in the general formula C_nH_{2n+Z}, fairly well classifies a compound type; that is, the hydrogen deficiency is caused by naphthenic rings (in the absence of olefins). Therefore, a deficiency of two hydrogens, generally written as $Z = 0$, signifies a monocycloalkane (mononaphthene); $Z = -2$ indicates a dicycloalkane (dinaphthene); $Z = -4$, a tricycloalkane (trinaphthene); $Z = -6$, a tetracycloalkane (tetranaphthene), and so on. However, with a nominal mass resolution NF-MS (e.g., FIMS or FDMS), a heptanaphthene ($Z = -12$) cannot be distinguished from a paraffin ($Z = +2$), an octanaphthene ($Z = -14$) from a

Table 7.1 Excerpt from *Mass and Abundance Tables for Use in Mass Spectrometry*

Mass	246			Cont			
C	H	N	O	mass	PM	Pmi	Ratio
14	16	1	3	0.113011	15.883	17.803	8.921
14	18	2	2	0.136820	16.258	16.393	9.917
14	20	3	1	0.160629	16.632	14.996	11.090
14	22	4	—	0.184438	17.006	13.614	12.491
14	30	—	3	0.219482	15.726	17.561	8.954
15	2	—	4	0.995306	16.397	20.587	7.964
15	4	1	3	0.019115	16.772	19.196	8.737
15	6	2	2	0.042924	17.146	17.819	9.622
15	8	3	1	0.066733	17.520	16.455	10.647
15	10	4	—	0.090542	17.895	15.106	11.845
15	18	—	3	0.125586	16.614	18.940	8.772
15	20	1	2	0.149395	16.989	17.557	9.676
15	22	2	1	0.173204	17.363	16.188	10.725
15	24	3	—	0.197013	17.737	14.833	11.958
16	6	—	3	0.031690	17.503	20.398	8.580
16	8	1	2	0.055499	17.877	19.048	9.385
16	10	2	1	0.079308	18.252	17.712	10.304
16	12	3	—	0.103117	18.626	16.390	11.364
16	22	—	2	0.161971	17.720	18.775	9.438
16	24	1	1	0.185780	18.094	17.433	10.379
16	26	2	—	0.209589	18.469	16.105	11.467
17	—	3	—	0.009221	19.515	18.027	10.825
17	10	—	2	0.068075	18.608	20.330	9.152
17	12	1	1	0.091884	18.983	19.022	9.979
17	14	2	—	0.115693	19.357	17.728	10.919
17	26	—	1	0.198355	18.825	18.732	10.050
17	28	1	—	0.222164	19.200	17.431	11.014
18	—	1	1	0.997988	19.871	20.690	9.604
18	2	2	—	0.021797	20.246	19.429	10.420
18	14	—	1	0.104459	19.714	20.386	9.670
18	16	1	—	0.128268	20.088	19.119	10.507
18	30	—	—	0.234740	19.931	18.811	10.595
19	2	—	1	0.010563	20.602	22.119	9.314
19	4	1	—	0.034372	20.977	20.885	10.044
19	18	—	—	0.140844	20.819	20.563	10.124
20	6	—	—	0.046948	21.708	22.394	9.693

Source: Ref. 1.

Table 7.2 Examples of Different Chemical Structures with the Same Mass and the Same or Different *Z*-Numbers

Structure	Rings	DB	Z	C	H	Mass
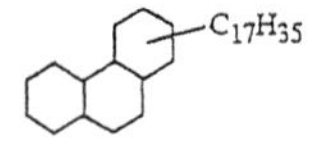 $C_{17}H_{35}$	3	0	−4	31	58	430
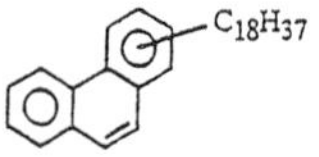 $C_{18}H_{37}$	3	7	−18	32	46	430
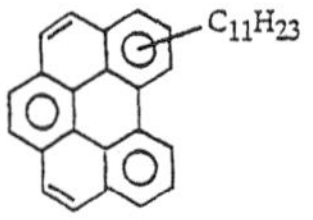 $C_{11}H_{23}$	6	11	−32	33	34	430
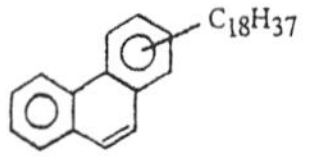 $C_{18}H_{37}$	3	7	−18	32	46	430
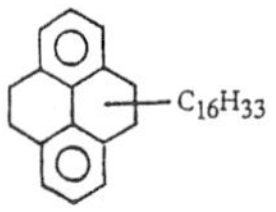 $C_{16}H_{33}$	4	6	−18	32	46	430
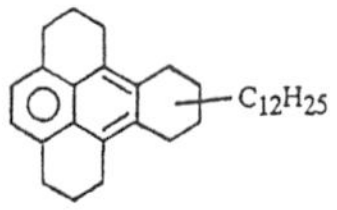 $C_{12}H_{25}$	5	5	−18	32	46	430
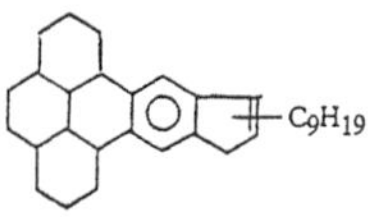 $C_{9}H_{19}$	6	4	−18	32	46	430
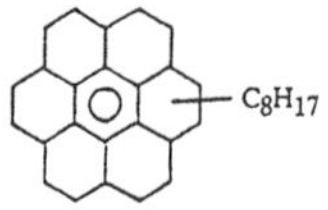 $C_{8}H_{17}$	7	3	−18	32	46	430
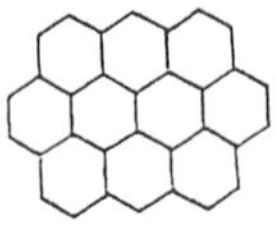	10	0	−18	32	46	430

mononaphthene ($Z = 0$), and so on, that is, from compounds differing by 14 Z-numbers (or multiples of 14). The resolution required to resolve those overlapping Z series equals $M/0.0939$. Thus, aliphatic hydrocarbons in a vacuum gas oil range (650–1000°F AEBP), which covers a molar mass (M) range of approximately 250 to 650 amu, require a medium resolution MS with a resolving power of at least 7000.

The above discussion is based on the assumption that a chromatographic separation is capable of producing a "pure" saturates fraction, which is free of olefins and aromatics. Olefins ($Z = 0$) would have the same exact molar mass as mononaphthenes; alkylbenzenes ($Z = -6$) would have the same exact molar mass as tetranaphthenes; mononaphthenobenzenes ($Z = -8$) as pentanaphthenes; and so on. Obviously, the unambiguous interpretation of the mass spectra depends on the effectiveness of a chromatographic separation. The problem may be even more complicated if aliphatic sulfur compounds are present in a sample. Those compounds are difficult to separate from aliphatic hydrocarbons and, if present, would require very high-resolution MS to resolve them from the hydrocarbon Z series. The interfering Z series (having the same nominal molar mass) are listed in Table 7.3.

The analysis of aromatic hydrocarbon fractions is even more complicated. Here, the Z-number is affected by both the total number of rings (R) and the number of double bonds (DB). The Z of hydrocarbons is defined by the formula $Z = -2[R + \text{DB} - 1]$. Table 7.4 gives examples of various hydrocarbon types and their Z-numbers.

Chromatographic separation by aromatic ring number greatly facilitates the MS analysis. However, for high-boiling petroleum fractions, "baseline" separations cannot be achieved, and adjacent fractions overlap. Often, abundant sulfur compounds further complicate the analysis because they coelute with aromatic hydrocarbons. Medium- to ultrahigh-resolution MS may be required to resolve sulfur compound series from hydrocarbon series, depending on compound type and molecular weight.

It is easy to understand that, with increasing boiling point of petroleum fractions and their concomitant increase in molecular weight and heteroatom content, the ability of MS to provide unambiguous compositional information is greatly reduced.

3. Response Factors

MS response factors depend on the ionization method as well as on the chemical nature of the molecules. Chasey and Aczel (6) summarized nicely their observations on molecular sensitivities for the EI mode. Their comments are of general interest although they may not be equally valid for FI, FD, and other MS modes. The trends are, at least in part, based on

Table 7.3 Examples of Hydrocarbons and Sulfur Compounds of Equal Nominal Mass

Aliphatic hydrocarbons				Aliphatic sulfur compounds		
Compound type	*Z* series	Formula	Mass	Compound type	*Z* series	Formula
Paraffins	+2 H	$C_{30}H_{62}$	422	Dicyclic	−2 S	$C_{28}H_{54}S$
Mononaphthenes	0 H	$C_{30}H_{60}$	420	Tricyclic	−4 S	$C_{28}H_{52}S$
Dinaphthenes	−2 H	$C_{30}H_{58}$	418	Tetracyclic	−6 S	$C_{28}H_{50}S$
Trinaphthenes	−4 H	$C_{30}H_{56}$	416	Pentacyclic	−8 S	$C_{28}H_{48}S$
Tetranaphthenes	−6 H	$C_{30}H_{54}$	414	Hexacyclic	−10 S	$C_{28}H_{46}S$
Pentanaphthenes	−8 H	$C_{30}H_{52}$	412	Acyclic	+2 S	$C_{28}H_{44}S$
Hexanaphthenes	−10 H	$C_{30}H_{50}$	410	Monocyclic	0 S	$C_{28}H_{42}S$

Table 7.4 *Z*-Numbers of Various Hydrocarbon Molecules. $Z = -2(R + DB - 1)$

Hydrocarbon	Formula	No. of rings	No. of double bonds	*Z*-number
Saturates				
n-octane	C_8H_{18}	0	0	2
Iso-octane	C_8H_{18}	0	0	2
Hexane	C_6H_{14}	0	0	2
Hexene	C_6H_{12}	0	1	0
Cyclohexane	C_6H_{12}	1	0	0
Cyclohexene	C_6H_{10}	1	1	−2
Decalin	$C_{10}H_{18}$	2	0	−2
Perhydrophenanthrene	$C_{14}H_{24}$	3	0	−4
Perhydropyrene	$C_{16}H_{26}$	4	0	−6
Perhydropicene	$C_{22}H_{36}$	5	0	−8
Perhydroperylene	$C_{20}H_{32}$	5	0	−8
Aromatics				
Benzene	C_6H_6	1	3	−6
Toluene	C_7H_8	1	3	−6
Hexyl benzene	$C_{12}H_{18}$	1	3	−6
Dodecyl benzene	$C_{18}H_{30}$	1	3	−6
Tetralin	$C_{10}H_{12}$	2	3	−8
Octahydrophenanthrene	$C_{14}H_{18}$	3	3	−10
Naphthalene	$C_{10}H_8$	2	5	−12
Tetrahydrophenanthrene	$C_{14}H_{14}$	3	5	−14
Fluorene	$C_{13}H_{10}$	3	6	−16
Phenanthrene	$C_{14}H_{10}$	3	7	−18
Tetrahydrochrysene	$C_{18}H_{16}$	4	7	−20
Pyrene	$C_{16}H_{10}$	4	8	−22
Chrysene	$C_{18}H_{12}$	4	9	−24
Perylene	$C_{20}H_{12}$	5	10	−28
Benzo[ghi]perylene	$C_{22}H_{16}$	6	11	−32
Coronene	$C_{24}H_{12}$	7	12	−36
Ovalene	$C_{32}H_{14}$	10	16	−50

the relationship between aromaticity and ionization potential. Chasey and Aczel make four main points:

1. Sensitivities increase
   (a) with the number of aromatic double bonds in the molecule
   (b) with the number of substituents (xylenes > toluene > benzene)

   (c) with decreasing length and decreasing branchiness of alkyl side chains (xylenes > ethylbenzene > *n*-propylbenzene > isopropylbenzene).

2. Sensitivities of aromatic thiophenes and aromatic furans are equivalent to those of the corresponding aromatic hydrocarbons (benzothiophenes = naphthalenes = benzofurans; dibenzothiophene = phenanthrenes = dibenzofurans). The examples given here are those of Chasey and Aczel. Molecules differing only by the size of an aromatic ring (five or six atoms as in indenes and naphthalenes) apparently have the same sensitivity.
3. Sensitivities of pyrrolic N compounds are higher than those of the corresponding hydrocarbons; those of pyridinic N compounds are lower (pyrroles > benzene > pyridines; indoles > naphthalenes > quinolines).
4. The effect of additional aromatic rings, alkyl substitution, heteroatoms, and so on, decreases with increasing molecular size (pyrroles >> benzenes, but carbazoles > phenanthrenes).

It is common practice to use the response factors published by Lumpkin (7) and to determine others as needed. All are fine-tuned "on the run" to make spectra come out right. As points 1(c) and 4 suggest, the differences in sensitivities are relatively small for the large molecules found in heavy petroleum fractions. The response factors are even more uniform for FIMS and FDMS (see below for definition and description) than for EIMS. Scheppele et al. (8), measured the sensitivities of 60 aromatic compounds and found those for FIMS varying only within about $\pm 20\%$.

In view of the complexity of petroleum fractions, the use of individual response factors would be difficult, or at least very tedious. Therefore, equal response factors are generally assumed for FIMS. This practice is acceptable when it is applied to narrow fractions of specific compound classes, for example, of monoaromatics or diaromatics, and so forth. In this case, the individual response factors are quite similar, and the remaining differences average out. The smoothing is helped by the high and fairly uniform degree of alkyl substitution prevalent even in heavy petroleum fractions if they contain only one compound class or very few closely related ones. Remember that, in most samples, the aliphatic carbon content is distinctly larger than the aromatic carbon content. Even so, the sensitivities of different compound classes can differ greatly. Therefore, in presenting quantitative data, and certainly for broader fractions, the neglect of response factors should be pointed out, leaving open the possibility of a minor tilt in the molar mass distribution.

B. High-Voltage Electron Impact MS

The "hot" ions produced by high-voltage (70–100 eV) EI-MS break up into fragments characteristic of the parent molecules. For instance, a common fragment from monoalkylbenzenes is that of mass 77, C_6H_5, the ring ion left over after dealkylation. Two other typical ones, 51 and 39 (for C_4H_3 and C_3H_3), arise from ring breakage after dealkylation and loss of C_2H_2 and C_3H_2 units. The mass 77 peak is the most intense of these, followed in sequence by those of 51 and 39 amu. Likewise, with monoalkylnaphthalenes, we find peaks of mass 127, 115, 102 (very small), 77, 63, and 51. The first one of these indicates the denuded naphthalene nucleus, the next two, fragments of the latter with one of the rings intact; and the last three, fragments of the ring system without any rings intact. Here, we recognize two of the benzene fragments (77 and 51 amu). Monoalkylphenanthrenes follow the same pattern: mass 177 for the denuded ring system, 151 for a fragment of the latter with two rings intact (loss of C_2H_2), 89 and 76 for one with one ring intact, and 63 and 51 for those with no ring intact. In reality, the fragmentation patterns are generally not as clean as those described here. Often multiple peaks are observed which arise from fragments with one or several H atoms more or less than expected, for example, 79, 78, 77 from benzene, and 77, 76, 75 from phenanthrene.

The alkyl chains break up into fragments, too. Loss of CH_3 gives the corresponding largest fragment; loss of C_2H_5 or C_2H_4, the next lower one. The loss of additional C_2H_4 units leads to more and smaller fragments. Typical ions of paraffinic chains are 15, 29, 43, 57, and so one, differing by 14 mass units (CH_2). In those of sufficient length, the most frequent fragment is the one with four carbon atoms (57), followed by C_3, C_2, and C_5. Thus, the peak of the distribution is C_4, and both sides fall off rapidly, but the high side can extend at a low level all the way to the parent peak. Monocyclic alkanes produce prominent fragments of 69, 83, 97, 111, and so on, amu, that is, homologous series generated by the splitting of a cyclohexane ring and differing by the mass of CH_2.

This brief description of the simplest fragmentation occurring in hydrocarbons may suffice to illustrate the complexity of the EI patterns obtained from petroleum fractions. As shown in Chapter 3, heavy petroleum fractions contain a broad range of molecules spanning MWs from a few hundred to several thousand daltons and chemical species from hydrocarbons to polar heterocompounds. Thus, fragmentation patterns of such fractions are hopelessly crowded, and nonfragmenting MS methods are preferred instead.

EI-MS of narrow fractions, preferably obtained by LC followed by GC, however, can indeed be most revealing. Such fractions contain molecules of similar molecular weight and chemical composition. Thus, their fragmentation patterns are simple enough for quantitative evaluation. They permit then, even in heavy fractions, the structural identification of multiple compounds contributing to a given peak in cases where this is not feasible with nonfragmenting MS. Examples are found mostly in papers with a geochemical orientation as, for example, by Payzant et al. (9–11) and by Zadro et al. (12).

C. Nonfragmenting MS Methods

The great advantage of NF-MS is the relative simplicity of its spectra. A drawback is their relatively low signal intensity. With low-voltage electron impact ionization, the number of parent ions formed increases rapidly as the ionizing voltage increases above the ionizing potential of the molecules in the sample. Thus, higher voltage gives a more intense signal (up to about 20–40 eV, depending on compound type). However, the higher ionizing voltage transfers more energy to the sample molecules which increases fragmentation. The surplus energy transferred to the molecule in excess of the ionizing potential is equilibrated and dissipated in various ways, for example, by increasing the internal energy of the molecule or by breaking atomic bonds. The challenge in NF-MS is to maximize the number of parent ions while keeping the number of fragment ions at acceptable levels.

Paraffins are highly susceptible to fragmentation. An ionizing voltage high enough to generate a good parent ion spectrum also breaks many (paraffinic) C–C bonds, producing a significant number of fragment ions. FIMS and FDMS are the only methods for generating reasonably clean parent ion mass spectra of saturates. Other classes of molecules, such as aromatics, have a wider energy difference between parent ion formation and fragmentation, that is, these classes of molecules can stabilize the excess energy in ways other than fragmentation. Thus, the ionizing voltage can be raised to about 16 eV to increase the number of parent ions formed without significantly increasing the production of fragment ions. They are well behaved and yield clean parent ion spectra with low-voltage MS (LV-MS).

1. FIMS and FDMS

We have had good success in using FIMS to produce parent ion peak spectra of heavy petroleum fractions (13,14). FIMS shows extremely low fragmentation because the ionization process imparts little excess energy to the formed ion. This is crucial when ionizing saturates. Both low-voltage electron impact (LV-EI) and chemical ionization (CI) transfer more excess energy than FIMS, leading to higher levels of fragmentation with saturates. As long as the analyte is volatile, the FIMS experiment is not sample limited. That is, additional sample can be vaporized as needed to obtain the required signal-to-noise ratio. FDMS can be considered a special form of FIMS where the sample material is placed on the ionizing surface of the probe before insertion into the sample chamber of the mass spectrometer. It, however, *is* sample limited and subject to experimental artifacts not normally present in FIMS.

In the FIMS technique, the emitting surface is a cathode consisting of an array of sharp tips on a support such as a wire, a grid, or a razor blade. This surface is located very close to the anode. The combination of an extremely small radius of curvature of the tips, the short distance, and the high electric field (5–10 kV) creates a very high field gradient (10^7–10^8 V/cm). The sample is introduced to this field gradient by evaporation and diffusion from a heated surface nearby. Because of sample losses due to low volatility during transfer from evaporator to ionizer, signal intensity can be lost for heavy molecules (≥1000 amu).

When a sample molecule comes close enough to the cathode or impinges on it, the high electric field removes an electron, converting it into a cation which is then repelled from the surface. The cation is accelerated through a slit into the mass spectrometer. The removal of an electron by the field is described as a "quantum mechanical tunneling effect." The energy transfer in this ionization process is only a few electron volts. Even when coupled with the internal thermal energy of the evaporated molecules, the energy available for bond breakage is limited; hence, little or no fragmentation ensues. A schematic of the ionization chamber of a FIMS spectrometer is shown in Fig. 7.2.

FDMS is a special form of FIMS. The cathode is the same, but here it is coated with the sample before it is placed in the mass spectrometer source. This eliminates the need to evaporate the sample and any losses

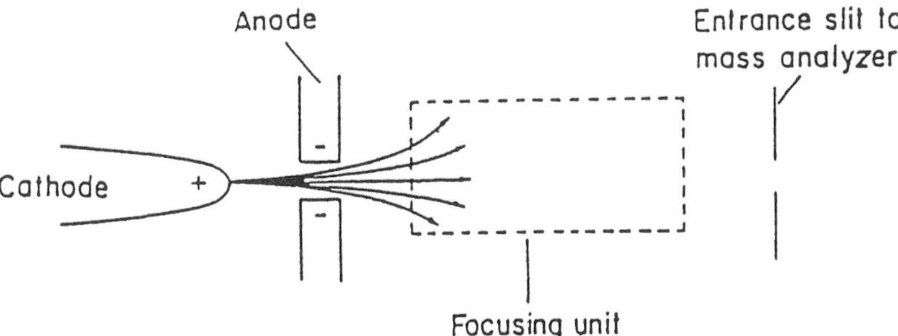

Figure 7.2 Schematic of a FIMS ion source. (From Ref. 15. Reproduced with permission of the publisher.)

on transport to the cathode. Instead, the ions are desorbed directly from the solid sample. The emitter is heated to the point where the sample melts and can be drawn by the surface tension to the tip where the molecules are converted to ions. As the ions are ejected from the surface into the mass spectrometer, more sample molecules can replace them at the tip. In this way, labile molecules can be handled without significant fragmentation, and even conventionally nonvolatile samples can be run. Thus, the method is especially suitable for the high-MW fractions of petroleum residues. Because of the low thermal energy to which the sample molecules are exposed, the total excess energy in this process is minimal. Thus, FDMS is an even milder method than FIMS and produces less fragmentation yet.

All this makes it look ideal. Regrettably though, FDMS also has its disadvantages. Preparing the emitter is an art which takes much experience. More importantly, the sample loading is quite small (less than a few μg), and it is rapidly used up, allowing little time for tuning the spectrometer on a sample before taking the spectrum. Multiple samples must be run and their results added together to minimize the limitations of the small single charge. This makes the technique tedious and time-consuming.

2. Other Soft Ionization Methods

Low-voltage MS is a "soft" electron impact MS method using 10–16 eV instead of the conventional 70–100 eV, thereby avoiding or greatly mitigating fragmentation for any petroleum compound classes other than saturates. Its sensitivity is about the same as that of FIMS, but compared to regular EI-MS, it is 10–100 times lower, and, therefore, the data acquisition time needed for the same signal-to-noise ratio increases proportionally.

Chemical ionization (CI), another soft method, has been used only occasionally for petroleum samples [e.g., by Dzidic et al. (16)]. In CI, a "high-pressure" (0.1–1 Torr) reagent gas (CH_4, CH_4H_{10}, NF_3, etc.) is ionized and allowed to contact the sample. At such pressures, the reagent ions undergo many collisions which help them lose excess energy left after ionization. The ions also exchange charges during collision with one another and with the sample molecules. The latter, of course, is the desired reaction. CI works best when the sample molecules retain the charge better than the reagent molecules.

With methane as reagent gas, CH_5^+ ions are formed, which react with sample molecules in the vapor phase according to the equation

$$CH_5^+ + M \rightarrow CH_4 + (MH)^+$$

The ion $(MH)^+$, which has a mass 1 amu greater than that of the molecular ion M^+, is called a quasi-molecular ion. Depending on the structure, sometimes $(M-H)^+$ ions are formed by elimination of H_2. This variability makes

CI less desirable as an ionization method for complex samples such as petroleum fractions. Other gases used as ionization reagents are ammonia, NF_3, and isobutane. They impart different energies to the sample molecules, for example, Ar about 40 eV and isobutane about 10–16 eV.

But there are other differences. In positive ion CI, three primary processes are normally observed:

- In *charge exchange*, the charge from the reagent is transferred. A portion of the excess energy of the reagent gas ions is also transferred. The latter can be as high as the difference between the ionization potentials of the sample and the reagent. This type of reaction is the only one occurring with rare gases (He, Ar, etc.) as reagents.
- In *proton transfer*, an H^+ ion is transferred between molecules. This is most common when strong bases are employed, for example, CH_5^+.
- In *alkyl transfer*, an alkyl fragment such as CH_3^+, $C_2H_5^+$, and so on, is transferred between the reagent and the sample.

An analogous set of processes exists for negative ion CI. Thus, different gases may react differently with the analyte. However, they all have the same problem of generating more than one type of ion (quasi-ions).

Recently, Dzidic et al. (17) reported an interesting use of this feature, exploiting it to differentiate between hydrocarbon compound classes. According to Wadsworth and Villalanti (18), Dzidic et al. have developed a method with NO ionization, which reacts with aliphatic hydrocarbons to give predominantly or exclusively the $(M - 1)^+$ ion, and with aromatics, the M^+ ion. This chemistry enables them to distinguish between aliphatic and aromatic hydrocarbons without prior separation from each other. Wadsworth and Villalanti combined this technique with GC and a suitable (proprietary) computer program in a method for rapid analysis of middle distillates and VGOs up to 950°F (about 510°C), which is suited even for on-line measurements.

Fast atom bombardment (FAB) MS is a method for special applications, namely, for polar, high-boiling samples when some fragmentation is desired for identification. Fan (19) used it for the characterization of naphthenic acids in six crude oils. The results permitted him to identify several types of naphthenic acids with different *Z*-numbers and to distinguish the oils by their naphthenic acid composition. This seems to be the only published application to petroleum samples to date. Rechsteiner (20) has used it with good success in our laboratories.

For FAB-MS, the sample is dissolved in a high-boiling polar liquid (the "matrix," e.g., glycerol or triethanolamine) and spread on a plate. Rare gas molecules (e.g., Xe or Ar) are ionized, accelerated, focused at the sample, and stripped of their charge before they hit their target. The impact of uncharged atoms sputters small amounts of the matrix and sample off the sample plate. The matrix evaporates, liberating charged sample ions, which are now propelled by the accelerating voltage into the mass spectrometer. FAB produces substantial amounts of fragments and rarely works for neutral compounds, two disadvantages for our purpose. Other problems and interferences with FAB are that the matrix material is ionized along with the sample, and that both sample and matrix material may form cluster ions (MH^+, M_2H^+, M_3H^+, etc.), which greatly complicate the spectrum.

D. Low-Voltage High-Resolution MS

The most useful application of LV-MS is in combination with high-resolution conditions. High resolution is achieved by narrowing the entry slit from the ion source to the mass spectrometer and also the exit slit before the detector. This narrows both the energy spread and the momentum spread in the ions. HR mass spectrometers with multiple sectors of high dispersion have an ultimate resolving power up to about 100,000. Upper resolving powers for useful work on samples in the 300–1000-MW range are about 50,000 or 60,000. For the sake of better signal-to-noise ratios, lower resolutions are employed whenever possible even with these instruments.

Single-sector machines achieve resolutions only up to about 5000. As a consequence of the narrow entry and exit slits, the signal intensity is drastically reduced. Older mass spectrometers, therefore, required very long scan times to produce reliable high-resolution data. New MS technology has shortened the scan times to only a few seconds for a wide mass range.

LVHR-MS not only provides a parent ion spectrum (for fractions free of saturates) as does FIMS or FDMS, but it can distinguish between molecules of equal nominal mass but different specific *Z* series. We saw two examples in Table 7.2. Another simple one would be alkylbenzenes, belonging to the series C_nH_{2n-6}, and alkyl-mononaphtheno-phenanthrenes, belonging to the series C_nH_{2n-20}. For example, decylbenzene and cyclopentylphenanthrene, both of 218 MW, are members of these two series. Having the same nominal mass but with an actual mass difference of 0.0939 amu, they can be differentiated by a mass spectrometer with a resolving power of ≥2325. "Routine" FIMS can be done up to about 5000 resolving power. Thus, our 218 peak is still within the limits of FIMS. But peaks requiring a resolution of >5000 must be left to LVHR-MS.

LVHR-MS can even distinguish between a nitrogen atom and a CH_2 group for molecules of moderate size. These two groups have a mass difference of only 0.0126 amu. A resolution of about 35,000 would be sufficient to distinguish ^{14}N from CH_2 in a molecule of mass 430, such as those in Table 7.2. Yet, even the resolving power of 50,000 is not enough to overcome some of the problems encountered with sulfur-containing compounds or with some of the pairs resolvable in small but not in large molecules. Other unresolvable cases are caused by the presence of ^{13}C isotopes or by the presence of more than one heteroatom per molecule.

Table 7.5 shows some more examples. It lists the maximum molar masses at which certain doublets encountered in petroleum can be distinguished with various resolutions. The first of the doublets, C–12H, indicates the amu difference of compounds whose composition differ by 1 carbon in one and 12 hydrogens in the other. Similarly, the second doublet indicates the mass difference between, for example, decalin, $C_{10}H_{18}$, and ethyl thiophenol, $C_8H_{10}S$, both having a nominal molar mass of 138, but an actual mass difference of 0.0905 amu. As we can see from Table 7.5, a minimal mass resolution MS cannot resolve any of these doublets in a heavy petroleum fraction (>650°F or >345°C AEBP, corresponding to a MW of ~300 daltons). A mass spectrometer with resolution of 1 in 10,000 does well only for the first two doublets. The other doublets need 1 in 50,000 or more.

Figure 7.3 shows a representative HR-FIMS mass spectrum. In this case, the sample was the vacuum gas oil (VGO) from Altamont crude. In Figures 7.4A and B, the mass scale was expanded 30- and 500-fold, respectively, over that in Figure 7.3. The partial spectrum in Figure 7.4B illustrates the power of high resolution (~1 in 25,000). The peak on the right has a subpeak on its right side which differs from the main peak by about 0.09 amu, corresponding roughly to the ^{12}C–12 H doublet in Table 7.5. Despite the obviously large difference in abundance, the two peaks are well resolved.

Table 7.5 Maximum Molar Mass at Which Doublets Are Resolved with Various Resolutions (R)

		Maximum molar mass		
Doublet	ΔM	R = 1000[a]	R = 10,000	R = 50,000
C–12H	0.0939	94	939	4695
C_2H_8–^{32}S	0.0905	91	905	4525
CH_4–O	0.0364	36	364	1820
C_3–$^{34}SH_2$	0.0165	17	165	825
CH_2–N	0.0126	13	126	630
^{13}CH–N	0.0081	8	81	405
C_3–$^{32}SH_4$	0.0034	3	34	170

[a]Nominal mass resolution.

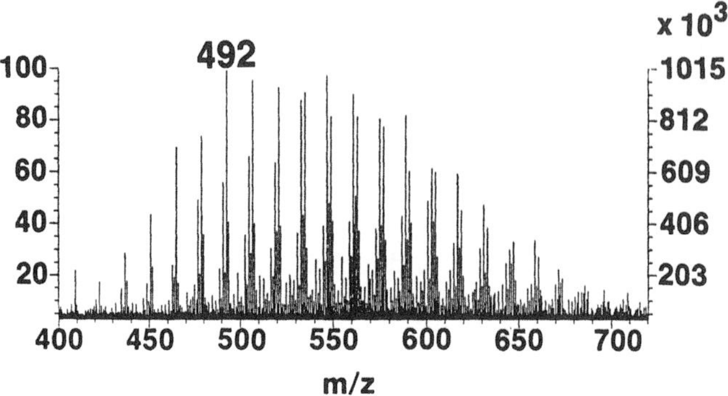

Figure 7.3 HR-FIMS spectrum of VGO from Altamont crude. (From Ref. 14.)

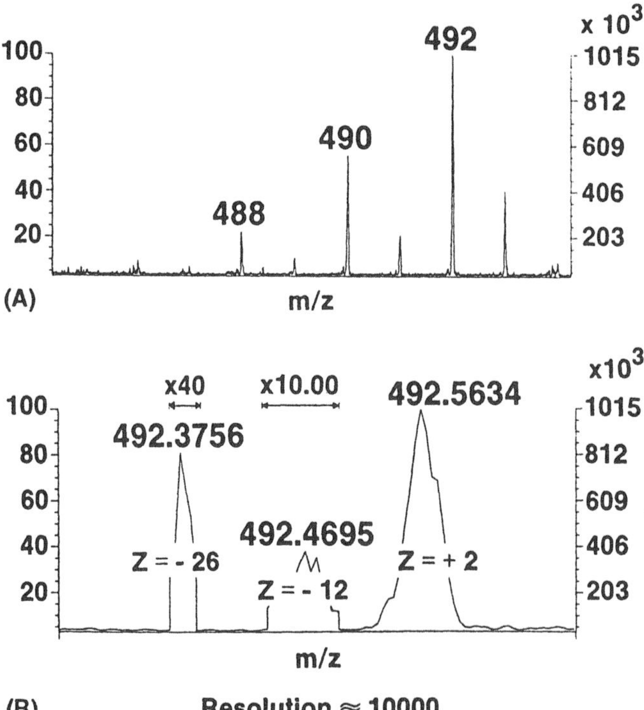

Figure 7.4 Sections of HR-FIMS spectrum of Fig. 7.3 with expanded mass scales. (A) 30-fold expansion, (B) 500-fold expansion. (From Ref. 14.)

Chasey and Aczel (6) discussed an application of LVHR-MS in combination with highly sophisticated evaluation methods in which they limited their instrument to a resolution of 15,000–20,000. They distinguished between species, unresolvable by HR-MS alone, by means of reasoning based on additional information, for example, from elemental analysis and simulated distillation. An example of their reasoning is the conclusion that mass peaks of 134–176 must represent benzothiophenes ($C_nH_{2n-10}S$) rather than mononaphthenophenanthrenes (C_nH_{2n-20}), which have the same nominal mass, but in higher boiling ranges. Because the lowest member of the latter has a MW of 190,* they are out of the disputed range and can be excluded from consideration. Table 7.6 illustrates this distinction and shows a number of examples for mononaphthenophenanthrene structures which would fit the pattern. Each of these, in turn, may have one or more alkyl substituents. This is another illustration of the variety of structures that may be represented by the same formula.

Chasey and Aczel (6) used their extended evaluation method to list the abundances of about 5000 carbon number homologues in 140 homologous series, including aromatics with one to seven rings, naphthenoaromatics, benzo- and higher aromatic thiophenes, benzo- and higher aromatic furans, aromatic nitrogen compounds, and miscellaneous SO-, S_2-, NO-, NO_2-, and N_2-containing compounds. Table 7.7 shows an example of their computer output. They further calculated the average carbon number in side chains for each molecule and also for the different *Z* series. This can be done for simple molecules from molar mass data and the number of aromatic and naphthenic rings with only minor assumptions. Large molecules, as found in nondistillable residues, would not lend themselves to this analysis without more information. This subject will be discussed later in more detail.

For high-resolution mass spectrometers, precise standardization of operating conditions is even more important for quantitative measurements than for regular mass spectrometers. Generally, the higher the resolution, the weaker the signal and the longer the data acquisition time. Special mass standards, mostly chlorinated hydrocarbon petroleum fractions, are employed for mass calibration. The parent petroleum fractions guarantee a complete mass sequence which is needed for the large number of het-

---

*Benzothiophenes would ordinarily coelute with the diaromatics fraction rather than with triaromatics as suggested by Chasey and Aczel's example. The distinction of benzothiophenes from alkylbenzenes, which have the same nominal mass but different *Z*-numbers ($Z = -6$; $Z = -10S$, where S stands for sulfur), by HR-MS is easy. Here, we have a good example of the value of HR-MS in complementing fractionation, whereas before we had shown how fractionation helps out in the interpretation of MS data.

Table 7.6 Several Benzothiophenes and Naphthenophenanthrenes as Examples of Molecules with Equivalent *Z*-Number (−10S and −20) But Different Mass Ranges in Chasey and Aczel's Reasoning (See Text)

$C_nH_{2n-10}S$					C_nH_{2n-20}			
Structure	C	H	S	Mass	Structure	C	H	Mass
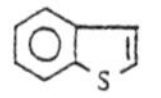	8	6	1	134				
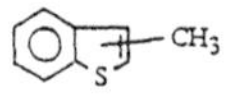 CH_3	9	8	1	148				
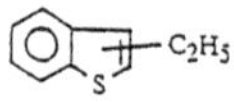 C_2H_5	10	10	1	162				
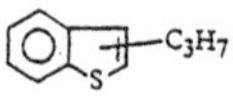 C_3H_7	11	12	1	176				
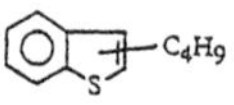 C_4H_9	12	14	1	190	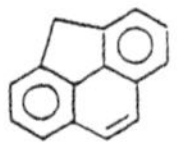	15	10	190
etc.					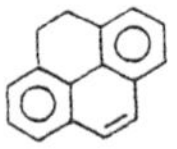	16	12	204
					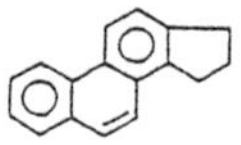	17	14	218
					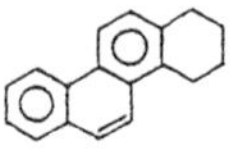	18	16	232
					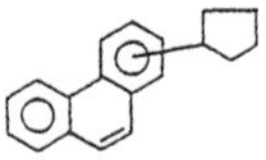	19	18	246
					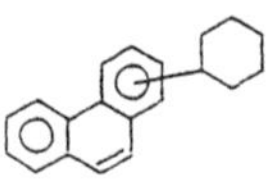	20	20	260
					etc.			

erocompound peaks. Even with such a standard there are still 10–20 heterocompound peaks between each pair of hydrocarbon peaks. The chloro compounds are preferred because of their higher sensitivities at the low ionization voltage compared to those of the fluorinated standards common with regular EI-MS.

The additional peaks in LVHR-MS require more intricate computer programs for data evaluation. Johnson and Aczel (21), Aczel et al. (22), and Aczel (23) discussed their computation technique in general terms but did not divulge details of their program. Now, Schmidt et al. (5) developed their own computer program based on Johnson and Aczel's general discussion and made it available to the general public.

E. Combined MS Methods

1. GC/MS

Gas chromatography coupled with MS (GC-MS) is a powerful method for petroleum distillates. Heavy petroleum fractions (from 650 to about 850°F, 345–450°C) are usually first separated by LC into compound-class fractions before application of GC, which separates the sample more or less by boiling point. MS, used as a detection method, gives the molecular weight and, with fragmenting ionization, the compound type of each peak. Zadro et al. (12), for instance, separated 263 peaks from several Australian crude oils by capillary GC and were able to identify more than 370 of the major alkane, naphthene, and aromatic hydrocarbons present in their samples up to C_{31}.

Petroleum samples boiling higher than about 850°F (450°C) AEBP, other than well-separated narrow fractions, are too complex for GC, which separates by boiling point and cannot differentiate between compound classes. Here, NF-MS must be combined with LC instead of GC. The most elegant way to do this is on-line, that is, in an integrated LC-MS unit. Otherwise, LC fractions are prepared separately, and each is submitted individually to NF-MS. Very high-boiling samples (>1200°F, >650°C AEBP) may not be seen with the same sensitivity as lower boiling ones by most MS methods. In part, this is due to slow or incomplete evaporation. For these cases, MS methods not limited by volatility are now available, for instance, FDMS, laser desorption (LS), fast atom bombardment (FAB), and electrospray (ES) techniques. These can be applied to very large (polar) molecules, even of several thousand molecular weight. However, because FAB and ES work only with highly polar, easily ionizable compounds, their application to petroleum fractions is limited.

GC/MS with its combination of high separation efficiency and molecular characterization has become a major tool in geochemistry where specific

Table 7.7 Example of Chasey and Aczel's Computer Output

Series	HRLV-MS aromatics analysis—carbon no. distribution, weight percent on stream										
carbon no.	C_nH_{2n}	C_nH_{2n-2}	C_nH_{2n-4}	C_nH_{2n-6}	C_nH_{2n-8}	C_nH_{2n-10}	C_nH_{2n-12}	C_nH_{2n-14}	C_nH_{2n-16}	C_nH_{2n-18}	C_nH_{2n-20}
1	0.0	0.0	0.0	0.0	0.0	0.0	0.0	0.0	0.0	0.0	0.0
2	0.0	0.0	0.0	0.0	0.0	0.0	0.0	0.0	0.0	0.0	0.0
3	0.0	0.0	0.0	0.0	0.0	0.0	0.0	0.0	0.0	0.0	0.0
4	0.0	0.0	0.0	0.0	0.0	0.0	0.0	0.0	0.0	0.0	0.0
5	0.0	0.0	0.0	0.0	0.0	0.0	0.0	0.0	0.0	0.0	0.0
6	0.0	0.0	0.0	0.0	0.0	0.0	0.0	0.0	0.0	0.0	0.0
7	0.0	0.0	0.0	0.0	0.0	0.0	0.0	0.0	0.0	0.0	0.0
8	0.0	0.0	0.0	0.0	0.0	0.0	0.0	0.0	0.0	0.0	0.0
9	0.0	0.0	0.0	0.0	0.0	0.0	0.0	0.0	0.0	0.0	0.0
10	0.0	0.0	0.0	0.0	0.0	0.0	0.0	0.0	0.0	0.0	0.0
11	0.0	0.0	0.0	0.0	0.0	0.0	0.0	0.0	0.0	0.0	0.0
12	0.0	0.0	0.0	0.0	0.0	0.0	0.0	0.006	0.002	0.0	0.0
13	0.0	0.0	0.0	0.0	0.0	0.0	0.0	0.011	0.007	0.0	0.0
14	0.0	0.0	0.0	0.0	0.0	0.0	0.0	0.017	0.018	0.011	0.0
15	0.0	0.0	0.0	0.0	0.0	0.0	0.024	0.028	0.032	0.019	0.0
16	0.0	0.0	0.0	0.0	0.0	0.0	0.028	0.031	0.044	0.042	0.005
17	0.0	0.0	0.0	0.0	0.0	0.0	0.025	0.044	0.055	0.116	0.011
18	0.0	0.0	0.0	0.0	0.0	0.0	0.032	0.057	0.086	0.151	0.073
19	0.0	0.0	0.0	0.024	0.043	0.038	0.037	0.073	0.121	0.150	0.135
20	0.0	0.0	0.0	0.26	0.053	0.044	0.065	0.087	0.180	0.175	0.203
21	0.0	0.0	0.0	0.054	0.069	0.065	0.087	0.090	0.243	0.213	0.239
22	0.0	0.0	0.0	0.091	0.152	0.153	0.152	0.214	0.325	0.258	0.234
23	0.0	0.0	0.0	0.163	0.214	0.164	0.143	0.205	0.330	0.268	0.184

24	0.0	0.0	0.0	0.232	0.312	0.266	0.197	0.328	0.360	0.291	0.182
25	0.0	0.0	0.0	0.291	0.419	0.332	0.306	0.381	0.364	0.281	0.179
26	0.0	0.0	0.0	0.310	0.453	0.300	0.268	0.312	0.290	0.264	0.311
27	0.0	0.0	0.0	0.291	0.331	0.433	0.353	0.420	0.417	0.247	0.096
28	0.0	0.0	0.0	0.257	0.328	0.338	0.239	0.228	0.292	0.222	0.002
29	0.0	0.0	0.0	0.251	0.261	0.271	0.190	0.203	0.223	0.144	0.040
30	0.0	0.0	0.0	0.206	0.256	0.202	0.156	0.160	0.188	0.162	0.176
31	0.0	0.0	0.0	0.117	0.141	0.129	0.112	0.135	0.145	0.149	0.087
32	0.0	0.0	0.0	0.087	0.126	0.117	0.104	0.111	0.108	0.087	0.030
33	0.0	0.0	0.0	0.060	0.073	0.080	0.070	0.102	0.122	0.077	0.021
34	0.0	0.0	0.0	0.033	0.051	0.057	0.051	0.056	0.071	0.057	0.033
35	0.0	0.0	0.0	0.034	0.051	0.041	0.036	0.041	0.044	0.046	0.019
36	0.0	0.0	0.0	0.011	0.033	0.019	0.025	0.041	0.043	0.038	0.029
37	0.0	0.0	0.0	0.010	0.023	0.028	0.020	0.018	0.036	0.025	0.019
38	0.0	0.0	0.0	0.009	0.011	0.010	0.019	0.017	0.026	0.022	0.013
39	0.0	0.0	0.0	0.0	0.0	0.0	0.0	0.013	0.020	0.020	0.0
40	0.0	0.0	0.0	0.0	0.0	0.004	0.0	0.008	0.0	0.009	0.017
41	0.0	0.0	0.0	0.0	0.0	0.0	0.0	0.0	0.0	0.0	0.013
42	0.0	0.0	0.0	0.0	0.0	0.0	0.0	0.0	0.0	0.0	0.0
43	0.0	0.0	0.0	0.0	0.0	0.0	0.0	0.0	0.0	0.0	0.0
44	0.0	0.0	0.0	0.0	0.0	0.0	0.0	0.0	0.0	0.0	0.0
45	0.0	0.0	0.0	0.0	0.0	0.0	0.0	0.0	0.0	0.0	0.0
46	0.0	0.0	0.0	0.0	0.0	0.0	0.0	0.0	0.0	0.0	0.0
47	0.0	0.0	0.0	0.0	0.0	0.0	0.0	0.0	0.0	0.0	0.0
48	0.0	0.0	0.0	0.0	0.0	0.0	0.0	0.0	0.0	0.0	0.0
49	0.0	0.0	0.0	0.0	0.0	0.0	0.0	0.0	0.0	0.0	0.0
50	0.0	0.0	0.0	0.0	0.0	0.0	0.0	0.0	0.0	0.0	0.0

Source: From Ref. 6. Reproduced with permission of the publisher.

single-biomarker compounds are of prime interest. Petroleum chemists use GC/MS and LC/MS to determine the numerous homologues of compound types in various petroleum-related samples: chromatographic fractions of petroleum distillates or residues (e.g., Refs. 24–28), pyrolysis products (10,29–31), and converted refinery streams or reactor products (32). Gallegos et al. (33) analyzed C_{40}–C_{60} saturates, high-MW polynuclear aromatics and metalloporphyrins in crudes and deposits by high-temperature GC-MS. Computers are employed for precise instrument control as well as for data evaluation.

2. LC/MS

LC/MS is more difficult than GC/MS in that the solvent needs to be removed before the eluate can be processed by the mass spectrometer. Hsu et al. (24) and McLean and Hsu (34) used a moving belt interface for this purpose. Here, the effluent is sprayed continuously onto a polyimide belt (3 mm wide) which traverses several heated and pumped chambers for solvent removal. Ultimately, it passes through the ionization chamber of the mass spectrometer where the sample is vaporized and can be ionized by any method, including FAB and FI. The principle of the moving belt interface is illustrated in Fig. 7.5.

Thermospray is another recent technique for the transfer of eluate without solvent into the mass spectrometer. Vestal et al. (36) reported a new improved version for use by itself or in combination with electron impact or chemical ionization. A schematic diagram is shown in Figure 7.6. The

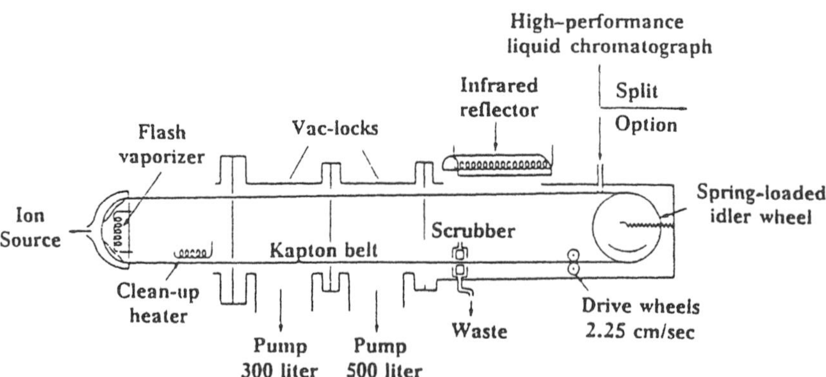

Figure 7.5 Schematic of a moving belt interface. (From Ref. 36. Reproduced with permission of the publisher.)

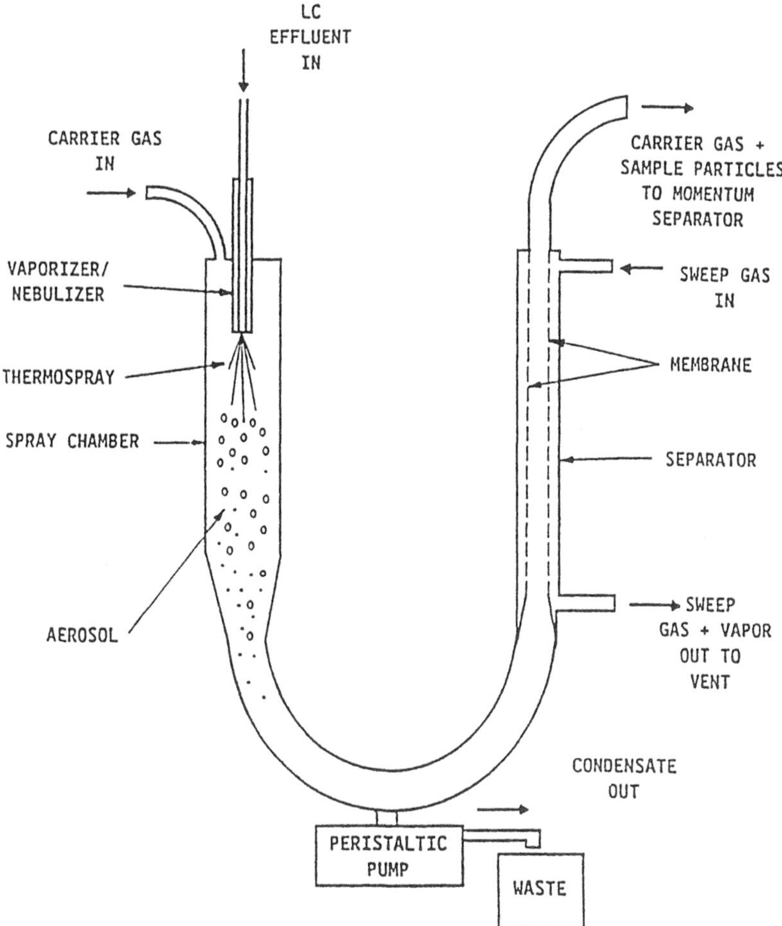

Figure 7.6 Schematic of a thermospray unit. (From Ref. 36. Reproduced with permission of the publisher.)

column eluate is nebulized by an inert carrier gas (helium) in the spray chamber. The solvent is vaporized by this process, and some of the solute is ionized. In the next section, most of the solvent is condensed on the walls and pumped out.

The remaining solvent is then removed in a countercurrent membrane separator. Finally, the dry sample aerosol is freed of excess helium in a momentum separator so that it can be introduced into the mass spectrometer without overwhelming the vacuum system. The thermospray method by itself produces protonated cations without fragmentation and, thus, a very clean parent ion spectrum from easily ionizable (very polar) compounds (36). Other samples need additional ionization by EI or CI. Covey et al. (37) warned that (earlier) thermospray techniques did not detect all the molecules present in a (biological) sample.

The great advantage of thermospray is the gentle heat treatment which makes it particularly useful for very high-boiling and temperature labile samples. Thus, for heavy (650–1000°F, 345–540°C) and very heavy (>1000°F) petroleum fractions, it seems better suited than the moving belt method. But, so far, we have not seen any application to petroleum samples. More information on present LC/MS interfaces in general terms is found in a review by Cairns and Siegmund (35).

A detailed survey of the state of the art in LC/MS was published by Covey et al. (37). Besides methodology, these authors also describe applications, though none in our field of petroleum chemistry. They discuss such interfaces as the moving belt, direct liquid introduction, thermospray, atmospheric pressure ionization, heated pneumatic nebulizer, electrospray, MAGIC (monodisperse aerosol generated for introduction of liquid chromatographic effluents), LC/MS/MS, and SFC/MS (supercritical fluid chromatography/MS). In an earlier review on LC/MS, Arpino and Guichon (38) discussed some principal subjects as the pros and cons of on-line LC/MS and those of a few interfaces then available, the choice of column size and running conditions, and some splitter designs.

The on-line LC-MS combination has several advantages over the separate applications of these techniques. It saves time and effort in that it automates solvent evaporation and minimizes the amount of solvent. In addition, because the column effluent is continuously monitored, it allows the distinction and quantitation of different overlapping compound series by the use of selected ion chromatograms.

Ion chromatograms are very useful as demonstrated in Fig. 7.7. Here we see two chromatograms, one with a solid line for all the molecules with mass 380.25 and one with a dotted line for those with mass 380.34 (~4000 resolution). Looking just at the individual mass spectra, it would be difficult to differentiate between the compound classes. The visual image of the ion chromatograms makes it easy to see the differences between these ions. Five examples of compound types distinguished by this technique are indicated in this figure.

Another set of examples of compounds with equal nominal mass but different structure are shown in Fig. 7.8. Again, LC/MS can easily differentiate the mono-, di-, and triaromatics because of the chromatographic separation. However, the molar masses of the two diaromatic compounds

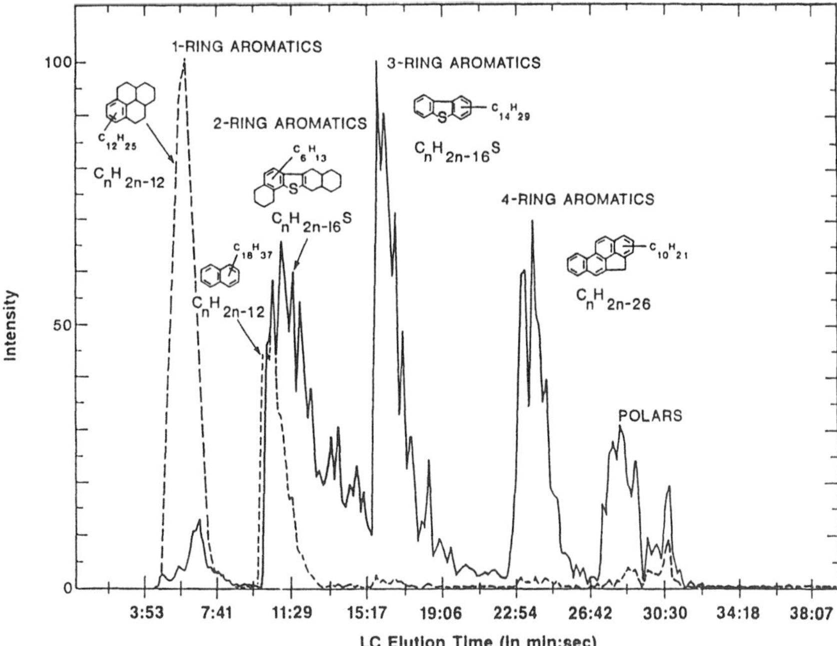

Figure 7.7 Selected ion chromatograms for masses 380.25 (solid line) and 380.34 (dotted line). (From Ref. 24. Reproduced with permission of the publisher.)

are too close for the resolving power of the mass spectrometer (the required resolution would be >125,000). Here, we have an example of the present limitation of this powerful technique.

3. MS/MS

For simplicity, in this discussion we will describe the MS/MS approach with a triple quadrupole system. Similar information may be obtained with other MS/MS configurations, but the outline of the experiment is less clear.

MS/MS, or tandem MS, is a way to combine the simple parent ion spectrum obtained by NF-MS with the discriminative power of fragmenting MS. In concept, the MS-MS experiment uses one stage of MS to select an ion of interest which is then reacted by collision with neutral molecules. The transferred translational energy causes the selected ion to break up into fragments similar to those obtained by EI ionization. The fragment ions are analyzed in the second MS stage.

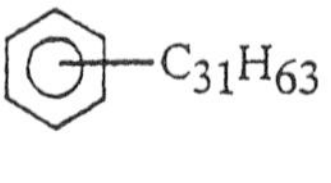

$-C_{31}H_{63}$

$C_{37}H_{68}$
MW 512.532

1-Ring aromatic

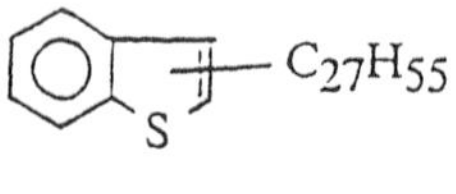

$-C_{27}H_{55}$

$C_{35}H_{60}S$
MW 512.442

2-Ring aromatic

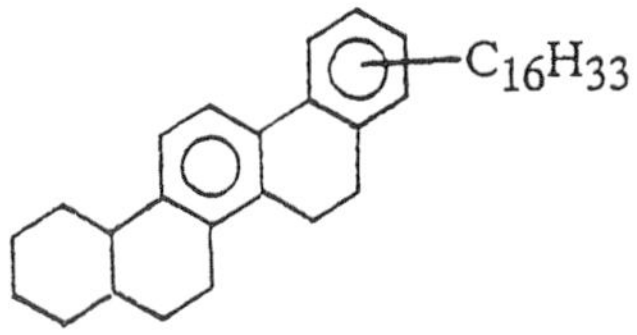

$-C_{16}H_{33}$

$C_{38}H_{56}$
MW 512.438

2-Ring aromatic

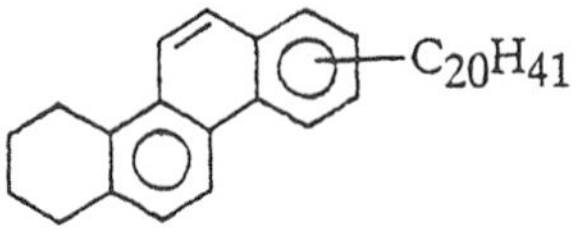

$-C_{20}H_{41}$

$C_{38}H_{56}$
MW 512.438

3-Ring aromatic

Figure 7.8 Possible structures of compounds with a nominal mass of 512. Examples from Ref. 24.

The data can be evaluated in three basic ways: (1) determining all or some of the fragments belonging to some parent ion; (2) determining all the parent ions belonging to some fragment ion; and (3) determining all parent ions which show the same neutral loss, that is, those parent ions which differ from a significant fragment ion by a fixed mass. Easiest to understand is option (1). The first MS stage gives a parent ion pattern, and the second, a series of fragmentation patterns, one for each ion selected from the first spectrum. Because the fragments all come from a single molecular composition, all isotope effects are eliminated. Each fragmentation pattern is, therefore, simple and readily interpreted. The measurement is not restricted to just one parent ion. Usually, it is performed for a set of parent ions, but only a few of these may be later selected for evaluation from the computer bank in which all the results were stored.

In option (2), the inverse measurement is performed. The data are collected from all of the parent ions which produce a specific fragment. The search can be repeated to find all the parent ions belonging to other fragment ions.

The same principle holds for option (3). The spectral data are collected by linking the parent ions to the fragment ions by a fixed mass which is chosen to be specific for a known substructure. For example, a mass of 63 units may be chosen as a diagnostic for catacondensed aromatics. Again, multiple fragments can be examined during an experiment.

F. MS Group-Type Methods

One of the applications of mass spectrometry is group-type MS. It uses fragment patterns and sets of empirical relations to sort the contributions from different molecules to certain selected peaks and to reconstruct the molecular distribution from these peak areas.

The ASTM D-2786 method, for instance, measures the concentration of eight compound types—paraffins, naphthenes with one to six rings, and monoaromatics—in the saturates fractions. The application of this method is restricted to petroleum fractions having an average carbon number between 16 and 32. The ASTM MS group-type method D-3239 determines aromatic compound types present in chromatographic fractions of aromatics separated from vacuum gas oils. Other ASTM group-type methods determine various aliphatic and aromatic compound types in middle distillates (D-2425, D-2625).

The 22-component hydrocarbon type analysis by Gallegos et al. (39; see also Ref. 40) covers 8 types of aliphatic hydrocarbons, 3 types each of mono- and di-aromatics, 2 types each of tri- and tetraaromatics, and 4 types of thiophenes. It was designed for the analysis of unfractionated petroleum distillates covering a boiling range from 350 to 1050°F AEBP (175–565°C). In contrast to the earlier group-type methods, it requires the use of a high-resolution mass spectrometer. Recently, Bouquet and Brument (41) disclosed a 33-component method for atmospheric and vacuum gas oil fractions boiling up to 1200°F (650°C) AEBP. Their method covers aliphatic hydrocarbons up to tetracycloalkanes ($Z = 2$ to $Z = -6$), aromatics up to $Z = -38$, including three types of tetraaromatics and two types of pentaaromatics, and S compounds up to $Z = -42$S including seven types of disulfides.

MS group-type methods are fast, relatively cheap, and require only a few milligrams sample. However, they have their limits. They are restricted to samples free of olefins and, ordinarily, with less than 2–5% S, N, or O compounds and in some cases, with much more stringent limits. Other restrictions, for example, to certain boiling (and MW) ranges also apply, depending on the method. Furthermore, only those compound types are considered in the calculation for which results are given in the output. If any other compound types are present in the sample to any large degree, which may easily be the case, major errors may result. The methods were developed for certain types of petroleum fractions as mentioned before. Different types, greater amounts of single hydrocarbons, thermally unstable materials, or other compound-type distributions can cause large deviations.

The MS group-type analysis makes use of average fragmentation patterns specific to the different compound types, that is, to certain groups of molecular structures. But the masses, as well as the abundances, of fragment ions are characteristic of the specific parent molecule. Although, as a class, monocyclic alkanes produce prominent fragments of 69, 83, 97, 111, etc. amu, different monocyclic alkane isomers may only show some of these peaks, and these with differing peak heights. The sum of all of the peaks gives the average distribution of the parent molecules. Generally, a limited set of peaks of a molecule is selected for evaluation. Calibration ensures that the system is operating in a reproducible fashion. A response factor matrix approach is used for quantitation and to compensate for overlaps in fragmentation, that is, for those fragment peaks composed of contributions from several different compound types (of the same molar mass). For example, the fragments characteristic for paraffins can also come from the fragmentation of "long" alkyl chains on aromatics.

Selected fragment peak intensities within the homologous series of compound types are added together, and the sums are evaluated with a matrix of response factors to give the concentration of each compound type in the sample. The more powerful MS group-type methods use several computer generated matrices based on a distillation curve of the sample. Ashe and Colgrove (42) described the principle of this evaluation in simple terms. More details of the procedure are found in a review by Teeter (40).

Several assumptions must be made. The most important one may be that the response matrix will cover all samples as, for instance, with D-2786. This is not always true, depending on the compound-type distribution. Another assumption concerns the compound types selected as present in a sample and those ignored.

MS group-type analyses can be expected to have errors in the order of ±20%. However, *relative* changes are ordinarily reported with better accuracy (40). For instance, two related samples containing 18% dicycloalkanes and 9% tricycloalkanes can reliably be considered to contain di- and tricycloalkanes in a ratio of 2:1 even though the absolute amounts may be off by 20% or more. If, however, the sample has a composition different from that assumed for the method, the errors may approach 100%. Occasional testing by the more tedious method of chromatographic separation combined with other independent techniques, for example, FIMS, will minimize misleading results.

A typical output sheet of the 22 × 22-component group-type method can be seen in Fig. 7.9. The results are normalized to 100%. The left-hand column lists the chemical formula of a species, the middle column its name (revealing the compound type), and the right-hand one its percentage in the sample. A number of compound types known to be present in petroleum-derived fractions are not included. For example, among "monoaromatics," trinaphthenobenzenes ($Z = -12$) such as monoaromatized steranes and/or decahydropyrenes, and tetranaphthenobenzenes ($Z = -14$) such as monoaromatized hopanes are missing. Similarly, among "diaromatics" only three Z series (−12, −14, and −16) are included. For "triaromatics," only two Z series are covered (−18 and −22), whereas a very likely one (see FIMS data for triaromatics in Section III.A) represented by mononaphthenophenanthrenes/anthracenes (i.e., −20) is missing. The "tetraaromatic" compound types include only Z series −24 and −28, but miss such important ones as −22 (e.g., pyrenes/fluoranthenes) and −26 (e.g., dihydrobenzopyrenes). Obviously, compounds other than those covered by the matrix make their contributions to measured fragment ions, and because all data are normalized to 100%, the results of an analysis may be very inaccurate.

Note the reminder at the bottom of a report sheet (Fig. 7.9), pointing out some of the limitations of the method.

III. APPLICATIONS

In this section, we describe two specific analytical methodologies, which show representative information accessible by NF-MS.

A. FIMS of VGO Compound-Class Fractions

This example illustrates the application of FIMS to the characterization of saturates, aromatics, and some nitrogen-compound fractions separated from vacuum gas oils (650–1000°F or 345–540°C AEBP). Let us first look at the FIMS spectra of three saturates fractions in Fig. 7.10.

Even at first glance, it is obvious that these samples are quite different from each other. Seven molar mass series can be discerned in each spectrum by their hydrogen deficiency, or Z-number. We will refer to them as the "apparent Z series," which we have assigned to specific compound types. The peaks marked with the open circles belong to the $Z = 2$ series,

```
        TWENTY-TWO COMPONENT HYDROCARBON TYPE ANALYSIS

RUN NO:   H18071                  ACQUIRED AT 15:58 ON 11/01/83
SER NO:   CH                      ANALYZED AT 15:50 ON 12/04/83
SAMPLE:   SGQ 4117
CONDS.:   MULT425, MAG975, SM214, HV294+57, REP432
SUBMITTED BY: RMTE                ANALYST: CRABS                ACCT. NO.: 58008

       C(N)H(2N+2)     PARAFFINS                        40.2
       C(N)H(2N)       MONOCYCLOPARAFFINS               16.9
       C(N)H(2N-2)     DICYCLOPARAFFINS                  7.3
       C(N)H(2N-4)     TRICYCLOPARAFFINS                 3.6
       C(N)H(2N-6)     TETRACYCLOPARAFFINS               0.0
       C(N)H(2N-8)     PENTACYCLOPARAFFINS               0.0
       C(N)H(2N-10)    HEXACYCLOPARAFFINS                0.0
       C(N)H(2N-12)    HEPTACYCLOPARAFFINS               0.0

      TOTAL NAPHTHENES                                  27.7
                                             SATURATES       68.0

    MONOAROMATICS
       C(N)H(2N-6)     ALKYLBENZENES                    11.7
       C(N)H(2N-8)     BENZOCYCLOPARAFFINS               3.0
       C(N)H(2N-10)    BENZODICYCLOPARAFFINS             0.8
    DIAROMATICS
       C(N)H(2N-12)    NAPHTHALENES                      3.3
       C(N)H(2N-14)                                      0.8
       C(N)H(2N-16)                                      1.5
    TRIAROMATICS
       C(N)H(2N-18)                                      1.1
       C(N)H(2N-22)                                      0.6
    TETRAAROMATICS
       C(N)H(2N-24)                                      0.4
       C(N)H(2N-28)                                      0.0
                                             AROMATICS       23.1

       C(N)H(2N-4)S    THIOPHENES                        0.0
       C(N)H(2N-10)S   BENZOTHIOPHENES                   4.6
       C(N)H(2N-16)S   DIBENZOTHIOPHENES                 4.4
       C(N)H(2N-22)S   NAPHTHOBENZOTHIOPHENES            0.0
                                       SULFUR COMPOUNDS       9.0

CALCULATED %C= 85.6   %H= 13.2   %THIOPHENIC S=  1.2

PLEASE NOTE--
   SAMPLE TYPES CORRECTLY ANALYZED
    1. C12 TO C36, 250 DEG F TO 1050 DEG F HYDROCARBONS
    2. OLEFIN-FREE HYDROCARBONS
    3. LESS THAN 5% OXYGEN, NITROGEN, OR SULFUR COMPOUNDS.
   ONLY LISTED TYPES ARE CONSIDERED, ALL OTHERS ARE IGNORED
   FACTORS WHICH CONTRIBUTE TO AN INCORRECT ANALYSIS
    1. LARGE AMOUNTS OF A SINGLE HYDROCARBON
    2. UNUSUAL DISTRIBUTION OF COMPOUNDS
    3. THERMALLY UNSTABLE MATERIALS
```

Figure 7.9 Typical output sheet of the 22 × 22-component group-type method. (From Ref. 40. Reproduced with permission of the publisher.)

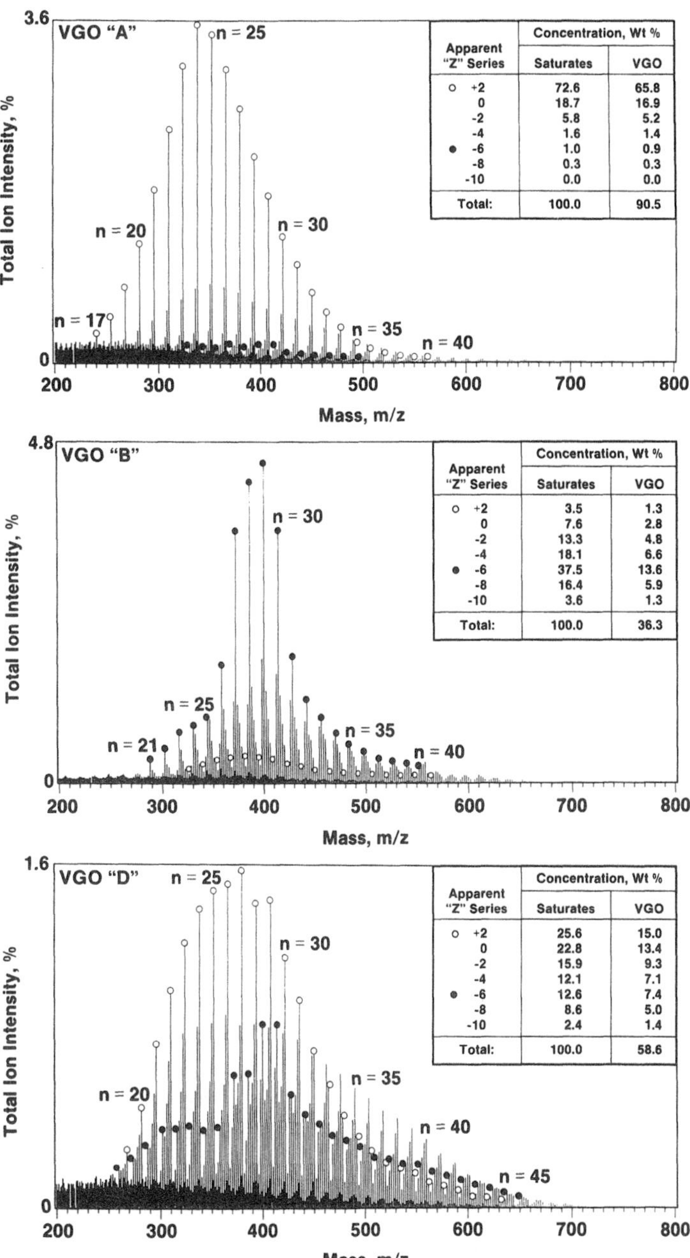

Figure 7.10 Field ionization mass spectra of saturates from three vacuum gas oils. (From Ref. 13. Reproduced with permission of the publisher.)

represented by paraffins. This series dominates in VGO "A"; it is hardly present in VGO "B." The nearest series of peaks to the left of the $Z = 2$ peaks have masses lower by two units and belongs to the $Z = 0$ series, represented by mononaphthenes. Moving leftward, the next series of peaks belongs to the $Z = -2$ series (dinaphthenes), and the following ones, to $Z = -4$ (trinaphthenes). The peaks to the left of those, $Z = -6$ (tetranaphthenes), are marked with filled circles. These dominate in VGO "B" and are also quite prominent in VGO "D." Two more Z series can be discerned from these spectra. These are $Z = -8$ and $Z = -10$, represented by pentanaphthenes and hexanaphthenes, respectively.

The FIMS spectra in Fig. 7.10 allow us to see the different molar mass distributions among the main aliphatic types in the three VGOs. For instance, in VGO "A" and "D," the mononaphthenes are the second most abundant species after the paraffins, but in VGO "B," the tri- and pentanaphthenes predominate after the tetranaphthenes. Without FIMS analysis, these differences would have escaped us.

Similar differences in abundance are obvious among the aromatic ring-number fractions of Fig. 7.11. Caution must be exercised when interpreting FIMS results for aromatics because these fractions, unlike saturates, may contain significant amounts of aromatic sulfur compounds which coelute with aromatic hydrocarbons during chromatographic separation. Sulfur compounds have the same integer (nominal) molar mass as hydrocarbons and, thus, cannot be distinguished by nominal mass resolution FIMS.

The aromatic fractions in our example (Fig. 7.11) did not exceed 0.05 wt% S for fractions from VGO "A" and 1.0 wt% for fractions from VGO "B." The concentration of 1.0 wt% sulfur in a fraction having an average molecular weight of 400 translates into 12.5% of sulfur compounds (assuming one sulfur atom per molecule). Thus, some of the apparent Z series in FIMS of aromatic fractions from VGO "B" are likely to be represented by both aromatic hydrocarbons and aromatic sulfur compounds. In our discussion of the results in Fig. 7.11, we will focus our attention on hydrocarbon compound types.

In the top spectra of Fig. 7.11, we can see that the alkylbenzenes ($Z = -6$) and mononaphthenobenzenes ($Z = -8$) predominate in the monoaromatics of VGO "A" with 39.1 and 25.0 wt%, respectively, whereas the dinaphthenobenzenes ($Z = -10$, 24.2%) and trinaphthenobenzenes ($Z = -12$) are the most abundant compound type in the monoaromatics from VGO "B." It is important to point out, however, that monoaromatics account for only 4.28 wt% of VGO "A" and 16.10 wt% of VGO "B" (24.2 and 22.0 wt%, respectively). The data illustrate the great complexity of these compound-class fractions. The complexity of the entire VGO is truly immense. For example, the alkylbenzenes in VGO "A" (accounting

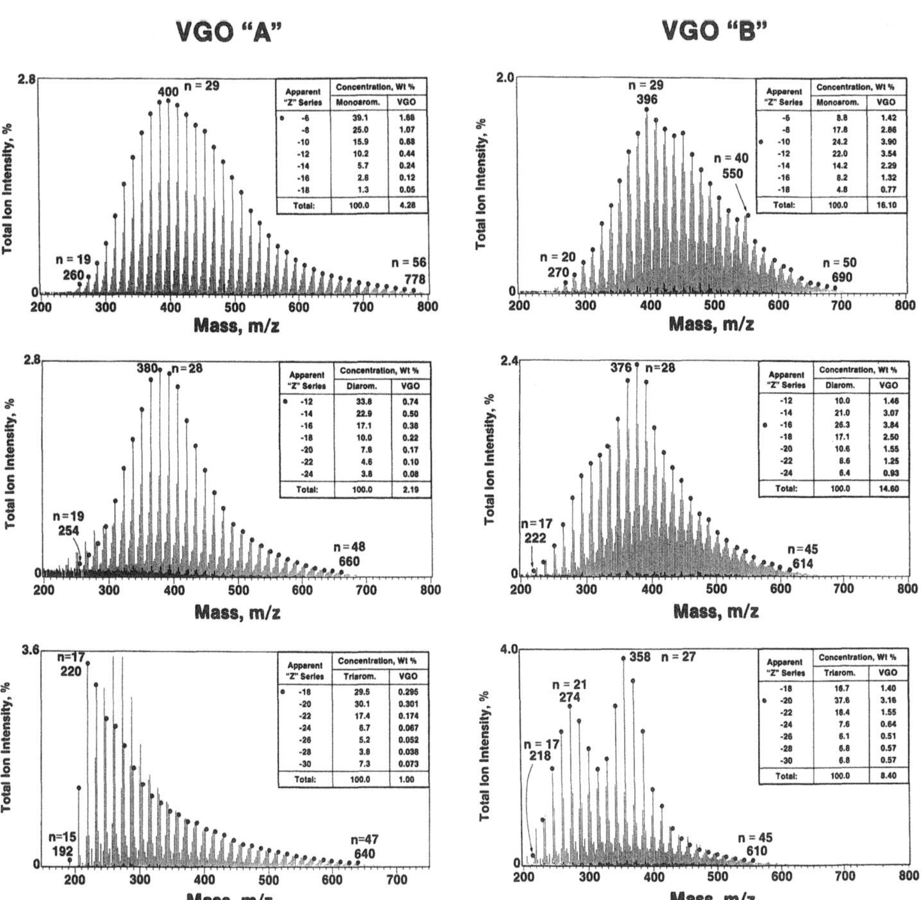

Figure 7.11 Field ionization mass spectrum of mono-, di-, and triaromatics from VGOs "A" and "B." (From Ref. 13. Reproduced with permission of the publisher.)

for only 1.68 wt% of the VGO "A") range from C19 to C56, that is, they encompass 37 carbon number members. The most abundant one, C29, constitutes for only about 0.12 wt% of VGO "A." All others account for even smaller fractions of the VGOs. Moreover, each carbon number member of this series may be represented by a large number of isomers.

The FIMS spectra of diaromatics (Fig. 7.11, center) are equally diverse as those of the monoaromatics. They reveal that the alkylnaphthalenes ($Z = -12$) are the most abundant compound type (33.8 wt%) in the diaromatics of VGO "A," whereas the dinaphthenonaphthalenes ($Z = -16$) are the major ones (26.3 wt%) in those from VGO "B." The concentrations of these two Z series in the parent VGOs are only 0.74 wt% and 3.84 wt%, respectively. Again, each series covers a broad carbon number range, and each carbon number member may be represented by a large number of isomers.

The triaromatics from both VGOs (Fig. 7.11, bottom) have distinctly narrower and lower molar mass ranges than the diaromatics and especially the monoaromatics. Interestingly, the $Z = -20$ series in the triaromatics from VGO "B" shows a distinctly bimodal distribution, which most likely indicates that two different homologous series may be present, for example, dihydropyrenes and triaromatic steranes. The $Z = -20$ series predominate in the triaromatics from both VGOs (30.1% in VGO "A" and 37.6% in VGO "B"). It is important to note, however, that the $Z = -20$ series accounts for only 0.3 wt% of VGO "A" and that it covers over 30 carbon number members (C15 to C47), with the most abundant one, C17, at only about 300 ppm. Once again, despite a certain ambiguity in terms of specific compound types and their molecular structure, the detail of compositional information provided by this analytical approach (i.e., HPLC and FIMS) is apparent. Group-type MS methods would not have given the correct information for either of these VGOs because of the limitations discussed in Section II.F.

Two examples, shown in Fig. 7.12, demonstrate the utility of FIMS for the characterization of nitrogen-compounds in VGO boiling range distillates. The two N-compound class fractions were separated from a VGO by liquid chromatography on basic and acidic alumina as described in Chapter 6, Section IV.C.3. The pyrrolic N-compound fraction produced by this separation method consists almost exclusively of five-membered nitrogen heterocycles with one pyrrolic >NH group. This composition was confirmed in three ways, by HPLC with model compounds, by FTIR, and by 14N NMR. The other N-compound fraction consists of basic N compounds with predominantly six-membered nitrogen heterocycles (again confirmed by HPLC with model compounds and by 14N NMR). Such complementary information is essential for the correct interpretation of the FIMS spectra of heterocompound fractions such as these two.

The first and perhaps the most significant observation is the fact that these spectra consist of odd-numbered peaks (both spectra have been corrected for 13C contributions), which are characteristic of nitrogen compounds with one nitrogen atom per molecule. (Strictly speaking, any odd number of nitrogen atoms per molecule could account for odd numbered peaks, but one nitrogen per molecule was consistent with the results of elemental analysis in this case.)

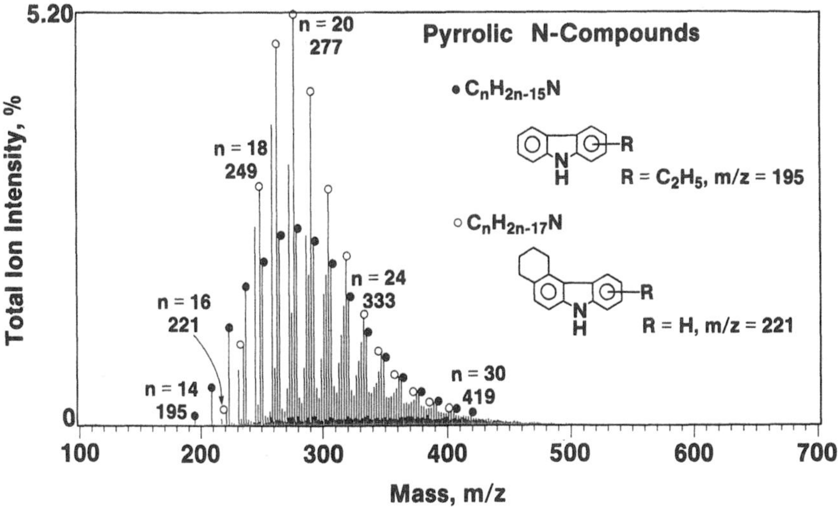

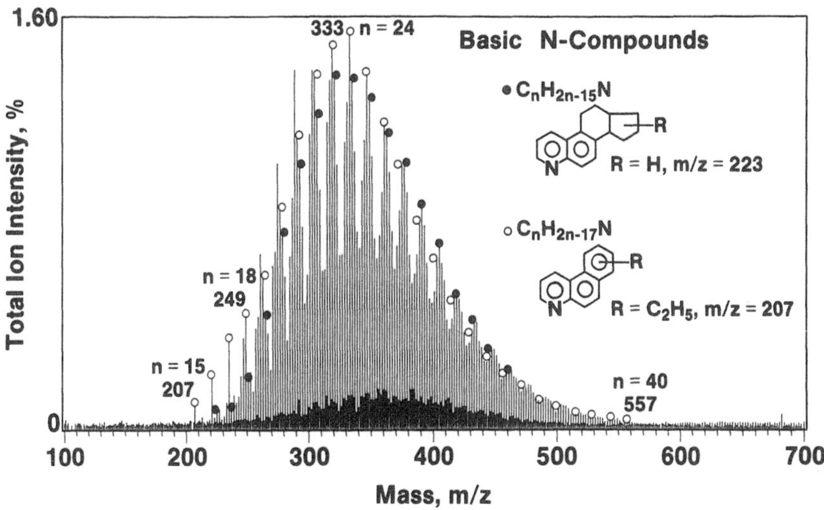

Figure 7.12 Field ionization mass spectra of pyrrolic and basic N-compounds from a VGO. (From Ref. 13. Reproduced with permission of the publisher.)

The second important observation is the predominance of the $Z = -17N$ and $Z = -15N$ series in both spectra. The ones in the FIMS spectrum of pyrrolic N compounds (Fig. 7.12, top) can be readily assigned to mononaphthenocarbazoles and carbazoles, respectively. The ones in the FIMS spectrum of basic N compounds represent nitrogen in six-membered rings such as the examples shown in the bottom part of Fig. 7.12. Without chromatographic separation and additional complementary information (elemental analysis, FTIR, and 14N NMR), the correct assignment of compound types to these series would have been impossible.

Third, note again the narrower molar mass distribution and the shift toward lower values compared with the spectra for saturates and aromatics. In contrast to lower-boiling fractions, VGOs contain molecules with more highly condensed ring structures and polar groups which, therefore, have lower molar masses for their boiling point than paraffins and molecules with simpler ring structures. These observations support our contention of increasing diversity with increasing AEBP, expressed in Chapters 2 and 3.

B. HRLV-MS of High-Boiling Fractions

The application of relatively simple accounting techniques assisted by computers allow the computation of average compositional data of a sample from the HRLV-MS spectrum (6,23). Each molar mass is assigned a chemical formula which determines its compound type within the limitations discussed before. Thus, the compound-type distribution can be calculated. Other data accessible in this fashion are as follows:

1. Weight percent of each homologous series and compound type (group-type determination)
2. Average MW and carbon number as well as carbon number in side chains for each homologous series and for the entire aromatic fraction, however broadly or narrowly this is defined
3. MW distribution; Z-number distribution; %C, H, S, N, O; and even simulated distillation results of the aromatic fractions

Within certain limits, this technique can be applied to whole distillation cuts. The errors caused by the fragmentation of saturates may, to some degree, be corrected by compensatory calculations.

The full power of this method is revealed in an example provided by Chasey and Aczel (6), which shows the effect of upgrading procedures on petroleum fractions. These authors presented data obtained on a lube oil base stock before and after three types of treatment:

1. Low severity extraction, low severity hydrotreatment
2. Low severity extraction, high severity hydrotreatment
3. High severity extraction, no hydrotreatment

Table 7.8 compares the ring number distributions of the original base stock fraction and the three refined products obtained from it.* The results clearly delineate the effects of the different treatments, whereas the average numbers, shown in Table 7.9, do not reveal these effects at all. Even more detail can be seen in Table 7.10, which lists the weight percent of numerous compound types in the four samples. Here the specific effects of the refining procedures on the diverse chemical groups become very evident. The impact of extraction on the number of carbon atoms in side chains is nicely illustrated in a plot like that of Fig. 7.13.

The advantages of on-line LC-MS are evident in a paper by Hsu et al. (24). This group analyzed a high sulfur VGO (650–950°F or 340–510°C

Table 7.8 Ring Number Distributions of a Lube Oil Base Stock and Three Refined Products Produced from It

	Weight percent[a,b]			
Component	Dewaxed distillate	Sample A	Sample B	Sample C
	Hydrocarbon ring systems			
1 ring	9.05	13.93	16.94	17.69
2 ring	10.37	11.48	13.25	11.56
3 ring	5.89	5.41	5.10	4.22
4 ring	2.49	1.76	2.51	1.39
5+ ring	0.80	0.31	0.81	0.63
Total	28.60	32.89	38.61	35.49
	Thiopene ring systems			
1 ring	Traces	Traces	Traces	Traces
2 ring	6.72	2.87	Traces	Traces
3 ring	6.99	5.49	2.80	1.62
4 ring	1.32	0.77	0.33	0.00
5+ ring	0.31	0.06	0.00	0.00
Total	15.34	9.19	3.13	1.62
	Furan ring systems			
Total	0.21	0.40	Traces	0.30
Grand total	44.15	42.47	41.74	37.40

[a]All values normalized to reflect concentrations in whole oils.
[b]Traces means less than 0.005%; 0.00 means not detected, less than 0.001% (10 ppm).
Source: Ref. 6.

*All samples were dewaxed before mass spectrometry.

Table 7.9 Average Compositional Values of Chasey and Aczel's Feed and Refined Products

	Feed (dewaxed distillate)	Refined products		
		A	B	C
Avg. molecular weight	333	340	330	353
Avg. *Z*-number	14.8	13.6	13.3	12.4
Avg. carbon number	24.8	25.5	24.9	26.5
Avg. carbon atoms in side chains	12.2	13.5	13.3	15.5
%C	86.58	87.33	88.53	88.35
%H	9.88	10.49	10.63	11.18
%S	3.51	2.13	0.84	0.43
%O	0.03	0.05	0.00	0.04

Source: Ref. 6.

AEBP) fraction on a dinitroanilino propyl column connected with a LVHR mass spectrometer by means of a moving belt interface. The column separated the sample by ring number, and the MS continuously monitored the effluent. The instrument was set at 200 ppm resolution (1 in 5000), high enough for the desired resolution of thiophenes from hydrocarbons of equal aromatic ring size and low enough for good sensitivity.

Selected ion chromatograms (SICs) permitted the monitoring of ions of the same nominal mass, for example, 380, but different structures. Figure 7.8 shows the SICs of masses 380.25 and 380.34. The former can be either $Z = -16$S dibenzothiophenes or $Z = -26$ or -12 hydrocarbons, for example, mononaphthenotetraaromatics and perhaps tetranaphthenotriaromatics, and the latter, mainly trinaphthenomonoaromatics and alkylnaphthalenes, both with $Z = -12$. A few representative structures are shown for the main peaks in Fig. 7.7. Another way of displaying the main components in the mono-, di-, tri-, and tetraaromatics is that of Fig. 7.14.

Although these compound types could be distinguished under the experimental conditions, there are others which cannot be differentiated this way. The distinction of $Z = -26$ hydrocarbons from $Z = -16$S compounds would require a resolution of about 1 in 50,000, and that of dinaphthenotriaromatic hydrocarbons and benzothiophenes, a resolution of about 1 in 150,000. Both of the latter compound types are triaromatic and, therefore, elute together.

These remarks are not meant to distract from the accomplishment of the authors or the tremendous potential of on-line LC-LVHR-MS. Con-

Table 7.10 Weight Percent and *Z*-Numbers of Compound Types in a Lube Oil Base Stock and Three Products Produced from It

			Weight percent[a,b]			
Z-no.	Possible structure(s)	Aromatic ring no.	Dewaxed distillate	Sample A	Sample B	Sample C
	Hydrocarbon ring systems					
8	Indans, tetralins	1	3.40	6.45	6.35	6.90
10	Dinaphthenobenzenes	1	3.09	4.30	5.55	5.23
12	Naphthalenes	2	2.74	3.48	4.13	3.92
14	Acenaphthenes, diphenyls	2	3.44	3.68	4.37	3.60
16	Fluorenes	3	4.19	4.32	4.75	4.03
18	Phenanthrenes	3	3.54	3.14	3.02	2.34
10	Naphthenophenanthrenes	3	2.35	2.26	2.08	1.89
22	Pyrenes, fluoranthenes	4	1.62	0.80	1.01	0.83
24	Chrysenes	4	0.70	0.89	1.13	0.42
26	Cholanthrenes, benzo[*h,i*]fluoranthenes	5	0.17	0.07	0.37	0.14
28	Dibenzopyrenes, benzofluoranthenes	6	0.19	0.05	0.64	0.48
30	Benzchrysenes	6	0.20	0.13	0.09	0.13
32	Benzo[*ghi*]perylenes	6	0.15	0.01	0.09	0.02
34	Dibenzopyrenes	6	0.12	0.03	0.00	0.01
36	Coronenes	7	0.08	0.06	0.00	0.00
38			0.04	0.01	0.00	0.00
40	Additional seven- and eight-ring condensed		0.03	0.01	0.00	0.00
42	aromatic hydrocarbons		Traces	Traces	0.00	0.00
44			Traces	0.00	0.00	0.00
	Thiophene ring systems					
8/S	Dinaphthenothiophenes	1	Traces	0.00	0.00	0.00
10/S	Benzothiophenes	2	3.48	1.24	0.00	0.00
12/S	Naphthenobenzothiophenes	2	1.98	1.46	0.00	0.00
14/S	Dinaphthenobenzothiophenes	2	1.34	0.17	0.00	0.00
16/S	Dibenzothiophenes	3	4.11	3.12	2.65	1.61

Table 7.10 (Continued)

Z-no.	Possible structure(s)	Aromatic ring no.	Weight percent[a,b]			
			Dewaxed distillate	Sample A	Sample B	Sample C
18/S	Thiophenoacenaphthenes	3	1.92	1.76	0.01	0.00
20/S	Thiophenofluorenes	4	0.96	0.61	0.14	0.01
22/S	Thiophenophenanthrenes	4	1.04	0.58	0.33	0.00
24/S	Thiophenonaphthenophenanthrenes	4	0.29	0.19	0.00	0.00
26/S	Thiophenopyrenes	5	0.10	0.02	0.00	0.00
28/S	Thiophenochrysenes	5	0.03	0.01	0.00	0.00
30/S	Thiophenocholanthrenes	6	0.13	0.03	0.00	0.00
32/S	Thiophenobenzpyrenes	6	0.04	Traces	0.00	0.00
34/S	Thiophenobenzochrysenes	6	Traces	0.00	0.00	0.00
	Furan ring systems					
8/O	Dinaphthenofurans	1	0.00	0.00	0.00	0.02
10/O	Benzofurans	2	Traces	0.00	0.00	0.05
12/O	Naphthenobenzofurans	2	Traces	0.00	0.00	0.00
14/O	Dinaphthenobenzofurans	2	0.01	0.00	0.00	0.03
16/O	Dibenzofurans	3	0.08	0.13	0.00	0.10
18/O	Naphthenodibenzofurans	3	0.05	0.12	0.00	0.07
20/O	Fluorenofurans	4	0.01	0.03	0.00	0.02
22/O	Phenanthrenofurans	4	0.03	0.07	0.00	0.00
24/O	Naphthenophenanthrenofurans	4	0.02	0.04	0.00	0.00
26/O	Pyrenofurans	5	0.01	0.01	0.00	0.00
28/O	Chrysenofurans	5	Traces	Traces	0.00	0.00
30/O	Cholanthrenofurans	6	0.00	Traces	0.00	0.00
32/O	Benzopyrenofurans	6	0.00	Traces	0.00	0.01
34/O	Benzochrysenofurans	6	Traces	0.00	0.00	0.00
36/O	Benz[*ghi*]perylenofurans	7	Traces	Traces	0.00	0.00

[a]All values normalized to reflect concentrations in whole oils.
[b]Traces means less than 0.005%; 0.00 means not detected, less than 0.001% (10 ppm).

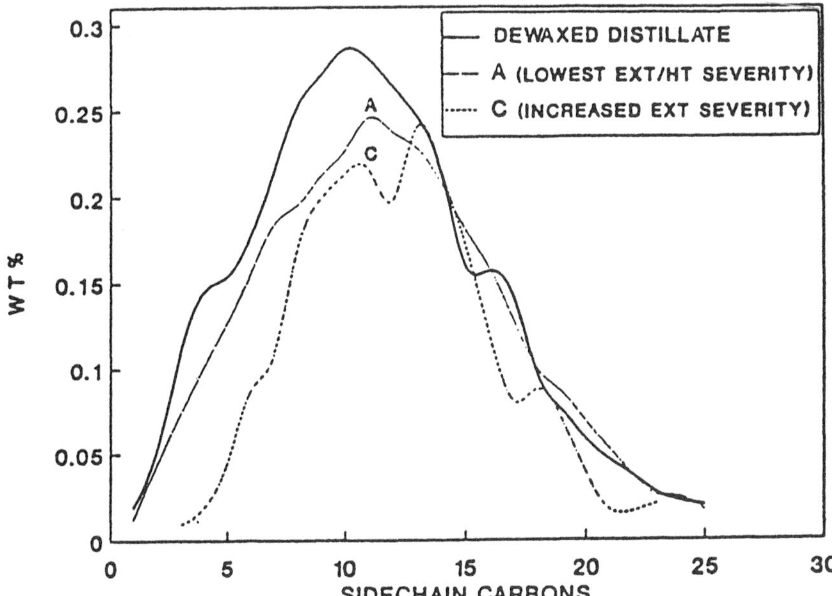

Figure 7.13 Abundance of compounds with different carbon numbers in side chains showing the impact of extraction. (From Ref. 6. Reproduced with permission of the publisher.)

sidering the speed of the analysis and the enormous wealth of data, this method is a remarkable tool. It is important, though, as with any method, to keep in mind its (present) limitations.

The immense complexity of the fraction is evident in Fig. 7.15 which shows the LV mass spectrum of one scan of the LC effluent in the diaromatic region, that is, of a rather narrow fraction (in structural rather than molecular-weight terms). This chromatogram spans a range from $<C_{19}$ to $>C_{60}$ and probably exhibits several thousand peaks, each of which is likely to contain several or even many isomers. Another scan in the same region may well display several other series of compound types. On the one hand, we have come a long way in the analysis of heavy petroleum fractions, in terms of both speed and detail, but obviously we are still far from being able to measure the entire composition of even the lightest of the heavy fractions.

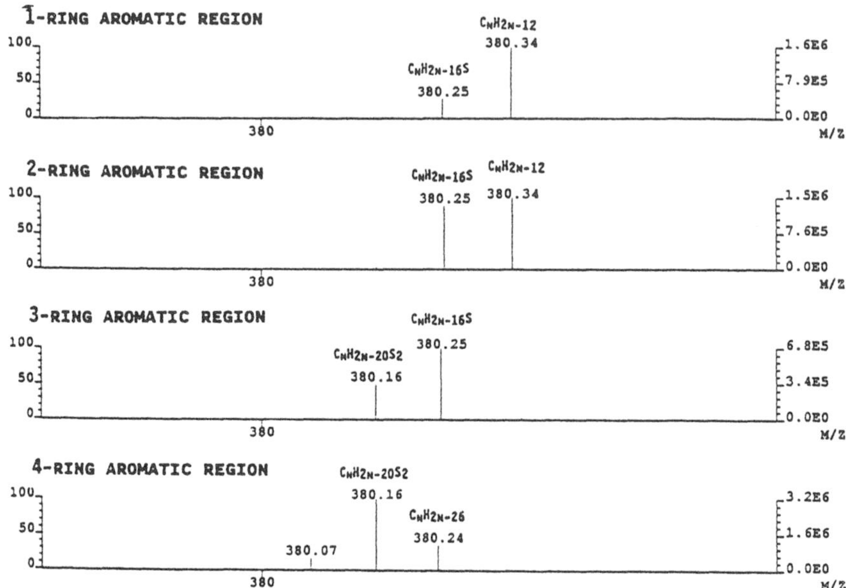

Figure 7.14 Distribution of molecular ions with a nominal mass 340 among the mono-, di-, tri-, and tetraaromatics from a VGO. (From Ref. 24. Reproduced with permission of the publisher.)

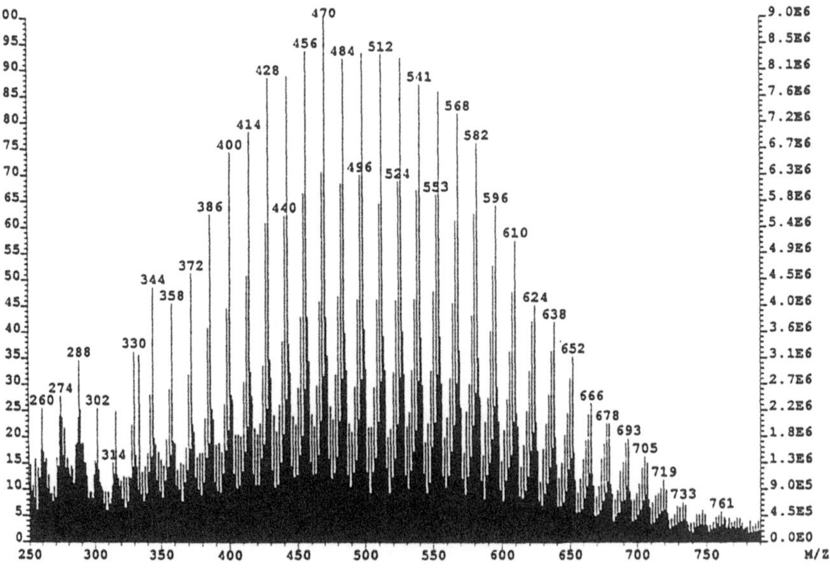

Figure 7.15 Low-voltage medium resolution (5000) mass spectrum of one scan in the diaromatics region of an on-line LC-MS experiment. (From Ref. 24. Reproduced with permission of the publisher.)

REFERENCES

1. Beynon, J. H., and Williams, A. E. 1963. *Mass and Abundance Tables for Use in Mass Spectrometry*. Elsevier, Amsterdam.
2. Petrakis, L., Allen D. T., Gavales, G. R., and Gates, B. C. (1983). Analysis of synthetic fuels for functional group determination. *Anal. Chem.* **55**:1557–1564.
3. Roboz, J. 1968. *Introduction to Mass Spectrometry, Instrumentation and Techniques*. Interscience, New York.
4. Watson, J. T. 1985. *Introduction to Mass Spectrometry*. Raven Press, New York.
5. Schmidt, C. E., Sprecher, R. F., and Batts, B. D. 1987. Low voltage, high resolution mass spectrometric methods for fuel analysis application to coal distillates. *Anal. Chem.*, **59**:2027–2033.
6. Chasey, K. L., and Aczel, T. 1991. Polycyclic aromatic structure distribution by high-resolution mass spectrometry. *Energy Fuels*, **5**:386–394.
7. Lumpkin, M. E. 1958. Low voltage techniques in high molecular weight mass spectrometry. *Anal. Chem.*, **39**:682–685.
8. Scheppele, S. E., Grizzle, P. L., Greenwood, G. J., Marriott, T. D., and Perreira, N. B. 1976. Determination of field-ionization relative sensitivities for the analysis of coal-derived liquids and their correlation with low-voltage electron-impact relative sensitivities. *Anal. Chem.*, **48**:2105–2113.
9. Payzant, J. D., Hogg, A. M., Montgomery, D. S., and Strausz, O. P. 1985. A field ionization mass spectrometric study of the maltene fraction of Athabasca bitumen. Part I—The saturates. *AOSTRA J. Res.*, **1**:175–182.
10. Payzant, J. D., Montgomery, D. S., and Strausz, O. P. 1988. The identification of homologous series of benzo[b]thiophenes, thiophenes, thiolanes and thianes possessing a linear carbon framework in the pyrolysis oil of Athabasca asphaltene. *AOSTRA J. Res.*, **4**:117–131.
11. Payzant, J. D., Lown, E. M., and Strausz, O. P. 1988. Structural units of Athabasca asphaltene: The aromatics with a linear carbon framework. *AOSTRA J. Res.*, **5**:445–453.
12. Zadro, S., Haken, J. K., and Pinczewski, W. V. 1985. Analysis of Australian crude oils by high resolution gas chromatography-mass spectrometry. *J. Chromatogr.*, **323**:305–322.
13. Boduszynski, M. M. 1988. Composition of heavy petroleums. 2. Molecular characterization. *Energy Fuels*, **2**:697–613.
14. Rechsteiner, C. E., and Boduszynski, M. M. 1990, 1991. Unpublished work.
15. Davis, R., and Frearson, M. 1987. *Mass Spectrometry*. John Wiley & Sons, Chichester.
16. Dzidic, I., Balicki, M. D., Petersen, H. A., Nowlin, G. D., Evans, W. E., Siegel, H., and Hart, H. V. 1991. A study of catalytic hydrodenitrification by the use of ammonia chemical ionization GC/MS and GC/MS/MS. *Energy and Fuels*, **5**:382–386.
17. Dzidic, I., Peterson, H. A., Wadsworth, P. A., and Hart, H. V. 1992. Town-

send Discharge Nitric Oxide Chemical Ionization Gas Chromatography/Mass Spectrometry (TDCINO) for Hydrocarbon Type Analysis of Middle Distillates. *Anal. Chem.*, **64**:2227–2232.

18. Wadsworth, P. A., and Villalanti, D. C. 1992. Pinpoint hydrocarbons types. *Hydrocarbon Processing*, 109–112 (May).
19. Fan, T.-P. 1991. Characterization of naphthenic acids in petroleum by fast atom bombardment mass spectrometry. *Energy Fuels*, **5**:371–375.
20. Rechsteiner, C. E. 1986, 1987. Unpublished work.
21. Johnson, B. H., and Aczel, T. 1967. Analysis of complex mixtures of aromatic compounds by high-resolution mass spectrometry at low ionizing voltages. *Anal. Chem.*, **39**:682–685.
22. Aczel, T., Allen, D. E., Harding, J. R., and Knipp, E. A. 1970. Computer techniques for quantitative high-resolution mass spectral analyses of complex hydrocarbon mixtures. *Anal. Chem.*, **42**:341–347.
23. Aczel, T. 1973. Anwendungen der hochaufloesenden massenspektrometrie bei der analyse von erdoel und kohlederivaten. *Erdoel Kohle*, **26**(1):27–31.
24. Hsu, C. S., McLean, M. A., Qian, K., Aczel, T., Blum, S. C., Olmstead, W. N., Kaplan, L. H., Robbins, W. K., and Schulz, W. W. 1991. On-line liquid chromatography/mass spectrometry for heavy hydrocarbon characterization. *Energy Fuels*, **5**:395–398.
25. Frakman, Z., Ignasiak, T. M., Montgomery, D. S., and Strausz, O. P. 1987. Nitrogen compounds in Athabasca asphaltene: The carbazoles. *AOSTRA J. Res.*, **3**:131–138.
26. Nishioka, M., Campbell, R. M., Lee, M. L., and Castle, R. N. 1986a. Isolation of sulfur heterocycles from petroleum- and coal-derived materials by ligand exchange chromatography. *FUEL* **65**:270–273.

26a. Nishioka, M., and Lee, M. L. 1986b. Determination and structural characteristics of sulfur heterocycles in coal- and petroleum-derived materials. *Amer. Chem. Soc. Div. Petr. Chem.* **31**(2):827–833.

27. Aczel, T., Colgrove, S. G., and Reynolds, S. D. 1985. High-resolution mass spectrometric analysis of coal liquids. *Amer. Chem. Soc. Div. Fuel Chem.*, **30**:209–220.
28. Aczel, T., Williams, R. B., Brown, R. A., and Pancirov, R. J. 1978. Chemical Characterization of Synthoil Feeds and Products. *Analytical Methods for Coal and Coal Products*, edited by C. Karr Jr., Academic Press, Vol. 1, Chapter 17, pp. 499–540.
29. Mojelsky, T. W., Montgomery, D. S., and Strausz, O. P. 1985. Catalyzed oxidation of high molecular weight components of Athabasca oil sand bitumen. Non-distillable ester residue. *AOSTRA J. Res.*, **2**(2):131–137.
30. Mojelsky, T. W., Montgomery, D. S., and Strausz, O. P. 1986. The side chains associated with the undistillable aromatic and resin components of Athabasca bitumen. *AOSTRA J. Res.*, **2**(3):177–184.
31. Mojelsky, T. W., Montgomery, D. S., and Strausz, O. P. 1986. Pyrolytic probes into the aliphatic core of oxidized Athabasca asphaltene non-distillable ester residue. *AOSTRA J. Res.*, **3**(1):43–51.

32. Dzidic, I., Balicki, M. D., Peterson, H. A., Nowlin, J. G., Evans, W. E., Siegel, H., and Hart, H. V. 1991. A study of catalytic hydrodenitrification by the use of ammonia chemical ionization GC/MS and GC/MS/MS. *Energy Fuels*, **5**:382–386.
33. Gallegos, E. G., Fetzer, J. C., Carlson, R. M., and Peña, M. M. 1991. High-temperature GC/MS characterization of porphyrins and high molecular weight saturated hydrocarbons. *Energy Fuels*, **5**:376–381.
34. McLean, M. A., and Hsu, C. S. 1990. Combination of Moving Belt LC/MS and Low Voltage Electron Impact Ionization for the Characterization of Heavy Petroleum Streams. *The 38th ASMS Conference on Mass Spectrometry and Allied Topics*, pp. 605–606.
35. Cairns, T., and Siegmund, E. G. 1990. Review of the Development of Liquid Chromatography/Mass Spectrometry. *Liquid Chromatography/Mass Spectrometry*, edited by M. A. Brown. ACS Symposium Series 420, American Chemical Society, Washington, DC, Chapter 1, pp. 1–11.
36. Vestal, M. I., Vestal, C. H., and Wikes, J. G. 1990. Application of Combination Ion Source to Detect Environmentally Important Compounds. *Liquid Chromatography/Mass Spectrometry*, edited by M. A. Brown. ACS Symposium Series 420, American Chemical Society, Washington, DC, Chapter 14, pp. 215–231.
37. Covey, T. R., Lee, E. D., Bruins, A. P., and Henion, J. D. 1986. Liquid chromatography/mass spectrometry. *Anal. Chem.*, **58**:1451A–1461A.
38. Arpino, P. J., and Guiochon, G. 1979. LC/MS coupling. *Anal. Chem.*, **51**:682A–701A.
39. Gallegos, E. J., Green, J. W., Lindeman, L. P., Letourneau, R. L., and Teeter, R. M. 1967. Petroleum Group-Type Analysis by High Resolution Mass Spectrometry. *Anal. Chem*, **39**:1833–1838.
40. Teeter, R. M. 1985. High-resolution mass spectrometry for type analysis of complex mixtures. *Mass Spec. Rev.*, **4**:123–143.
41. Bouquet, M., and Brument, J. 1990. Characterization of heavy hydrocarbon cuts by mass spectrometry. Routine and quantitative measurements. *Fuel Sci. Tech. Int.*, **8**(9):961–886.
42. Ashe, T. R., and Colgrove, S. G. 1991. Petroleum mass spectral hydrocarbon compound type analysis. *Energy Fuels*, **5**:356–360.
43. Kemp, W. 1991. *Organic Spectroscopy*, 3rd ed. W. H. Freeman and Company, New York, Chapter 2, pp. 19–99.

8

Structural Group Characterization by Advanced NMR Techniques

I. WHAT ADVANCED NMR TECHNIQUES AND WHY?

In Chapter 5, we discussed the basic NMR techniques and some of their applications to the analysis of petroleum fractions. Detailed band assignments were presented in Tables 5.5 and 5.6 of Chapter 5 and the overlap between many neighboring bands was evident. Spectral editing and 2D techniques, developed during the last 15 years, have made it possible to resolve many of these bands which in regular ^{13}C-NMR spectra overlap. In Section II of this chapter we sketch the basic principles of these techniques. Very few applications of the advanced techniques to petroleum fractions have been published. Therefore, to show the potential as well as the limitations, we deviate here from our restriction to deal only with heavy petroleum fractions in this book and discuss some applications of these techniques to coal liquid fractions.

The results obtained from conventional, and even more so those from advanced, NMR techniques can be further enhanced by combination with data from elemental and other analyses. This is best done by means of linear programming and similar computer methods. Brown and Ladner (1) introduced the first, still simple version of combining such analytical methods. The advent of ^{13}C-NMR and the use of numerous rigorous mathematical relations among the structural parameters of hydrocarbon molecules made it possible to extend Brown and Ladner's initial concept. Additional empirical relations and certain boundary conditions now permit a fairly detailed description of heavy petroleum fractions on the basis of ^{1}H- and ^{13}C-NMR measurements together with data from elemental analysis.

Gavales and co-workers (2) introduced a different way of describing complex molecules in terms of what they call "functional groups" and made use of these new descriptors in their "functional group analysis" (for references, see Section III.C). These composite methods, that is, the Brown–Ladner methods and functional group analysis, are the subject of Section III of this chapter.

In Chapter 7, we had demonstrated that MS, in combination with appropriate fractionation, can give us the composition of petroleum distillation cuts in terms of most or all of the significant compound types present. This combination, in our view, is the core of current compositional analysis. We consider other methods, as useful as they may be, as secondary or complementary. This is the case even for the powerful NMR techniques.

Advanced NMR techniques add two additional pieces of information to the results from our core methods. First, they can be applied to solubility fractions of nondistillable residues which are outside the range of MS. Although they do not provide the molecular characterization of which MS is capable, they can give us the concentration of numerous structural groups in a sample and the extent to which multiring molecules consist of large ring systems or contain several smaller interconnected aromatic ring clusters or single rings. This information is most detailed and most accurate if the sample is a narrow fraction.

Second, advanced NMR techniques can furnish supplementary data on petroleum distillates and their chromatographic fractions in addition to those obtained by FIMS. Examples are the (average) nature of the alkyl chains, such as branching and, to a degree, their average number and size distribution; further, the degree of aromatic alkyl substitution and the extent of aromatic, naphthenic, and hydroaromatic ring fusion. Again, samples containing molecules with single large ring systems can be differentiated from those containing several smaller interconnected ring clusters. NMR can also distinguish between certain (average) heteroatom positions in petroleum fractions.

II. THE TOOLS

A. Spectral Editing

During the last 15 years, several kinds of pulsed techniques, called spectral editing, were developed which allow the separation of primary, secondary, tertiary, and quaternary carbons in CH_n configuration, that is, CH_3, CH_2, CH, and C_q. A few of these are sensitive enough for heavy petroleum fractions. Here, the Overhauser effect (see Chapter 5) is overcome by effective timing, for instance, by turning the decoupler on for a very short time, namely, just at the last radiation pulse before data acquisition and off again right afterward.

One group of editing techniques takes advantage of "scalar coupling" between carbon and proton nuclei, which brings about a $^1H-^{13}C$ *polarization transfer*. This is an exchange of spin populations between the more sensitive ^1H nuclei and the less sensitive ^{13}C nuclei, and it causes a fourfold increase in the ^{13}C resonance intensity. Even more importantly, the sequence can be repeated as soon as the *protons* have reached their equilibrium rather than after the longer relaxation time of the ^{13}C nuclei. The increased sensitivity decreases the number of spectra necessary for the desired quality, and the reduced waiting time between spectra reduces the total time expended.

The pulses in polarization transfer techniques are specifically timed as to their duration as well as to their intervals (delay times). In addition, their phases may be varied for certain sequences. The delay times can be chosen in such a way that, with some manipulation, spectra containing only CH, or only CH_2, or only CH_3 are obtained. The resonances always appear as sharp singlets, enabling a one-to-one comparison with the normal spectrum.

With different sets of pulses, delay times, and decoupling procedures, the method DEPT (distortionless enhancement by polarization transfer) evolved from this principle. Other combinations of delay times between pulses and broad-band decoupling led to the spin-echo methods such as GASPE (gated spin echo) and PCSE (part-coupled spin-echo technique) which, however, do not have the advantage of polarization transfer with its signal enhancement and accelerated relaxation. In a brief, yet easy-to-read discussion, Shoolery (3) explained the underlying model for both. Other good reviews for the user—as opposed to the expert—are those by Dennis and Pabst (4) and Petrakis and Allen (5).

The two techniques mostly employed for fossil fuel samples at this time are the spin-echo technique GASPE (or the practically identical PCSE) and the polarization transfer technique DEPT. Both give comparable in-

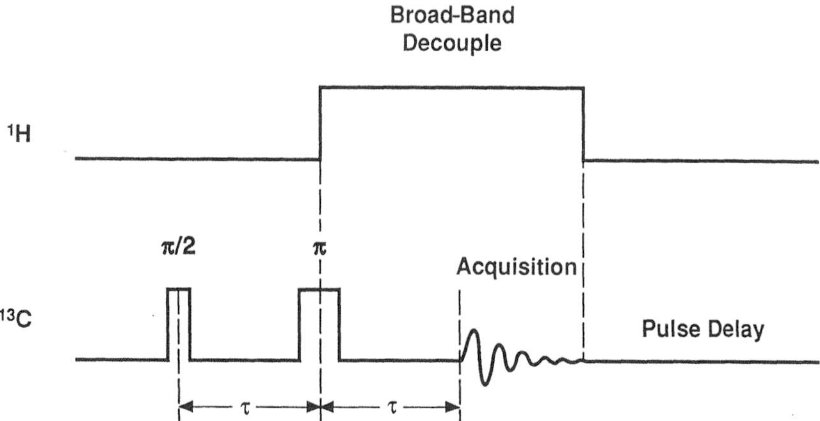

Figure 8.1 GASPE pulse sequence. (From Ref. 6. Reproduced with permission of the publisher.)

formation about carbon-type distribution, but they differ in theory as well as in complexity and speed of operation.

The spin-echo techniques use delay times, τ, to select the CH_n type for a spectrum. The pulse sequence of GASPE (and PCSE) is shown in Fig. 8.1. The delay times are chosen in terms of multiples of J, the coupling constant between ^{13}C and ^{1}H nuclei. J depends on carbon type, for example, aliphatic or aromatic, and can be either calculated or determined by experiment. Depending on τ, some of the CH_n spectra appear as upright, others as inverted (upside down), and others not at all (zero values). This makes it possible to distinguish and isolate the CH, CH_2, and CH_3 subspectra from each other.

In the polarization transfer methods, spin-echo sequences and gated broad-band decoupling are used along with additional pulses in a more complex arrangement as illustrated for DEPT in Fig. 8.2. We recognize the 90° and 180° pulses applied to the ^{13}C nuclei, each followed by a delay time, τ. Here the ^{1}H nuclei are not only exposed to the regular broad-band decoupler but also irradiated with three carefully timed and phased pulses.* The delay times between the pulses are set to $\frac{1}{4}J$ s. The first two

*Pulse phases are usually given in parentheses or as subscripts. Here, we omit phase designation of pulses which are of the same phase (x), but add it in parentheses if a pulse has a 90° phase difference, for example, $\pi/2(y)$. Pulses and receiver phases are often reversed on

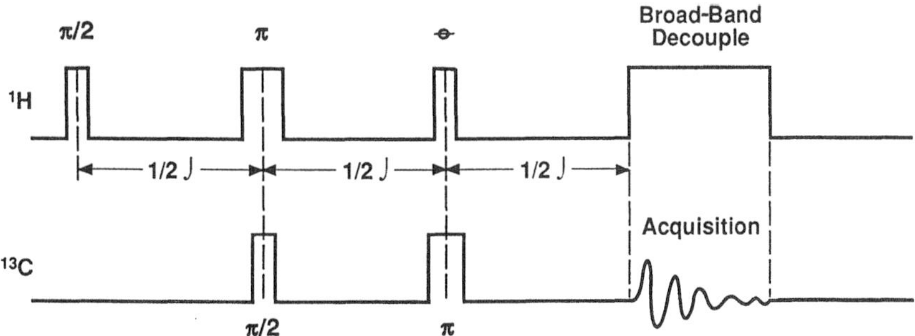

Figure 8.2 Pulse sequences used for the polarization transfer method DEPT.

pulses are of constant duration, the third one, $\Theta(y)$, is variable. Its choice determines whether the signal amplitude will be positive, zero, or negative. Thus, signal intensity in DEPT is controlled by the choice of pulse angle, Θ, that is, by the pulse length needed to change the spin by the angle $\Theta°$.

Individual CH, CH_2, and CH_3 subspectra are generated by appropriate combination of spectra recorded at 45°, 90°, and 135°. The spectrum taken at $\Theta = 90°$ can be used as the CH subspectrum without further manipulation, and the CH_2 and CH_3 subspectra are obtained in their final form by linear combination of the spectra taken at 45° and 135°:

$$CH_2 = \tfrac{1}{2}(45° - 135°) \quad \text{and} \quad CH_3 = \tfrac{1}{2}(45° + 135° - 0.707 \cdot 45°)$$

In reality, minor correction factors must be applied to compensate for instrumental inadequacies.

The DEPT sequence was first succinctly described by Doddrell, Pegg, and Bendall (7). According to Bendall and Pegg (8) and to Barron et al. (9), DEPT is better than the spin-echo methods (GASPE and PSCE) for two reasons: Its CH and CH_3 subspectra are more accurate and it is considerably faster because of its stronger signal. However, it can only measure CH, CH_2, and CH_3; quaternary C needs to be determined in a separate measurement. Such a method, later called QUAT, was made available by Bendall et al. (10). It superseded SEBBORD, a spin-echo method for determining quaternary carbons, reported by Cookson and Smith (11), which, however, is not as good in the aliphatic range.

---

alternate accumulations to suppress unwanted effects, but these techniques will not be discussed here.

B. 2D Methods

Two principle types of 2D methods can be distinguished: heteronuclear and homonuclear correlation spectroscopy (sometimes abbreviated as HETCOR and HOMCOR). Within these, there are several specific techniques, but their discussion would be beyond the scope of our book. 2D methods are qualitative rather than quantitative and allow only rough estimates of the amounts of functional groups in a sample. However, they allow us to pinpoint peaks in conventional NMR spectra for more accurate peak assignments and quantitative evaluation.

In heteronuclear correlation spectroscopy, the correlation between protons and ^{13}C nuclei is exploited. A spectrum is generated by collecting a number of ^{1}H and ^{13}C-NMR data, using the spin-echo sequence with monotonically increasing delay times. The ^{13}C spectrum is plotted by computer in one dimension and spread in the other according to the chemical shift of the protons to which they are coupled. The two-dimensional result can be displayed either as a contour plot or as a stacked plot. Examples of the two are given in Figs. 8.3 and 8.4. The contour plot (Fig. 8.3) gives a clear picture of the position of a peak in the

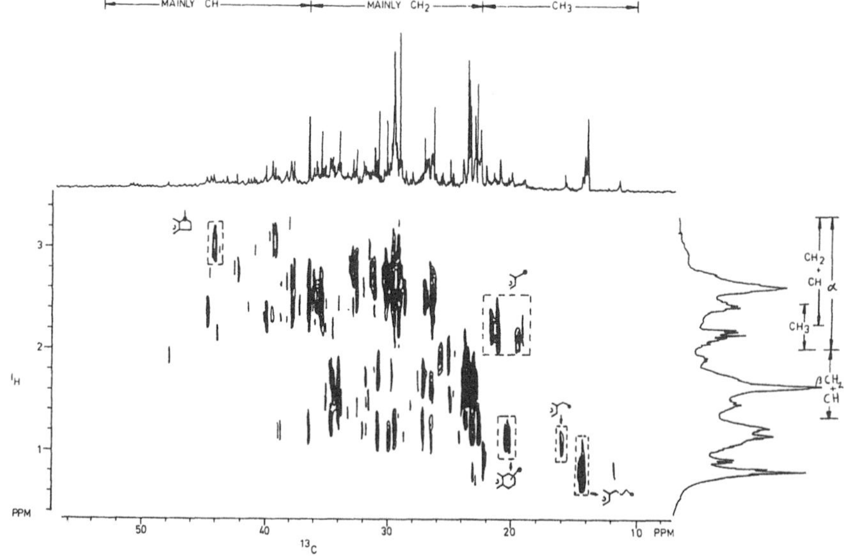

Figure 8.3 2D heterocorrelated NMR spectrum in the contour-mode of a monoaromatic coal liquifaction fraction; Aliphatic region. (From Ref. 12. Reproduced with permission of the publisher.)

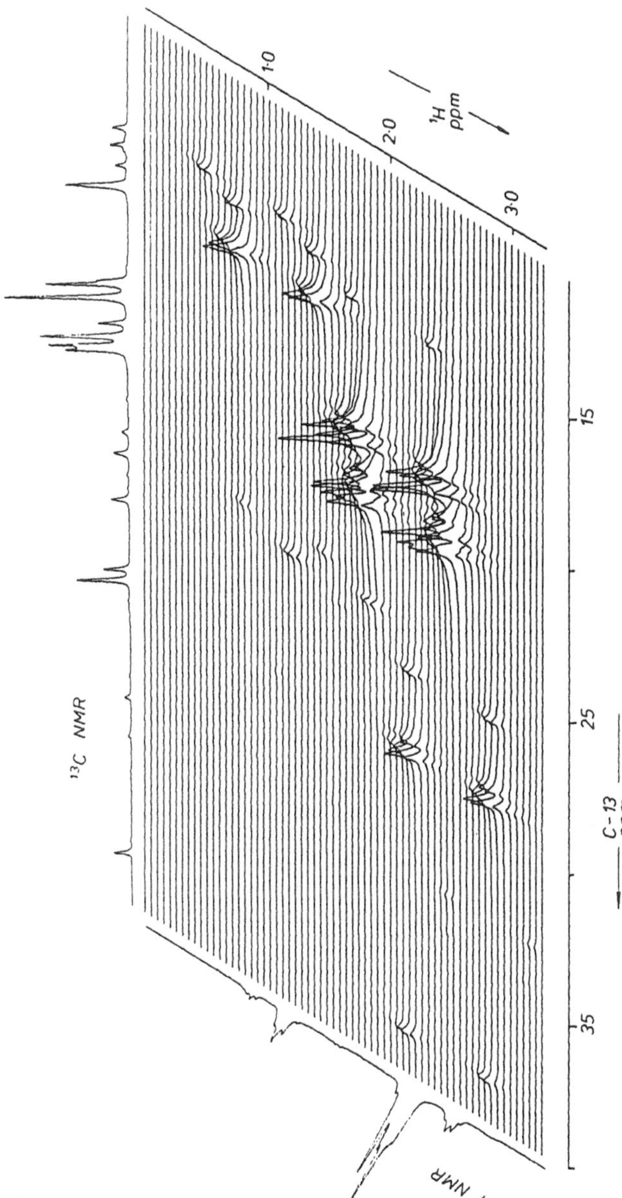

Figure 8.4 2D heterocorrelated NMR spectrum in the stacked-mode display of an aromatic petroleum solvent (medium naphtha). (From Ref. 13. Reproduced with permission of the publisher.)

scheme but little indication of its height. The stacked plot (Fig. 8.4) shows the peak heights but makes their precise coordination more difficult. For a review of the principles involved, we refer again to Shoolery (3), Dennis and Pabst (4), or Petrakis and Allen (5).

In Fig. 8.4, the different spectra taken in the heterocorrelated experiment are clearly displayed. Taking so many spectra, of course, is time-consuming even though 2D measurements generally do not require the same high-quality data (number of measurements) as spectral editing. Another example, presented in Fig. 8.5, demonstrates the power of heteronuclear correlation spectroscopy by the clean separation of olefinic from aromatic carbons which, in a regular ^{13}C-NMR spectrum, would have been hopelessly mixed up. The main types of olefinic structural groups in the sample are identified. The unlabeled peaks represent aromatic structures. Similar unraveling can be done with other carbon types.

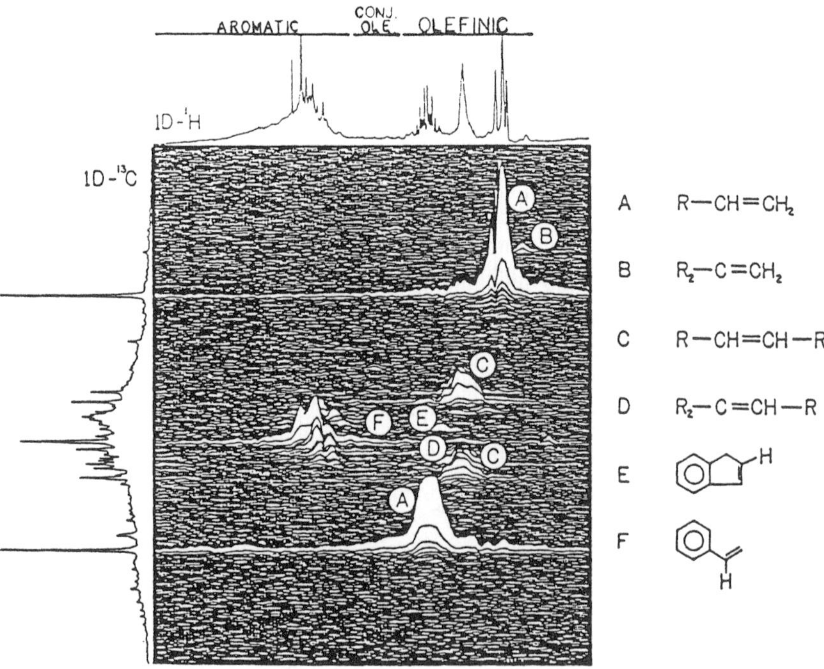

Figure 8.5 2D heterocorrelated NMR spectrum, aromatic region, in the stacked-mode display of a fossil fuel sample. (From Ref. 4. Reproduced with permission of the publisher.)

The other main 2D method, homonuclear correlation spectroscopy, makes use of the cross-correlation between coupled protons. Again, the resonance signals fall on the diagonal of the plot, and correlations are indicated by off-diagonal peaks. As in heteronuclear correlation spectroscopy, these additional correlations afford peak assignments even for complex spectra (in this case only ^{1}H-NMR spectra). Because of the high sensitivity and short relaxation time of protons, these measurements take much less time than *hetero*correlated 2D methods.

Snape et al. (12) used COSY, a version of homonuclear correlation spectroscopy, for the structural analysis of two compound-class fractions. These were the monoaromatics and diaromatics of a coal liquefaction solvent of bp = 250–450°C (480–840°F), that is, a boiling range spanning that of atmospheric to light vacuum gas oils. Figures 8.6 and 8.7 show the two-dimensional contour pattern obtained by COSY for the two fractions. The structural elements displayed in the upper right corner were identified from the Hα-Hβ correlations and their placements in the correlation patterns.

C. Applications

1. DEPT

Very few applications of spectral editing methods to heavy petroleum samples have been published so far. Most of the work on fossil fuels was done on coal oils.

Barron et al. (9) used DEPT for the analysis of several light fossil fuel samples with generally good results. Their samples consisted of two hydrotreated coal-derived naphthas, two coal-derived diesel fuels, and a shale oil extract. They point out three potential sources for error:

1. Differences in the extent of polarization transfer resulting from incorrectly chosen J values; for instance, an average J_{CH} value instead of two different ones for aliphatic and aromatic CH.
2. Variable relaxation during the pulse sequence in samples with high-molecular-weight species which tumble more slowly and, therefore, have shorter relaxation times. They warn that relaxation agents might exacerbate the problem and should be avoided.
3. Pulse strength inhomogeneities and weak rf fields. These effects can, to some extent, be corrected by calibration with a known sample.

Dereppe and Moreaux (14) applied DEPT to a heavy commercial isoparaffinic oil (Soltrol, 350–400°C = 660–750°F boiling range) with excellent results and to a petroleum asphaltene sample (no MW or other data

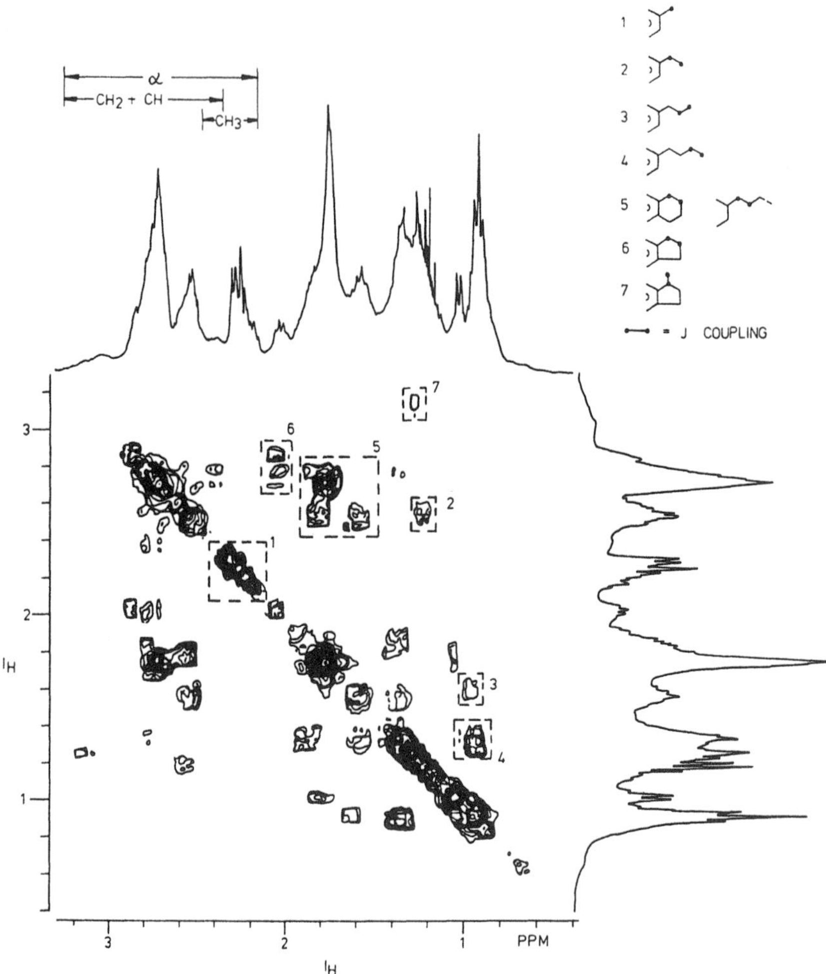

Figure 8.6 2D homocorrelated NMR spectrum of a monoaromatic coal oil fraction (AEBP = 250–450°C = 480–840°F). (From Ref. 12. Reproduced with permission of the publisher.)

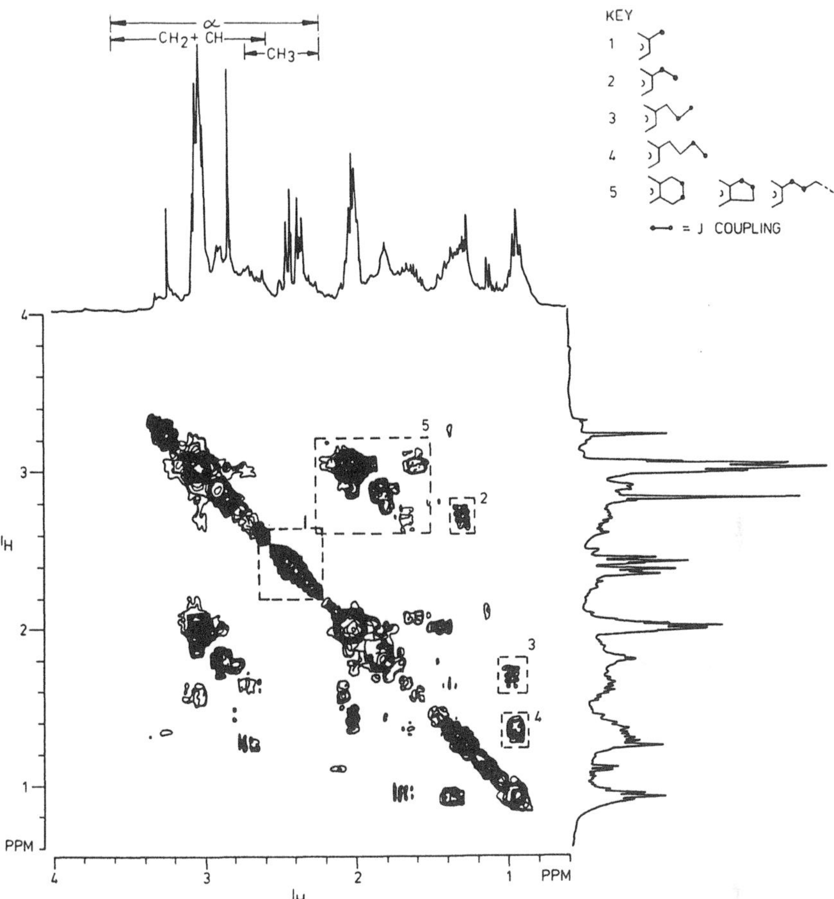

Figure 8.7 2D homocorrelated NMR spectrum of a diaromatic coal oil fraction (AEBP = 250–450°C = 480–840°F). (From Ref. 12. Reproduced with permission of the publisher.)

given) with good results. Measuring relaxation times, T_2,* of the different types of C nuclei in the asphaltene sample, they found 40–50 ms for its aromatic carbons and 100–500 for the aliphatic ones. Even with a pulse sequence as long as 10 ms, they argue, the difference in relaxation times

*The spin relaxation times of protons are designated as T_1, those of C nuclei, as T_2, and the relaxation times of entire molecules as t_2.

Table 8.1 Fraction of Carbon Types from DEPT and QUAT Special Data for a Hydrocarbon Mixture

	Fraction of carbon type in mixture		
Carbon type	Exptl.	Theor.	% Diff.
C			
Aromatic	0.161	0.167	3.6
Aliphatic	0.008	0.010	20
CH			
Aromatic	0.341	0.333	2.4
Aliphatic	0.010	0.010	0.0
CH_2	0.329	0.322	2.2
CH_3	0.151	0.156	3.2

Source: Ref. 16.

between the carbon types should lead to an error of no more than 10%, which is roughly the error of the integral intensity measurements of this type of sample. Thus, the larger average size of the molecules in this sample should only marginally contribute to the experimental error.

We will see later that Snape and Marsh (15) disagree with this statement. Their T_2 values, found with several asphaltene samples, were only 20 ms. They believe that the experimental errors caused by large molecules (with their PSCE technique) can, therefore, be distinctly greater with the customary delay times (τ) of 6–10 ms. However, Dereppe and Moreaux's DEPT results agreed well with those from elemental analysis and conventional ^1H- and ^{13}C-NMR. All in all, these authors concluded that DEPT was well suited even for the analysis of asphaltene samples.

More recently, Netzel (16) applied the combination of DEPT and QUAT to two sets of model compounds* and to four medium fossil fuel samples: a tar sand distillate (ambient—316°C), a tar sand pyrolytic oil, a shale oil distillate, and a shale oil pyrolizate. In all cases, the results were good to excellent. Table 8.1 shows the close agreement of his experimental data with those calculated from their model compounds. The results demonstrate the great power of this technique, at least for such small molecules and their mixtures.

*1. *o*-Ethyltoluene in $CDCl_3$. 2. Toluene, 2,2,4-trimethylpentane, *o*-ethyltoluene, acenaphthene, 2,3-dimethylnaphthalene, 1-methylnaphthalene, tetralin, cyclohexane, heptane, and tetradecane in $CDCl_3$.

A more critical test is provided by Netzel's measurements on his tar sand and shale oil samples. In Table 8.2, these are compared with corresponding data he obtained by independent techniques. The agreement in both cases is quite satisfactory.

In addition to applying the conventional corrections to the original DEPT equations, Netzel corrected both the initial 90° DEPT spectrum of the hydrocarbon mixture and the QUAT spectrum for residual methylene carbons.

Table 8.2 Structural Parameters of Oils from Tar Sands and Oil Shales Measured by DEPT and Directly (by Combustion and by ^{1}H-NMR)

Structural parameter	Tar sand distillate	Tar sand pyrolytic oil	Shale oil MBFA[a]	(Tube reactor)[b]
Atomic H/C				
Combustion	1.75	1.80	1.76	1.54
^{13}C DEPT NMR	1.78	1.73	1.73	1.45
% Aromatic H				
^{1}H-NMR, direct	4.5	4.4	4.3	12.2
^{13}C DEPT NMR	4.8	4.6	4.4	12.4
% Olefinic H				
^{1}H-NMR, direct	0.0	2.0	2.9	0.9
^{13}C DEPT NMR	0.0	1.3	2.4	0.7
% Aliphatic H				
^{1}H-NMR, direct	95.5	93.6	92.8	86.9
^{13}C DEPT NMR	95.2	94.1	93.2	87.0
% Aromatic C				
^{1}H-NMR, direct	17.6	20.5	23.6	43.7
^{13}C DEPT NMR	16.8	20.2	24.9	41.2
% Olefinic C				
^{1}H-NMR, direct	0.0		2.6	
^{13}C DEPT NMR	0.0	1.7	2.6	0.6
% Aliphatic C				
^{1}H-NMR, direct	82.4	79.5	76.4	56.3
^{13}C DEPT NMR	83.3	79.9	75.2	58.7

[a]From the material balance Fischer assay of Green River oil shale.
[b]From the tube reactor pyrolysis of New Albany oil shale.
Source: Ref. 16.

His normal and DEPT spectra of the tar sand pyrolysis oil are shown in Fig. 8.8. These plots nicely demonstrate the advantage of spectral editing over conventional ^{13}C-NMR spectroscopy. Although the CH and CH_3 regions are separated in their main sections, that of CH_2 grossly overlaps with both of the others. The overlap between aromatic C and CH appears

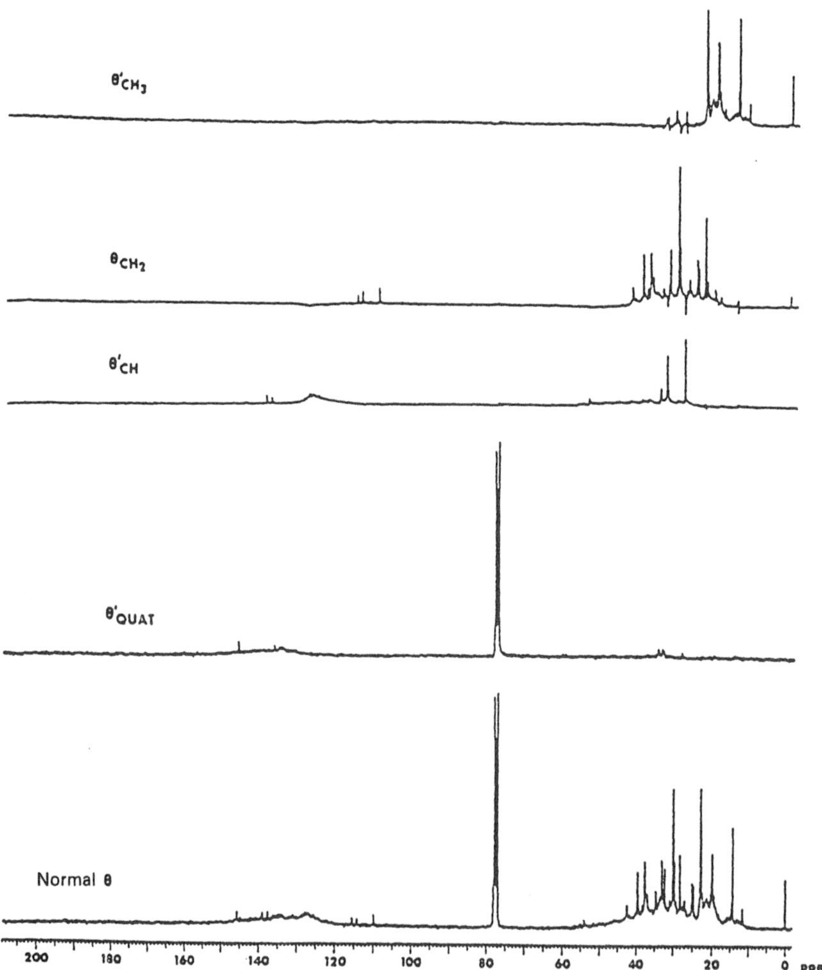

Figure 8.8 Normal and DEPT ^{13}C-NMR spectra of a pyrolytic oil from an asphalt ridge tar sand sample. (From Ref. 16. Reproduced with permission of the publisher.)

to be minimal, but it would have been more severe with a sample of greater complexity than the one chosen for this spectrum.

Another convincing demonstration of the value of spectral editing are the normal and the DEPT spectra of the 200–425°C saturates fraction from Cerro Negro crude oil (17) shown in Fig. 8.9. Here, the overlap among the three main CH_n types is massive.

Figure 8.9 is interesting for another reason, too. Note the large humps in the normal and the CH and CH_2 spectra of this highly naphthenic crude. The hump seems to extend down to about 22 ppm rather than to 24 ppm as assumed before (e.g., Ref. 18). Likewise, the hump for CH_3 next to naphthenic rings seems to reach to 25 ppm rather than 23 ppm. Netzel and Guffey found by GC/MS that this fraction contained only about 1 wt% paraffins. The remainder consists of alkylnaphthenes. The low peaks around 30 ppm further indicate a dearth of long alkyl groups.

The authors evaluated their spectra in terms of carbon and hydrogen distributions. They determined 23.8 (atom)% of the carbon to be in CH_3 groups, 42.5% in CH_2, 28.5% in aliphatic CH, and 5.2% in aliphatic C. They also calculated the number of branches per molecule and the average number of naphthenic rings per molecule. For the latter, however, they used a formula which did not perform well with model compounds (19). They seem to be reluctant to make use of the naphthenic hump for calculating the amount of naphthenic carbon which is a cautious and, possibly, wise attitude.

Other examples of DEPT spectra overcoming the overlap among the various CH_n regions in ^{13}C-NMR spectra of fossil fuel samples are found, for example, in papers by Dennis and Pabst (4), Dreeskamp et al. (20), and Abu-Dagga and Rüegger (21).

2. Spin-Echo Techniques

As mentioned earlier, PCSE and GASPE are virtually identical. Snape chose the name PCSE (part-coupled spin-echo technique), and Cookson and Smith, the name GASPE (gated-spin echo).

Snape applied the technique to coal liquifaction products ranging from highly aliphatic oils (22) to aromatic oils (23–25), to various other light to intermediate samples (24–26), and to several low-molecular-weight (≤610) asphaltenes (23–26). Later, Snape and Marsh (15) investigated another set of coal-derived samples and also some petroleum fractions, both of which included asphaltenes of up to 3100 effective molecular weight (presumably measured by VPO). They concluded that, even for the high-molecular-weight asphaltenes, the PCSE peaks, including the more difficult CH_2 signals, were well separated and reasonably quantitative except for those of the aromatic carbons, which were probably too low.

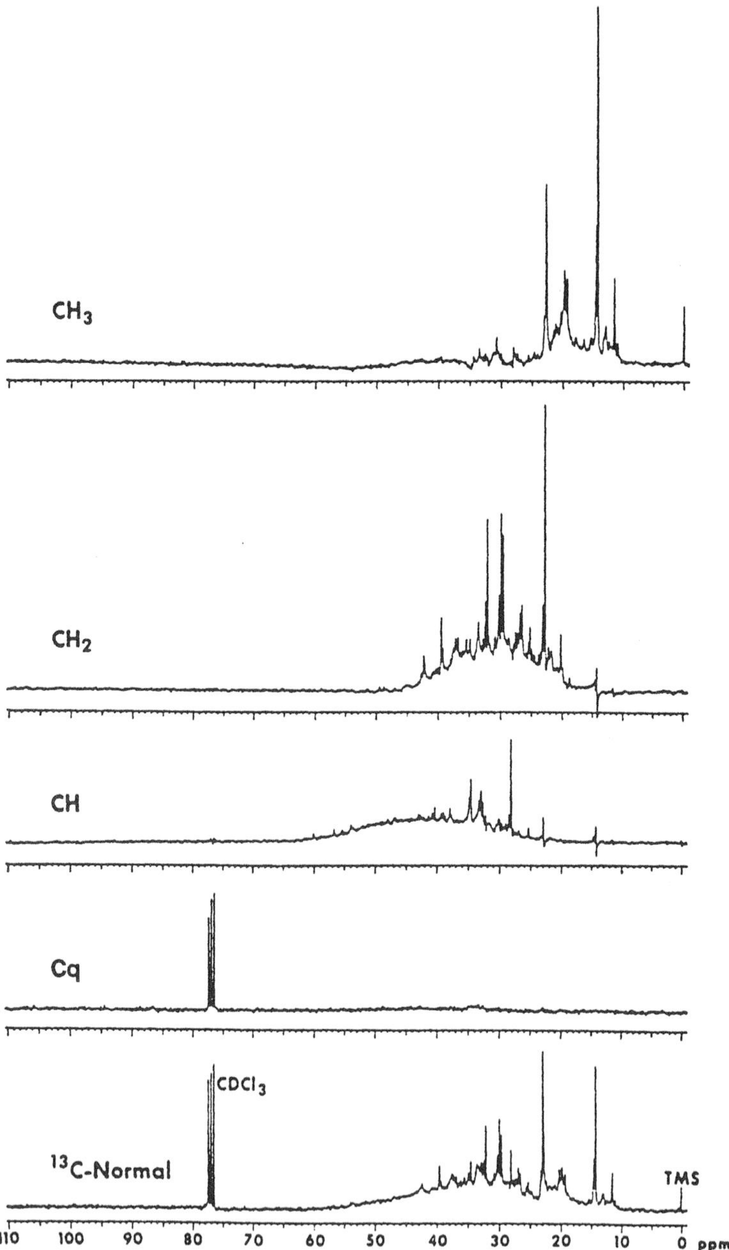

Figure 8.9 Normal and DEPT ^{13}C-NMR spectra of the 200–425°C (390–800°F) saturates fraction from Cerro Negro crude. (From Ref. 17. Reproduced with permission of the publisher.)

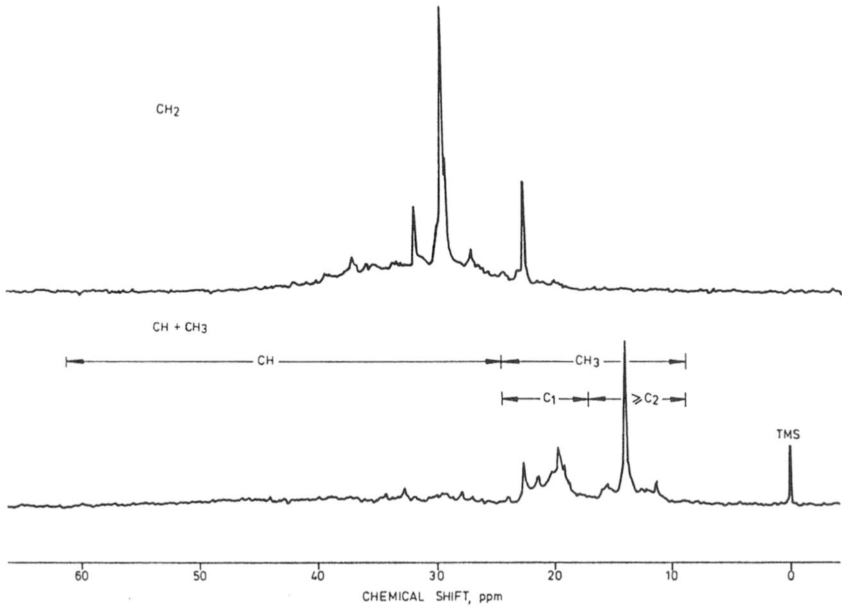

Figure 8.10 PCSE aliphatic spectra, aliphatic region, of an asphaltite aromatic oil sample. (From Ref. 15. Reproduced with permission of the publisher.)

Figure 8.10 shows the aliphatic CH_2 and CH/CH_3 subspectra of one of Snape and Marsh's aromatic oils (280 MW), and Fig. 8.11, those of their Athabasca asphaltenes (1740 MW). The peaks in both figures are remarkably narrow for both materials, although, of course, not as sharp as those of single compounds.

Snape and Marsh (15) took spin-echo spectra of two asphaltene samples (one from a coal extract and of 420 MW, the other from Athabasca bitumen and of 1750 MW). Varying the pulse lengths from 100 down to 1 ms in roughly geometric sequence, they found that the longer pulses (>1 ms) were insufficient for both samples. In addition, the size of the hump under the peaks increased distinctly with lower pulse length down to 4 or even 1 ms. The latter observation calls for caution in the evaluation of the humps for the determination of naphthenic carbon (27–29). It suggests that short pulses (≤4 ms) should be used for asphaltene samples.

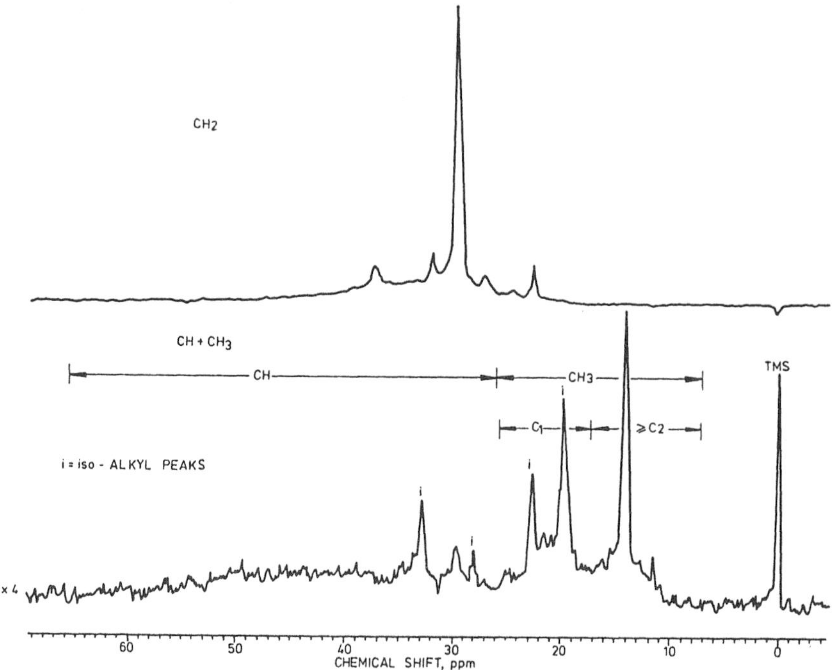

Figure 8.11 PCSE aliphatic spectra, aliphatic region, of an Athabasca bitumen asphaltene sample. (From Ref. 15. Reproduced with permission of the publisher.)

Snape and Marsh also measured spin-spin relaxation times, T_2, of various CH_n groups in these and two other samples; see Table 8.3. Long relaxation times assure quantitative measurements at the conventional pulses of 4–10 ms. Short relaxation times as, for instance, those found for aromatic CH (20 ms) in some samples, may lead to substantial underestimation of the amounts of the corresponding groups. According to Snape and Marsh, even the *naphthenic* CH and CH_2 groups of two of their asphaltene samples, which had relaxation times of about 40 ms, may lose 20% of their signal at the usual 8-ms pulses. These conclusions are less favorable than those drawn by Dereppe and Moreaux (14). Both groups agree that ^{13}C spectral editing methods are quite accurate when applied to relatively light oils whose carbons have longer relaxation times, but Snape and Marsh feel they could give low results with asphaltenes, especially with those of higher MW, which have very short relaxation times.

Table 8.3 Spin Relaxation Times, T_2, of CH_n Groups in Various Asphaltene and Oil Samples

	T_2 (ms)		
	Asphaltenes		Aromatic oil
CH_n group	High MW	Low MW	Low MW
Aromatic C_q	60	—	—
Aromatic CH	20	—	—
Long alkyl CH_2	150	130	500
Terminal CH_3 in long alkyls	150	180	250
Cyclic CH_2 and CH	40	70	100

Source: Ref. 15.

Snape and Marsh (15) conclude:

1. Provided that T_2 are not prohibitively short ($< \sim 20$ ms), spin-echo ^{13}C-NMR techniques can be used to estimate concentrations of C, CH, CH_2, and CH_3 groups in heavy fossil fuel fractions such as aromatic oils and even asphaltenes
2. ^{13}C-NMR data yield much more reliable and detailed information on the nature of aliphatic groups than that derived previously by indirect methods

Cookson and Smith applied their spin-echo technique (GASPE) to a number of fossil fuel distillates with "reasonably quantitative results" (13, 30–32). They warn, however, that "all multiplet selection procedures are limited by transverse relaxation effects, and thus difficulties are expected for very heavy materials." In other words, very large molecules do not tumble fast enough, and, therefore, their relaxation times may be too short for accurate measurements, especially for the advanced techniques used in spectral editing.

Initially, Cookson and Smith (30) tried the GASPE procedure with a petroleum-derived aromatic solvent (120–190°C, 250–375°F AEBP) and two coal-derived oils (120–350°C, 250–660°F and 210–380°C, 410–715°F, AEBP). Figure 8.12 shows the aliphatic region of their CH_n subspectra along with the conventional spin-echo NMR spectra of one of their coal oils. Figure 8.13 displays the corresponding aromatic spectra. In both regions the comparison of the subspectra with regular ones is very favorable. Even though there are small visible artifacts in the CH and CH_3 spectra near 27 and 30 ppm, the authors expect their accuracy to be between 3%

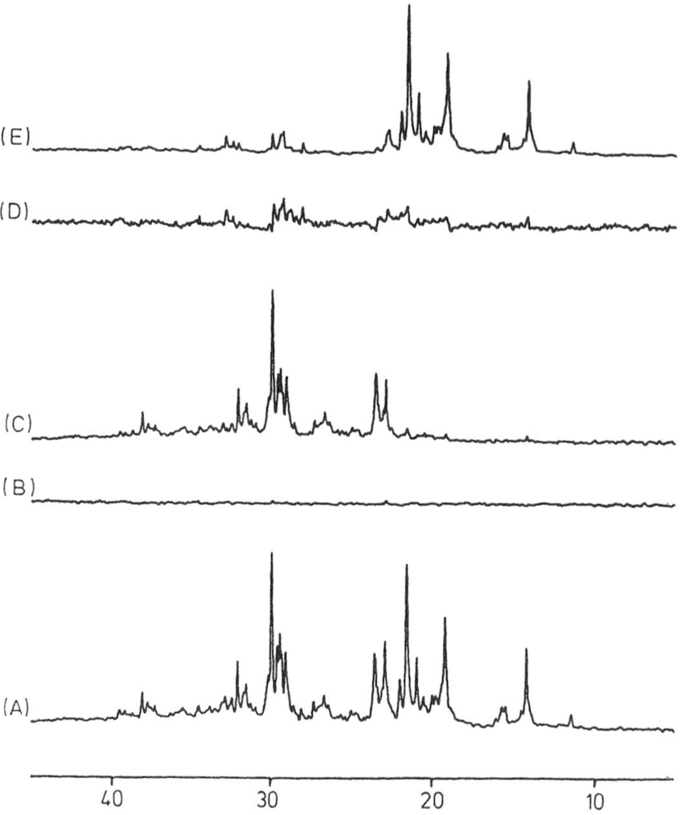

Figure 8.12 Conventional spectrum (A) of coal recycle oil, aliphatic region, and C, CH_2, CH, and CH_3 subspectra (B, C, D, E) obtained by GASPE. (From Ref. 30. Reproduced with permission of the publisher.)

and <1% absolute, respectively. C/H ratios of the three samples measured by elemental analysis were about 5% lower than those calculated from the subspectra.

Figure 8.12 again demonstrates the advantage of spectral editing over conventional ^{13}C-NMR spectroscopy. Here, the CH_2 and CH_3 regions are fairly well separated, but that for CH overlaps with both of the others. Especially in the 22- and 29-ppm bands, there would have been no way to distinguish CH from CH_3 and CH_2, respectively, in a conventional spectrum. Little if any aliphatic quaternary C was detectable in the sample.

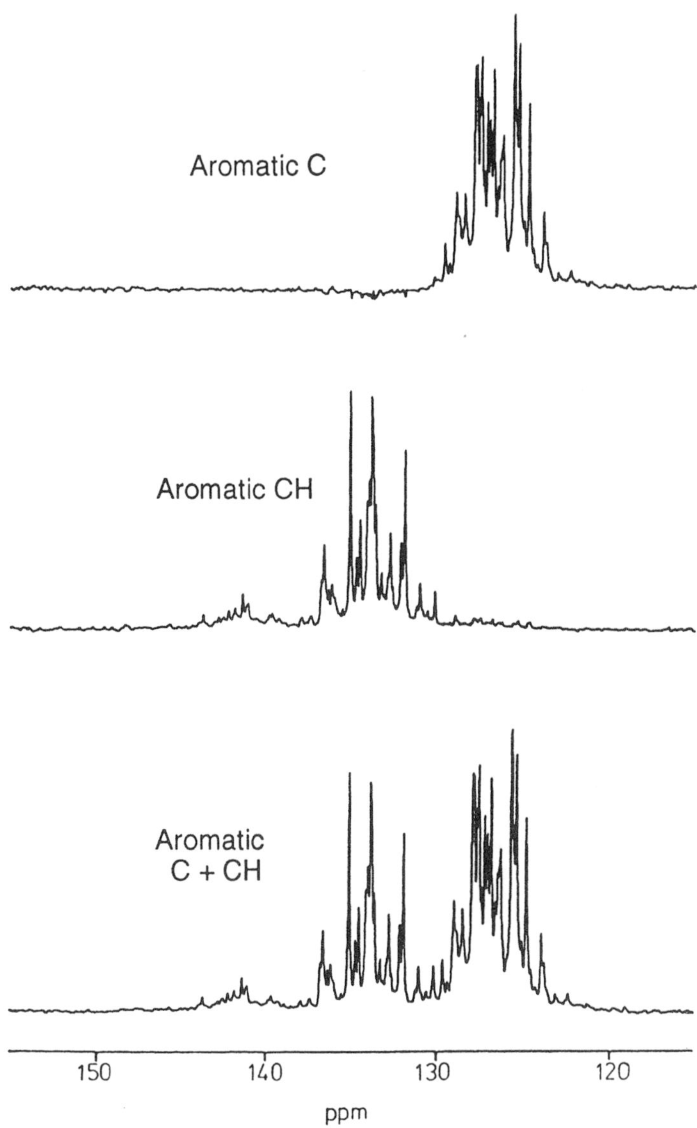

Figure 8.13 C and CH subspectra, aromatic region, of coal recycle oil. (From Ref. 30. Reproduced with permission of the publisher.)

In a paper accompanying the previous one, Cookson and Smith (31) reported measurements with SEBBORD, a procedure which provides only quaternary C signals. Appropriate scaling and comparison with a regular NMR spectrum allowed them the quantitative determination of protonated and unprotonated aromatic carbons. Application to light petroleum and coal-derived samples of up to 240°C (465°F) boiling point gave good results. With heavy oil samples they observed deviations of up to 15%, which is in accord with Snape's (23) earlier observation with SEBBORD. Cookson and Smith (32) compared GASPE and SEBBORD results obtained on nine light oil samples with those from a combination of conventional NMR with elemental analysis as an independent reference. They concluded that SEBBORD was somewhat better than GASPE in the quantitation of aromatic C and CH.

3. 2D Methods

We have already referred to the application of homonuclear (^1H-^1H) 2D measurements to coal-derived fractions of monoaromatics and diaromatics (boiling range 250–450°C) by Snape et al. (12). This group also applied heteronuclear correlated 2D NMR, HETCOR, to the same two fractions. The resulting two-dimensional patterns are reproduced in Figs. 8.3 and 8.14. In this application, Snape et al. identified four kinds of CH_3 groups in different placements (α, β, γ^+) with respect to the aromatic rings and one kind of CH. The main experimental conditions for their two methods are listed in Table 8.4.

Snape et al. (12) point out that "the number of structural possibilities were limited in our samples enabling detailed structural information to be obtained on the aliphatic groups present." This implies that with heavier, more complex fractions these two methods, HOMCOR and HETCOR, might not have worked so well. Cookson and Smith (33) warn "the more detailed the spectral subdivisions attempted, and the more complex the sample under study, the greater is the possibility that ambiguities will be significant."

Nevertheless, it is remarkable what can be done with modern sophisticated NMR techniques, at least for gas oil fractions and lighter ones (see also refs. 61 and 62). A final assessment of their efficacy for heavier fractions will have to wait for a study comparing the results from advanced NMR spectroscopy with those from independent measurements such as extensive fractionation combined with FIMS. Hoping for a positive outcome, we have collected in Table 8.5 all the information accessible so far by spectral editing, and in Table 8.6 that provided by the 2D methods.

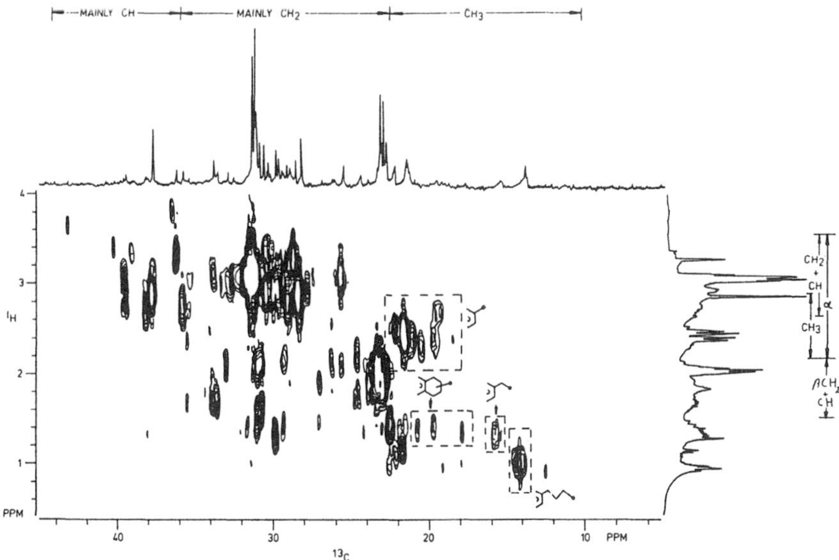

Figure 8.14 2D heterocorrelated NMR spectrum, aliphatic region, of a diaromatic coal liquifaction fraction. (From Ref. 12. Reproduced with permission of the publisher.)

Table 8.4 Conditions for 2D NMR Experiments on a 300 MHz Spectrometer

Settings	Nuclei	HETCOR	COSY
90 pulse widths (μs)	^1H	30	13.5
	^{13}C	17	—
Dwell time (μs)	^1H	165	333
	^{13}C	17	—
Sweep width (Hz)	^1H	—	1000
	^{13}C	12.048	—
Assay size ($t_1 \cdot t_2$)		$64 \cdot 2k$	$256 \cdot 2k$
t_1 increment (μs)		165	333
No. of scans per t_1 value		460	32
Acquisition time (ms)		85	170.5
Total acquisition time (h)		16	1
Choice of J_{CH} (Hz)		125	

Source: Ref. 12.

Table 8.5 Structural Elements Identifiable by Spectral Editing

Carbon type	Symbol	Chemical shift (ppm)
Aromatic quaternary C	CA0	120–160
Aromatic CH	CA1	115–130
Aliphatic + naphthenic C_q	$CL_q + CN_q$	20–60
Internal (bridgehead) naphthenic C_q	CNI_q	37–60
Aliphatic + naphthenic CH	CL1 + CN1	32–60
Internal (bridgehead) naphthenic CH	CNI	37–60
CH in alkyl side chains except isopropyl or isobutyl	CL1g	37–60
β-CH intetralin or indan structures	CN1β	27.5–37
Aliphatic + naphthenic CH_2	CL2 + CN2	20–56
CH_2 in alkyl side chains adjacent to CH	CL2ad1	37–60
Ring-joining CH_2	CL2αα	32–43
CH_2 for ε-carbon in alkyl chains $\geq C_8$	CL2ε	29.7
CH_2 α to aromatic ring intetralin or indan structures and in $Ar{-}CH_2{-}CH_2{-}Ar$ structures	CLN2αβ	27.5–37
CH_2 in naphthenic structures	CN2	24.0–37
Shielded CH_2α to an aromatic ring	CL2αshld	24–27.5
β-CH_2 in indan structures	CN2βind	24–27.5
β-CH_2 in propyl side chains	CL2prop	24–27.5
β-CH_2 in tetralin structures	CN2β	22.5–24
CH_2 next to terminal CH_3 in chains ≥4	CL2n3t	22.5–24
Aliphatic CH_3	CL3	20–45
β-CH_3 in isopropyl side chains	CL3iprop	24–27.5
CH_3 on naphthenic and hydroarom. groups	CL3αN	18–24
α-CH_3 not shielded by adjacent groups	CL3α	20.5–22.5
α-CH_3 shielded by 1 adj. ring or group	CL3αshld1	18–22.5
CH_3 β to an aromatic ring	CL3β	15–18
Ch_3 gamma or farther from an arom. ring	CL3β	11–15
α-CH_3 shielded by 2 adj. rings or groups	CL3αshld2	11–15

Table 8.6 Structural Elements Identified by 2D Methods in Crude Oil and Coal Oil Fractions (No Quantitative Data)

Carbon type	Symbol	Correlation method
α-CH_3	CL3α	Homo, hetero
β-CH_3	CL3β	Hetero
γ^+-CH_3	CL3γ	Hetero
CH_3 α to naphth. ring fused to aromatic ring	CL3αN	Hetero
Aliphatic α-C_2H_5		Homo
Aliphatic β-C_2H_5		Homo
Aliphatic γ-C_2H_5		Homo
Aliphatic C_2H_4 α to a naphthenic ring		Homo
α- or β-C_2H_4 in 6-membered naphthenic ring		Homo
Naphthenic α-C_2H_4 in 5-membered naph. ring		Homo
Naphthenic α-C_2H_5 in 5-membered naph. ring		Homo

III. NMR-BASED METHODS

A. The Brown-Ladner Method

1. The Original Brown-Ladner Method

Brown and Ladner (1,34) described the combination of ^{1}H-NMR data with elemental analysis for the purpose of calculating structural parameters which cannot be determined by either measurement alone. This method gave them access (before the advent of ^{13}C-NMR) to the aromaticity, f_a, that is, the ratio of aromatic to total carbons; the degree of aromatic ring substitution, σ, which they defined as the ratio of substituted *aromatic* to all the aromatic carbons; and the ratio of peripheral aromatic to total aromatic carbons, CAP/CA, in our nomenclature. They called this third parameter the "hydrogen to carbon ratio of the hypothetical unsubstituted aromatic material" and gave it the symbol H_{aru}/C_{ar}. Brown and Ladner (1) expressed their three new parameters in terms of ratios of carbon, oxygen, and specific hydrogen species to total hydrogen, all of which are directly measurable quantities:

Aromaticity:

$$f_a = \frac{(C/H) - (H_\alpha/xH) - (H_0/yH)}{C/H} \tag{8.1}$$

Degree of substitution:

$$\sigma = \frac{(H_\alpha/xH) + (O/H)}{(H_\alpha/xH) + (O/H) + (H_{ar}/H)} \tag{8.2}$$

$$\frac{\text{Peripheral } C_{ar}}{\text{Total } C_{ar}}:$$

$$\frac{H_{aru}}{C_{ar}} = \frac{(H_\alpha/xH) + (H_{ar}/H) + (O/H)}{(C/H) - H_\alpha/xH - (H_o/yH)} \tag{8.3}$$

The various symbols and their meanings are displayed in Table 8.7. In this section, which is more of historical than of practical interest, we have used Brown and Ladner's original nomenclature rather than the more efficient one we chose for Section III.A.2.

Brown and Ladner's parameters added valuable new information to the structural data heretofore available and have been widely used by many researchers in our field. However, they require several assumptions:

1. All the oxygen is attached directly to the aromatic ring systems and is not shared between them as, for instance, in diphenylether.
2. Aromatic rings must not be linked by C–C bonds as, for instance, in biphenyl.
3. Values must be assumed for x and y.

The first assumption seems fully justified for petroleum samples and even for coal liquids and shale oil. The second assumption is more ques-

Table 8.7 Symbols Used in Brown and Ladner's Parameters

Symbol[a]	Meaning
C	Total number of carbon in the sample
H	Total number of hydrogen in the sample
C/H	Ratio of all carbon to all hydrogen in the sample
H_α/H	Ratio of hydrogen α to an aromatic ring to all H
x	Average number of H atoms per α-carbon
H_α/x	Average number of C atoms α to an aromatic ring
H_o/H	Ratio of other aliphatic H, farther from an aromatic ring than α, to all H atoms
y	Average number of H atoms per (β and β$^+$) carbons
H_β/y	Average number of C atoms β and farther from an aromatic ring
O/H	Ratio of oxygen to all H
H_{ar}/H	Ratio of aromatic H to all H
f_a	Ratio of aromatic C to all carbons
σ	Ratio of substituted aromatic C to all aromatic carbons
H_{aru}/C	Hydrogen to carbon ratio of the hypothetical unsubstituted aromatic material. Same as peripheral aromatic C to all aromatic carbons.

[a]Brown and Ladner's original symbols.

tionable though not too serious. The occurrence of biphenyl type linkages would increase both σ and H_{aru}/C. The major consideration is in the third assumption, the choice of x and y. Their true values are inaccessible, but at least their limits can be fixed reasonably well. Brown and Ladner (1) assumed x and y to be equal and close to 2, which is a good approximation for unprocessed petroleum fractions, but not always for process streams. Today, with the help of ^{13}C-NMR, we can estimate x and y more precisely as will be discussed next.

2. An Extended Brown-Ladner Method

With the greater wealth of experimental NMR data available today, we can also use a larger set of relations between intramolecular hydrocarbon groups and calculate more structural parameters. Such relations have been described, for instance, by Williams (35), Van Krevelen (36), Hirsch and Altgelt (37), Haley (38), Aczel et al. (39), Herod et al. (40), Abu-Dagga and Rüegger (21), and most recently by Altgelt (41). It is now possible to determine or estimate from NMR measurements and elemental analysis almost the entire average molecular composition of a heavy distillate or refinery stream. There are now more than 30 independent rigorous molecular relations available. In addition, Altgelt developed an empirical relation for estimating CAII, one of the more important parameters of his scheme and defined later. Several other, minor ones need to be estimated a priori, but "what if" calculations can be used to sharpen the initial assumptions.

The first of these assumptions concerns the distribution of naphthenic rings between those with five and six members. The second determines the distribution of quaternary carbon (if any) between paraffinic and naphthenic carbon. The third handles the degree of pericondensation in the aromatic and naphthenic ring parts of the ring systems. The fourth decides whether internal shared carbon atoms between aromatic and naphthenic rings (very rare) belong to the aromatic or to the naphthenic ring system. The fifth one finally fixes the number of ring clusters per molecule. Very large molecules may contain several ring clusters separated by alkyl chains, by direct linkages as in biphenyl, or by sulfur atoms. However, for petroleum distillates and refinery streams this number is always very close to 1. Therefore, 1 is the initial choice, but it can be changed in subsequent calculations. Errors in this guess affect mainly molecular parameters of secondary interest and only minimally the important ones. The other assumptions are of even lesser significance.

More difficult is the decision of how to assess the distribution of sulfur and nitrogen among the various functional groups and other positions in a petroleum molecule. These choices must be based on experience and are, thus, somewhat arbitrary.

The most serious problem is the fact that the naphthenic carbon content and the number of shared aromatic–naphthenic carbon atoms at the boundary between such ring systems must be taken directly from the spectrum. Both are less accurate than most of the other primary values. Especially the shared aromatic–naphthenic carbons (CANS for Carbons Naphthenic–Aromatic Shared) may have errors as large as ±50%. For single compounds, more rigorous relations apply which would allow the calculation of CANS as well as of some of the assumed values from other spectral peaks. However, here we deal with mixtures (not even with average molecules) which exclude the use of these additional equations.

Examples of rigorous relations between molecular parameters range from the trivial $C_{CH_3} = 3H_{CH_3}$ to less obvious ones such as

$$\mathrm{RA} = \frac{\mathrm{CA} + \mathrm{CAI} + \mathrm{CAII}}{6} \tag{8.4}$$

which says the number of aromatic rings, RA, equals one-sixth the sum of aromatic carbon, CA, the number of internal aromatic carbons, CAI (also called bridgehead carbons*), and the number of internal internal carbons, CAII. Two types of internal ring carbons can be distinguished. We designate those connecting *two* rings (as in naphthalene) with the symbol CAI (for internal aromatic C or, rather, C Aromatic Internal). CAII denotes the internal ring atoms connecting *three* aromatic rings in pericondensed systems such as pyrene. Figure 8.15 shows examples of these internal aromatic carbon types in various molecular structures.

Another important equation relates the number of aromatic ring systems, SA, with CA and CAI or with RA and CAI in a molecule:

$$\mathrm{SA} = \tfrac{1}{6}(\mathrm{CA} - 2\mathrm{CAI} + \mathrm{CAII}) \tag{8.5a}$$

or

$$\mathrm{SA} = \mathrm{RA} - \tfrac{1}{2}\,\mathrm{CAI} \tag{8.5b}$$

Simple examples for molecules with more than one aromatic ring system are biphenyl, dinaphthylmethane, and 9,10-dihydroanthracene. All of these have two aromatic ring systems (SA = 2). Those in the last example are separated by a naphthenic ring. On the other hand, octahydroanthracene, has two naphthenic rings separated by an aromatic ring. Here SA = 1,

*We prefer the term "internal ring carbons" to the term "bridgehead carbons" because it is more precise and not as easily confused with other structural elements.

	CA •	CA I ○	CA II ◉	RA
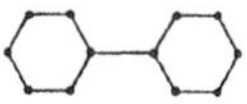	12	0	0	2
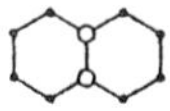	10	2	0	2
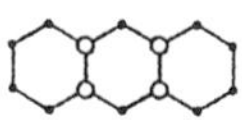	14	4	0	3
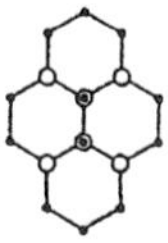	16	6	2	4
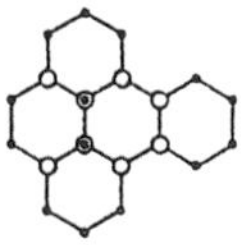	20	8	2	5
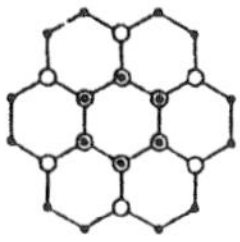	24	12	6	7

$$CA = 6RA - CA\,I - CA\,II$$

Figure 8.15 Examples of fused 6-membered rings and their carbon carbon numbers in various positions. CA = total aromatic C atoms; CAI = internal aromatic C atoms; CAII = internal internal aromatic C atoms; RA = number of aromatic rings.

but the number of naphthenic ring systems is two (SN = 2).

biphenyl 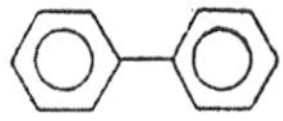 dinaphthylmethane

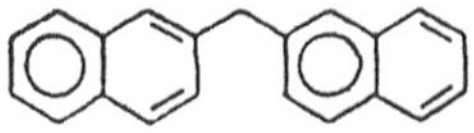

9,10 dihydroanthracene 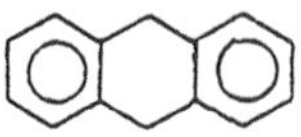 octahydroanthracene

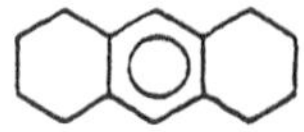

There is even an equation relating the total number of rings, aromatic and naphthenic, in a hydrocarbon molecule to the total number of carbon atoms, C, the number of aromatic carbons, CA, and the number of hydrogen atoms, H:

$$R(A + N) = C + 1 - \tfrac{1}{2}(H + CA) \qquad (8.6)$$

More equations are shown in Table 8.9. The empirical relation is necessary to estimate CAII, which was already defined. It has an average error of about ±6%, which is within the experimental error of most of the ^{13}C-NMR measurements.

CANS is the number of the carbon atoms shared by aromatic and naphthenic rings. CANSI is the number of the internal carbons shared by aromatic and naphthenic rings. These, as CNII and even more so CANSII, are quite rare in straight-run petroleum fractions. In catacondensed systems, such as tetrahydrophenanthrene, they are nonexistent. CANSI, are found only in pericondensed hydroaromatic molecules, for example, in decahydropyrene:

decahydropyrene

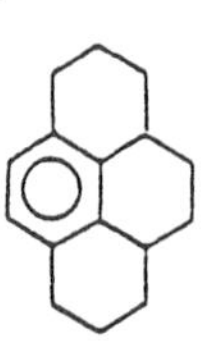

In this molecule there are four shared aromatic-naphthenic carbons, CANS. The two in the center are internal, CANSI, the other two are peripheral, CANSP. CNII is the number of internal internal naphthenic carbons and CANSII that of the internal internal shared aromatic–naphthenic carbons in pericondensed mixed aromatic–naphthenic ring systems such as dihydropyrene:

dihydropyrene

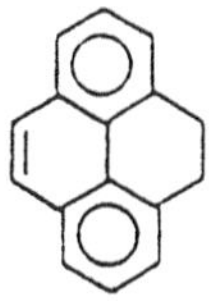

The difference to CANSI in decahydropyrene is that here the two internal CANS protrude into the aromatic structure.

In Table 8.8, we have collected all the data currently accessible by measurement. Here, we also introduce our nomenclature for the symbols we use for designating these parameters. We tried to choose simple, descriptive terms which are easy to understand and to remember and which can be used directly in computer programming.

The measured values are generally in terms of wt% of sample, for example, wt% C, wt% H, wt% aliphatic H, wt% aromatic C, and so on. The percentages can either be used directly, or they can be converted into numbers of atoms per molecule by equations like the following:

$$C = \frac{MW \cdot wt\%\ C}{1201} \tag{8.7}$$

$$C = \frac{MW \cdot wt\%\ H}{100.8} \tag{8.8}$$

Dealing with numbers of atoms or of atomic groups is simpler than dealing with percentages. It results in average parameters, which must be viewed as average parameters of the corresponding groups in the sample rather than those of an average molecule. The calculated numbers of C and H in the various structural parameters can also be reconverted to wt% of C and H by solving the above equations for wt% C and wt% H. As always, extended Brown-Ladner calculations should be applied to data obtained from narrow fractions rather than to those from broader samples.

Most of the structural parameters, which can be measured or calculated by Altgelt's version of the extended Brown-Ladner method, are listed in Table 8.9. Several of the calculated parameters can now be checked to some extent by 2D in combination with regular ^{13}C-NMR measurements as indicated, for example, in Figs. 8.3 and 8.14.

Altgelt (41) has tested his method so far only by model calculations using molecular structures. These consisted of very diverse types, ranging from alkyl naphthenes to highly pericondensed hydroaromatics consisting of several aromatic and naphthenic ring systems found only in specific refinery streams. Those calculations performed with the correct input data

Table 8.8 Molecular Parameters of Hydrocarbons Accessible by Direct Measurement

Parameter	Symbol	Method	Reference
%C	%C	Elemental analysis	
%H	%H	Elemental analysis	
Molecular weight	M	VPO, SEC, MS	
Total C (atoms)	CT	From %C and M	
Total H (atoms)	HT	From %H and M	
Aliphatic H	HL	Proton NMR	42
Aromatic H	HA	Proton NMR	42
Monoaromatic H	HAm	Proton NMR	16
(Diaromatic H)	HAd	Proton NMR	16
(Triaromatic H)	HAtr	Proton NMR	16
Aliphatic H	HL	Proton NMR	
Hα to an aromatic ring	HLα	Proton NMR	42
Hβ to an aromatic ring	HLβ	Proton NMR	42
Hγ to an aromatic ring	HLγ	Proton NMR	42
Aromatic C	CA	^{13}C-NMR	42
Aromatic unsubstituted peripheral C + internal CA	CAPu + CAII	^{13}C-NMR	18,27,43–45
Aromatic quaternary C (internal + unsubstituted peripheral)	CAq	Spectral edit ^{13}C-NMR	13,23,24
CAq substituted by CH or CH_2	CAq,12	Spectral edit ^{13}C-NMR	18,21,44
CAq substituted by CH_3 + internal CA	CAq,3,I	Spectral edit ^{13}C-NMR	18,21,44
CAq substituted by CH_3 + CAP fused to naphthenic ring	CAq,3+CANS	Spectral edit ^{13}C-NMR	12,43,44,45

Aliphatic C	CL	^{13}C-NMR	42
C in CH_3	CL3	Spect. edit ^{13}C-NMR	13,23,24
C in α-CH_3	CL3α	Spect. edit ^{13}C-NMR, 2D-NMR	12,13,23,24
C in β-CH_3	CL3β	Spect. edit ^{13}C-NMR, 2D-NMR	13
C in γ-CH_3 and farther away	CL3γ	2D-NMR	13
C in CH_2	CL2	Spect. edit ^{13}C-NMR	13,23,24
C in α-CH_2	CLα2	Spect. edit ^{13}C-NMR	13,23,24
C in β-CH_2	CLβ2	Spect. edit ^{13}C-NMR	13,23,24
C in $\geq\gamma$-CH_2	CLγ2	^{13}C-NMR	18
C in CH	CL1	Spect. edit ^{13}C-NMR	13,23,24
Quaternary aliph. C	CLq	Spect. edit ^{13}C-NMR	13,23,24
C in naphthenic rings, qualitative	CN12	^{13}C-NMR	28
α- or β-C_2H_4 in 6-membered naphthenic ring	C6N2$\alpha\beta$	2D-NMR	21
α-C_2H_4 in 5-memb. naphth. ring	C5N2α	2D-NMR	12
α-C_2H_3 in 5-memb. naphth. ring		2D-NMR	12
C in alkyl CH_2, qualitative	CL2ch	^{13}C-NMR	45
ε-CH_2 in long chains	CL2lch	^{13}C-NMR	42,44
CH_2 in long chains 1 or 2 removed from end CH_3	—	^{13}C-NMR	42,44
Olefinic C	Co	^{13}C-NMR	16
C in olefinic CH_2	Co2	^{13}C-NMR	16
C in olefinic CH	Co1	^{13}C-NMR	16
C in olefinic Cquat	Coq	^{13}C-NMR	16

Table 8.9 Determination of Main Structural Parameters

Parameter	Symbol		Obtained from
Aromatic C	CA		From ^{13}C-NMR
Unsubstituted CA	CAPH		From ^{1}H-NMR
CA substituted with CH or CH_2	CAPS12		From ^{13}C-NMR
CA substituted with CH_3	CAPS3		From ^{13}C-NMR
Substituted CA	CAPS	=	CAPS12 + CAPS3
Peripheral CA	CAP	=	CAPS + CAPH
(Unsubst. monoarom. CAP)	CAPu,m		From ^{1}H-NMR
(Unsubst. diarom. CAP)	CAPu,d		From ^{1}H-NMR
(Unsubst. triarom. CAP)	CAPu,tr		From ^{1}H-NMR
Internal CA	CAI	=	CAPH + CAI (from ^{13}C-NMR) CAPH (from ^{1}H-NMR)
Internal internal CA	CAII	=	From ^{1}H- and ^{13}C-NMR, checked by empirical rel.
Number of aromatic rings	RA	=	(CA + CAI + CAII)/6
Number of arom. ring systems	SA	=	(CA − 2CAI + CAII)/6
Aromaticity	f_a	=	CA/CL from ^{13}C-NMR
Degree of substitution of aromatic rings			CAPS/CAP
Aliphatic C	CL		From ^{13}C-NMR
CL in CH_3	CL3		From ^{13}C-NMR
CL in CH_2	CL2		From ^{13}C-NMR
CL in CH	CL1		From ^{13}C-NMR
Quaternary CL	CLq		From ^{13}C-NMR

CL3 in α position	CL3α		From ^{13}C-NMR
CL3 in γ and farther positions	CL3γ		From ^{1}H-NMR
CL3 in β position	CL3β		CL3 − CL3α − CL3γ
CL12 in α position	CL12α		CAPS12 (see above)
CL2 in long chains (≥ 8 C)	CL2lch		From ^{13}C-NMR
Naphthenic C	CN		From ^{13}C-NMR
Total aliphatic C in alkyl chains	CLch	=	CL − CN
Number of alkyl chains attached to aromatic rings = periph. arom. C carrying alkyl chains	CAPSch	=	CL3′ − Cl1 − 2CLq + CNI
Shared aromatic/naphthenic carbons	CANS	=	Empirical
Internal shared aromatic/naphthenic carbons	CANSI	=	Empirical
Internal aromatic/naphthenic ring carbons	C(AN)I	=	2(RA + RN − SAN)
Naphthenic carbons	CN	=	Empirical or from ^{13} C-NMR
Average number of CN/RN	m_n	=	Empirical
Naphthenic carbons	CN	=	m_nRN − (CNI + CANS)
Sum of internal naphthenic C and shared aromatic/naphthenic C	CNI + CANS	=	m_nRN − CN
Naphthenic internal ring carbons	CNI	=	(CNI + CANS) − CANS
Peripheral naphthenic carbons	CNP	=	CN − CNI
Number of aromatic + naphthenic rings	R(A+N)	=	CT + 1 − (HT+CA)/2
Number of naphthenic rings	RN	=	R(A+N) − RA
Number of naphthenic ring systems	S(N)	=	RN − (CNI + CANS)/2
Number of arom + naphth. ring systems	S(A+N)	=	SA + SN
Number of fused arom + naphth. ring systems	SAN	=	Empirical

Source: Ref. 41.

taken from the "paper molecules" gave good results for CAII, SN, and other parameters which depend on assumptions and empirical relations. As long as the initial guesses were correct, most parameters were found with zero error, the others with errors smaller than 15%. The deviations increased, of course, when incorrect input data were used, for example, data that would ordinarily be obtained from ^{13}C-NMR and would have experimental errors of between 10 and 25%.

Some of the earlier equations used, for example, by Charlesworth (46) and Dickinson (47), for the calculation of the number of naphthenic rings or that of naphthenic carbon atoms from NMR data, gave grossly wrong results when applied to model compounds (41,48). Obviously, some of these relations were approximations, but even as such they did not work very well.

The extended Brown-Ladner method as described here ignores the presence of heteroatoms in the samples. If heteroatoms are also to be accounted for, several more assumptions are needed about their distribution (e.g., of sulfur among thiophenic and cyclic and acyclic sulfidic groups). With and without the consideration of heteroatoms, the method offers the possibility of monitoring in great detail the average changes occurring during chemical reactions, specifically during those used in upgrading heavy ends of petroleum such as cracking, hydrocracking, desulfurization, and denitrogenation. It is reasonably fast, although, for maximum output, it requires NMR spectral editing techniques. It gives supplementary information to that obtained by MS, primarily the average number and length of alkyl chains and their degree of branching and the distribution of aromatic and naphthenic rings among hydroaromatic ring systems. The latter may be of particular value for the description of certain very complex molecules occurring in refinery streams.

B. Computer-Assisted Molecular Structure Construction

When NMR results were presented in the form of average molecular structures, it was found that more than one such structure might be compatible with a set of data. In these cases, it was difficult to find all the possible versions. Oka and co-workers (2,49) developed an algorithm, their "Computer-Assisted Molecular Structure Construction" (CAMSC), which allowed the calculation of all the molecular structures consistent with the experimental data. Their program reduced the tedium involved with the manual solution of the task and ensured that no feasible structure was overlooked. The same group (50) published an expanded version which includes functional groups containing nitrogen and which, by making use of MS data, generates several formulas with different molecular weights rather than just one.

Today, we realize that average structures are unlikely to represent the real molecules in a sample. Nevertheless, CAMSC is an interesting and powerful technique which may find other applications. The main reason for discussing it here in some detail is that it provides the basis for the linear programming method described in the following section.

Oka et al. (2) defined a set of functional groups as suitable building blocks for the construction of complex molecules of the type found in coal extracts, but the concept can be equally well applied to petroleum samples. The basic elements of their system, nine aromatic skeletons and one olefinic one, are listed in Table 8.10. Table 8.11 displays 24 aliphatic groups which were designed to attach to these skeletons or, in some cases, to each other. A clever bookkeeping system ensures that all atoms changing place when the building blocks are put together are accounted for. The parameters used for this purpose are explained in Table 8.12.

The equation

CH_3--$(CH_2)_8$-[tetralin] = [benzene] + [cyclic fragment] + 8--CH_2-- + o--CH_3

gives an example of how a molecule is built up from such building blocks. More aromatic skeletons than given in Table 8.10 can be formed by suitable combinations. For instance, the structure of biphenyl results from combining two number 2 groups with one number 13 group. Similarly, various kinds of alkyl groups can be formed by combining certain aliphatic groups, for example, an *n*-propyl group from groups 15 and 18. Oxygen atoms can be inserted in appropriate positions at the last stage after the hydrocarbon part of the structure has been established. In their first program the authors did not consider sulfur and nitrogen. They also limited the number of catacondensed aromatic and saturated rings to six and four, respectively, and that of pericondensed aromatic and naphthenic rings to four and zero.

CAMSC utilizes six experimental input parameters obtained from proton and ^{13}C-NMR spectra, elemental analysis, and molecular-weight determinations:

$H\alpha$	Hydrogen on α-carbons
$H\beta$	Hydrogen on $\beta^{(+)}$-carbons
H_{ar}	Aromatic hydrogen including phenolic OH
C_{ar}	Aromatic carbon

Table 8.10 Aromatic and Olefinic Functional Group Types

i	Aromatic groups	NC_i	AS_i
1	C = C	2	4
2		6	6
3	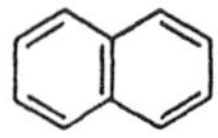	10	8
4	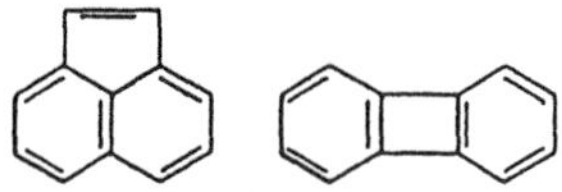	12	8
5	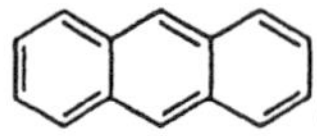 (2)[a]	14	10
6	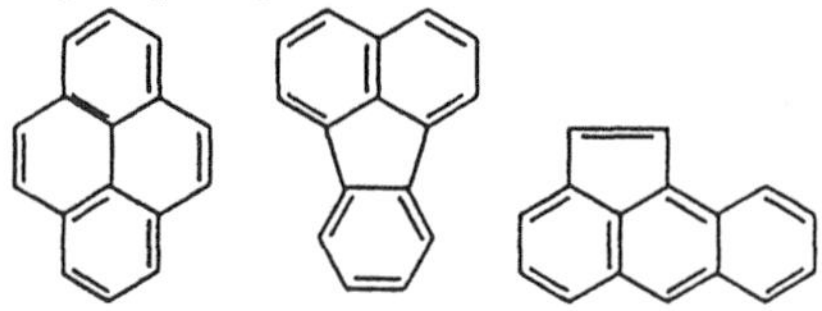	16	10
7	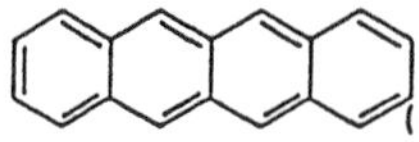 (5) 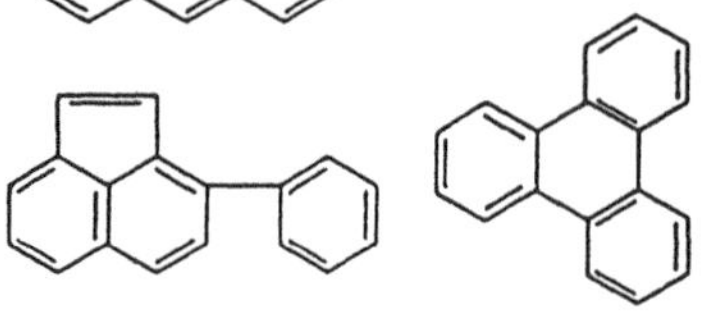	18	12
8	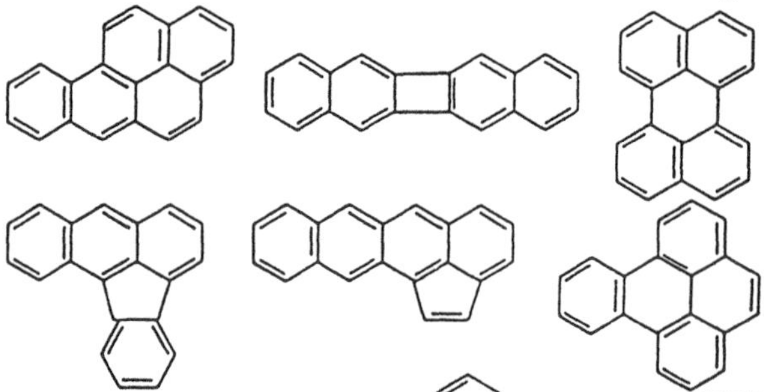	20	12
9	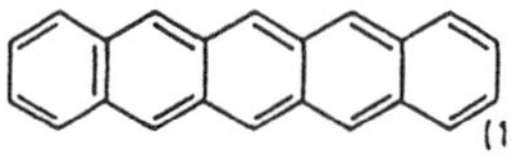 (15) 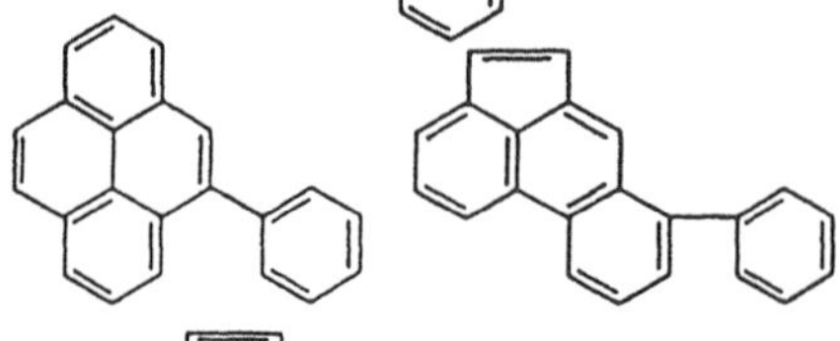	22	14
10	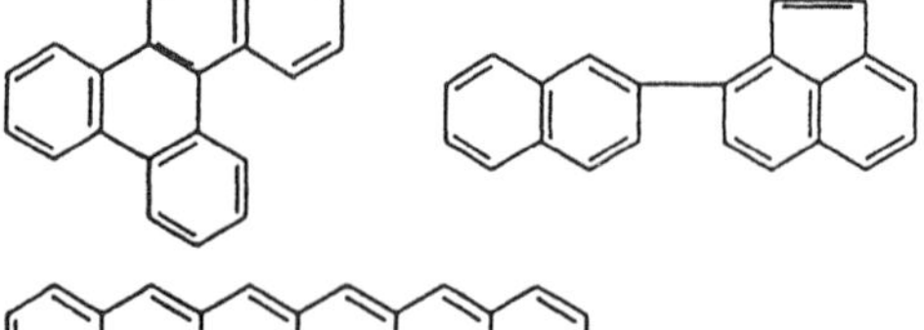 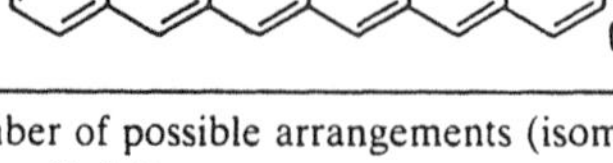 (>15)	26	16

[a]Number of possible arrangements (isomers).
Source: Ref. 2.

Table 8.11 Aliphatic Functional Group Types

i	Aliphatic groups	HA_j	S_j	H_j	C_j
11	● − CH_2 − ●	2	2	2	1
12	● − CH_2CH_2 − ●	4	2	4	2
13	●——●	0	2	0	0
14	● − CH_3	3	1	3	1
15	● − CH_2CH_3	2	1	5	2
16	● − $CH(CH_3)_2$	1	1	7	3
17	⊗ − CH_2	−1	0	2	1
18	○ − CH_2	0	0	2	1
19	●——⊗	−1	1	−1	0
20	●——○	0	1	−1	0
21	⊗——⊗	−2	0	−2	0
22	⊗——○	−1	0	−2	0
23	○——○	0	0	−2	0
24			2	4	2
25		4 (5 or 6)[a]	2	6	3
26		4 (6)	2	8	4
27	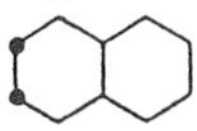	4 (5)	2	14	8
28	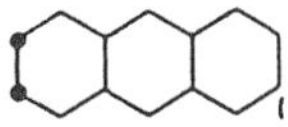 (2)[b]	4 (5)	2	20	12
29	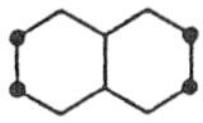	8 (9)	4	10	6
30	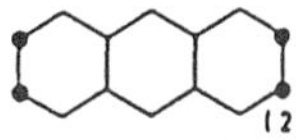 (2)	8 (10)	4	16	10
31	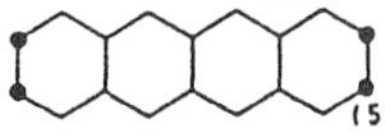 (5)	8 (10)	4	22	14
32	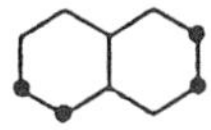	7 (9)	4	10	6
33	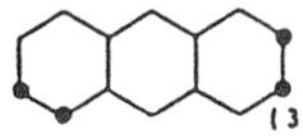 (3)	7 (11)	4	16	10
34	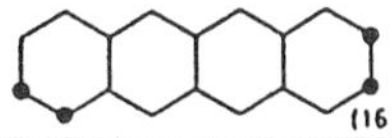 (16)	7 (11)	4	22	14

[a]Estimated correct values for alicyclic groups since the H_β hydrogens in the alicyclic groups become H_α hydrogens in some degree, as far as the ^{1}H-NMR chemical shift is concerned.
[b]Number of possible arrangements (isomers).
Key: ● aromatic carbons; ⊗ C_α carbons; ○ carbons in β or β⁺ positions.
Source: Ref. 2.

Table 8.12 Functional Descriptors Used by Oka et al.

Functional descriptor	Function
NC_i	No. of C in an aromatic or olefinic group.
AS_i	No. of ligands on periphery of unsaturated groups or aromatic ring.[a]
C_j	No. of aliphatic C.
H_j	No. of aliphatic H.
S_j	No. of C replacing aromatic H.
HA_j	No. of α-H in an aliphatic group.
i	Elements in aromatic or olefinic groups.
j	Elements in aliphatic groups.

[a]AS_i for ethylene is 4, that of aromatics equals the number of peripheral C.
Source: Ref. 49.

C_{al} Aliphatic carbon
O Oxygen in phenolic (O_p) and ether (O_e) groups

Oka et al. had considered the use of additional input from NMR spectra, for example, H3γ (from γCH_3), H from methylene bridges, H from olefinic carbons, carbonyl and ether carbons, but found it unnecessary for samples of low and moderately high molecular weights.

An example may demonstrate how CAMSC works. Input data are collected from NMR and elemental analysis, for example, $H_\alpha = 7$, $H_\beta = 3$, $H_{ar} = 8$, $C_{ar} = 16$, $C_{al} = 4$, $O_p = 1$, and $O_e = 1$. Appropriate output data are chosen from Tables 8.10 and 8.11, for example, benzene skeleton, naphthalene skeleton, methylene bridge, methyl group, ethyl group, and direct bond. Both data sets are fed into the computer which proceeds to construct the hydrocarbon structure according to the flow chart shown in Fig. 8.16. In this case, Structure 1 results:

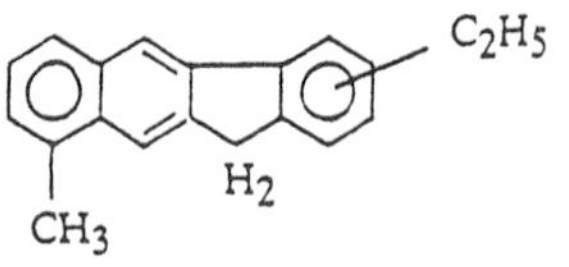

Structure 1

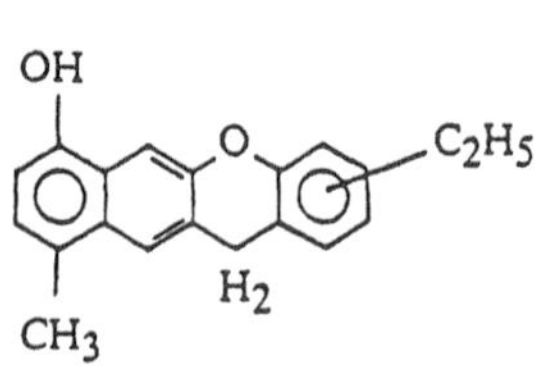

Structure 2

Then the oxygen atoms are inserted by hand where deemed appropriate and consistent with the ^{1}H-NMR spectrum. The final result is Structure 2.

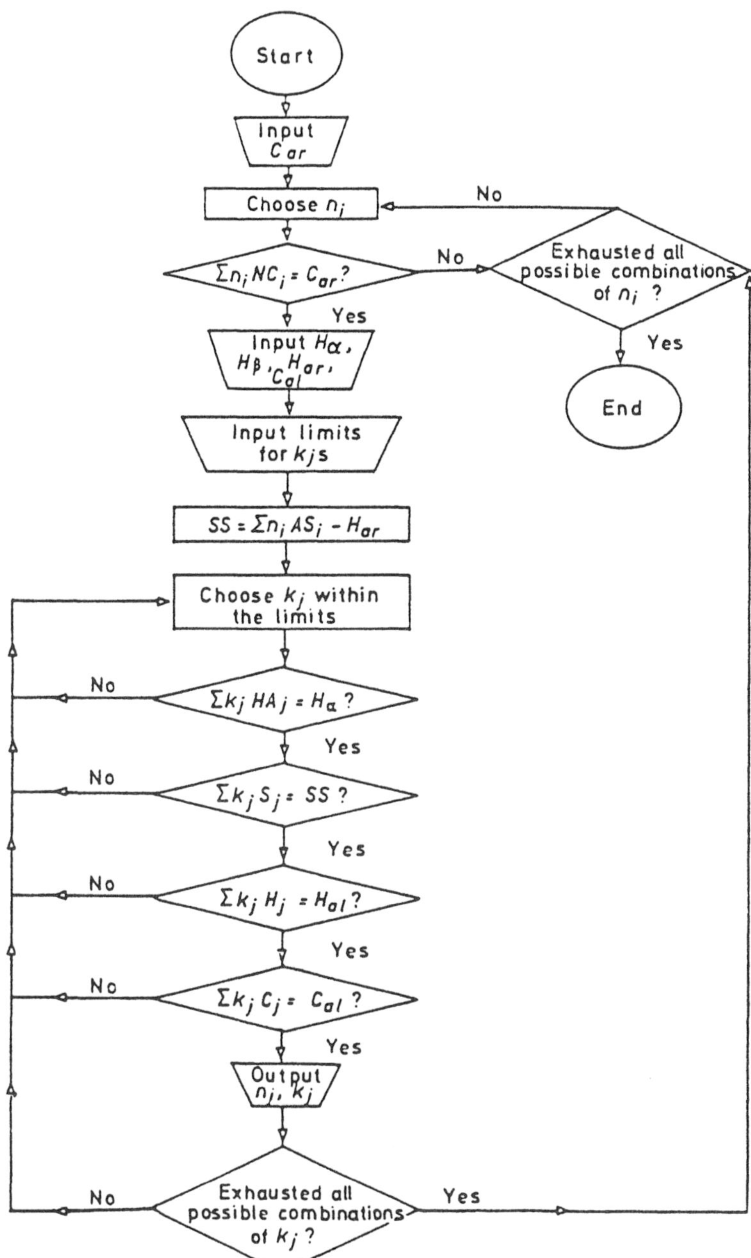

Figure 8.16 Flow chart of computer program used for CAMSC. (From Ref. 2. Reproduced with permission of the publisher.)

In this simple example, the output data allow only one hydrocarbon structure. However, the number of feasible structures increases with molecular weight. The number of permissible structures can be reduced by utilizing more analytical information and also by chemical reasoning and experience.

Oka et al. (2) applied their method with good success to the data from fractions of coal extracts published by several research groups. Their results, obtained with CAMSC and sharpened with keen reasoning, were at least as good as those of the original researchers and in some cases more complete.

C. Functional Group Analysis by Linear Programming

If the number of structural parameters or functional groups obtainable is limited because of insufficient experimental data, linear programming can be used to estimate certain parameters which are not accessible to measurement or direct calculation. Early attempts at applying mathematical optimization techniques to the structural analysis of petroleum and other fossil fuel fractions by several groups* did not work too well, mainly because they overstressed the limited NMR capabilities available at their time and because their methods treated the samples as consisting only of average structures rather than of diverse components in broad mixtures.

The approach by Gavales (54), later improved by Petrakis et al. (55), is fundamentally different from those early versions. These workers developed a linear programming routine based on the set of functional parameters displayed in Tables 8.10 and 8.11 which they had originally selected for their CAMSC method. They start out with a set of balance equations

$$\sum_{j=1}^{n} A_{ij}y_j = b_i \qquad (i = 1, \ldots, m) \tag{8.9}$$

where vector y represents the unknown concentrations of the functional groups; vector b, the experimental data; and where the elements a_{ij} of matrix A are the stochiometric coefficients. The subscript i refers to an experimental data point and j to a functional group. The element a_{ij} is then the contribution of data point i to functional group j. For example, if data point 2 is the concentration of aromatic hydrogens, and functional group 3 is the naphthalene ring as in Table 8.10, then, in this case, $a_{23} = 8$. A chemically feasible solution must have all functional group concentrations, y_i, positive or zero, that is,

$$y_i \geq 0 \tag{8.10}$$

*Bestougeff and Pierre (51), Hirsch and Altgelt (37), Haley (38), Katayama et al. (52), and Kiet et al. (53).

If the number of experimental data, m, is smaller than the number of functional groups, n, Eq. (8.9), with the restraint of Eq. (8.10), has either no solution or more than one (mathematically an infinite number). Linear programming provides a systematic procedure for finding such solutions. The textbooks by Gass (56) and by Hillier and Lieberman (57) are two of many useful guides to the practical application of this technique.

The best non-negative solution of Eq. (8.9) can be found by maximizing the function J:

$$J = \sum_{i=1}^{n} c_i y_i \tag{8.11}$$

Treating the problem as one of optimization ensures that a solution can be generated and that this solution is the best in the sense of maximizing J among all feasible solutions of Eq. (8.9). The coefficients c_i are chosen to express qualitative information or estimates about the relative importance or abundance of any of the functional groups. For example, if we would expect that naphthalene is a more abundant ring system than phenanthrene in the sample, we might choose 2 for the c_i of naphthalene and 1 for that of phenanthrene. The solution may still contain phenanthrene groups, but now most likely of lower concentration than for the case of equal C_i's. Each set of c_i yields a solution. Some of these solutions may be identical even for different sets of c_i.

Not all of the functional groups originally defined and shown in Tables 8.10 and 8.11 can be used for a linear programming run. Their number must not be too much greater than the number of experimental input data. Therefore, they must be carefully chosen to be sufficient and concise. This does not seem to be a problem with coal-derived samples (2), but it could be quite difficult for the analysis of some refinery streams which may have rather large aromatic ring clusters and, thus, presumably, a greater variety in their structure than straight-run petroleum fractions or coal extracts. Gavales et al. (54) added some more experimental input data to those chosen by Oka et al. (2).

Oka et al. (49), on whose paper much of the above discussion is based, give an example of linear programming applied to a coal asphaltene fraction. The experimental input data are listed in Table 8.13 and the functional groups used in the analysis, in Table 8.14. Table 8.15 shows the matrix A, constructed from the data of Table 8.14, and the values of vector b, taken from Table 8.13. The negative numbers for some of the aromatic H atoms result from the definition of the filled circles in groups 6–9 and 12–14. These circles represent peripheral aromatic carbons, which lost their hydrogens when they were connected to aliphatic substituents.

Table 8.13 Experimental Input Data for Numerical Example

Elemental analysis %	^{1}H-NMR fraction	Concentration in sample Moles per 100 g of sample	
C = 803	H_A = 0.313	C_A = 4.97	H_F = 0.422
H = 6.6	H_{OH} = 0.054	C_{ali} = 1.75	H_{OH} = 0.356
O = 10.8	H_F = 0.064	H_A = 2.07	O = 0.0675
N = 1.05	H_α = 0.288	H_α = 1.90	N = 0.0750
S = 1.05	H_β = 0.222	H_β = 1.47	S = 0.0328
	H_γ = 0.059	H_γ = 0.389	

Source: Ref. 49.

Table 8.14 Functional Groups Chosen for Numerical Example

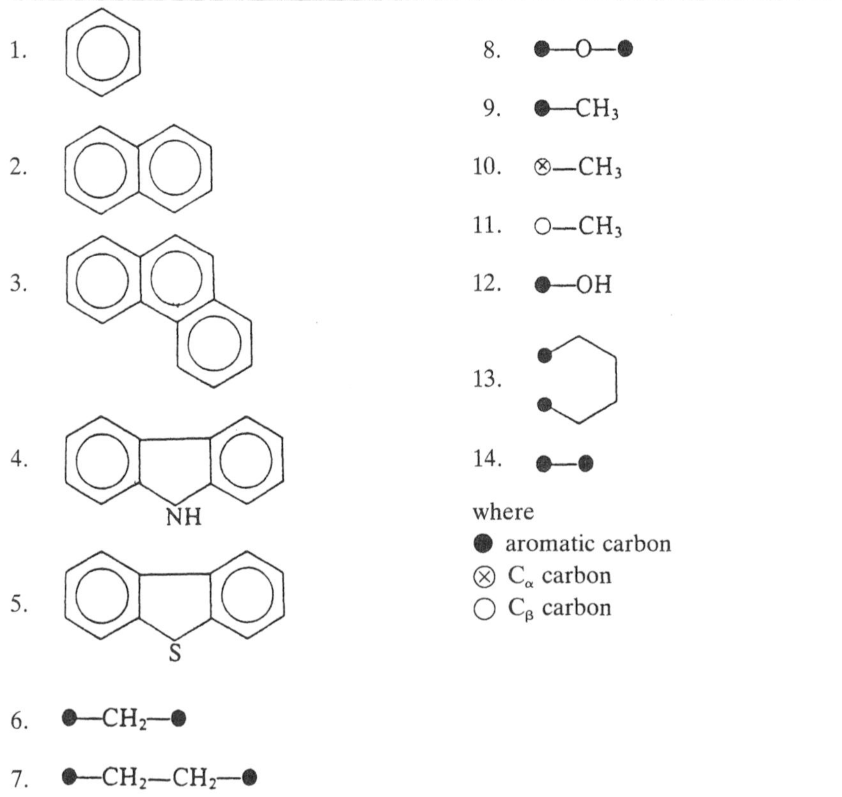

Source: Ref. 49.

Table 8.15 Matrix *A* Constructed for Numerical Example

	A Matrix														*b* Vector
Functional group	1	2	3	4	5	6	7	8	9	10	11	12	13	14	b
Aromatic C	6	10	14	12	12	0	0	0	0	0	0	0	0	0	4.97
Aliphatic C	0	0	0	0	0	1	2	0	1	1	1	0	4	0	1.75
Aromatic H	6	8	10	8	8	−2	−2	−2	−1	0	0	−1	−2	−2	2.07
α-H	0	0	0	1	0	0	4	0	3	−1	0	0	4	0	1.90
β-H	0	0	0	0	0	0	0	0	0	3	−1	0	4	0	1.47
γ-H	0	0	0	0	0	0	0	0	0	0	3	0	0	0	0.389
Phenolic H	0	0	0	0	0	0	0	0	0	0	0	1	0	0	0.356
Methylene Bridge H	0	0	0	0	0	2	0	0	0	0	0	0	0	0	0.422
Oxygen	0	0	0	0	0	0	0	1	0	0	0	1	0	0	0.675
Nitrogen	0	0	0	1	0	0	0	0	0	0	0	0	0	0	0.0750
Sulfur	0	0	0	0	1	0	0	0	0	0	0	0	0	0	0.0328

Source: Ref. 49.

The results of the linear programming procedure, that is, the functional group concentrations, are presented in Table 8.16 for three sets of c_i. In the first set, all c_i values were made equal, that is, all the functional groups listed in Table 8.14 were assumed to have the same probability or basic abundance. Interestingly, this led to zero concentration for the naphthalene ring structure in the mixture. In the second set of c_i, the relative probability of groups 2 and 13, denoting the naphthalene group and a saturated ring fused to an aromatic ring (tetralin type), respectively, was chosen five times as large as the rest. With this choice, the linear programming procedure gave a nonzero abundance of both of these groups in the average structure. The third set of c_i, a 1:1 mixture of the previous two, gave intermediate results. This example graphically illustrates the importance of additional information for the choice of the c_i values. It also shows that, in the absence of such information, almost any result can be obtained by a judicious (though otherwise arbitrary) choice of c_i.

Petrakis et al. (55) modified and refined in several ways the linear programming method developed by Gavalas et al. (54).

1. Instead of maximizing the linear function J, they minimize a function P, which is the square of the deviation between an input and a calculated (concentration) value, y_j.
2. More input data were used in the balance equations, $A_{ij}y_j = b_i$.
3. Additional secondary, less precise, data were incorporated, together with estimates as auxiliary information (the former c_i's), in a separate function P.

Petrakis et al. distinguish among three kinds of minimization functions. The most basic "general minimization function" is used if no secondary information on the relative abundance of the various functional groups is available and all functional groups are assumed to be equally probable. This corresponds to all the c_i values of the earlier approach being equal to 1.

A concrete example may serve to illustrate how P is constructed in this case. Let us consider the distribution of carbons among the functional groups in Table 8.16. Of the 14 functional groups, 11 were found to be represented by the experimental data. If each of these is assumed to be equally probable, each contains 1/11 of the total carbon. The general minimization function for carbon ($i = 1$) is then

$$P_1 = \sum_{A_{1j} \sim 0} = \left(\frac{b_1}{11A_{1j}} - y_j \right)^2 \tag{8.12}$$

Table 8.16 Results of Numerical Example

Functional group	Concentration (moles/100 g of liquid)		
	All $C_i = 1$	$C_2 = C_{13} = 5$; all other $C_i = 1$	1/2 solution 1 + 1/2 solution 2
1. 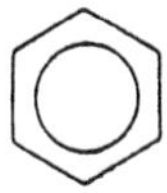	0.618	0.395	0.506
2. 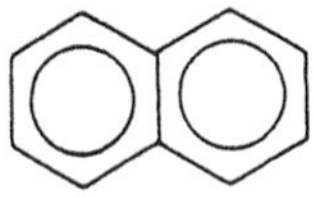	0	0.133	0.0665
3. 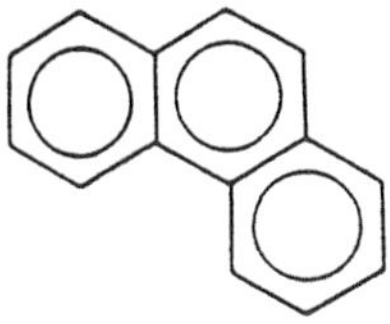	0	0	0
4. 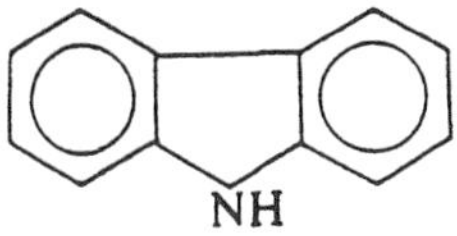 NH	0.0750	0.0750	0.0750
5. 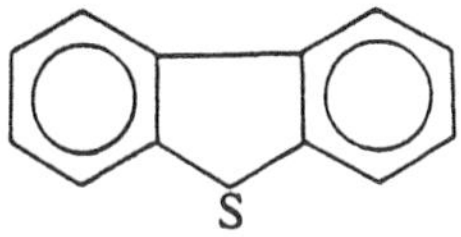 S	0.0328	0.0328	0.0328
6. ●—CH_2—●	0.211	0.211	0.211
7. ●—CH_2—CH_2—●	0.0901	0	0.0450
8. ●—O—●	0.319	0.319	0.319
9. ●—CH_3	0.666	0.546	0.606
10. ⊗—CH_3	0.533	0.353	0.443
11. ○—CH_3	0.130	0.130	0.130
12. ●—OH	0.356	0.356	0.356
13. 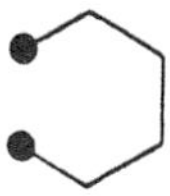	0	0.135	0.0675
14. ●—●	0.118	0	0.0590

All concentrations, y_j, must be non-negative, that is, $y_j \geq 0$, as before. Also as before, the constraints of Eq. (8.9) must be obeyed. For instance, their general minimization function for the distribution of aromatic hydrogen is

$$P_2 = \sum_{A2j\sim 0} = \left(\frac{1}{A_{2j}}(b_2 + 2y_6 + 2y_7 + 2y_8 + y_9 + y_{12} + 2y_{13} + 2y_{14}) - y_j\right)^2 \quad (8.13)$$

The terms $2y_6$, $2y_7$, and so on must be added to b_2 because the respective functional groups (6, 7, etc.) displace the corresponding numbers of hydrogens from the aromatic structures to which they attach.

Petrakis et al. have similar expressions for the remaining types of atoms known from experimental input, for example, alpha, beta,* and gamma hydrogen, oxygen, and sulfur. The final minimization function is the sum of those individual functions (in this case 7):

$$P_8 = \sum_{i=1}^{7} = P_i \quad (8.14)$$

To repeat, this is the general minimization function for seven input data and no additional secondary information or guesses.

Next, Petrakis et al. (55) incorporate high-resolution nonfragmenting MS data, and finally also data from low-resolution fragmenting mass spectrometry. Because several assumptions must be made in the evaluation of the mass spectra of heavy fractions for this purpose, the authors consider these data less accurate than the elemental analysis and NMR data (a debatable point, depending on the sample). Therefore, they use the MS data as secondary information in specific minimization functions as a means to estimate the relative abundances of the functional groups of interest.

In their utilization of mass spectra, they first assign structures to each of the major peaks. From these molecular structures and peak heights, they calculate the concentration of the various functional groups of interest present in the compound which is represented by a peak. Summation over all appropriate peaks gives the relative abundances, f_i, of these functional groups in the sample. Neglecting or reassigning those present at very low concentration helps to keep the number of functional groups low.

*Beta hydrogen is defined here as H in CH_n groups, two or more places removed from an aromatic ring. Gamma is defined as H in CH_3 groups, three or more places removed from an aromatic ring.

Minimizing the function

$$P = \sum_{i=1}^{n} (y_j - f_i)^2 \tag{8.15}$$

subject to the constraints of Eqs (8.9) and (8.10) gives the desired set of functional group concentrations, y_i. This set satisfies the seven primary input data from elemental composition and NMR and also, to the greatest extent possible, the secondary input data, f_i, from MS.

For the third minimization, Petrakis et al. extract additional secondary information from high-voltage–low-resolution MS. According to these authors, this method allows the estimation of the nature and amount of ring systems as they would be without their alkyl substituents.

They tested their technique with two mixtures of 11 model compounds including tetralin, biphenyl, diphenyl ethane, ethyl naphthalene, and dinaphthyl methane in slightly different proportions. The results, shown in Table 8.17, are fairly close to the known true average values for most functional groups, especially when MS data were included in the analysis.

They also applied their method to four real samples: a coal liquid and three hydrodesulfurization products obtained from it by treatment for different lengths of time. Petrakis et al. could show how some of the functional groups were used up by the treatment, whereas others remained intact (e.g., benzene, naphthalene, and phenanthrene rings). Allen et al. (58) applied this technique to several coal-derived oil and asphaltene fractions. Later, Allen (59) and Allen et al. (60) further modified the procedure and used it to compare the structural profiles of the heavier fractions of crude oils, shale oil, tar sands, and coal liquids.

In their modification Allen et al. (60) expanded the set of functional groups to a total of 25, which is more than can ordinarily be used in one operation because of the limited input information. Eight of these groups contain heteroatoms, for example, dibenzothiophene, quinoline, and carbazole. He considers this assembly a pool from which he selects those groups which, from experience, he thinks he needs for a specific sample. For petroleum and shale oil distillates, for instance, he ignores bridges between ring systems. He also utilized more ^{13}C-NMR data than before, thus increasing the number of input data. In addition, he incorporated data from analytical separations. The greater input allows him to utilize a larger assortment of functional groups from his pool, sometimes even up to its entirety. His article demonstrates the great flexibility which the method of functional group analysis has now attained.

A good source for more detail about this method is a small book by Petrakis and Allen (5). Finally, a word of caution. This method works best

Table 8.17 Functional Group Concentrations Calculated by Linear Programming from Two Model Compound Mixtures

A. Mixture composition

	Mixture			Mixture	
	#1 (g)	#2 (g)		#1 (g)	#2 (g)
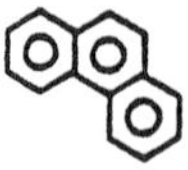	13.2	13.2	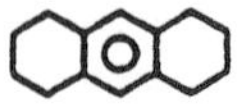	13.1	13.1
				31.4	14.3
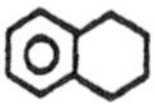	3.3	3.3			
	14.0	14.0	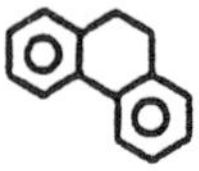	6.7	6.7
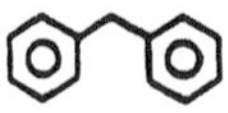	7.4	7.4	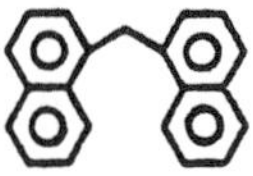	7.8	7.8
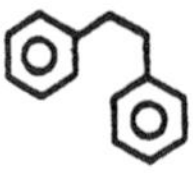	6.9	6.9	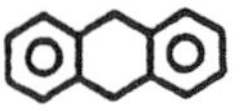	6.3	6.3
	7.0	7.0			

B. Functional group concentrations

	Mixture #1		Mixture #2			
	True conc.	Result of algorithm Eqs. (8.1), (8.2), (8.10)	True conc.	Result of algorithm Eqs. (8.1), (8.2), (8.10)	Concentration estimated from mass spectrum	Result of algorithm Eqs. (8.1) (8.2), (8.10)
	0.585	0.463	0.585	0.420	0.503	0.516
	0.309	0.279	0.199	0.245	0.211	0.215
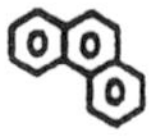	0.075	0.148	0.075	0.112	0.098	0.093
—CH_3	0.251	0.250	0.141	0.140	0.177	0.140
—CH_3	0.202	0.240	0.092	0.150	0.108	0.098
	0.129	0.000	0.129	0.000	0.113	0.096
	0.143	0.153	0.143	0.135	0.097	0.106
	0.076	0.109	0.076	0.138	0.063	0.099
	0.165	0.137	0.165	0.122	0.121	0.162

Source: Ref. 55.

with an exactly defined, or better yet, an overdefined system, that is, a set of terms where the number of experimental data is at least as large as the number of parameters to be calculated. With an underdefined system, there is no guarantee that the results are accurate; they cannot even be checked for accuracy except by independent means. An exactly defined or an overdefined system would allow the introduction of experimental error terms in the calculation and, thus, an estimate of the precision. Eventually, a balance will have to be found between reliability and maximum output data.

IV. SUMMARY

Spectral editing and 2D NMR techniques have made it possible to distinguish between many structural parameters which were not accessible before because their peaks overlapped so severely with others that they could not be evaluated. Now subspectra of CH_3, CH_2, CH, and C_q can be obtained separately. Hetero- and homocorrelated 2D measurements allow the assignment of specific functional groups to a number of peaks in otherwise overcrowded spectra.

Computer-assisted methods such as the extended Brown-Ladner method and functional group analysis can combine the information from NMR and other results for an integrated structural analysis, yielding a rather detailed description of the average composition of a sample. These methods, of course, are limited by the constraints of the NMR and other primary results and, to repeat, they give only average results.

Altgelt's version of the extended Brown-Ladner method utilizes about 30 rigorous mathematical relations between structural parameters and an additional approximate one. For very large pericondensed ring systems and for heterocompounds several assumptions must be made. So far, his scheme has only been tested with model calculations and not with real data from spectra of model compound mixtures. In principle, it can provide information on the nature of the paraffinic chains and complex ring systems in petroleum fractions which is not available from the combination of chromatography and FIMS.

Allen's functional group analysis has been applied to mixtures of model compounds and to a number of fossil fuel fractions, including heavy petroleum fractions, apparently with good success.

REFERENCES

1. Brown, J. K., and Ladner, W. R. 1960. A study of the hydrogen distribution in coal-like materials by high-resolution nuclear magnetic resonance spec-

troscopy II—A comparison with infra-red measurement and the conversion to carbon structure. *Fuel*, **39**:87–96.
2. Oka, M., Chang, H.-C., and Gavales, G. R. 1977. Computer-assisted molecular structure construction for coal-derived compounds. *Fuel*, **56**:3–8.
3. Shoolery, J. N. 1984. Recent developments in ^{13}C- and proton-NMR. *J. Nat. Prod.*, **47**(2):226–259.
4. Dennis, L. W., and Pabst, R. E. 1987. Polarization Transfer and 2-D Experiments in NMR of Solutions of Humic Materials and Fossil Fuel Liquids. *NMR of Humic Substances and Coal*, edited by A. L. Warshaw and M. A. Mikita. Lewis Publishers, Chelsea, Chap. 6, pp. 99–128.
5. Petrakis, L., and Allen, D. T. 1987. *NMR for Liquid Fossil Fuels*. Analytical Spectroscopy Library, Volume 1. Elsevier, Amsterdam.
6. Collin, P. J., and Wilson, M. A. 1983. Use of INEPT and GASPE N.M.R. pulse sequences. *Fuel*, **62**:1243–1246.
7. Doddrell, D. M., Pegg, D. T., and Bendall, M. R. 1982. Distortionless enhancement of NMR signals by polarization transfer. *J. Magn. Reson.*, **48**:323–327.
8. Bendall, M. R., and Pegg, D. T. 1983. Complete accurate editing of decoupled ^{13}C spectra using DEPT and a quaternary-only sequence. *J. Magn. Reson.*, **53**:272–296.
9. Barron, P. F., Bendall, M. R., Armstrong, M. J., and Atkins, A. R. 1984. Application of the DEPT pulse sequence for the generation of $^{13}CH_n$ subspectra of coal-derived oils. *Fuel*, **63**:1276–1280.
10. Bendall, M. R., Pegg, D. T., Dodrell, D. M., Johns, S. R., and Willing, R. I. 1982. Pulse sequence for the generation of a ^{13}C subspectrum of both aromatic and aliphatic quaternary carbons. *J. Chem. Soc. Chem. Commun.* 1982, 1138–1140.
11. Cookson, D. J., and Smith, B. E. 1981. Improved resolution of low field quaternary carbon resonances in ^{13}C N.M.R. spectroscopy. *J. Chem. Soc. Chem. Commun.*, 1981, 12–13.
12. Snape, C. E., Ray, G. J., and Price, C. D. 1986. Two-dimensional N.M.R. analysis of aromatic fractions from a coal liquifaction solvent. *Fuel*, **65**:877–880.
13. Cookson, D. J., and Smith, B. E. 1982. ^{1}H and ^{13}C N.M.R. spectroscopic methods for the analysis of fossil fuel materials: Some novel approaches. *Fuel*, **61**:1007–1013.
14. Dereppe, J. M., and Moreaux, C. 1985. Measurement of CH_n group abundances in fossil fuel materials using DEPT ^{13}C N.M.R. *Fuel*, **64**:1174–1176.
15. Snape, C. E., and Marsh, M. K. 1985. Structural analysis of heavy fossil fuel fractions using ^{13}C NMR spectral editing. *Am. Chem. Soc. Pet. Chem. Prepr.*, **30**(2): 20–25.
16. Netzel, D. A., and Miknis, F. P. 1981. NMR study of US and western shale oils produced by pyrolysis and hydropyrolysis. *Fuel*, **61**:1101–1109.

17. Netzel, D. A., and Guffey, F. D. 1989. NMR and GC/MS investigation of the saturate fractions from the Cerro Negro heavy petroleum crude. *Energy Fuels*, **3**:455–460.
18. Bartle, K. D., Ladner, W. R., Martin, T. G., Snape, C. E., and Williams, D. F. 1979. Structural analysis of supercritical-gas extracts of coals. *Fuel*, **58**:413–422.
19. Altgelt, K. H. 1992. To be published.
20. Dreeskamp, H., Potemka, T., and Müller, A. 1989. Aspects of quantitative ^{13}C-N.M.R.-spectroscopy of high boiling oil residues and coal tar pitches. *Fuel*, **68**:972–977.
21. Abu-Dagga, F., and Rüegger, H. 1988. Evaluation of low boiling crude oil fractions by N.M.R. spectroscopy. Average structural parameters and identification of aromatic components by 2D N.M.R. spectroscopy. *Fuel*, **67**:1255–1262.
22. Snape, C. E. 1982. Spin echo ^{13}C NMR: A valiable aid to the characterization of coal liquefaction products. *Fuel*, **61**:775–777.
23. Snape, C. E. 1982. Estimation of quaternary and tertiary aromatic carbon in coal liquefaction products by spin echo ^{13}C N.M.R. *Fuel*, **61**:1164–1167.
24. Snape, C. E. 1983. Estimation of aliphatic C/H ratios for coal liquefaction products by spin-echo ^{13}C N.M.R. *Fuel*, **62**:621–624.
25. Snape, C. E. 1983. Spin Echo ^{13}C NMR: A valiable aid to the characterization of coal liquefaction products. *Proc. Int. Conf. Coal Sci.*, 624–627.
26. Snape, C. E., Ladner, W. R., Petrakis, L., and Gates, B. C. 1984. The chemical nature of asphaltenes from some coal liquifaction processes. *Fuel Proc. Tech.*, **8**:155–168.
27. Galya, L. G., and Young, D. C. 1983. A new carbon-13 NMR method for determining aromatic, naphthenic, and paraffinic carbon. *Am. Chem. Soc. Pet. Chem. Prepr.*, **28**(5):1316–1318.
28. Young, D. C., and Galya, L. G. 1984. Determination of paraffinic, naphthenic and aromatic carbon in petroleum derived materials by carbon-13 NMR spectrometry. *Liq. Fuels Technol.*, **2**(3):307–326.
29. Yamashita, G. T., Saetre, R., and Somogyvari, A. 1989. Evaluation of integration procedures for PNA analysis by ^{13}C-NMR. *Am. Chem. Soc. Pet. Chem. Prepr.*, **34**(2):301–305.
30. Cookson, D. J., and Smith, B. E. 1983. Determination of carbon C, CH, CH_2 and CH_3 group abundances in liquids derived from petroleum and coal using selected multiplet ^{13}C N.M.R. spectroscopy. *Fuel*, **62**:34–38.
31. Cookson, D. J., and Smith, B. E. 1983. Investigation of aromatic carbon sites in materials derived from petroleum and coal using ^{13}C N.M.R. methods. *Fuel*, **62**:39–43.
32. Cookson, D. J., and Smith, B. E. 1983. Quantitative estimation of CH_n group abundances in fossil fuel materials using ^{13}C N.M.R. methods. *Fuel*, **62**:986–988.
33. Cookson, D. J., and Smith, B. E. 1987. An Investigation of the Utility of ^{1}H and ^{13}C NMR Chemical Shift Data When Applied to Fossil Fuel Products.

Coal Science and Chemistry, edited by A. Volborth. Elsevier Science Publishers, Amsterdam, pp. 31–60.
34. Brown, J. K., and Ladner, W. R. 1960. A study of the hydrogen distribution in coal-like materials by high-resolution nuclear magnetic resonance spectroscopy I—the measurement and interpretation of the spectra. *Fuel*, **36**:79–86.
35. Williams, R. B. 1958. *Symposium on Composition of Petroleum Oils, Determination and Evaluation*. ASTM Spec. Tech. Publ. 224, p. 168.
36. Van Krevelen, D.W. 1961. *COAL*. Elsevier Publishing Company, Amsterdam, pp. 322, 435–452.
37. Hirsch, E., and Altgelt, K. H. 1970. Integrated structural analysis. A method for the determination of average structural parameters of petroleum heavy ends. *Anal. Chem.*, **42**:1330–1339.
38. Haley, G. A. 1972. Unit sheet weights of asphalt fractions determined by structural analysis. *Anal. Chem.*, **44**:580–585.
39. Aczel, T., Williams, R. B., Brown, R. A., and Pancirov, R. J. 1978. Chemical Characterization of Synthoil Feeds and Products. *Analytical Methods for Coal and Coal Products. Vol. 1*, edited by C. Karr, Jr., Academic Press, New York. Chapter 17, 499–540.
40. Herod, A. A., Ladner, W. R., and Snape, C. E. 1981. Chemical and Physical Structure of Coal. *Philos. Trans. Roy. Soc. London, Ser. A*, **300**:3–14.
41. Altgelt, K. H. 1992. Manuscript in preparation.
42. Snape, C. E., Ladner, W. R., and Bartle, K. D. 1983. Structural Characterization of Coal Extracts by NMR. *Coal Liquifaction Products. Vol. 1. NMR Spectroscopic Characterization and Production Processes*, edited by H.D. Schultz. John Wiley & Sons, New York, Chap. 4, pp. 69–84.
43. Fischer, P., Stadelhofer, J.W., and Zander, M. 1978. Structural investigation of coal-tar pitches and coal extracts by ^{13}C N.M.R. spectroscopy. *Fuel*, **57**:345–352.
44. Maekawa, Y., Yoshida, T., and Yoshida, Y. 1979. Quantitative ^{13}C N.M.R. spectroscopy of a coal-derived oil and the assignment of chemical shifts. *Fuel*, **58**:864–872.
45. Cookson, D. J., Latten, J. L., Shaw, I. M., and Smith, B. E. 1985. Property-composition relationships for diesel and kerosine fuels. *Fuel*, **64**:509–519.
46. Charlesworth, J. M. 1980. Influence of temperature on the hydrogenation of Australian Loy-Yang Brown Coal. 2. Structural analysis of the asphaltene fractions. *Fuel*, **59**:865–870.
47. Dickinson, E. M. 1980. Structural composition of petroleum fractions using proton and ^{13}C N.M.R. spectroscopy. *Fuel*, **59**:290–294.
48. Shenkin, P. S. 1983. Hidden assumptions in the average structure methods. *Amer. Chem. Soc. Div. Petr. Chem.*, **28**(5); 1387–1375.
49. Oka, M., Hsia, Y.-P., and Gavales, G. R. 1983. Computer-Assisted Molecular Structure Construction. *Coal Liquifaction Products. Vol. 1. NMR Spectroscopic Characterization and Production Processes*, edited by H.D. Schultz. John Wiley & Sons, New York, Chap. 9, pp. 187–213.

50. Chang, P., Oka, M., and Hsia, Y.-P. 1982. A computer program, AMCLA, to analyze aromatic mixture and coal liquid. *Energy Sources*, **6**:67–83.
51. Bestougeff, M. and Pierre, M. 1968. Recherche de formules developées D'hydrocarbures de poids moléculaire élevé. *Ann. Chim.* **3**:481–490.
52. Katayama, Y., Hosoi, T., and Takeya, G. 1975. A method of structure determination for aromatic heavy oils with a computer. *Nippon Kagaku Zasshi*, **1**:127–134.
53. Kiet, H. H., Malhorta, S.-L., and Blanchard, L.-P. 1978. Structure parameter analyses of asphalt fractions by a modified mathematical approach. *Anal. Chem.*, **50**:1212–1218.
54. Gavalas, G. R., Allen, D., and Oka, M. 1980. Unpublished results, quoted by Oka et al. (49).
55. Petrakis, L., Allen D. T., Gavales, G. R., and Gates, B. C. 1983. Analysis of synthetic fuels for functional group determination. *Anal. Chem.*, **55**:1557–1564.
56. Gass, S. I. 1958. *Linear Programming—Methods and Applications*. McGraw-Hill Book Co., New York.
57. Hillier, F. S., and Lieberman, G. J. 1984. *Operations Research*. Holden-Day, San Francisco.
58. Allen, D. T., Petrakis, L., Grandy, D. W., Gavales, G. R., and Gates, B. C. 1984. Determination of functional groups of coal-derived liquids by n.m.r. and elemental analysis. *Fuel*, **63**:803–809.
59. Allen, D. T. 1985. NMR characterization and property estimation in heavy fuels. *Am. Chem. Soc. Pet. Chem. Prepr.*, **30**:270–273.
60. Allen, D. T., Grandy, D. W., and Petrakis, L. 1985. Heavier fractions of shale oils, heavy crudes, tar sands, and coal liquids: Comparison of structural profiles. *Ind. Eng. Chem. Process Des. Dev.*, **24**:737–742.
61. Cookson, D. J., and Smith, B. E. 1987. One- and two-dimensional NMR methods for elucidating structural characteristics of aromatic fractions from petroleum and synthetic fuels. *Energy Fuels*, **1**:111–120.
62. Netzel, D. A. 1987. Quantitation of carbon types using DEPT/QUAT NMR puls sequences: Application to fossil-fuel-derived oils. *Anal. Chem.*, **59**:1775–1779.

9

Complementary Characterization Methods

I. GENERAL CONSIDERATIONS

We believe highly useful information for the petroleum chemist on the composition of heavy petroleum fractions can be provided by the right combination of separation and spectroscopic methods. In our experience, chromatographic compound-class separation and nonfragmenting MS analysis of the fractions is usually the most revealing combination as discussed in Chapters 6 and 7. In those cases, when more detail on the nature of the aromatic and naphthenic ring systems and their alkyl substitution is desired, NMR can provide most of the answers. Additional methods are needed for certain specific problems, for example, the determination of polar functional groups or the elucidation of the way vanadium and nickel are bound to the molecules. This is where IR, UV, and x-ray spectroscopy can help. Nondistillable residues and alkane-insoluble petroleum fractions (asphaltenes) can be analyzed by NMR, IR, and x-ray spectroscopy which give average concentrations of certain structural parameters and functional groups. Considerable additional detail is accessible by the application of mild degradation methods such as selective oxidation and mild pyrolysis.

We restrict our discussion of these chemical degradation methods mainly to the analysis of the otherwise almost intractable nondistillable (>1300°F, >700°C AEBP) and alkane-insoluble petroleum fractions (SEF fractions, asphaltenes). Most of the molecules in these fractions contain several heteroatoms, often different ones. This fact prevents their further separation into simple chemically defined compound classes. Because of their low volatility, these materials are difficult to analyze by MS. The average results provided by NMR and IR are of limited value because of the enormous complexity of these fractions in terms of both MW and chemical composition.

A better method for their analysis is their mild degradation into fragments large enough and sufficiently unaltered to enable the researcher to draw conclusions about their relation to the original molecules. In Section III, we touch on some of these methods and their application to insoluble fractions and nondistillable residues.

The most important of these complementary methods—in our biased view—are briefly described in this chapter. We realize the somewhat arbitrary selection, but we did not intend to write a handbook with a complete array of all the methods which have been applied in the analysis of heavy petroleum fractions.

II. SPECTROSCOPIC METHODS

A. Infrared Spectroscopy

1. Basics

Infrared spectroscopy is a well-established method which used to be employed mainly for comparative, semiquantitative analysis. With presently available instruments, truly quantitative work can be done. The molecular vibrations causing IR absorption, the resulting spectrum, and band assignments have been expertly described, for example, by Bellamy (1), Nakanishi and Solomon (2), and recently, by Lin-Vien et al. (3). Extensive lists of spectra are available in print (Aldrich, Ref. 69) and for computer access (Sadtler, Ref. 70). Kemp's book, now in its third edition (4), gives a brief and simple but solid introduction into this field with good practical hints and guidance. Instead of covering the basics here again, with few exceptions we defer to his excellent text.

Spectra are displayed either as percent transmittance or as absorptivity versus frequency (cm^{-1}). Transmittance, T, defined as the ratio of transmitted light over incident light, or percent transmittance ($100T$), usually shows more detail over the entire range and is generally the preferred display. Absorbance, A, on the other hand, is proportional to the concentration and is, therefore, used for quantitative measurements. The two are related by the equation

$$A = -\log_e (T)$$

It is more common now to record the apparent absorptivity, A_{app},

$$A_{app} = -\log_{10} (T)$$

At sufficiently low concentration, Beer's law is valid:

$$-\log_{10} (T)_\nu = c\, l\varepsilon_\nu$$

Here ν is the wavelength of the light, c the sample concentration, l the sample thickness (path length), and ε_ν the (apparent) molecular absorption coefficient at that wavelength. The true molar absorption coefficient is proportional to the natural logarithm of the transmittance.

IR spectroscopy is relatively inexpensive, simple, and fast. Although it is ordinarily used for the identification of single compounds, it can contribute also to our field of interest by giving quick information on the distribution of several structural and functional groups in a sample. IR measurements may complement NMR data. In combination with regular NMR spectroscopy, it may provide quick yet fairly detailed information on the distribution of CH_n groups. However, in our context, the main application of IR spectroscopy is the determination of polar groups such as OH, NH, and the various CO groups.

2. Modern IR Spectroscopy

Two important recent developments are the combination of IR instruments with computers and the introduction of Fourier transform (FT) techniques. In the older, dispersive IR spectroscopy, the light is refracted by a prism or a grating and scanned by a moving slit which takes several minutes for one measurement. In FT-IR the entire spectrum is obtained by an interferometer in a fraction of a second. Thus, several hundred measurements can be taken in 5 min and averaged by computer. This multiplexing, together with a higher-energy throughput, leads to greatly increased (about 100-fold) sensitivity and precision over those achievable with dispersive instruments. Even wave-number resolution is enhanced.

Built-in computers—in FT as well as in dispersive instruments—permit immediate (on-line) data manipulation, for example, the subtraction of background, solvent, and reference spectra, baseline corrections, "zooming" (expansion of signal intensity or wave-number range), and curve resolving. With these improvements, good quantitative determinations have now become routine. However, absolute quantitative data require rigorous standardized experimental conditions or added standards.

New diffuse reflectance infrared (DR-IR) techniques [e.g., Fuller and Griffiths (5), Yang and Mantsch (6), Christy et al. (8)] seem to give equally good spectra as obtained in conventional ways from solutions in cells or from KBr pellets but make sample preparation easier and safer than before (7). Another synonym for this technique is DRIFT (for diffuse reflectance infrared Fourier transform). The sample is deposited from solution onto finely ground KBr in a small cup which is placed into a diffuse reflectance accessory after removal of the solvent in a vacuum oven. A related technique, variable angle specular reflectance, allows rotation of the sample holder for optimization. Bukka et al. (9) applied this technique to fractions obtained from tar sand bitumen.

Increased resolution by band narrowing techniques (10–12) in conjunction with reference spectra allows the distinction of CH_2 groups next to other groups such as alkyls, aromatic rings, carbonyls, or alkoxies. Figure 9.1 shows an example of IR spectra with and without band narrowing. CH_n groups and aromatic C can be identified, measured, and ratioed. The ratio of the band at 1602 cm^{-1} (aromatic C–C stretching) to that at 2920 cm^{-1} (aliphatic hydrogen–carbon stretching in CH_2 in the β^+ position of an aromatic ring) is a good relative measure for the aromaticity of a sample.

FT-IR is so sensitive that it can be used for detection in HPLC and even in GC and SFC. The special flow cells required and other, more intricate interfacing devices have been reviewed by Norton et al. (13).

3. Band Assignments

General band assignments were established several decades ago. Those shown in Table 9.1 were collected for coal fractions, but they apply also to heavy petroleum fractions. Painter et al. (17) clarified an uncertainty surrounding the absorption near 1600 cm^{-1}. In heavy petroleum fractions, and especially in coal samples, the absorption in this region is much more intense than in aromatic model compounds. Painter et al. proved that phenols and aromatic structures in large molecules, which are predominant in these materials, absorb much more strongly than aromatic hydrocarbons of low molecular weight. They also warned of complications due to interactions in the range from 1000 to about 1350 cm^{-1}, which make earlier assignments here (e.g., to ethers) quite uncertain.

Resolution enhancement by Fourier self-deconvolution (band narrowing) led to the recognition and distinction of bands which could previously not be separated. Wang and Griffiths (11) and Yang et al. (7) tentatively assigned specific functional groups to these bands in spectra obtained from coal samples. Examples of such new assignments are CH_3 next to aromatic rings (2948 cm^{-1}) and next to alkyl groups (2960 cm^{-1}); CH_2 (together with some CH_3) next to aromatic rings (2916 cm^{-1}) and in aliphatic chains

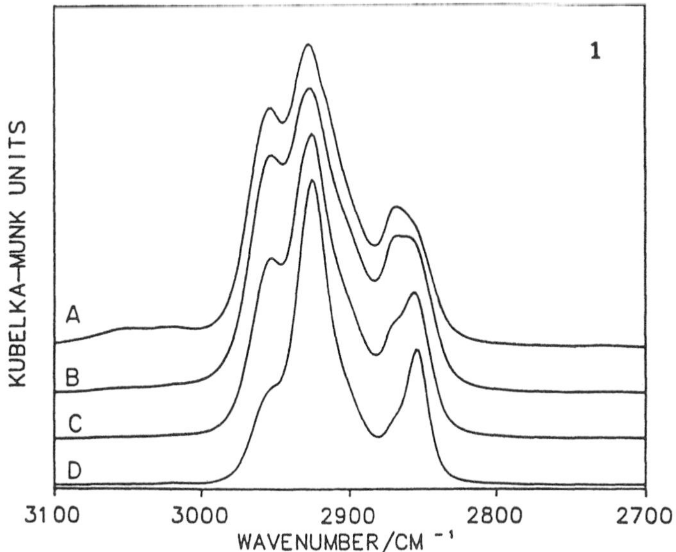

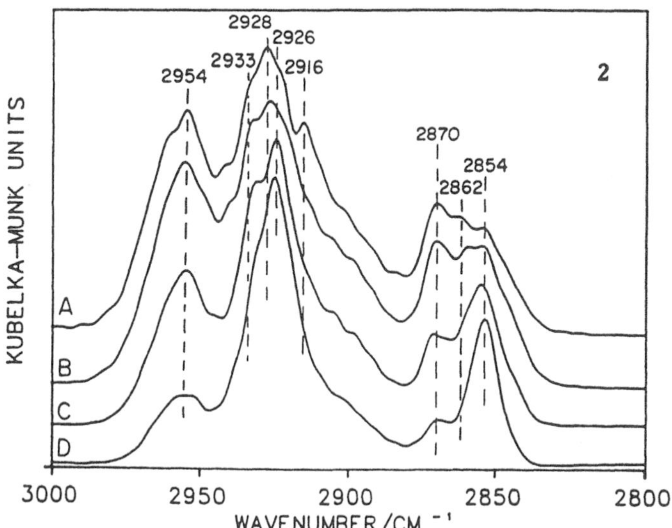

Figure 9.1 IR spectra in the C–H stretching region (1) without and (2) with band narrowing. The four curves in each spectrum indicate SEC fractions of different M_w: (A) 182, (B) 294, (C) 660, and (D) 4890. (From Ref. 7. Reproduced with permission of the publisher.)

Table 9.1 IR Band Assignments

Aliphatic and aromatic groups		Oxygen-containing function groups	
Wave number (cm^{-1})	Assignment	Wave number (cm^{-1})	Assignment
		3300	Hydrogen bonded
3030	Aromatic C—H		
2950 shoulder	CH_3		
2920, 2850	Aliphatic—CH: CH_2 and CH_3		
		1835	C=O, anhydride
		1775–1765	C=O, ester with electron withdrawing group attached to single bonded oxygen Ar—O—C(=O)—R
		1735	C=O, ester
		1690–1720	C=O, ketone, aldehyde and, —COOH
		1650–1630	C=O highly conjugated Ar—C(=O)—Ar

1600	Aromatic ring stretch	~1600	High, conjugated hydrogen bonded C═O
		1560–1590	Carboxyl group in salt from —COO^-
1490 shoulder	Aromatic ring stretch		
1450	CH_2 and CH_3 bend, possibility of some aromatic ring modes		
1375	CH_3 groups		
		1300–1110	C—O stretch and O—H bend in phenoxy structures, ethers
		1100–1000	Aliphatic ethers, alcohols
900–700	Aromatic C—H out-of-plane bending modes		
860	Isolated aromatic H		
833 (weak)	1,4 Substituted aromatic groups		
815	Isolated H and/or 2 neighboring H		
750	1,2 Substituted, i.e., 4 neighboring H		

Source: Ref. 17.

(2926 cm^{-1}); CH_2 next to an alkoxy (2928 cm^{-1}) and next to a carbonyl group (2933 cm^{-1}); and CH in unspecified environments (2905 and 1897 cm^{-1}). The rocking vibration band of CH_2 in straight aliphatic chains varies slightly with their length. For *n*-pentyl groups, it is located at 726 cm^{-1}, and with longer chains, it shifts to lower frequencies until it stabilizes at 720 cm^{-1} for chains with more than nine carbon atoms (7).

4. Applications to Petroleum Analysis

In our field, IR spectroscopy has been mostly employed for measuring oxygen- and nitrogen-containing groups and for evaluating shifts in certain bands due to aggregation or other interactions. McKay et al. (18,19) succeeded in using IR spectroscopy, by itself as well as in combination with other methods, for the quantitative determination of acids and bases in high-boiling petroleum fractions. They distinguished between carboxylic acids, phenols, carbazoles, and cyclic amides among the acids, and pyridine and acridine variants among the bases. Their main solvent was methylene chloride. Occasionally, they used THF to minimize association between carboxylic acids and to distinguish them from amides. Extinction coefficients were determined from model compounds.

By means of IR spectroscopy, Jacobson and Gray (16) determined phenol, pyrrole, carboxyl, ketone, amide, and sulfoxide groups in two bitumen fractions, namely, the pentane insolubles and the most polar fraction of the pentane solubles. The use of two solvents, methylene chloride and THF, permitted them to observe frequency shifts of the carboxylic acid, ketone, and amide absorption bands in going from one to the other and, thus, to clearly identify these groups. The response factors published by Bunger et al. (14) and by Petersen (20) served for quantitation. Table 9.2 shows the absorption ranges, response factors, and the concentrations of these groups from the two fractions in both solvents.

Green et al. (21) established the calibration curves for OH compounds in terms of % transmittance versus O content, shown in Fig. 9.2. The nonaromatic hydroxyl compounds display remarkably little scatter, that of the phenolic compounds is greater with about ±3%. The model compounds were appropriately selected for syncrude fuels; for our purpose, for example, for petroleum samples, we would have liked to see some compounds with longer aliphatic chains. According to Green et al., the pyrrolic benzologues showed less scatter in their calibration than the phenolics, but the amides and other carbonyls had considerably more.

Green's band assignments, shown in Table 9.3, are slightly different from those by Jacobson and Gray (Table 9.2). The deviation may in small part be due to the different solvent, dichloroethane versus dichloromethane (methylene dichloride). Note that the absorption maxima of some of their

Table 9.2 IR Absorption Data of Polar Groups in Peace River Bitumen Fractions

Structure	Range of absorption (cm^{-1})[a]		10^{-4} *B* (L mol^{-1} cm^{-2})	
	CH_2Cl_2	C_4H_8O	CH_2Cl_2[a]	C_4H_8O[b]
Aromatic hydroxyl	3540–3600		0.5	
Indole[c]	3455–3465		0.7	
Carboxylic acid	1700–1745	1720–1735	1.5	1.2
Ketone	1690–1700	1695–1705	0.7	0.7
Amide	1625–1690	1630–1695	1.5	
Sulphoxide	1025–1040		0.6	

[a]Values previously reported by Bunger et al. (14).
[b]Values previously reported by Petersen (15).
[c]Indole represents benzologues of pyrrole.
Source: Ref. 17.

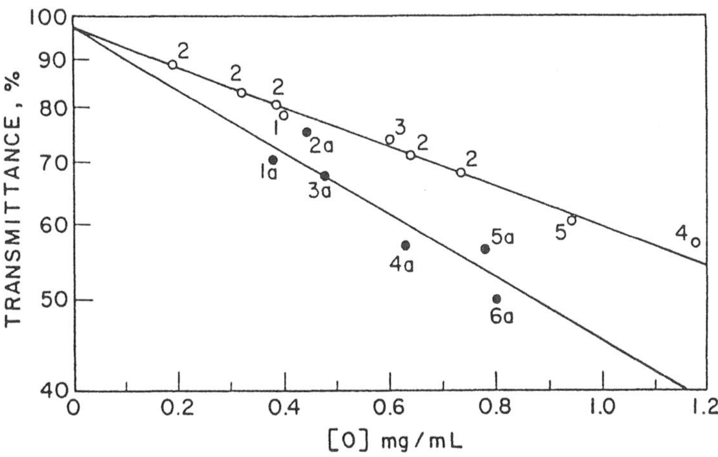

Figure 9.2 Calibration curve for the IR absorption of OH compounds. Key to compounds. Alcohols: 1—acenaphthol, 2—9-hydroxyfluorene, 3—benzylalcohol, 4—*n*-butanol, 5—water. Hydroxyaromatics: 1a—2-naphthol, 2a—2,4-dimethyl-6-*tert*-butylphenol, 3a—3,4,5-trimethylphenol, 4a—4-hydroxybiphenyl, 5a—*m*-cresol, 6a—2,3,5-trimethylphenol. (From Ref. 21. Reproduced with permission of the publisher.)

Table 9.3 IR Bands Assignments by Green et al.

Class	Frequencies (cm^{-1})
Hydroxyaromatics	3580–3560
(Water)[a]	3680, 3600
Pyrrolic benzologs	3465–3450
Amides, C = 0[b]	1730, 1690, 1660, 1630

[a]The 0–H stretching band at 3680 was used as a water indicator. If more than a trace of water was indicated, the sample was dried and rerun.

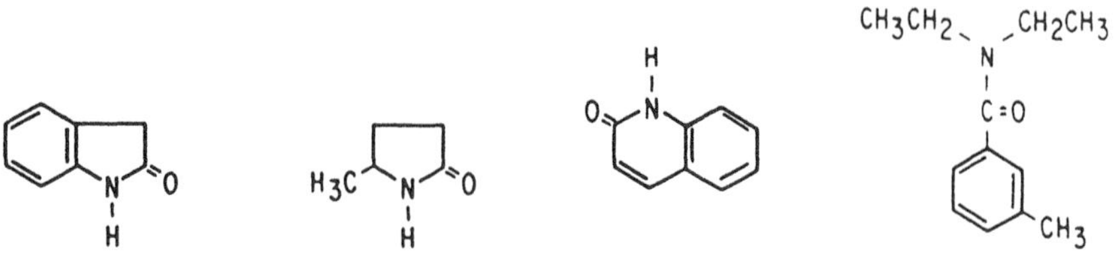

1733 cm^{-1}	1700 cm^{-1}	1662 cm^{-1}	1628 cm^{-1}
Benzo-5-member lactam	5-member lactam	Pyridone, quinolone, etc.	*N, N*-Disubstituted amide

[b]Work with model compounds suggests structures analogous to the following for the various amide bands (shown above).
Source: Ref. 21

amide model compounds, displayed in Table 9.3, span the entire range of the carbonyl bands. This overlap prevents the distinction between amides and other carbonyl groups without prior separation, the use of other, independent data, or some other means. Carboxylic acids, for instance, can be made visible among other carbonyl compounds by esterification with trifluoroethanol (22) which shifts their absorption from the region between 1710 and 1750 cm^{-1} to 1760 cm^{-1}, well outside that of the amides.

Acylation with trifluoroacetic acid of the OH groups, also advocated by Yu and Green (22), has the additional advantages of enhanced sensitivity, reduced or eliminated hydrogen-bonding, and no more interference from water. Table 9.4 demonstrates the band shifts and changes in molar absorptivities from the underivatized to the derivatized functional groups. Figure 9.3 shows the spectra of a Cerro Negro acid concentrate in the original (A) and the derivatized (B) forms.

Peterson's investigations of hydrogen bonding (15) in regular and oxidized asphaltic residues by IR spectroscopy added greatly to our knowledge in this field as well as to our list of peak assignments and response factors. By measuring the IR absorption of these materials at different concentration, for example, in films (c = 100%) and in CCl_4 at 0.07 and 0.004

Table 9.4 Band Maxima (ν_{max}) and Molar Absorptitives (ϵ) of IR Bands from Underivatized and Derivatized OH and COOH Groups (for Dilute Dichloromethane Solutions, ~1–10 mg/ml)

Functional group	Underivatized $\bar{\nu}_{max}$ (cm^{-1})[a]	Underivatized ϵ ($L\ mol^{-1}\ cm^{-1}$)[b]	Derivatized $\bar{\nu}_{max}$ (cm^{-1})	Derivatized ϵ ($L\ mol^{-1}\ cm^{-1}$)
		A. From pure compounds		
OH (alcoholic)	3610 ± 10	60 ± 10	1785 ± 5	420 ± 40
OH (phenolic)	3590 ± 10	120 ± 10	1800 ± 5	610 ± 50
C(O)OH (aliphatic)	1750 ± 5 (monomer) 1715 ± 5 (dimer)	580 ± 40	1755 ± 5	330 ± 20
C(O)OH (aromatic)	1740 ± 10 (monomer) 1700 ± 10 (dimer)	610 ± 40[c]	1740 ± 5	370 ± 30[c]
		B. From petroleum fractions		
OH	3600 ± 10	nd[d]	1800 ± 5	600 ± 40
C(O)OH	1750 ± 5 (monomer) 1710 ± 5 (dimer)	470 ± 40	1760 ± 5	280 ± 14

[a]Median ± range for petroleum and related pure compounds.
[b]Mean ± SD.
[c]Results for para alkyl and 2-naphthoic acids were excluded from the mean. ε's for para alkyl and 2-naphthoic acids were approximately 50% greater than those of other aromatic acids.
[d]Not determined.
Source: Ref. 22.

g/ml, he demonstrated that OH and NH groups undergo extensive hydrogen bonding at high concentrations (a broad shoulder extending to almost 3000 cm^{-1}). At low concentrations, hydrogen bonding is much less pronounced, and the peaks of free OH (3610 cm^{-1}) and NH (3480 cm^{-1}) groups become visible. Later, Petersen extended these studies to the identification of 2-quinolones in asphalts, their association with each other and with carboxylic acids in CCl_4 solutions (15), and their interactions in asphalts (23).

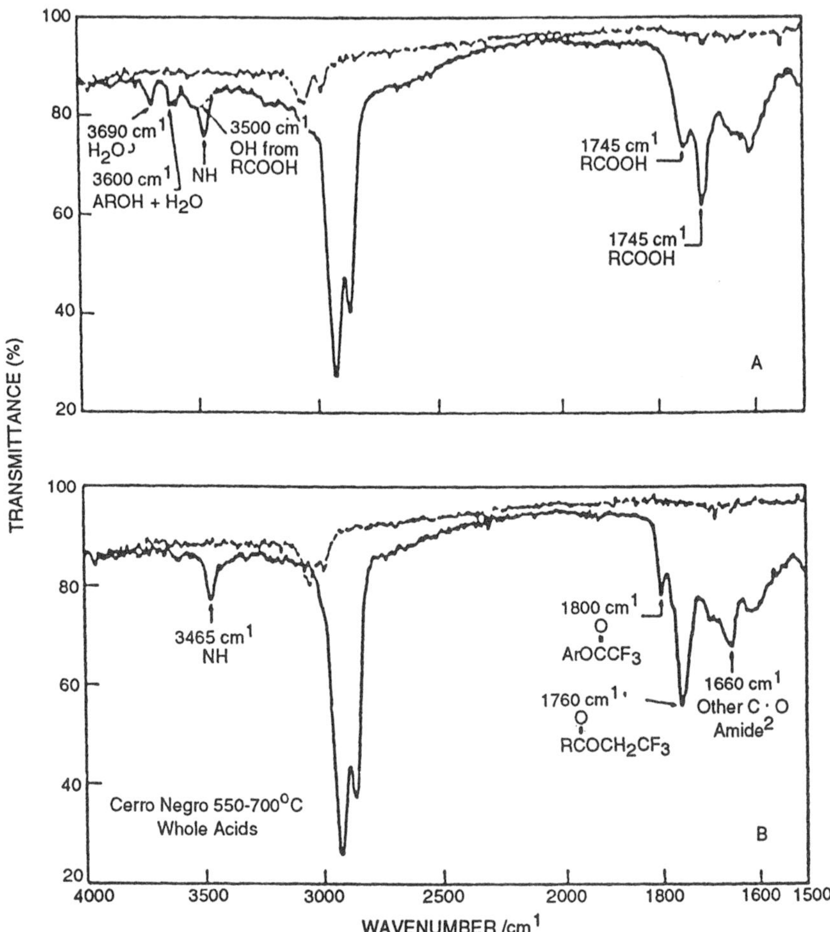

Figure 9.3 IR spectra of an underivatized (A) and a derivatized (B) Cerro Negro acid fraction in methylene chloride. (From Ref. 22. Reproduced with permission of the publisher.)

In less complex, lower-boiling petroleum fractions, parts of the hydrocarbon skeleton can be assessed by IR. Specifically, the alkyl substitution of aromatic rings can be determined from the out-of-plane C–H deformation bands. For example, Jokuty and Gray (24) distinguished three substitution types of carbazoles in the GC fractions of a hydrotreated gas oil:

Type I. One band. All aromatic hydrogen atoms are in adjacent groups of three or four. Representative isomers: 4 alkyl, 4,5-dialkyl.

Type II. Two bands. Some aromatic hydrogens are in groups of two, and some are in groups of three and four. Representative isomers: 3,4-dialkyl, 3,4,5-trialkyl.

Type III. Three bands. Some aromatic hydrogens are in groups of three and four, some in groups of two, and some are isolated. Representative isomers: 3-alkyl, 3,5-dialkyl.

B. UV/Visible Spectroscopy

The UV/visible spectrum, although not as specific for chemical group types as NMR and IR, can distinguish between aromatic compounds with different ring numbers and configurations. The spectra of some model compounds are presented in Fig. 9.4. The patterns are not distinct enough to recognize or distinguish these compounds in complex mixtures, but they can be useful for their identification in narrow fractions. For additional helpful background information without overwhelming detail, we refer again to Kemp's book (4).

UV/visible photodiode arrays are employed as detectors for the fractionation of petroleum samples, for example, for the chromatographic separation of aromatics by ring number. As mentioned in Chapter 6, these detectors can record the entire spectrum many times every second. The spectra, plotted in three dimensions versus time, give a clear picture of how a separation progresses. In a trade brochure, Owen (25) described the principle of this method and the basic instrumentation.

UV/visible spectra can be useful also in studies of refining processes. An example is given in Fig. 9.5 which shows two 3-dimensional chromatograms obtained in our laboratory (26) by ring number fractionation with a UV/Vis detector. The upper left chromatogram (a) is that of a sample taken at the beginning of a desulfurization run when the catalyst was fresh. The right one (b) is of a sample taken toward the end of a run when the catalyst was nearly spent and the temperature had been raised to maintain constant conversion. Both are first-stage hydrocracker (HCR) feeds. The lower chromatogram is that of the second-stage hydrocracker recycle stream (c). Comparison of the first two with the third chromatogram clearly demonstrates that large aromatic molecules are generated in the second-stage hydrocracker. Even specific aromatic ring types could be distinguished in narrow fractions by UV spectra. FIMS spectra provided molecular detail. The combination of fractionation, UV detection, and FIMS made it possible to deduce a novel reaction sequence leading from small naphthalenes to such large aromatic ring structures as coronenes and ovalenes.

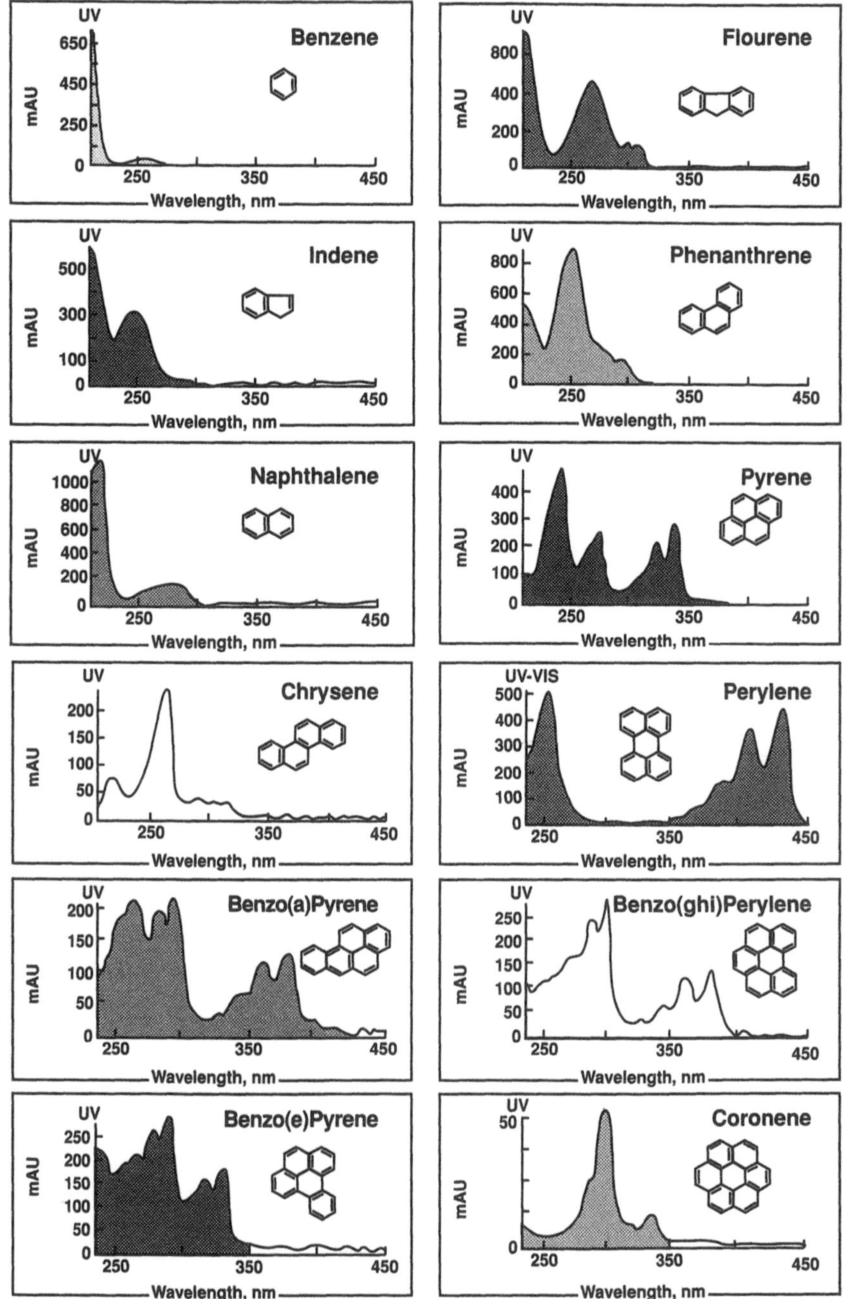

Figure 9.4 UV/Vis spectra of aromatic compounds.

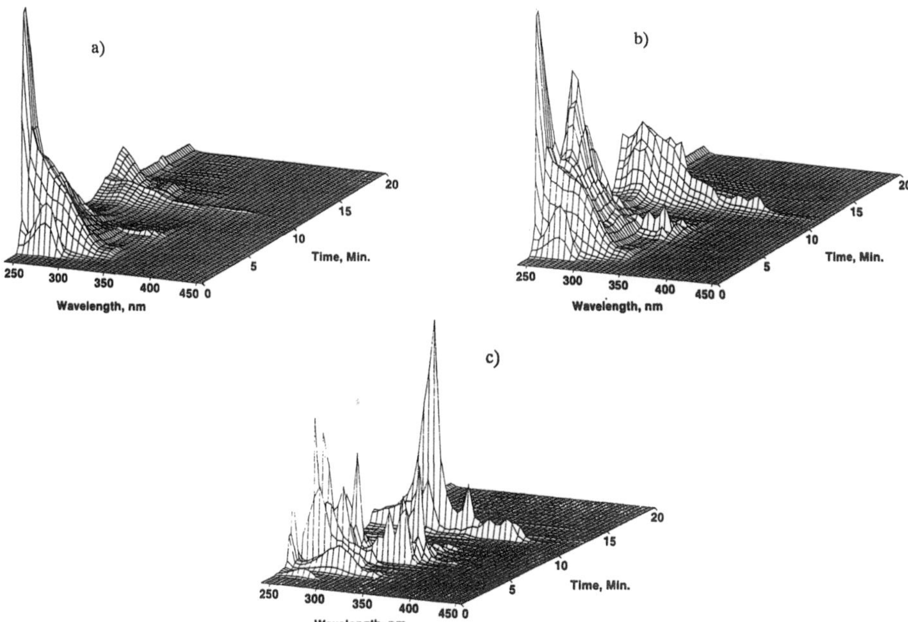

Figure 9.5 Three dimensional chromatograms with UV/Vis photodiode array detection of VGOs from hydrodesulfurization reactor with fresh and nearly spent catalyst. (a) Start of run, first-stage HCR feed; (b) end of run, first-stage HCR feed; (c) end of run, second-stage HCR recycle stream. (From Ref. 26. Reproduced with permission of the publisher.)

In contrast to the UV/Vis spectra of narrow fractions, those of broad or high-boiling samples, are featureless. An exception is the Soret band which may be prominent in those samples which have sufficient amounts of porphyrin. Figure 9.6 shows a UV/Vis map comparable to that in Fig. 9.5, but taken from a SEF fraction. Figure 9.7 may illustrate the point even more clearly.

C. X-Ray Techniques

1. X-Ray Absorption Spectroscopy

Extended x-ray absorption fine structure (EXAFS) and x-ray absorption near-edge structure (XANES) spectroscopy are tools for the investigation of the immediate chemical environment of x-ray absorbing elements such as metals and sulfur. Goulon et al. (29) applied these two methods to Boscan asphaltenes to determine the ways in which vanadium is bound in

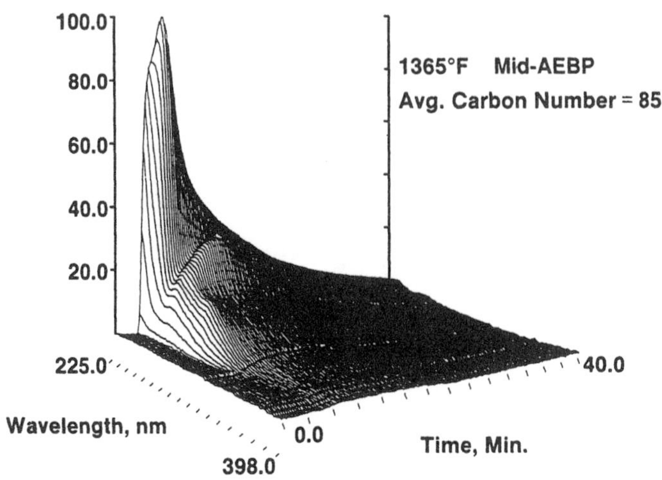

Figure 9.6 UV/Vis map of a SEF fraction, mid-AEBP = 1365°F. (From Ref. 27. Reproduced with permission of the publisher.)

these samples; see Chapter 10. Recent applications of XANES and x-ray photoelectron spectroscopy (XPS) for the determination of sulfur compounds in petroleum samples were described by Kelemen et al. (30,31) and Waldo et al. (32). Zhang et al. (34) applied XANES to several narrow fractions of Maya atmospheric residues to study the distribution *V* and Ni in these crude oils as a function of AEBP. Again they found that prior fractionation led to much cleaner spectra than otherwise attainable, which allowed them to claim with certainty the near or complete absence of nonporphyrinic metal structures in these samples. More results of this study are presented in Chapter 10.

Because these measurements require rather high x-ray densities and a broad smooth spectral range, they are usually performed with synchroton radiation, for example, at the Brookhaven National Synchroton Light Source or the French National Synchroton Radiation Facility. The technique involves several adjustments, subtraction of atomic absorption background, and corrections for self-absorption and, in case of XANES, for multiple scattering. The interpretation depends on the choice of model compounds, which is often not straightforward because of the great variety of structural features and the present dearth of model compounds as, for example, in the case of S compounds (31). The basic instrumentation and principles of this technique have been described for chemists by Fay et al. (35).

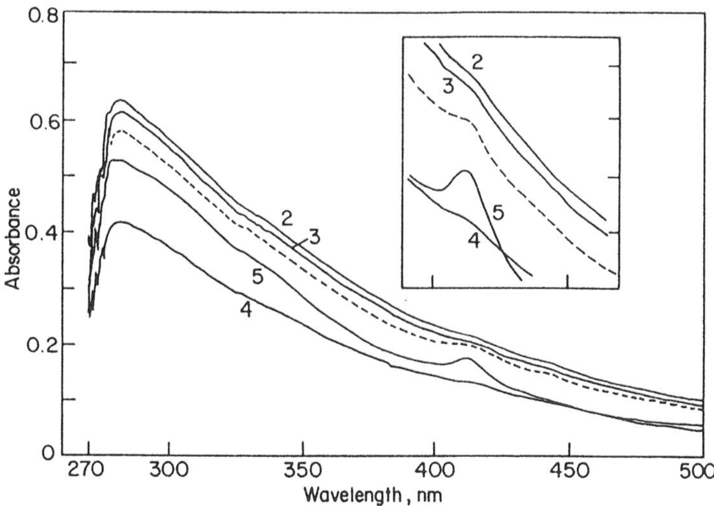

Figure 9.7 UV/Vis spectra of an Athabasca asphaltene and four SEC fractions derived from it. (From Ref. 28. Reproduced with permission of the publisher.)

2. X-Ray Diffraction

X-ray diffraction had been used by Yen, Erdman, Pollack (36) for the determination of f_a, the fraction of aromatic over total carbon in a molecule. These days, f_a is easily and precisely obtained by NMR spectroscopy or even by IR. Its determination by XRD is even inappropriate as Ebert (37) pointed out. X-ray data can be very misleading (38–43) if they are used to translate geometric data—measurements of (aromatic) sheet diameter—into structural information. On the one hand, not all aromatic atoms contribute to the stack diameter seen by x-ray diffraction, whereas, on the other hand, nonaromatic atoms such as hydroaromatic carbons and other substituents on aromatic rings may contribute to the diffraction pattern. Thus, the interpretation of these measurements is quite arbitrary.

III. CHEMICAL METHODS

A. Pyrolysis

In pyrolysis, the simplest degradation method, the sample is rapidly heated under exclusion of air to a temperature high enough to break some of the chemical bonds. Typical reactions are dealkylation and the breakage of (aliphatic) S–S and S–C bonds. The problem here is to find a compromise between specificity and yield, which means that the temperature and other reaction conditions must be carefully chosen.

A common procedure is to heat the sample in a tube to the desired temperature for several hours. The tube may be evacuated (44,45), or swept with an inert gas [e.g., (46–48)], or just sealed [e.g., (49)]. Typical temperatures reported for these procedures are 300–400°C (570–750°F), and sample sizes are in the order of 100 mg or less. The reaction products are collected in various ways and usually analyzed afterward. Only with gas-swept arrangements can the products be analyzed on-line by GC.

Flash pyrolysis on a thin, electrically heated wire with properly chosen Curie temperature, for example, 610°C (1130°F), ensures rapid warm-up and brief exposure time for suppression of side reactions [e.g., (50–53)]. In this procedure, a very small amount (about 100 μg) of analyte is pressed onto the wire. Warm-up time is 0.1 s, and the Curie point guarantees constant temperature for the duration of the process, usually a few seconds. The gaseous products are directly transferred to a gas chromatograph, preferably with MS as detector.

Sinnenghe Damsté et al. (51,52) applied this technique to kerogen samples and identified numerous alkylthiophenes, alkylbenzothiophenes, and alkyl thiolanes among the fragments. The content of thiophenic moieties in the pyrolysis products was considerably smaller than that in the analyte. Thus, the results are qualitative rather than quantitative.

Another variation of flash pyrolysis was recently described by Payzant et al. (54). Here, the sample solution is dripped from a dropping funnel at a slow rate (<1 ml/min) to the bottom of a flask which is heated by a temperature-controlled sand bath, typically at 800°F (about 430°C). A flow of nitrogen sweeps the reaction products and the solvent to a cooled collection flask for subsequent concentration and analysis. There was no evidence for degradation of the toluene. Several grams of product for further analyses can be obtained this way. Payzant et al. separated the pyrolysis oil into a number of compound groups by a combination of LC with an oxidation/reduction method for the sulfur compounds. GC-FTD of these fractions allowed them to identify numerous compound types and their homologues. Their results are described in Chapter 10.

B. Selective Oxidation

1. Ruthenium Ion Catalyzed Oxidation

Selective oxidation with $NaIO_4$ in $H_2O/CCl_4/CH_3CN$, catalyzed by Ru(VIII) ions, can give more specific information on the nature of insoluble petroleum fractions. This method selectively oxidizes alkyl-substituted aromatic carbons to CO_2, leaving behind alkyl substituents as monocarboxylic acids, and polymethylene bridges between aromatic rings as α,ω-dicarboxylic acids. Some of the condensed aromatic rings survive the treatment as benzene polycarboxylic acids (Fig. 9.8). The various acids are made volatile by esterification with diazomethane, and the esters are analyzed by GC-MS.

Mojelski et al. (56) named this technique RICO (for Ru ion catalyzed oxidation) and described in detail the processes involved. When they applied it to Athabasca asphaltenes (56,57), about 40% of their oxidation product remained unvolatilized. Further esterification, volatilization by pyrolysis, and GC-MS analysis of the resulting esters led to the conclusion that this residue consisted primarily of substituted naphthenic rings. It contained 4.7% S, 0.8% N, about 22% O, and less than 3% aromatic carbon.

RICO provides the distribution of alkylsubstituents attached to aromatic rings, that of paraffin chains connecting two aromatic rings or ring systems, and information on the nature of the aromatic and naphthenic ring systems in the original sample. The amounts of alkyl chains found this way seem to be reasonably quantitative. Those of aromatic and naphthenic rings appear more tentative because the fate of the aromatic heterocompounds is not known except that some of the sulfur compounds are recovered in the form of sulfones (56). Because we know that the insoluble fractions (asphaltenes) and nondistillable residues may contain as many as 10 het-

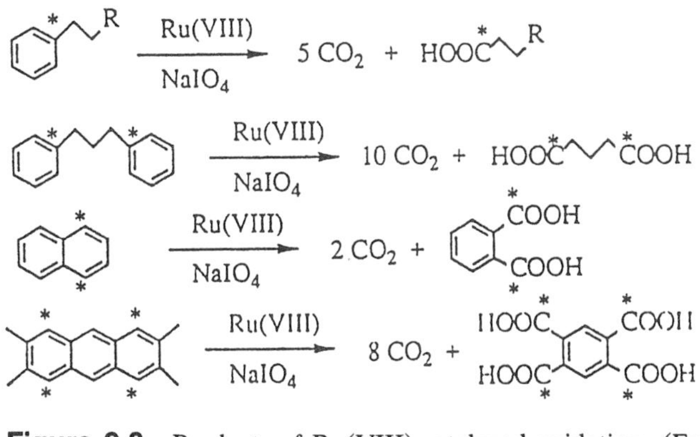

Figure 9.8 Products of Ru(VIII)-catalyzed oxidation. (From Ref. 55. Reproduced with permission of the publisher.)

eroatoms per molecule on average (see Chapter 4), the information provided by RICO is valuable but incomplete. Mojelski's et al. (57) results will be further discussed in Chapter 10.

2. Oxidation of Sulfur Compounds

Oxidation is also used for the separation of neutral sulfur compounds from hydrocarbons [e.g., (58–60)]. The S compounds are converted to sulfoxides and sulfones which are strongly retained on chromatographic columns and can, thus, be separated from the hydrocarbons. After their recovery, they can be reconverted to the sulfides by reduction with $LiAlH_4$ for further analysis. By combining such oxidation and reduction procedures with ESCA measurements, Ruiz et al. (59) could distinguish between thiophene derivatives and three different types of sulfides. Others, for example, Arpino et al. (60), described similar oxidation–reduction procedures for subsequent characterization and identification of the isolated S compounds.

Strausz et al. (61) warn of side reactions which may occur in such oxidative procedures. They developed a scheme based on the selective oxidation of aliphatic sulfides to sulfoxides with tetrabutylammonium periodate which leaves the thiophenes unchanged. In addition, tetrabutylammonium periodate does not oxidize the sulfides beyond the sulfoxide stage to sulfones, even when used in excess, as do most of the other oxidizing agents. This has the advantage of easier and more quantitative reconversion of the products to sulfides after their separation from the hydrocarbons.

The thiophenes are oxidized to their sulfones under anhydrous conditions with *m*-chloroperbenzoic acid, which stops at this stage instead of continuing the reaction to quinones and other products. Both the sulfoxides and sulfones can be reduced again with $LiAlH_4$ after chromatographic separation from the hydrocarbons. The method is relatively quick and easy to perform, but, according to the authors, it is not always quantitative. The original thiophenes may not be completely recovered; in fact, occasionally they may not even be detected at all.

C. Selective Reduction

1. Reduction by Raney Nickel

Reduction by Raney nickel is another mild degradation method. It may be performed, for example, by reacting a 10-mg sample with a suspension of 0.5 mg of catalyst in 2.5 ml of refluxing absolute alcohol for 1–2 h under nitrogen. However, according to Sinnenghe Damsté et al. (62), it can only be applied to molecularly dissolved samples. Because pentane-insoluble petroleum fractions form aggregates which interfere with this method and direct methods are available for the characterization of pentane soluble fractions, we see little incentive for its use in the analysis of heavy petroleum fractions. Furthermore, comparison with results from x-ray spectroscopy suggests that the sulfur found after Raney nickel reduction in very high AEBP samples is less than 20% of the sulfur present.

2. Reductive Alkylation

Reductive alkylation is ordinarily used to make pentane- or heptane-insoluble petroleum fractions more soluble and less prone to aggregation in hydrocarbon solvents. The sample is dissolved in THF and treated with an excess of metallic potassium under exclusion of air. After 24 h, the remaining K is removed for weighing, and the mixture is quenched with alkyl iodide at 0°C. This produces the final alkylated product. Sternberg and co-workers (63,64), who developed this method, added naphthalene as an electron transfer agent to the mixture. Ebert (43,68), however, demonstrated that petroleum residues contain sufficient aromatic carbon to make the reaction proceed well without additional naphthalene, at least with the two residue samples he worked with (one was a vacuum residue, the other an atmospheric residue). According to Ebert, it is even preferable to leave out this agent because side reactions are then mitigated.

Ignasiak et al. (65), in contrast, claimed that the presence of naphthalene in the reaction mixture of their Athabasca asphaltenes doubled the number of S atoms removed and also the number of the alkyl groups introduced. However, the absence or presence of naphthalene was not the only difference in the reaction conditions of these two groups of workers, so that their different results may have been caused by other experimental details.

A principal reaction of this procedure is the conversion of aromatic double bonds via dianions to naphthenic ones with alkyl groups inserted onto the corresponding carbon atoms. Two other important reactions are the desulfurization of thiophenic structures and the cleavage and transalkylation of (aliphatic) sulfides. For instance, dibenzothiophene is transformed to biphenyl (64), and diphenylsulfide is converted mainly to phenylalkylsulfide and benzene, and, to a minor extent, to biphenyl (65).

In petroleum residues, surprisingly, only about one-third of the thiophenic sulfur is taken out by the reaction (43). Subsequent treatments remove similar though decreasing amounts of sulfur. Very likely, the first anion created in a molecule prevents another one from forming (43). Only after its removal by alkylation can the next anion be brought forth. Ebert (66) observed this obstructional behavior with perylene, which readily forms the dianion with K in THF, but not the tetraanion.

Francisco et al. (67) developed an interesting two-step reaction sequence of cleaving carbon–sulfur bonds (and also some C–C bonds) in heavy oils with alkali metal, followed by selectivity labeling the resulting thiols with

^{13}C- and ^{2}H-enriched methyl groups. Two-dimensional ^{13}C and ^{2}H-NMR spectroscopy of the samples before and after cleavage gave clearly recognizable chemical shift data of the methyl labels enabling identification of several types of S compounds in three main classes:

1. Diaryl sulfides ranging from dibenzyl to bis(1-naphthylmethyl) sulfide
2. S compounds that form thiols on reductive protonation, namely, alkylaryl- and aryl-arylsulfides as well as certain thiophene derivatives
3. Those S compounds that resist cleavage by reductive protonation (about 60% of all the sulfur in the two samples investigated by these authors), for example, dibenzothiophene derivatives and dialkylsulfides

IV. SUMMARY OF CHARACTERIZATION METHODS

The auxiliary methods described in this chapter are used for two purposes. First, they give information on petroleum distillate fractions which the main methods cannot provide, primarily on the nature of heterocompounds. Second, they can provide data on the nondistillable residue fractions which are not amenable to useful fractionation into compound classes and to characterization by MS as are the distillates.

IR spectroscopy permits the determination of many heteroatom containing functional groups, primarily the various carbonyls, OH and NH. It also can be used for finding the distribution of H atoms in the aromatic rings and that of CH and CH_2 groups in the aliphatic part of the hydrocarbon structure. UV/Vis and x-ray spectroscopy help in special situations, for example, in the study of porphyrin structures. In narrow aromatic compound fractions, UV/Vis can identify the predominating aromatic ring system.

Mild degradation methods, such as pyrolysis, selective oxidation, and reduction, break up the otherwise intractable molecules of pentane-insoluble petroleum fractions into representative smaller fragments which now can be characterized by chromatographic separation and MS. This allows us to draw conclusions about the structure of the parent molecules. For example, selective ruthenium ion catalyzed oxidation (RICO) provides the distribution of alkyl substitution to aromatic rings of both kinds, regular alkyl groups and chains connecting two aromatic rings. In addition, it gives information about the nature of the aromatic and naphthenic ring systems in the original sample. Reductive alkylation affords us insight into the types of sulfur groups in such complex fractions.

REFERENCES

1. Bellamy, L. J. 1975. *The Infra-Red Spectra of Complex Molecules*, 3rd ed. Chapman and Hall, London; John Wiley & Sons, New York.
2. Nakanishi, K., and Solomon, P. H. 1977. *Infrared Absorption Spectroscopy*, 2nd ed. Holden-Day, San Francisco.
3. Lin-Vien, D., Colthup, N. B., Fateley, W. G., and Grasseli, J. G. 1991. *Handbook of Infrared and Raman Characteristic Frequencies of Organic Molecules*. Academic Press, San Diego,
4. Kemp, W. 1991. *Organic Spectroscopy*, 3rd ed. W.H. Freeman and Company, New York, Chap. 2, pp. 19–99.
5. Fuller, M. P., and Griffiths, P. R. 1978. Diffuse reflectance measurements by infrared fourier transform spectrometry. *Anal. Chem.*, **50:**1906–1910.
6. Yang, P. W., and Mantsch, H. H. 1987. *Appl. Opt.*, **26:**326. Cited by Yang et al. (7).
7. Yang, P. W., Mantsch, H. H., Kotlyar, L. S., and Woods, J. R. 1988. Characterization of Athabasca tar sand maltenes by diffuse reflectance infrared spectroscopy. *Energy Fuels*, **2:**26–31.
8. Christy, A. A., Dahl, B., and Kvalheim, O. L. 1989. Structural features of resins, asphaltenes and kerogen studied by diffuse reflectance infrared spectroscopy. *Fuel*, **68:**430–435.
9. Bukka, K., Hanson, F. W., Miller, J. D., and Oblad, A. G. 1992. Fractionation and characterization of Whiterocks tar sands bitumen. *Energy Fuels*, **6:**160–165.
10. Fuller, M. P., Hamadeh, I. M., Griffiths, P. R., and Lowenhaupt, D.E. 1982. Diffuse reflectance infrared spectroscopy of powdered coals. *Fuel* **61:**529–536.
11. Wang, S. H., and Griffiths, P. R. 1985. Resolution enhancement of diffuse reflectance I.R. spectra of coals by Fourier self-deconvolution. *Fuel*, **64:**299–306.
12. Chen, P., Yang, P. W., and Griffiths, P. R. 1985. Effect of preheating on chemical structure and infrared spectra of Yanzhou coal. *Fuel*, **64:**307–312.
13. Norton, K. L., Lange, A. J., and Griffiths, P. R. 1991. A unified approach to the chromatography–FTIR interface: GC-FTIR, SFC-FTIR, and HPLC-FTIR with subnanogram detection limits. *J. High Res. Chromat.*, **14:**225–229.
14. Bunger, J. W., Thomas, K. P., and Dorrence, S. M. 1979. Compound types and properties of Utha and Athabasca tar sand bitumens. *Fuel*, **58:**183–195.
15. Petersen, J. C. 1967. An infrared study of hydrogen bonding in asphalt. *Fuel*, **46:**295–305.
16. Jacobson, J. M., and Gray, M. R. 1987. Use of I.R. spectroscopy and nitrogen titration data in structural group analysis of bitumen. *Fuel*, **66:**749–752.
17. Painter, P. C., Snyder, R. W., Starsinic, M., Coleman, M. M., Kuehn, D. W., and Davis, A. 1981. Concerning the application of FT-IR to the study

of coal: A critical assessment of band assignments and the application of spectral analysis programs. *Appl. Spectrosc.*, **35:**475–485.
18. McKay, J. F., Cogswell, T. E., Weber, J. H., and Latham, D. R. 1975. Analysis of acids in high boiling petroleum distillates. *Fuel*, **54:**50–61.
19. McKay, J. F., Weber, J. H., and Latham, D. R. 1976. Characterization of nitrogen bases in high boiling petroleum distillates. *Anal. Chem.*, **48:**891–898.
20. Petersen, J. C. 1971. A thermodynamic study by infrared spectroscopy of the association of 2-quinolone, some carboxylic acids, and the corresponding 2-quinolone–acid mixed dimers. *J. Phys. Chem.*, **76:**1129–1135.
21. Green, J. B., Stierwalt, B. K., Green, J. A., and Grizzle, P. L. 1985. Analysis of polar compound classes in SRC-II liquids. Comparison of non-aqueous titrimetric, i.r. spectrometric and h.p.l.c. methods. *Fuel*, **64:**1571–1580.
22. Yu, S. H.-T., and Green, J. B. 1989. Determination of total hydroxyls and carboxyls in petroleum and syncrudes after chemical derivatization by infrared spectroscopy. *Anal. Chem.*, **61:**1260–1268.
23. Petersen, J. C., Barbour, R. V., Dorrence, S. M., Barbour, F. A., and Helm, R. V. 1971. Tentative identification of 2-quinolones in asphalt and their interaction with carboxilic acids present. *Anal. Chem.* **43:**1491–1496.
24. Jokuty, P. L., and Gray, M. R. 1991. Resistant nitrogen compounds in hydrotreated gas oils from Athabasca bitumen. *Energy Fuels*, **5:**791–795.
25. Owen, A. J. 1988. *The Diode Array Advantage in UV/Visible Spectroscopy*. Hewlett Packard, Waldbronn, Germany.
26. Sullivan, R. F., Boduszynski, M. M., and Fetzer, J. C. 1989. Molecular transformations in hydrotreating and hydrocracking. *Energy Fuels*, **3:**603–612.
27. Boduszynski, M. M. 1988. Composition of heavy petroleums. 2. Molecular characterization. *Energy Fuels*, **2:**697–613.
28. Yokota, T., Scriven, F., Montgomery, D. S., and Strausz, O. P. 1986. Absorption and emission spectra of Athabasca asphaltene in the visible and near ultraviolet regions. *Fuel* **65:**1142–1149.
29. Goulon, J., Esselin, C., Friant, P., Berthe, C., Muller, J.-F., Poncet, J.-L., Guilard, R., Escalier, J.-C., and Neff, B. 1984. Structural characterization by x-ray absorption spectroscopy (EXAFS/XANES) of the vanadium chemical environment in various asphaltenes. *Collect. Collog. Semin.* (*Inst. Fr. Pet.*), **40:**153–157.
30. Kelemen, S. R., George, G. N., and Gorbaty, M. L. 1990. Direct determination and quantification of sulfur forms in heavy petroleum and coal. 1. The X-ray photoelectron spectroscopy (XRA) approach. *Fuel*, **69:**939–944.
31. Kelemen, S. R., George, G. N., and Gorbaty, M. L. 1990. Direct determination and quantification of sulfur forms in heavy petroleum and coal. 2. The sulfur K edge X-ray photoelectron spectroscopy approach. *Fuel*, **69:**945–949.
32. Waldo, G. S., Carlson, R. M. K., Moldowan, J. M., Peters, K. E., and Penner-Hahn, J. E. 1991. Sulfur speciation in heavy petroleums: Information

from x-ray absorption near-edge structure. *Geochim. Cosmochim. Acta*, **55**:801–814.

34. Zhang, G., Ziemer, J. N., Boduszynski, M. M., and Biggs, W. R. 1993. The Binding of Metals in Maya Crude Oil: Looking for the Elusive "Metallo-nonporphyrins." *Energy Fuels*, manuscript submitted for publication.
35. Fay, M. J., Proctor, A., Hoffmann, D. P., and Hercules, D. M. 1988. Unravelling EXAFS spectroscopy. *Anal. Chem.*, **60**:1225A–1243A.
36. Yen, T. F., Erdman, J. G., and Pollack, S. S. 1961. Investigation of the structure of petroleum asphaltenes by x-ray diffraction. *Anal. Chem.*, **33**:1587–1597.
37. Ebert, L. B. 1990. Comment on the study of asphaltenes by x-ray diffraction. *Fuel Sci. Tech. Int.*, **8(5)**:563–569.
38. Ebert, L. B., Scanlon, J. C., and Mills, D. R. 1984. X-ray diffraction of *n*-paraffins and stacked aromatic molecules: Insight into the structure of petroleum asphaltenes. *Liq. Fuels Tech.*, **2**:257–286.
39. Ebert, L. B., Mills, D. R., and Scanlon, J. C. 1987. Reductive alkylation of aromatic hydrocarbons: Petroleum resid vs. model compounds. *Amer. Chem. Div. Petr. Chem.*, **32(2)**:419–425.
40. Ebert, L. B., Kastrup, R. V., and Scanlon, J. C. 1988. The disruption of aromatic stacking order in mesophase coke via reductive alkylation. *Mater. Res. Bull.*, **23**:1757–1763.
41. Ebert, L. B., Scanlon, J. C., and Clausen, C. A. 1988. Combustion tube soot from a diesel fuel/air mixture: Issues in structure and reactivity. *Energy Fuels*, **2**:438–445.
42. Ebert, L. B., Rose, K. D., and Scanlon, J. C. 1989. Reductive alkylation of petroleum residua using potassium metal and tetrahydrofuran at room temperature. *Fuel*, **68**:935–937.
43. Ebert, L. B., Kastrup, R. V., and Scanlon, J. C. 1989. Characterization of petroleum residua through reductive chemistry. *Fuel Sci. Techn. Int.*, **7(4)**:377–397.
44. Rubinstein, I., and Strausz, O. P. 1987. Thermal treatment of the Athabasca oil and bitumen and its component parts. *Geochim. Cosmochim. Acta*, **43**:1887–1893.
45. Rubinstein, I., Spyckerelle, C., and Strausz, O. P. 1979. Pyrolysis of asphaltenes: a source of geochemical information. *Geochim. Cosmochim. Acta*, **43**:1–6.
46. McIntire, D. D., Montgomery, D. S., and Strausz, O. P. 1987. Asphaltene pyrolysis distillates from Athabasca bitumen and cretaceous heavy oils related to depth of burial and degree of wheatering. *AOSTRA J. Res.*, **2**:251–265.
47. Behar, F., and Pelet, R. 1984. Characterization of asphaltenes by pyrolysis and chromatography. *J. Anal. Appl. Pyrol.*, **7**:121–135.
48. Eglinton, T. I., Philip, R. P., and Rowland, S. J. 1988. Flash pyrolysis of artificially matured kerogen from the Kimmeridge Clay, U.K. *Organ. Geochem.*, **12(1)**:33–44.

49. Fowler, M. G., and Brooks, P. W. 1987. Organic Geochemistry of Western Canada Basin tar sands and heavy oils. 2. Correlation of tar sands using hydrous pyrolysis of asphaltenes. *Energy Fuels*, **1**:459–467.
50. Van de Meent, D., Brown, S. C., Philp, R. P., and Simoneit, B. R. T. 1980. Pyrolysis–high resolution gas chromatography and pyrolysis gas chromatography–mass spectrometry of kerogens and kerogen precursors. *Geochim. Cosmochim. Acta*, **44**:999–1013.
51. Sinnenghe Damsté, J. S., Van Dalen, A. C. K., De Leeuw, J. W., and Schenk, P. A. 1988. Identification of homologous series of alkylated thiophenes, thiolanes, thianes and benzothiophenes present in pyrolysates of sulphur-rich kerogens. *J. Chromatog.*, **435**:435–452.
52. Sinnenghe Damsté, J. S., Eglington, T. I., De Leeuw, J. W., and Schenk, P. A. 1989. Organic sulfur in macromolecular sedimentary organic matter: I. Structure and origin of sulfur-containing moieties in kerogen, asphaltenes and coal as revealed by flash pyrolysis. *Geochim. Cosmochim. Acta*, **53**:873–889.
53. Eglinton, T. I., Sinnenghe Damsté, J. S., Kohnen, M. E. L., and De Leeuw, J. W. 1990. Rapid estimation of the organic sulfur content of kerogens, coals and asphalt by pyrolysis–gas chromatography. *Fuel*, **69**:1394–1404.
54. Payzant, J. D., Lown, E. M., and Strausz, O. P. 1991. Structural units of Athabasca asphaltene: The aromatics with a linear carbon framework. *Energy Fuels*, **5**:445–453.
55. Strausz, O. P., and Lown, E. M. 1991. Structural features of Athabasca bitumen related to upgrading performance. *Fuel Sci. Tech. Int.* **9(3)**:269–281.
56. Mojelsky, T. W., Ignasiak, T. M., Frakman, Z., McIntyre, D. D., Lown, E. M., Montgomery, D. S., and Strausz, O. P. 1992. Structural features of Alberta bitumen and heavy oil asphaltenes. *Energy Fuels*, **6**:83–96.
57. Mojelsky, T. W., Montgomery, D. S., and Strausz, O. P. 1985. Ruthenium (VIII) catalyzed oxidation of high molecular weight components of Athabasca oil sand bitumen. *AOSTRA J. Res.* **2**:131–137.
58. Drushel, H. V. 1972. Analytical characterization of residua and hydrotreated products. *Am. Chem. Soc. Div. Petr. Chem.*, F92–F101.
59. Ruiz, J.-M., Carden, B. M., Lena, L. J., Vincent, E.-J., and Escalier, J.-C. 1982. Determination of sulfur in asphalts by selective oxidation and photoelectron spectroscopy for chemical analysis. *Anal. Chem.*, **54**:689–691.
60. Arpino, P. J., Ignatiadis, I., and De Rycke, G. 1987. Sulphur-containing polynuclear aromatic hydrocarbons from petroleum. *J. Chromatog.*, **390**:329–348.
61. Strausz, O. P., Lown, E. M., and Payzant, J. D. 1990. Isolation of Sulfur Compounds from Petroleum. *Geochemistry of Sulfur in Fossil Fuels*, edited by W. L. Orr and C. M. White. ACS Symposium Series 429. American Chemical Society, Washington, DC: Chap. 5, pp. 83–92.
62. Sinnenghe Damsté, J. S., Eglington, T. I., Rijpstra, W. I. C., and de Leeuw, J. W. 1990. Characterization of Organically Bound Sulfur in High-Molecular-Weight, Sedimentary Organic Matter Using Flash Pyrolysis and Raney Ni Desulfurization. *Geochemistry of Sulfur in Fossil Fuels*, edited by W. L. Orr

and C. M. White. ACS Symposium Series 429. American Chemical Society Washington, DC, Chap 26.

63. Sternberg, H. W., Delle Donne, C. L., Pantages, P., Moroni, E. C., and Markby, R. E. 1971. Solubilization of an lvb coal by reductive alkylation. *Fuel*, **50:**432–442.
64. Sternberg, H. W., and Delle Donne, L. C. 1974. Solubilization of coals by reductive alkylation. *Fuel*, **53:**172–175.
65. Ignasiak, T., Kemp-Jones, A. V., and Strausz, O. P. 1977. The of molecular structure of Athabasca asphaltene. Cleavage of the carbon–sulfur bonds by radical ion electron transfer reactions. *J. Organ. Chem.*, **42:**312–320.
66. Ebert, L. B. 1986. The potassium metal reduction of perylene in tetrahydrofuran: Evidence for the prylene dianion. *Tetrahedron*, **42:**497–500.
67. Francisco 1988. Personal communication.
68. Ebert, L. B., Milliman, G. E., Mills, D. R., and Scanlon, J. C. 1988. Reductive Alkylation of Aromatic Hydrocarbons, Prylene, Decacyclene, and Dibenzothiophene. *Polynuclear Aromatic Compounds*, edited by L.B. Ebert. Advances in Chemistry No. 217. American Chemical Society of Washington, DC, Chap. 7, pp. 109–126.
69. Pouchert, C. J. 1975. The Aldrich Library of Infrared Spectra. The Aldrich Chemical Company, Inc., 940 W. St. Paul Ave., Milwaukee, Wisconsin, 53233.
70. Sadtler Research Laboratories, a subsidiary of BIO-RAD, 3316 Spring Garden Street, Philadelphia, PA 19104.

10

Composition of Heavy Petroleum Fractions

I. OVERVIEW

A. What Compositional Information Do We Need?

Until the 1970s, the heavy ends of crude oils were mainly used as fuel oils and for making paving and roofing asphalts. Now they are becoming increasingly important as feed for making gasoline, diesel, and jet fuel. In this conversion, large molecules are broken down to smaller ones, heteroatoms must be removed, the aromatics are hydrogenated to naphthenes, and these are cracked to paraffins and preferably isoparaffins. Information on the feed composition helps the refiner in choosing the right processing route, and the process chemist to improve and optimize existing processes and to design new ones. Among the important compositional details to be known are the yields of high-boiling and residual fractions (i.e., the AEBP distribution curve), concentration of heteroatoms (S, N, O, V, and Ni), concentration of various compound groups and classes in a given fraction (e.g., "acids," "bases," "aromatics," "saturates"), heterocompound types (e.g., carboxylic acids, amides, pyrrolic N-compounds, thiophene types, etc.), concentration and type of aromatics (e.g., mono-, di-, tri-, etc.), concentration and type of aliphatic hydrocarbons (e.g., paraffins, mononaphthenes, dinaphthenes, trinaphthenes, etc.), and the carbon-number distribution of these compound types. Availability of this kind of detailed compositional information decreases rapidly with increasing boiling point of the fractions.

The detailed analytical information is relatively easy to get for light (<400°F, <200°C AEBP) and middle distillates (400–650°F, 200–345°C AEBP), but increasingly harder for higher-boiling fractions. Deep distillation of atmospheric residues (>650°F, >345°C AEBP) with high-vacuum short-path stills facilitates the analysis of high-boiling fractions (650–1300°F, 345–700°C AEBP) by eliminating interferences from more refractory "nondistillable" residues (>1300°F, >700°C AEBP). Light vacuum gas oils (LVGOs), with a nominal boiling range of approximately 650–800°F (345–430°C) AEBP, can be analyzed directly by MS group-type methods, which determine concentrations of various compound types. The LVGOs can also be fairly readily separated by liquid chromatography into well-defined compound-class fractions such as mono-, di-, tri-, and tetraaromatics, which can then be further characterized in terms of specific compound types (chemical formula) and carbon-number distribution. Besides the new, more complex compound types in LVGOs, most or all of the simpler ones found in middle distillates are also present here, except with higher carbon numbers.

Heavy vacuum gas oils (HVGOs) with a nominal boiling range of approximately 800–1000°F (430–540°C) AEBP are distinctly harder to analyze than LVGOs. The main reason is a much higher concentration of heterocompounds. In addition, both concentration and diversity of aromatics in HVGOs are much greater than in LVGOs. The chromatographic separations of HVGOs are less clean and produce overlapping compound-group and compound-class fractions. The chromatographic fractions of saturates and aromatics can be analyzed by MS group-type methods (e.g., D2786 and D3239, respectively). However, the MS group-type methods have limited application, particularly if the HVGO boiling range extends beyond 1000°F (540°C) AEBP. The presence of relatively large amounts of heterocompounds in HVGOs requires the use of special chromatographic separations to produce sufficiently narrow fractions for analysis by high-resolution MS.

Super-heavy vacuum gas oils (SHVGOs) with a nominal boiling range of approximately 1000–1300°F (540–700°C) AEBP are already immensely complex. Although they can be completely volatilized during MS analysis, the resulting spectra are very complex and difficult to interpret. The bulk of SHVGO components involve sulfur, nitrogen, and oxygen compounds. These very high-boiling oils also contain metalloporphyrins. Saturates in SHVGOs seldom account for more than 10–20 wt%, and aromatics for more than 50–80 wt%. The latter usually contain significant amounts of neutral heterocompounds, which are very difficult to separate from the aromatic hydrocarbons. Chromatographic separations of SHVGOs into concentrates of certain groups or classes of compounds facilitate their further analysis. However, severe overlapping of adjacent fractions is to be expected. No MS group-type methods are available for the analysis of these fractions. FIMS and/or FDMS of saturates separated from SHVGOs can provide information in terms of Z series (paraffins, mononaphthenes, dinaphthenes, etc.) and carbon-number distribution. Analysis of aromatics and other chromatographic fractions separated from SHVGOs would require ultra-high-resolution MS, which is not readily available. Complementary information on the composition of these fractions can be obtained by using NMR, FTIR, and other spectroscopic techniques.

The "nondistillable" residues (>1300°F, >700°C AEBP) are the most complex of all crude oil fractions. They consist almost exclusively of heterocompounds, many of which involve several heteroatoms per molecule. These residues typically contain 30–50% of the total sulfur, 70–80% of the total nitrogen, and 80–90% of the total vanadium and nickel present in the crude oil. Saturates and aromatics, together, account for only a few weight percent, at most, of the "nondistillable" residue. Chromatographic separations of nondistillables are complicated by their limited solubility in solvents which are typically used as mobile phases in liquid chromatography. Irreversible adsorption on column packings can also severely limit the use of chromatography. Frequently, precipitation of insolubles (asphaltenes) by a low-boiling paraffin such as *n*-pentane or *n*-heptane, is used to facilitate the chromatographic separation of the soluble part of the sample (the "maltenes"). It has been demonstrated, however, that direct chromatographic separation of "nondistillable" residues on ion exchange resins without prior precipitation of asphaltenes produces compound-group fractions such as "acids," "bases," and "neutrals" with reasonably good recovery of the material (>90–95%). The "neutrals" can then be further separated into saturates and aromatics. The separations in this range are fairly poor, and adjacent fractions overlap severely. We advocate, instead, the separation of nondistillables by "sequential elution fractionation" (SEF), which produces operational but reasonably well-defined and distinctly different, fractions with progressively higher average molecular weights and mid-AEBPs. In other words, we advocate their separation by AEBP rather than by chemical composition. The AEBP fractions can then be subjected to further characterization.

B. Overall Composition

In this section, we present a brief overview of the composition of heavy ends with emphasis on its change with boiling point (AEBP). For the sake of clarity and easy reading, only the most important features are sketched. Many details are left to Sections II and III, in which our cursory discussion given here will be fleshed out, substantiated, and properly referenced. In Section II, the subject will be treated from a different point of view, subdivided into the main compound classes. In Section III, the composition of the main distillation cuts of a crude oil will be described in terms of the main compound groups, ring types, and carbon numbers.

To distinguish among the great variety of petroleum components, we make extensive use of the terms "compound groups," "compound classes," and "compound types," which were defined in Chapter 2. Occasionally, we will have to abandon our preferred nomenclature and refer to other traditional, but less well-defined terms such as, for example, asphaltenes and maltenes. The reason for yielding here is our extensive reference to the literature where these terms are common. In some cases, we have tried to replace them with more precise language, but other times this was not possible because of insufficient data or the sake of proper reference. In Section III, we discuss the topic of asphaltenes in detail.

It is customary to distinguish between light and heavy crude oils and bitumens. The light crude oils give high yields of low-boiling distillates which are mostly paraffinic. Even their heavy ends have a very high proportion of paraffinic carbon atoms, partly in large paraffin molecules or else in alkyl chains connected to naphthenic and aromatic rings. Heavy crude oils contain greater amounts of naphthenic and aromatic structures and of heterocompounds. Not only do they contain less low-boiling material, but generally they also have less paraffinic character. Bitumens lack low-boiling fractions, but may encompass a wide range of paraffinic, naphthenic, and aromatic carbon distributions. Both heavy crude oils and bitumens have high concentrations of heterocompounds.

Most petroleum compounds boiling below 650°F (345°C) AEBP are primarily paraffins, mono- and dinaphthenes, mono- and diaromatics. There are some sulfur compounds and very few other heterocompounds. The composition of these distillates is relatively simple and generally well known. In higher-boiling fractions (>650°F), naphthenes and aromatics with higher ring numbers quickly become more prominent, and the concentration of heterocompounds increases rapidly with rising boiling point.

In the range of 800–1000°F (425–540°C) AEBP, we encounter the same compound types found in the lower boiling ranges but now with more alkyl substituents and, thus, increased complexity. In addition to these, new compound types show up. Not only does the ring number increase, but heterocompounds account now for more than half of all molecules. The diversity of molecular structure and functionality and the increased alkyl substitution make it increasingly difficult to separate these high-boiling fractions into compound groups and compound classes.

In the super-heavy VGO range (1000–1300°F, 540–700°C AEBP), the overlap between the various compound-class and compound-group fractions is quite severe. Here, "polars" make up the majority of constituents, and even the "aromatics" may consist of as much as 80% neutral heterocompounds. Heterocompounds with several heteroatoms, for example, metalloporphyrins appear. The great majority of compounds in this range contain one or more heteroatoms.

In the nondistillable residues, almost all the molecules are polar heterocompounds, and many comprise several functionalities. Despite much work, little is known about the composition. Molecular level analyses (by MS) are very difficult, and chromatographic separations are inefficient. NMR, IR, and some other techniques give average information on ring structure, side-chains, polar functional groups in the mixture. Destructive methods, pyrrolysis and oxidation, have provided some access to their composition in terms of ring structure and alkyl chain distribution.

Figure 10.1 illustrates the progression of ring types with increasing boiling point in general terms and only for the unsubstituted parent molecules (with one exception). Alkyl substitution increases the molecular weight more than the boiling point, shifting the position in the plot upward as demonstrated by the C_{10}-substituted pyrene. Heteroatoms in polar functional groups have the opposite effect, increasing the boiling point much more than the MW. They would shift the position of the corresponding molecules, toward the right or toward the bottom in a graph like that of Fig. 10.1. Examples for the latter case are the following compounds:

Compound	Formula	Boiling point		MW
		°F	°C	
n-Pentadecane	$C_{15}H_{32}$	518	270	212
2,3-Dimethylnaphthalene	$C_{12}H_{12}$	516	269	156
2-Methylindole	C_9H_9N	523	273	131
Benzoic acid	$C_7H_6O_2$	518	270	122

Little has been published on the composition of heavy and super heavy vacuum gas oils (800–1300°F, 430–700°C AEBP). There is mainly the work by the API-60 program [summarized by McKay et al. (1–3)], the NIPER work [summarized by Green et al. (4)], and our own (5,6). Thus, little information is available on the changes in composition of a crude oil with AEBP. The most detailed study of this kind to date was performed on Kern River crude (5,6). Even though this heavy naturally biodegraded crude oil is quite different from most other crude oils, it is a good sample to demonstrate the changing composition with increasing AEBP. Figure 10.2 shows the results in terms of 6 chromatographic fractions—saturates,

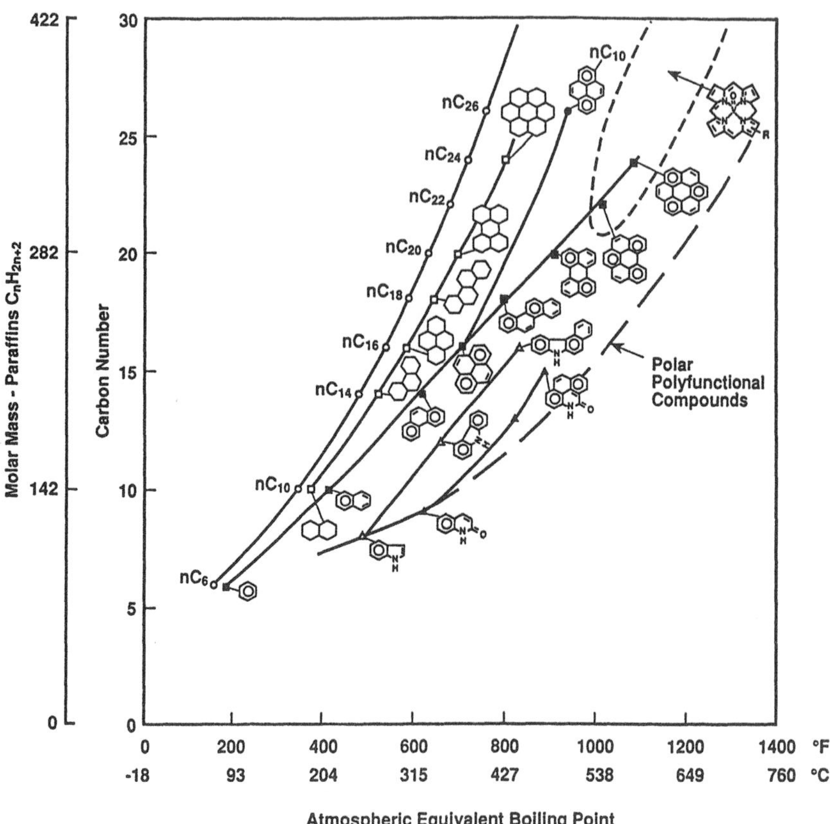

Figure 10.1 Distribution of compound types and compound classes in petroleum as a function of boiling point.

mono- to pentaaromatics, and 3 polar heterocompound fractions—for each of 10 distillates. Four solubility fractions were also obtained from the "non-distillable" residue (>1300°F AEBP) but only one of these, the pentane-soluble SEF-1 fraction (1365°F, 735°C mid-AEBP) was successfully separated by liquid chromatography. The concentrations of these compound classes and groups are indicated by the width of their bands. The gradual and consistent change in their distribution with AEBP is clearly manifested here.

The first fraction, with AEBP ≤500°F (260°C) and amounting to about 7% of the crude oil, consists of 88% saturates, 8% monoaromatics, and

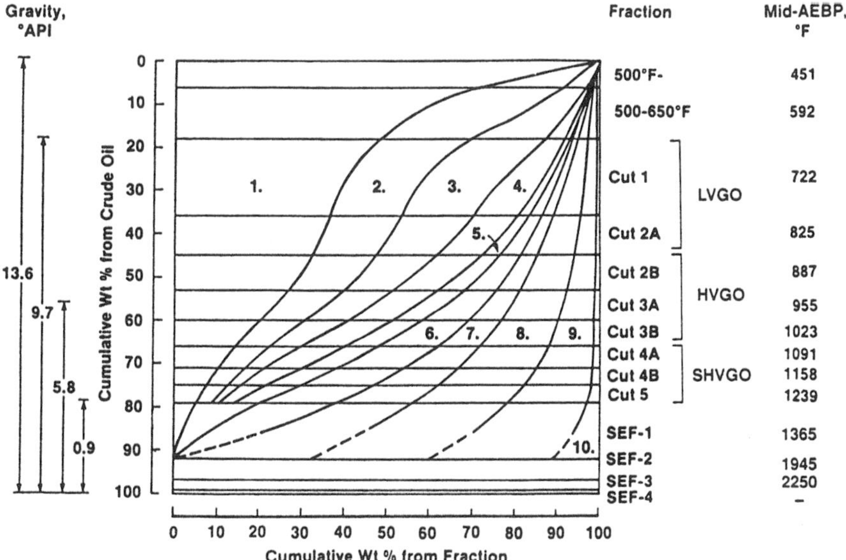

Figure 10.2 Distribution of saturates, aromatic compound classes, and polar compound fractions in Kern River petroleum. Key to HPLC fractions: (1) saturates, (2) monoaromatics, (3) diaromatics, (4) triaromatics, (5) tetraaromatics, (6) pentaaromatics and greater, (7) "basic" compounds, (8) "pyrrolic" N-compounds, (9) "acidic" compounds. (From Ref. 6. Reproduced with permission of the publisher.)

4% diaromatics. In the 500–650°F distillate, the saturates have dropped to under 55%, the monoaromatics are up to almost 35%, whereas the diaromatics remain about the same; here we also see 2% triaromatics and 3% polar heterocompounds. With increasing AEBP, the saturates' content continues to drop and that of the polar heterocompound classes rises almost in (inverse) proportion. The total aromatics remain fairly constant up to about 60% of total crude (near 1000°F or 540°C AEBP), but the tetraaromatics and, even more so, the pentaaromatics increase strongly at the expense of the mono- and diaromatics. The increase in the aromatics, in general, and especially in the "pentaaromatics," arises to a great extent from neutral heterocompounds.

The compositional changes between cuts are gradual, not abrupt. FIMS spectra (5), not shown here, demonstrate that this is the case not only with compound groups and compound classes, in general, but with the compound types as well.

Not only does the amount of the aromatic compound classes change with the boiling point but so does their internal composition, namely, the proportion of alkane structure within their molecules. Both their naphthenic ring structure and the carbon number of paraffinic substituents increases. The increasing paraffinicity makes the higher-boiling aromatic compound types (and the heterocompound types) harder to separate from each other than the lower-boiling ones.

More than half of the truly nondistillable residue consists of the pentane-soluble fraction (SEF-1) with mid-AEBP of 1365°F (741°C). In this fraction, the saturates content is down to about 1% and that of the total aromatics, to about 8%. The remainder of the SEF-1 fraction consists of polar heterocompounds, about evenly divided between basic, pyrrolic, and acidic ones. The pentane-insoluble/cyclohexane-soluble SEF-2 (mid-AEBP of 1945°F, ~1060°C) and cyclohexane-insoluble/toluene-soluble SEF-3 (mid-AEBP of 2250°F, ~1230°C) fractions together constitute less than 9% of this crude oil. Assuming the trends of the lower-boiling fractions continue, we can expect almost half of this material to belong to the acidic heterocompound group, another 40% to the pyrrolic, and only 15% to the basic one.

Without deep distillation, 82% of this crude oil would be atmospheric residue (>650°F or >345°C). But high-vacuum short-path distillation yields three-fourths of this material as distillates, leaving only 21% of the crude oil as truly nondistillable residue (>1300°F or >700°C). From the chromatographic separations of the eight distillation cuts, spanning the AEBP range from 650 to 1300°F (345–700°C), we can see clearly and in reasonable detail the changes in composition which occur with increasing boiling point.

In principle, the scheme of Fig. 10.2 is representative for most crude oils although Kern River crude is not only very heavy, containing little low-boiling material, but also very naphthenic. Therefore, although the relative amounts of the nine compound classes will vary significantly in other oils, their respective changes with AEBP are likely to be similar. This statement is borne out by the (much more limited) data on six other crude oils in Tables 10.1–10.3. Those in Tables 10.1 and 10.2 confirm the sharp increase in the amount of heterocompounds, acids, bases, and neutral heterocompounds, with boiling point and the corresponding decrease of the hydrocarbons. Table 10.3 shows that, in addition to the increase in the amounts of acids and bases with increasing AEBP, the S, N, and O contents in these fractions generally rise too, and so does the hydrogen deficiency as indicated by the falling H/C ratio. We will come back to the plot in Fig. 10.2 later to see that more information can be extracted from it.

The picture we painted here is oversimplified. In reality, the separations become more and more difficult with the higher boiling fractions; and in

Table 10.1 Distribution of Compound Classes in Three AEBP Fractions from Four Crude Oils

Crude oil	Boiling range °F	Boiling range °C	Acids	Bases	Neutral N compounds	Aliphatic hydrocarbons	Aromatic hydrocarbons
Wilmington, California	700–995	370–535	5.6	6.8	4.2	36.9	46.5
	995–1245	535–675	9.3	12.7	21.3	20.8	36.0
	>1245	>675	18.0	19.0	41.0	4.0	15.0
Gach Saran, Iran	700–995	370–535	1.7	2.1	2.3	48.5	46.5
	995–1245	535–675	5.4	8.7	8.9	31.6	45.4
	>1245	>675	12.0	25.0	14.0	8.0	30.0
South Swan Hills, Alta.	700–995	370–535	1.8	2.2	1.9	65.9	29.7
	995–1245	535–675	3.5	4.5	5.6	57.5	28.9
	>1245	>675	12.0	13.0	10.0	34.0	27.0
Recluse, Wyoming	700–995	370–535	1.4	1.1	0.9	74.1	22.5
	995–1245	535–675	2.9	3.3	3.0	68.8	21.9
	>1245	>675	10.0	9.0	8.0	44.0	26.0

Source: Ref. 3.

Table 10.2 Weight Percent Acids, Bases, and Neutrals in Three Crude Oils

Crude oil	AEBP range °F	AEBP range °C	Acids	Bases	Neutrals
Cerro Negro	800–1025	425–550	6.9	4.1	89.2
	1025–1300	550–700	12.3	10.3	78.1
	>1300	>700	32.6	33.9	31.5
Wilmington	800–1000	425–540	10.8	5.2	80.1
	>1000	>540	35.7	28.5	36.5
Mayan	750–925	400–500	4.0	1.7	94.0
	>925	>500	29.0	22.0	48.2

Source: Data adapted from Ref. 4.

the high AEBP ranges, the compound class fractions are no longer as clean as their names imply. The overlap between fractions becomes worse, and the hydrocarbon fractions contain significant and increasing amounts of heterocompounds (see, e.g., Table 10.4). These are hard to separate from the hydrocarbons because of their often large alkyl substituents which shield the heteroatoms and hinder them from interacting with the active

Table 10.3 C, H, S, N, and O Distributions in Various Fractions of Cerro Negro Crude as a Function of AEBP

Fraction	AEBP range °F	AEBP range °C	C	H	H/C	S	N	O
Whole distillate	800–1025	425–550	84.58	10.92	1.538	3.46	0.34	0.45
	1025–1300	550–700	84.35	10.77	1.521	4.01	0.50	0.55
	>1300	>700	83.65	9.28	1.322	4.56	1.16	0.89
Acid fraction	800–1025	425–550	79.09	9.60	1.447	2.43	1.25	6.21
	1025–1300	550–700	81.09	9.17	1.347	3.42	1.33	3.95
Strong acids	>1300	>700	81.49	8.18	1.197	4.47	1.71	2.21
Weak acids	>1300	>700	80.59	8.52	1.260	4.42	1.22	4.09
Base fraction	800–1025	425–550	82.91	9.66	1.388	3.05	3.09	1.33
	1025–1300	550–700	80.04	9.38	1.381	3.49	2.22	1.58
Strong bases	>1300	>700	83.09	8.36	1.199	4.59	1.80	1.05
Weak bases	>1300	>700	81.95	9.20	1.338	5.25	1.23	1.49
Neutrals	800–1025	425–550	84.92	11.11	1.560	3.72	0.14	0.14
	1025–1300	550–700	84.63	11.09	1.567	4.01	0.17	0.18
	>1300	>700	84.50	9.05	1.533	4.18	0.33	0.17

Source: Data adapted from Ref. 4.

Table 10.4 C, H, S, N, and O Distributions in Some Compound Classes of Two >1250°F (>675°C) Residues

Crude oil Sample	C	H	H/C	S	N	O
Wilmington						
Whole nondistillable residue	84.5	9.7	1.378	2.57	1.62	1.5
Saturates	85.5	12.7	1.782	0.82	0.11	0.8
Aromatics	85.3	10.8	1.519	2.37	0.63	0.7
Neutral N-compounds	83.0	9.4	1.359	2.71	1.78	1.3
Base fraction	83.2	9.6	1.385	2.31	2.24	2.7
Acid fraction	80.2	9.1	1.362	2.53	2.33	5.5
Recluse						
Whole nondistillable residue	86.5	11.6	1.609	0.37	0.56	0.8
Saturates	—	—	—	0.12	0.11	0.3
Aromatics	—	—	—	0.37	0.62	1.5
Neutral N-compounds 1	—	—	—	0.52	0.77	1.9
Neutral N-compounds 2	—	—	—	0.22	0.77	2.5
Base fraction	—	—	—	0.45	1.23	1.5
Acid fraction	—	—	—	0.36	1.10	3.6

Source: Data adapted from Ref. 2.

sites on the chromatographic packings. On the other hand, most S compounds behave chromatographically very similar to hydrocarbons even without steric hindrance.

Heterocompounds, particularly sulfur compounds, can be found in almost all petroleum fractions, even in the lower-boiling ones. The sulfur concentration increases only moderately and in roughly linear fashion with the boiling point. Consequently, much of the sulfur in crude oils resides in the distillates. Nitrogen and oxygen have much lower concentrations in the low-boiling fractions which initially increase only moderately with rising boiling point until near 650°F (345°C), where their concentration curves turn up sharply; see Chapter 4, Figs. 4.14 and 4.15. Metals are absent in distillates below 1000°F (540°C). Their distributions are bimodal with a first hump around 1200°F (650°C) AEBP (in super-heavy vacuum gas oils) and a steady increase beyond 1500°F (815°C) (in the solubility fractions of nondistillable residues); see Fig. 4.13 of Chapter 4.

Numerous compound types have been identified among the simpler heterocompounds which carry just one or two heteroatoms. Most of these have been found in the boiling range of about 500–700°F (260–370°C). Monoheterocompounds are also present in higher-boiling fractions, but because of their greater complexity in terms of ring structure as well as number and length of substituents, they are harder to detect. Fractions with higher AEBPs comprise increasing amounts of compounds with two or more heteroatoms.

The distribution of heteroatoms and functionalities among compound groups is much more complex than one might expect from the names of these fractions (saturates, aromatics, acids, bases). Note, for instance, the heteroatom distribution in the nondistillable residues of two crude oils shown in Table 10.4. As expected, the acid fraction of the two residues is very rich in oxygen; but substantial amounts of O are also found in the base fraction as well as in the neutral N-compounds fraction and even in the saturates and aromatics fractions. Similarly, S and N are almost evenly distributed among all the fractions except for the saturates where they are less abundant though still present.

Little is known about the heteroatom-containing structural elements in the nondistillable fractions. Almost all we know about them comes from averaging measurements of the main functional polar groups and some sulfur configurations. These will be discussed in Section II. The concentration of heteroatoms in nondistillable residues is so high that the majority of molecules carry multiple heteroatoms. One easily identifiable multi-heteroatom compound type are the porphyrins which have four nitrogen atoms. Even these may contain additional heteroatoms. Generally, only the main functionalities such as >CO, –OH, >NH, and certain other structural features, for example, aromatic C–C and C–H, and the average number, length, and branching of paraffinic chains can be determined directly.

For all compositional features, including the heterocompounds, the changes from one fraction to the next are gradual. In fact, the composition of a higher-boiling cut can to a great extent be predicted from that of the next lower-boiling cut. This principle still holds even as we come to the end of the distillable constituents and enter the domain of the nondistillables around 1300°F (700°C) AEBP. All the main features found in high-vacuum short-path distillates are present also in the nondistillable pentane-soluble SEF-1 fraction (1365°F or 741°C mid-AEBP) with only gradually increased complexity. However, in the higher AEBP solubility fractions (SEF-2:1945°F, ~1060°C, mid-AEBP, and SEF-3:2250°F, ~1230°C, mid-AEBP) the rate of change becomes so great that it often masks the gradual progression.

In the nondistillable fractions, particularly in the pentane-insoluble ones (SEF-2, SEF-3, SEF-4), the structural complexity becomes immense. Molecular weights range from approximately 500 to possibly several thousand daltons. The exact range is not known. At the low end, fragmentation products may obscure the parent ion distribution in FIMS and FDMS spectra. At the high end, low sample volatility limits the MS methods.

Aggregation effects interfere with SEC and VPO measurements. FIMS/FDMS profiles can, however, be accepted as fairly representative in the 500–2000 dalton range. Among the larger molecules, we must expect assemblies of several ring clusters linked by paraffinic chains and sulfur as proposed, for example, by Ignasiak et al. (7), Strausz and Lown (8), and Mojelsky et al. (9).

In the discussion in Section II, this untidiness needs to be remembered. Although we must often simplify the issues when we try to explain the compositional principles involved, the true complexity of heavy petroleum fractions should always be kept in mind.

C. Molecular Fossils

Certain petroleum compounds are found in distinctly higher concentration than others of similar structure. These survived the diagenesis of petroleum with fewer changes than the remainder. They are called "biomarkers" or "molecular fossils" (10). Many hydrocarbons, oxygen compounds, and, more recently, some sulfur and nitrogen compounds have been identified as molecular fossils [see, e.g., (11,12)]. Among the main fossil molecules are diterpanes, triterpanes, steranes, and hopane derivatives. Most are paraffinic and naphthenic, some were partially or completely aromatized, and some contain heteroatoms. A few examples are displayed in Fig. 10.3. These molecules are present in higher concentrations than other ones because they did not suffer the random degradation as those did. But even most of the degraded molecules still carry more or less obvious remnants of the original structural features of their biogenic parent material.

Although the subject of molecular fossils is outside the scope of our book, it is worthwhile to point out their significance to our subject. Their prime importance is as biomarkers, enabling geochemists to relate crude oils to their parent kerogen and, thus, to draw conclusions about the origin of a crude oil, its genesis, and migration. They also help in the identification of oil samples in terms of their oil fields.

Their significance to the petroleum chemist lies in their structural similarity with most other petroleum components. These other components were altered much more extensively during the conversion of the biogenic material to oil. Most lost methyl groups and other alkyl substituents; some were partially or completely aromatized. Some rings may have been closed and others opened, and former ring carbons may have been lost. Yet, the overall ring structure of many petroleum constituents still bears distinct similarity to the original molecules.

A large part of the petroleum naphthenes, from mono-ring compounds to those with six rings, can be assumed to be directly derived from biological matter. Even many of the aromatic compounds show such a resemblance

Saturates

acyclic

monocyclic

tricyclic

Tricyclahexaprenane

tetacyclic

Cholestane (C_{27})

pentacylic

Hopane (C_{30} H_{32})

hexacylic

R

Aromatics

monoaromatic

triaromatic

triaromatic

R

hopanoids

steroid

O Compounds

OH

COOH

H

COOH

N Compounds

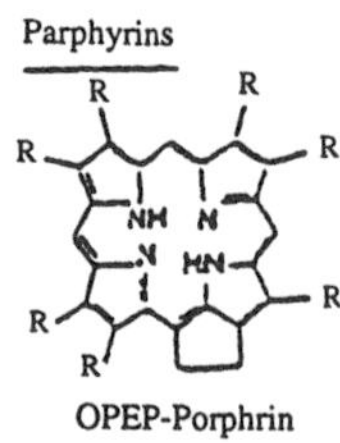

S Compounds

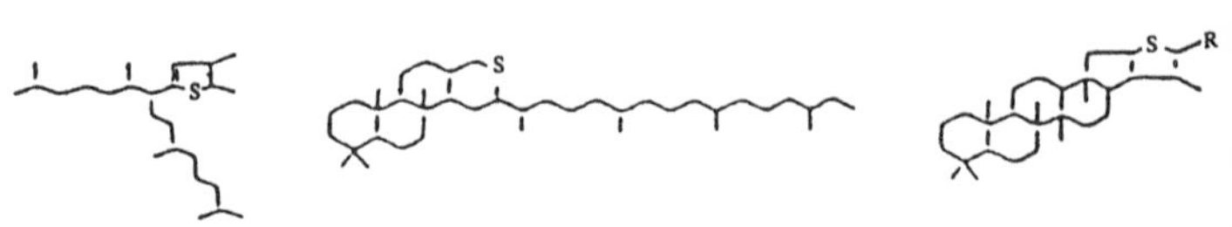

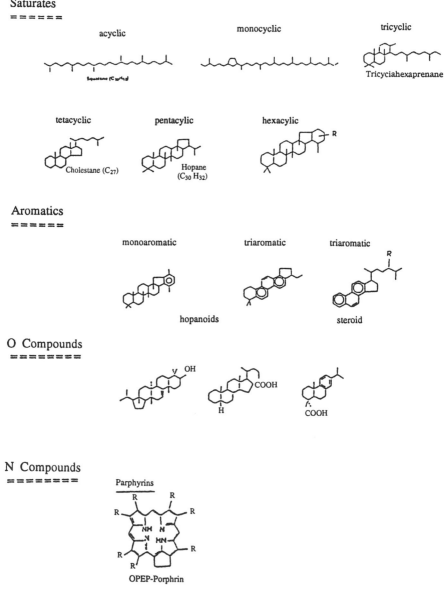

Figure 10.3 Examples of molecular fossils found in petroleum. (From Refs. 10, 13, and 14.)

in their ring structure with biogenic compounds, except for their aromaticity, that we may presume them to be their direct diagenetic dehydrogenation products. This similarity is preserved also in many heterocompounds, especially in many sulfur compounds, although the sulfur was not part of the biogenic material but was incorporated during diagenesis [see, e.g., (12,14) and references therein]. The composition of *high*-boiling fractions is to a great extent related to the molecular fossils found in the same crude oil. The survival and destruction of molecular fossils in refinery operations were discussed by Peters et al. (11).

D. A Cautionary Note

We need to mention here another point. In the literature, we find a number of detailed reports on the composition of heavy petroleum fractions. In some cases, entire (unfractionated) atmospheric or vacuum residues were a subject of detailed analyses. A closer inspection of the reported data reveals that the molecules identified are of such low molecular weight that they belong to distillate fractions preceding the residues. Thus, the reader is led to believe that these structures are part of truly residual material when, in reality, they are not.

The problem most likely arises from the use of insufficiently defined samples. For instance, a so-called vacuum residue typically contains some material <1000°F (535°C) AEBP. The main source of error may be low-efficiency distillation, producing distillates (and residues) with excessive overlap from neighboring fractions, especially from lower-boiling ones. Such a lower-boiling contaminant may be accepted as a legitimate part of the residue even if the molar mass (or carbon number) of the identified molecules clearly indicates that it should not be. Deep distillation of a residue into several narrow cuts by using high-vacuum short-path stills avoids or greatly mitigates these shortcomings. Unambiguous definition of the sample in terms of a boiling-point distribution as determined by a simulated distillation method would be another important step.

The following discussion must be understood with the caveat that reliable detailed information is available at best for the light (650–800°F AEBP) and heavy VGOs (800–1000°F AEBP). Information on the super-heavy (1000–1300°F AEBP) VGOs and, in particular, on nondistillable (>1300°F AEBP) residues is sketchy and averaged. Some of it can be augmented by extrapolation from results obtained on lower-boiling fractions, but, in general and despite much work in this area, little is truly known.

II. THE BASIC COMPOUND GROUPS AND THEIR DISTRIBUTION

A. Aliphatic Hydrocarbons

Although the aliphatic hydrocarbons, or saturates, in light and middle distillates are mainly paraffins with some mono- and dinaphthenes, in most *heavy* petroleum fractions, they consist primarily of naphthenes having up to six or more rings. These, however, bear varying and sometimes large amounts of alkyl carbon in side-chains. True paraffins are much less abundant in the heavy fractions of most crude oils. Olefins are rarely found in straight-run petroleum fractions; however, they may be present in substantial amounts in certain refinery streams, for example, in coker and catcracker products.

Figure 10.4 shows the distribution of seven compound types in the saturates of Kern River crude, namely, the paraffins and the mono- to hexanaphthenes, as a function of AEBP. The concentrations of these compound types in the 10 distillation cuts were determined by FIMS of the respective

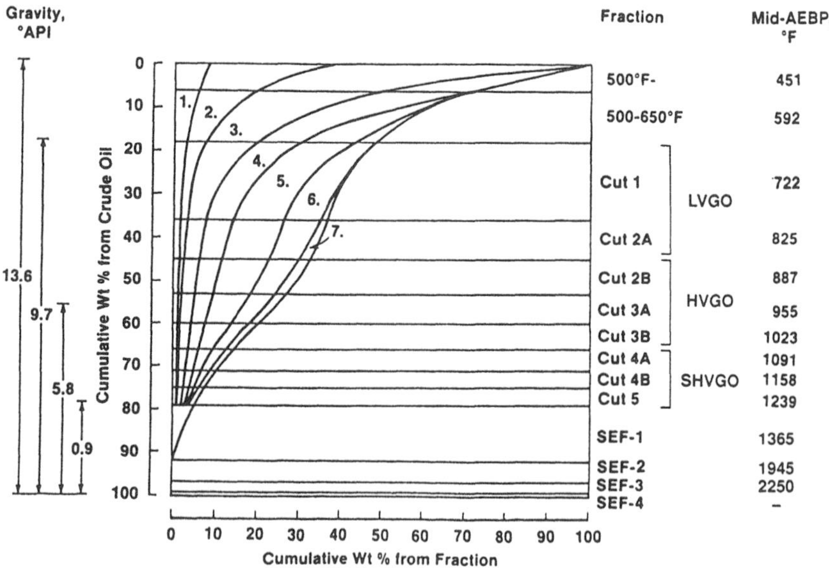

Figure 10.4 Distribution of alkane compound types in Kern River crude. Key to compound types: (1) paraffins, (2)–(7) monocyclic to hexacyclic alkanes. (From Ref. 6. Reproduced with permission of the publisher.)

saturates fractions. As mentioned earlier, this biodegraded crude oil is atypical.* Here, the paraffins are distinctly scarcer than in other crudes. The saturates consist predominantly of naphthenes, of which the tetranaphthenes are the most abundant ones in the heavy fractions (>650°F, 345°C).

The plot demonstrates a rapid decrease of the paraffins and mononaphthenes with increasing boiling point and an increase of the trinaphthenes and the higher ones. Cut 1, the first short-path distillate (mid-AEBP = 722°F or 383°C), has an average molecular weight of 335 daltons which corresponds to a carbon number of about 24. It contains about 2% paraffins, 3% mononaphthenes, 6–8% each of di-, tri-, and pentanaphthenes, and just a trace of hexanaphthenes. Tetranaphthenes, with 12% the most abundant type, brings the total saturates of this cut to about 35%. Assuming no pericondensation and no 5-membered rings (this exercise is done only to demonstrate the principle, not to give quantitative data), the carbon number of the ring clusters would be 6, 10, 14, 18, 22, and 26 for the mono- to hexanaphthenes, respectively. On average then, the mononaphthenes in this cut would have 18 carbon atoms in side-chains; the dinaphthenes, 14; the trinaphthenes, 10; the tetranaphthenes, 6; and the pentanaphthenes, 2. Since Kern River crude contains a fair amount of 5-membered rings, these numbers are on the low side. The chemical formula and actual carbon number distribution of each type can be determined by FIMS or FDMS.

The total amount of saturates in Kern River crude oil drops sharply with increasing boiling point, but their internal distribution remains much the same: roughly equal amounts of di-, tri-, and pentanaphthenes, up to twice as much tetranaphthenes, and much less paraffins and mono- and hexanaphthenes. FIMS and FDMS do not distinguish paraffins from heptanaphthenes (or mono- from octanaphthenes). But, in our experience, the concentrations of the compound types represented by $Z = 2$ and/or -12 continuously and smoothly decrease with increasing AEBP. Therefore, heptanaphthenes, if present at all in higher AEBP fractions, would account for very little material.

The higher-boiling members of an alkane-compound type are different from their lower-boiling ones by their greater content of carbon atoms in paraffinic side-chains. Thus, the paraffinicity of the naphthenic-compound types increases with increasing molecular weight and boiling point. In Table

*It was chosen specifically for its paucity of paraffins, which makes it easier to study the cycloalkane distribution and to look for heptanaphthenes (which in other crude oils, if present, are hidden in the FIMS spectra by the paraffins). Kern River crude also shows the relation of many of its components with molecular fossils more clearly than other crudes.

10.5 we illustrate this situation by showing the average carbon number of the six naphthene-compound types in some of the short-path distillates of Kern River crude oil. These numbers are given only to demonstrate the principle, not as a reliable picture of the chemical composition of the naphthenes.

The higher-boiling members of a given alkane compound type (e.g., the tetranaphthenes) are different from their lower-boiling ones by their higher number of carbon atoms in paraffinic side-chains. Thus, the paraffinicity of each of the naphthenic-compound types increases with increasing molecular weight and boiling point.

The distributions of aliphatic-compound types in the light VGO fractions of three naphthenic Yugoslav crude oils reported by Svob and Jambrec (15) were similar to those of our Kern River crude. However, in other crude oils large variations in the distribution of the saturates have been observed. In most of these, but not in all, the compound types of greatest abundance in the heavy fractions were the paraffins and the mono- and dinaphthenes (6,16). Three examples are shown in the FIMS spectra of three VGOs in Fig. 10.5. The differences are striking if one focuses on the corresponding compound types, that is, if one compares, for example, the paraffins (open circles), or any one of the other types, in the three VGOs. Thus, the most abundant species in the saturates of VGOs A and D are the paraffins, followed by the mono- and dinaphthenes. Tetranaphthenes (closed circles) are very low in VGO A, whereas they are close to the dinaphthenes in VGO D. VGO B is very different in every respect from the two other oils. The striking irregularities in the mass spectra of VGOs B and D are caused by the presence of "molecular fossils" such as steranes and hopanes (tetra- and pentanaphthenes, respectively).

Table 10.5 Estimated Average Carbon Numbers in Alkyl Side-Chains of the Main Naphthene Types in High-Boiling Distillates of Kern River Crude Oil

				Naphthenic ring number					
Dist. cut	Mid-AEBP (°F)	Av. MW	Estimated avg. C no.	1	2	3	4	5	6
				Estimated avg. C numbers in alkyl side chains					
1	722	335	24	18	14	10	6	2	0
2A	825	420	30	24	20	16	12	8	4
3A	955	525	37	31	27	23	19	15	11
4A	1090	680	48	42	38	34	30	26	21
5	1240	880	63	57	53	49	45	41	37

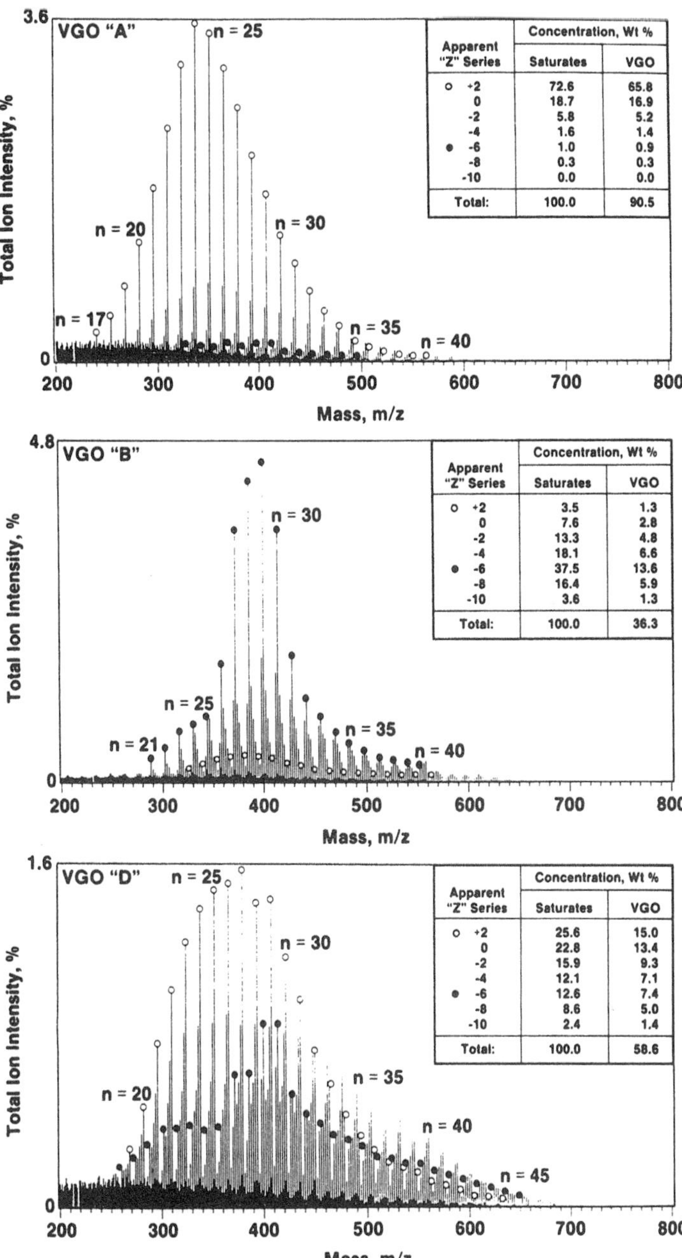

VGO "A"

Apparent "Z" Series	Concentration, Wt %	
	Saturates	VGO
○ +2	72.6	65.8
0	18.7	16.9
-2	5.8	5.2
-4	1.6	1.4
● -6	1.0	0.9
-8	0.3	0.3
-10	0.0	0.0
Total:	100.0	90.5

VGO "B"

Apparent "Z" Series	Concentration, Wt %	
	Saturates	VGO
○ +2	3.5	1.3
0	7.6	2.8
-2	13.3	4.8
-4	18.1	6.6
● -6	37.5	13.6
-8	16.4	5.9
-10	3.6	1.3
Total:	100.0	36.3

VGO "D"

Apparent "Z" Series	Concentration, Wt %	
	Saturates	VGO
○ +2	25.6	15.0
0	22.8	13.4
-2	15.9	9.3
-4	12.1	7.1
● -6	12.6	7.4
-8	8.6	5.0
-10	2.4	1.4
Total:	100.0	58.6

Figure 10.5 Distribution of alkane compound types in VGOs from three crude oils. (From Ref. 6. Reproduced with permission of the publisher.)

For the structures of some prototypes of naphthenes we refer to Chapter 2, Fig. 2.2. The degree of branching in the paraffins and the alkyl chains of the naphthene molecules can be determined by ^{13}C-NMR.

The chain length distribution of the alkyl groups in heavy naphthenes has not been widely studied to our knowledge. We are only aware of the work of Payzant et al. (16,17) who investigated the saturates in the pentane-soluble portion (a very broad fraction) of Athabasca bitumen by GC and FIMS. The carbon number distribution of their tetra- and pentanaphthenes had very distinct peaks at C_{29} and C_{30}, respectively, with steeply descending amounts to about C_{50}. A tetranaphthene with C_{29} could well be a sterane derivative, for example, stigmastane. Steranes are tetranaphthenes with two methyl groups and a longer branched chain of variable length. Their ring skeleton consists of 17 carbons. A sterane derivative with 50 C would, therefore, have 23 carbon atoms in side-chains. The structures of several sterane and hopane derivatives found and identified in crude oil are shown in Fig. 10.6.

A likely pentanaphthene type would be hopane which has six methyls and an isopropyl group with a total number of 30 carbon atoms. A hopane derivative with 50 C would have 20 additional carbon atoms in side-chains. Although we cannot assume that all the additional carbon atoms in these tetra- and pentanaphthenes are contained in only one side-chain, the probability of only a few but fairly long ones is high.

The MW and carbon number distributions of the tetranaphthenes in three short-path distillation cuts from Kern River crude oil are shown in Fig. 10.7. For Cut 1 (722°F mid-AEBP), the range is from C_{19} to C_{36} with a steep maximum at C_{28}. These numbers would correspond to sterane ring skeletons with 2, 19, and 11 carbons, respectively, in side-chains. The carbon number range for Cut 3B (1025°F, 552°C mid-AEBP) is C_{26}–C_{62}, with a maximum at C_{44}, corresponding to sterane ring skeletons with 9, 45, and 27 C atoms, respectively, in side-chains. For the highest-boiling short-path distillate, Cut 5 (1239°F, 671°C mid-AEBP), the lowest carbon number observed was 44 (27 C atoms in side-chains), 74 at the maximum (57 C atoms in side-chains), and 86 for the upper limit (69 C atoms in side-chains).

B. Aromatic Hydrocarbons

Aromatic hydrocarbons are defined as those hydrocarbons with at least one aromatic ring. Examples of different aromatic compound types had been presented in Chapter 2, Table 2.2. Most aromatic molecules, in heavy as well as in light petroleum fractions, contain many aliphatic carbon atoms. Indeed, their aliphatic carbon content is usually greater than their aromatic

Steranes (tetracyclic)

Cholestane (C_{27})

Ergostane (C_{28})

Stigmastane (C_{29})

Diacholestane (C_{27})

24-n-propylcholestane (C_{30})

Hopanes (pentacyclic)

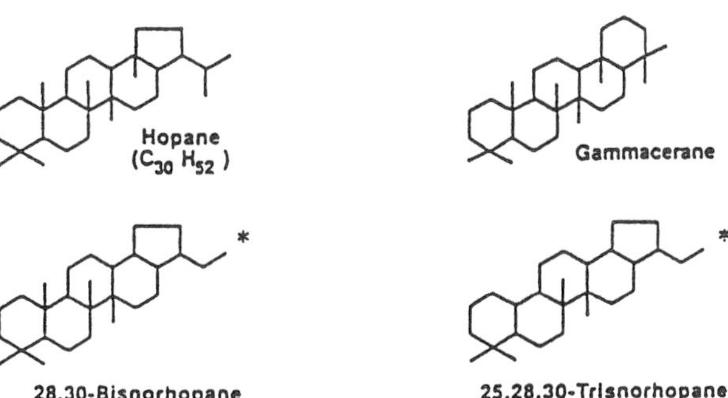

Figure 10.6 Structures of several sterane and hopane derivatives found and identified in crude oil. (From Ref. 12.)

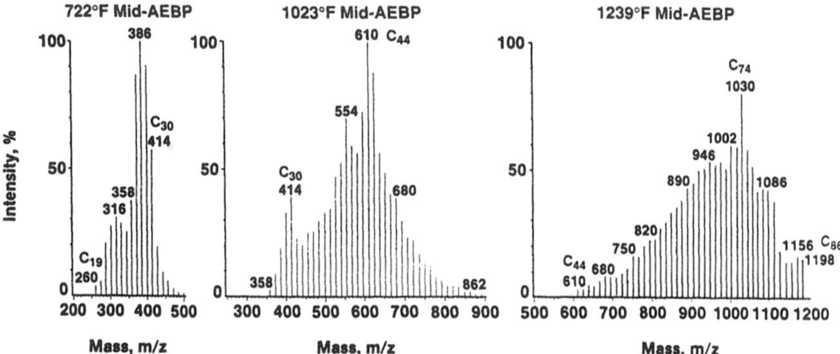

Figure 10.7 MW and carbon number distributions of tetranaphthenes in three short-path distillation cuts from Kern River crude oil. (From Ref. 6. Reproduced with permission of the publisher.)

one. As to be expected, the aliphatic character is most prevalent in monoaromatics and least in those of high aromatic ring number. The situation is similar to the paraffinicity in the naphthenes, shown in Table 10.5, except for the fact that aromatics not only bear paraffinic side chains but may also be fused or otherwise linked to naphthenic rings.

Aromatics, in contrast to aromatic hydrocarbons, are defined by their chromatographic separation from other compound classes. Thus, aromatics may and do contain substantial amounts of heterocompounds. Some of these can be removed by ligand exchange chromatography.

Boduszynski's (6) analysis of Kern River crude, shown earlier in this chapter in Fig. 10.2, gives the distribution of aromatics and other constituents as a function of AEBP for the entire crude. The data were generated by a combination of distillation, two-stage chromatographic fractionation (described in Chapter 6), and characterization by FIMS. Cut 1, the first short-path distillation cut having a mid-AEBP of 722°F (383°C) and, thus, being essentially a light VGO, contains about 50 wt% aromatics. Cut 3 B, a short-path distillate of 1023°F (550°C) mid-AEBP, corresponding to the lowest-boiling fraction of a vacuum residue, contains about 40 wt% aromatics. In the following higher-boiling cuts, the aromatics decrease rapidly with further increasing AEBP until they are down to about 8% in SEF-1 (1365°F, 740°C mid-AEBP) and to zero in SEF-2 (1945°F, 1065°C mid-AEBP). In the nondistillables, all, or almost all, the aromatic molecules also contain heteroatoms, most with polar functionalities.

Not only does the *amount* of aromatics change with AEBP but so does their composition. In Cut 1, we see about the same amount of mono- as diaromatics, about half as much triaromatics, and much smaller quantities of tetraaromatics and higher aromatics. This distribution shifts toward higher amounts of the higher ring number aromatics with increasing AEBP. Thus, in the fractions >1000°F (540°C) AEBP, all the aromatics with (aromatic) ring numbers from 1 to 4 are about equally abundant by weight, whereas the pentaaromatics increasingly predominate until, in Cut 5 (1239°F AEBP), they account for almost as much material as the lower aromatics (1–4 rings) together. By about 1500°F AEBP (815°C) the aromatic hydrocarbons have vanished, and the heterocompounds have taken over completely.

The "aromatics" may contain substantial amounts of heterocompounds. For example, about 40% of the aromatics fraction of Cut 1 of Kern River crude can be assumed to be neutral heterocompounds, containing sulfur, nitrogen, or oxygen! Most of these are in the tetra- and pentaaromatics fractions, but even the monoaromatics comprise some sulfur compounds. The heterocompound content of the higher-boiling aromatics is even higher and may reach about 80% of the total aromatics in Cut 5, the highest-boiling short-path distillate. The aromatic ring number of heterocompounds in any boiling range will be the same or slightly lower (on average) than that of the hydrocarbons.

By inference from smaller molecules in lower-boiling fractions, we can assume that the hydrocarbon structure of the heterocompounds is quite similar to that in the aromatics and that the trend observed in the hydrocarbons is continued, namely, the trend toward increased complexity, though not necessarily toward larger ring clusters.

Amounts of unsubstituted aromatic hydrocarbons are very low in straight-run petroleum fractions. Even alkylaromatics without naphthenic rings are in the minority. Most aromatics are fused to at least one naphthenic ring and contain substantial amounts of carbon in alkyl chains.

Few quantitative data have been published on the alkyl chain length distribution in the aromatics of heavy petroleum fractions. MS can, at best, give the total number of carbon in side-chains, but not the number of alkyl groups nor their carbon number distribution. Hydrocracking of distillates often produces waxy products which is an indication of the presence of fairly long linear (unbranched) alkyl groups in the feed molecules. Chain lengths differ from one crude oil to the next and also with AEBP. Let us take alkylbenzenes as an example for the increase with AEBP. Obviously, a higher-boiling alkylbenzene must have more paraffinic carbon than a lower-boiling one. Although it may simply have more substituents, it is more likely that it also has longer ones. The same holds true for other compound types.

Some clues for the alkyl side-chain distribution in heavy petroleum fractions may be gleaned from those of "molecular fossils." Most of the primary (aliphatic) molecular fossil molecules contain several methyl groups besides one longer chain. During aromatization, the methyls attached to quarternary naphthenic ring carbons are lost. Aromatized molecular fossils, thus, contain fewer methyl groups than aliphatic ones. We can expect a similar shift in regular, nonmolecular-fossil constituents.

Even so, petroleum molecules contain many more methyl groups and other short alkyl groups than longer ones. Singh et al. (19) observed in short-path and supercritical fluid extraction (SFE) fractions of a vacuum residue (North Gujarat Mix) about equally strong ^{13}C-NMR absorption for methyl substituted as for CH- and CH_2-substituted aromatic carbon atoms. The ratio of these two absorptions increased slightly with increasing AEBP (except for the SFE fraction for which it was lower).

A more detailed study based on ruthenium ion catalyzed oxidation of Athabasca bitumen asphaltenes [(8,9); see also Chapter 9] revealed that, here too, methyl groups and other short alkyl groups are strongly favored over longer ones. The carbon number distribution in these alkyl groups follows a steeply descending curve from a high maximum for methyls down to about C_7 where its slope drastically diminishes to extend to about C_{32}. These results concern the number of substituents with different length. A distribution of the same data in terms of carbon atoms in such alkyl groups would look different, favoring more the longer chains. In comparing the number of carbon atoms in methyl groups, short chains (2–4 C), medium chains (5–8 C), and long ones (9 and higher), we would find most of the carbon in medium chains.

C. Sulfur Compounds

The sulfur content in crude oils ranges from about 0.05 to about 14%, but few presently produced commercial crudes exceed 4% (14). Sulfur is broadly distributed across the range of petroleum fractions. From low levels in the naphthas, the S content rises gently in about linear fashion to high levels in the nondistillable fractions. The average number of sulfur atoms *per molecule* increases sharply in the higher-boiling ranges, mainly because of the increase in MW.

We can distinguish among five major compound classes: thiols, sulfides, disulfides, sulfoxides, and thiophenes. The first four can be subdivided into cyclic and acyclic species, and into alkyl-, aryl-, and alkylaryl-derivatives. The thiophenes are fused to aromatic rings. Thus, we have benzo-, dibenzo-, naphthobenzo-thiophenes, and other derivatives. The chemical structures of some S-compound prototypes are shown in Fig. 10.8. More examples were presented by Orr and Sinnenghe Damsté (14) and by Strausz et al. (20).

Both groups of authors also describe the recent advances made in the identification of S compounds, mainly of molecular fossils (biomarkers).

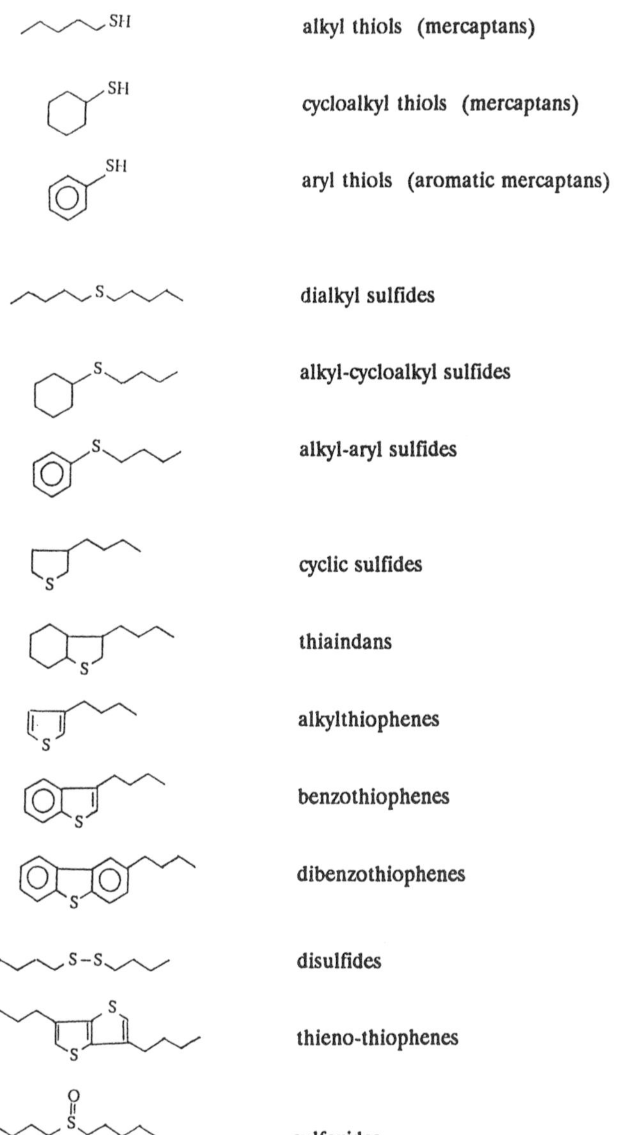

Figure 10.8 Examples of sulfur-compound types found in petroleum.

Some rather large S compounds, up to about C_{40}, have been found (20), ranging from aliphatic bi- and tricyclic terpenoid (18,21) and hexacyclic hopanoid sulfides (22,23) to thiophene derivatives (16). The corresponding sulfoxides, too, occur in crude oils and bitumens (18,21), but they may also be formed in the laboratory by inadvertant oxidation during sample work-up (24).

The nature of the sulfur compounds (and other heterocompound types) in very high-boiling and nondistillable petroleum fractions seems to be much the same as that of the lower-boiling fractions. We know that usually more than half the S in heavy petroleum fractions is present in the form of thiophene derivatives. The remainder consists primarily of sulfides (cyclic and acyclic) and a minor amount of sulfoxides [e.g., 24–28)]. Even such limited information as the distribution of sulfur among these groups can be quite useful for the process chemist, especially because some of the sulfidic sulfur seems to connect ring systems in the refractory pentane-insoluble fractions (7,29), and because molecules containing such sulfur bridges break down more easily on desulfurization than others.

By mild degradation, Sinnenghe Damsté et al. (30) demonstrated that much of the sulfur in very high-boiling materials is in the same kind of structures as in lower-boiling fractions, except that these are bound to other parts of the molecules by single or multiple sulfur bridges. They detected linear isoprenoid, steroid, hopanoid, and carotenoid carbon skeletons as well as thiophene derivatives among the fragments, some of them with 35 and more carbon atoms, many of them with long paraffinic chains. The methods for this study were flash pyrolysis and Raney Ni hydrogenation. Just how much of the total sulfur in the very high AEBP samples is seen this way is not clear, but comparison with results from x-ray spectroscopy suggests that it is less than 20%.

Chromatographic separation on conventional columns of S compound classes found in petroleum fractions from the related hydrocarbons is not possible, not even for low-boiling fractions. Aliphatic sulfides co-elute with aliphatic hydrocarbons and thiophenes with aromatic ones, for instance, benzothiophene together with diaromatics, dibenzothiophenes with triaromatics, and so on. But ligand exchange chromatography (LEC) has been reported as useful for separating sulfides and thiophene derivatives from hydrocarbons in middle distillates and somewhat higher-boiling cuts [e.g., (4,31)]. Concentrated fractions suited for characterization by GC/MS were obtained this way. Apparently two steps are needed to isolate sulfides and thiophenes separately. For optimum resolution, these compound classes are separated by aromatic ring number, for example, by supercritical fluid chromatography (SFC). This technique allowed Nishioka et al. (32) the identification and quantitation by GC/MS of about 100 thiophenic compounds ranging from methyldibenzothiophenes to alkylchrysenothiophenes and alkyldinaphtho-thiophenes, all in the VGO boiling range.

There are now several direct ways to measure the amounts of sulfur in thiophenic, sulfidic, and oxygen-containing structures in heavy petroleum fractions, even in nondistillable residues. K edge x-ray absorption spectroscopy (XANES) (33,34) and x-ray photoelectron spectroscopy (XPS) (27) look promising. With a number of model compounds, mostly unsubstituted molecules of low MW, these researchers could clearly differentiate the three main S compound classes. They reported sulfidic and thiophenic S contents in several residues (no further identification) and their heptane asphaltenes with an experimental error of only ±10%. The two methods, applied to the same samples, correlated quite well within a given sample type. However, the correlations deviated for different sample types, namely, petroleum residues, asphaltenes, and especially coal liquids. Gorbaty et al. (34) state that more work is needed to assess the value of these techniques. In particular, the effect of long alkyl substituents must be examined.

Waldo et al. (24) further improved the XANES method for better quantitation. Applying it to 29 whole petroleum samples and to 10 broad fractions (asphaltenes, maltenes, etc.), they found thiophenic compounds to account for 55% to almost 90% of the sulfur. Between 4 and 40% was in the form of sulfides and about 10% in sulfoxide structures. They distinguish between sulfide-rich (>30% sulfide, <60% thiophene) and thiophene-rich (<15% sulfide, >75% thiophene) crude oils. In several of their low-sulfur oils, the relative sulfoxide contents were very high (20–30% of total S). However, they attributed these high values to autoxidation during storage and handling rather than to the true nature of the original crude oil.

Several chemical methods have been described which show greater differentiation than the x-ray methods, but they appear to be tedious and possibly less reliable. Rose and Francisco (35) and Francisco et al. (28) developed an interesting two-step reaction sequence of reductively cleaving carbon–sulfur bonds—as well as C–C bonds—in vacuum residues and heptane insolubles. They treated the samples with alkali metal, and selectively labeled the resulting thiols with ^{13}C- and ^{2}H-enriched methyl groups. Two-dimensional ^{13}C and ^{2}H-NMR spectroscopy of the samples before and after cleavage gave clearly recognizable chemical shift data of the methyl labels enabling identification of several types of S compounds in three main classes:

1. Diaryl sulfides ranging from dibenzyl to bis(1-naphthylmethyl) sulfide
2. S compounds that form thiols on reductive protonation, namely, alkyl-aryl- and aryl-arylsulfides as well as certain thiophene derivatives

3. Those S compounds that resist cleavage by reductive protonation (about 60% of all the sulfur in the two samples investigated by these authors)

Dibenzothiophene derivatives and dialkylsulfides are examples of this third category. In a vacuum residue (>950°F, >500°C), Francisco et al. distinguished

- Benzylic sulfides (about 12%)
- Benzyl-alkylsulfides (1–2%)
- Mostly unhindered alkyl-arylsulfides, diarylsulfides, and benzothiophenes (8–12%)
- Hindered alkyl-arylsulfides, hindered diarylsulfides, and hindered benzothiophenes (about 15%)
- Dibenzothiophene and dialkylsulfide types (about 60%)

Ruiz et al. (36) studied the S distribution in an Arabian Light asphalt by combining oxidation and reduction procedures with ESCA measurements. Distinguishing between thiophene derivatives and three different types of sulfides, they found 51% of the S in thiophene structures, 33% in alkylalkyl- and arylalkyl-sulfides, and 13% in diarylsulfides. Strausz et al. (20) pointed out the danger of side reactions in such oxidative procedures and suggested ways to overcome them (see also Section III.C.1 of Chapter 9).

Sinnenghe Damsté et al. (30) described two methods for analyzing nondistillable petroleum fractions. One uses flash pyrolysis to sever sulfur-bound fragments from the matrix and GC/MS for their analysis. The other splits the molecules by Raney nickel desulfurization, a sulfur-selective reduction method. Both degradation techniques are mild and presumed to yield representative fragments. However, Raney nickel desulfurization does not work well with pentane insoluble fractions.

An important function of sulfur in very high AEBP fractions is that it links smaller aromatic ring clusters into larger structures (7,29).

D. Nitrogen Compounds

The nitrogen content of crude oils is quite low, between a few ppm and 0.8 wt%, mostly below 0.4%. Nitrogen is found at ppm levels in light and middle distillates. Its concentration increases significantly around 650°F (345°C) and continues to rise with increasing AEBP (see Chapter 4, Tables 4.19–4.21 and Fig. 4.12).

We distinguish between basic, neutral, and acidic nitrogen compounds in petroleum. The basic ones are mainly pyridine derivatives, also called aza compounds (medium basic), primary amines (strong bases), N-alkylindoles (weak bases), and alkylarylamines (weak bases). The neutral, or

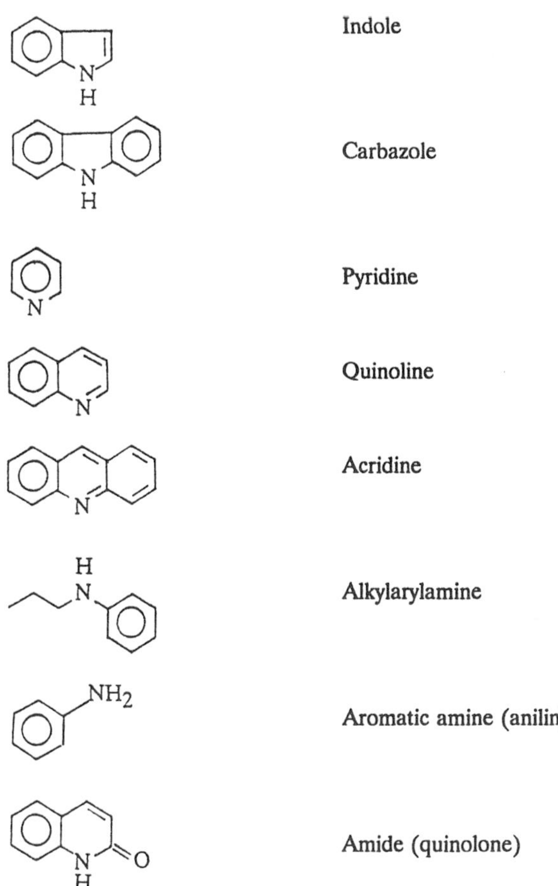

Figure 10.9 Examples of nitrogen compound types found in petroleum.

very weakly basic ones, include alkylindoles, alkylacridines, certain amides, and alkylhydroxypyridines. The acidic ones consist of indoles, carbazoles, (nonmetallic) porphyrins, and amides. Figure 10.9 lists some prototypes of these compound types.

Table 10.6 shows the distribution of N compounds in the heavy distillates of four crude oils as determined in two studies by McKay and co-workers (37,38). In the distillation cut of 1000–1250°F (535–675°C) AEBP, up to 7 wt% pyrrolic compounds, another 3 wt% diaza compounds (also basic) was found. Formulas, *Z* series, and carbon-number ranges of many N

Table 10.6 Distribution of N Compounds Found in (A) Acid and (B) Base Concentrates of Heavy Distillates from Four Crude Oils

	wt% in distillate					
Crude oil	Pyridine benzologs	Diaza compounds	Pyrrole derivatives		Amides	
Boiling range	B	B	B	A	B	A
Wilmington						
370–535°C	3.5	1.16	0.5	0.9	1.7	0.12
535–675°C	7.2	2.9	0.5	—	2.0	—
Gach Saran						
370–535°C	1.2	0.54	0.04	0.53	0.27	0.34
535–675°C	3.9	2.6	0.17	—	2.0	—
Recluse						
370–535°C	0.5	0.37	0.02	1.0	0.20	0.05
535–675°C	1.9	0.86	0.03	1.7	0.53	0.24
Swan Hill						
370–535°C	0.7	0.57	0.07	0.8	0.81	0.29
535–675°C	2.2	1.31	0.18	—	0.81	—

Temperature ranges in °C and °F: 370–535°C = 700–1000°F; 535–675°C = 1000–1250°F.
Source: Averaged IR and titration data adapted from Refs. 37 and 38.

compounds in the base concentrates are displayed in Table 10.7. The *Z*-numbers indicate not only compound types with different aromatic ring number to be present in these fractions but molecules also with various naphthenic rings. Some examples of likely structures are shown in Fig. 10.10.

The distribution of the N compounds from the same distillates in terms of strong, weak, and nontitrable bases is presented in Table 10.8. These numbers do not include N compounds from neutral or acid fractions, but only those from the base concentrates as described by McKay et al. (37). Although the total base content increases with AEBP, there is no uniform pattern for the distribution of the base types except for the predominance of strong bases.

In an earlier study of heterocompounds, Snyder and co-workers (39,40) measured the heterocompound distribution in three heavy distillates from a California crude oil. Their nitrogen data are displayed in Table 10.9. These results favor the pyrrole derivatives as the most abundant N-compound types in these three distillates. Their separation scheme with a silica column as the first device may have eliminated some of the very basic N compounds by irreversible adsorption and caused the pyrrole derivatives to gain such a prominent placing.

Table 10.7 Formulas, *Z* Series, and Carbon-Number Ranges of Some N Compounds in the Base Concentrates of Heavy Distillates from Four Crude Oils

General formula $C_nH_{2n+Z}X$	Z-number	Carbon-number range
$C_nH_{n-7}N$	−7	15–30
$C_nH_{n-9}N$	−9	15–30
$C_nH_{n-11}N$	−11	15–34
$C_nH_{n-13}N$	−13	15–33
$C_nH_{n-15}N$	−15	15–35
$C_nH_{n-17}N$	−17	15–33
$C_nH_{n-19}N$	−19	15–34
$C_nH_{n-21}N$	−21	16–31
$C_nH_{n-23}N$	−23	18–31
$C_nH_{n-25}N$	−25	18–33
$C_nH_{n-29}N$	−29	19–33
$C_nH_{n-17}NS$	−17	14–29
$C_nH_{n-19}NS$	−19	17–31
$C_nH_{n-12}N_2$	−12	15–31
$C_nH_{n-14}N_2$	−14	15–31
$C_nH_{n-16}N_2$	−16	16–31
$C_nH_{n-8}N_2S$	−8	12–29
$C_nH_{n-10}N_2S$	−10	12–29

Source: Data adapted from Refs. 37 and 38.

Holmes' (41) results on the abundance of N-compound types in a tar sand bitumen, Table 10.10, are similar to McKay's. Most of the N in this bitumen is again tied up in pyridinic structures, followed in sequence by pyrrolic, amide, and primary amine N. Pyridine stands here again for the class of compounds with at least one N in a six-membered aromatic ring including, for example, quinolines and acridines. The same holds for the indoles (containing at least one five-membered ring with one N). All of the listed groups of compound types may contain more than one N and, possibly, also some O or S. Holmes' method involved LC separations on basic alumina and silica, followed by analysis of the fractions by HR-MS and NMR. His distinction of amides I and II, for example, indicates different chromatographic behavior by members of the same compound type, presumably because of steric hindrance.

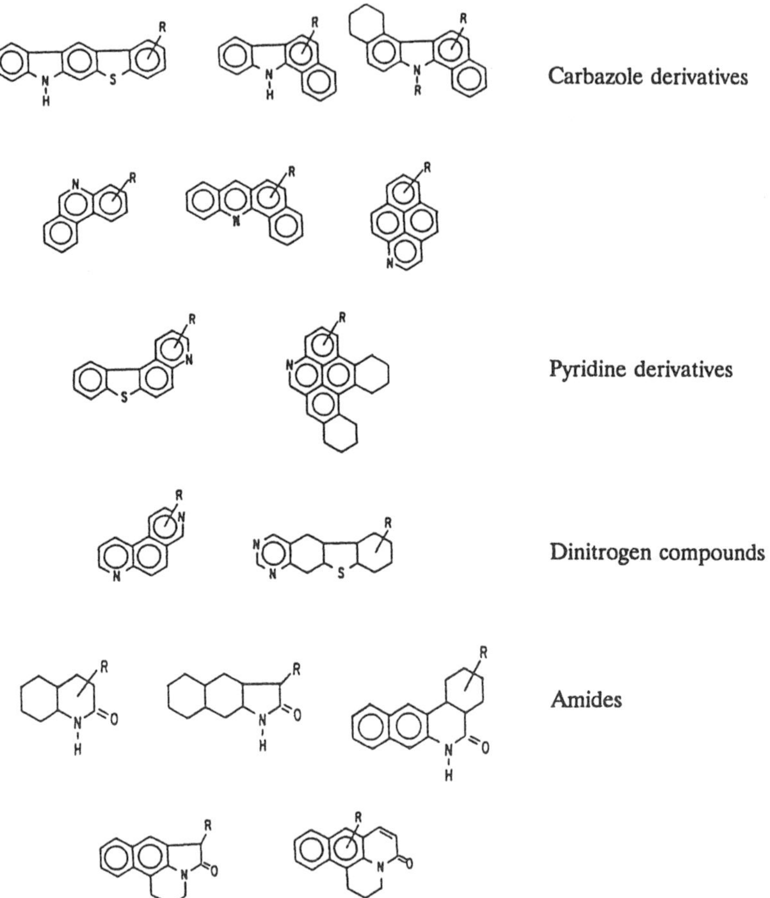

Figure 10.10 Nitrogen compounds suggested as being representative in crude oils by McKay et al. (37).

Among the acidic N compounds, the carbazoles are the most abundant species (Table 10.9 and Fig. 10.11). In Fig. 10.11 we show some representative examples of different types of carbazoles found in petroleum VGOs (650–1000°F; 345–540°C). Of these, the alkylbenzocarbazoles ($Z = -21$ N) were the most prominent type in a Wilmington VGO, followed by those with Z-numbers -17 N and -23 N (42). These findings are consistent with unpublished data (5,6,43) for VGO from San Joaquin Val-

Table 10.8 Nitrogen Distribution in Base Fractions of Heavy Distillates

Crude oil Boiling range	wt% Basic compounds	Percent of total base fraction Averaged IR and titration data Strong bases	 Weak bases	 Nontitratable bases
Wilmington				
370–535°C	6.8	58	24	18
535–675°C	12.7	65	15	20
Gach Saran				
370–535°C	2.1	74	8	18
535–675°C	8.1	50	28	22
Recluse				
370–535°C	1.1	62	18	20
535–675°C	3.3	65	18	17
Swan Hill				
370–535°C	2.2	55	28	15
535–675°C	4.5	63	15	22

Source: Averaged IR and titration data adapted from Ref. 37.

Table 10.9 Distribution of N Compounds Found in Three Heavy Distillates from a California Crude Oil

	wt% in distillate		
Boiling range, °C °F	400–700 205–370	700–850 370–455	850–1000 455–540
Indoles	0.07	0.59	0.75
Carbazoles	0.28	3.40	4.08
Benzcarbazoles	0.00	0.50	1.28
Pyrrol derivatives	0.35	4.49	6.11
Pyridines	0.35	0.66	1.3
Quinolines	0.21	1.74	2.0
Benzoquinolines	0.03	0.26	1.6
Pyridine derivatives	0.59	2.66	4.9
Pyridones, quinolones	0.2	1.2	2.0
Azaindoles	0.0	0.1	0.4

Source: Data adapted from Ref. 39.

Table 10.10 Distribution of Nitrogen Compound Types in Tar Sand Bitumen

Tentative identification of compound types[a]	pKa range	% of total N in Amines	Pyridines	Pyrroles	Amides
Strong base	9–11				
Alkylamines		1			
Weak base I	7–9				
Alkylpyridines I			5		
Alkylpyridines II			4		
Alkylpyridines III			12		
n-Alkylindoles				8	
Weak base II	2–7				
Unknown				0.2	
Alkylarylamines		0.3			
Very weak base					
Unknown				1	
Alkylindoles				3	
Alkylacridines			1		
Amids I					8
Alkylhydroxypyridines			12		
Nonbasic					
n-Alkylcarbazoles				2	
Alkylindoles/Carbazoles				12	
Amids II					3
Alkylcarboxamids					20
Unidentified			(7)		
		1.3	33	26.2	31

[a]Benzologues and cycloalkane derivatives are included in the definitions (e.g., "pyridines" include quinolines, etc.).
Source: Data adapted from Ref. 41.

ley crude. In Fig. 10.11, the compound types are marked with their respective Z-numbers.

The API-60/NIPER ABN method of preparing heterocompound concentrates by chromatography on ion exchange resins, as practiced by McKay et al. (37) and by Green and co-workers [(44); see also (42,45)], is the mildest way to isolate the N compounds from a petroleum distillate. It can be assumed to have much smaller losses due to irreversible adsorption and structural changes than these, which may occur on the more traditional

Z-Number	Compound Type	Representative Structures
-15N	Alkylcarbazoles	
-17N	Alkylnaphthenocarbazoles and/or alkylcyclohexyl-carbazoles	
-19N	Alkyldinaphthenocarbazoles	
-21N	Alkylbenzocarbazoles	
-23N	Alkylphenylcarbazoles and/or Alkylnaphtheno-benzocarbazoles	
-25N	Alkyldinaphtheno-benzocarbazoles	
-27N	Alkyldibenzocarbazoles	

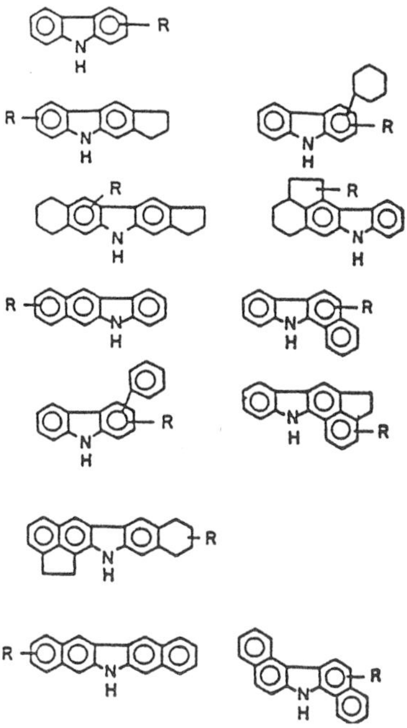

Figure 10.11 Examples of carbazoles found in petroleum. (From Ref. 42. Reproduced with permission of the publisher.)

column packings. Subsequent separation of the acid and base concentrates by LC can provide a number of fractions ranging from basic to weakly acidic N compounds and with different degrees of steric shielding (see Chapter 6). High-resolution MS, augmented by IR and NMR spectroscopy, can then be used for the identification and measurement of compound types and their MW distributions.

E. Oxygen Compounds

Oxygen, too, is present in petroleum in small amounts—about 1% or less, again concentrated in the >650°F (>345°C) fractions. Oxygen has been shown to occur primarily in the form of hydroxyl groups in phenols, carboxyl groups in carboxylic acids and esters, carbonyl groups in ketones, cyclic and acyclic ethers, and, in combination with N, as amides, and with S, as sulfoxides [e.g., (1–3,39,40,46–53)].

Seifert and co-workers [e.g., (46–50)] identified numerous phenols and carboxylic acids—often collectively called naphthenic acids—in petroleum. Figure 10.12, reproduced from Seifert and Teeter (49), gives an impression of the immense variety of compound types carrying the carboxyl group. Several phenolic and carboxylic acid compound types, found by McKay et al. (38) in two high-boiling petroleum fractions are shown in Fig. 10.13. McKay et al. (1–3) reported roughly equal amounts of carboxylic acids and phenols in a Wilmington nondistillable residue and only half as much amides; see Table 10.11.

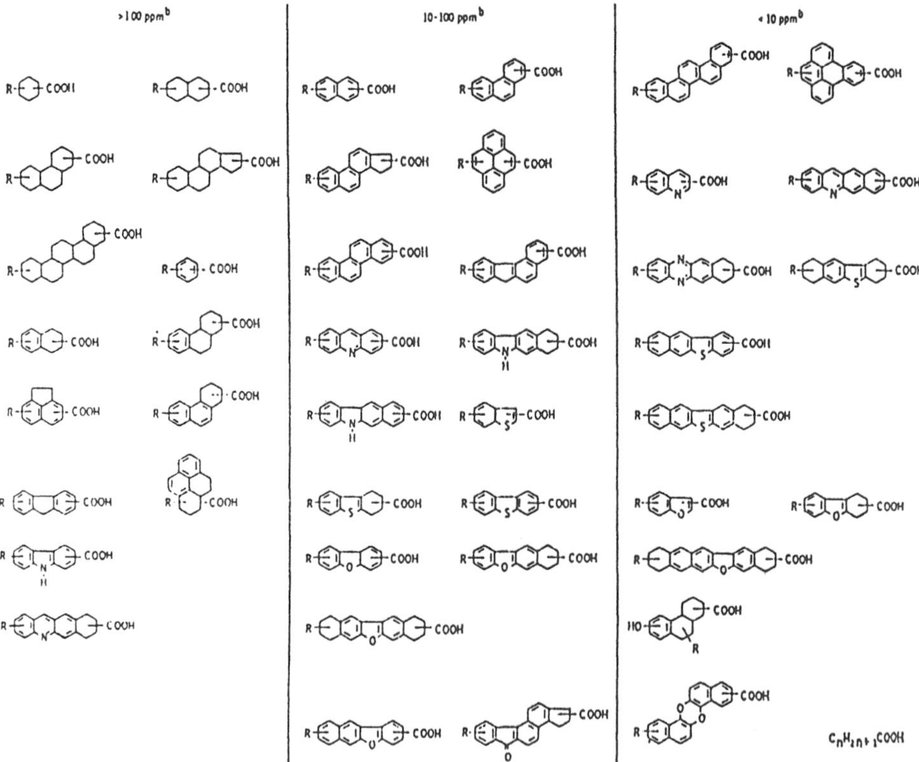

Figure 10.12 Examples of carboxylic acid compound types in petroleum. (From Ref. 49. Reproduced with permission of the publisher.)

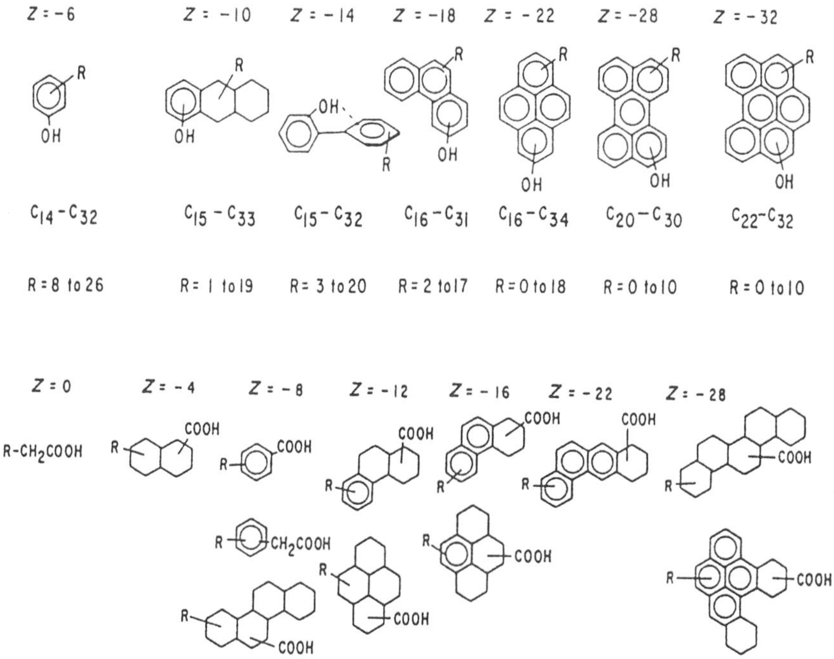

Figure 10.13 Oxygen compounds suggested as being representative in crude oils by McKay et al. (38).

Table 10.11 Acid Compound Types in Wilmington Residue (>1250°F, >675°C)

Compound type	Approximate wt% in acid fraction
Carboxylic acids	28
Phenols	28
Pyrroles	28
Amides	16

Source: Data adapted from Refs. 1–3.

Two relatively rapid preparative LC methods for the isolation of hydroxyl compounds in petroleum fractions have been developed recently. One of these (54) uses normal phase HPLC on silica with solvents containing tetramethylammonium hydroxide. Methylation of the fractions, followed by GC/MS analysis, allows the differentiation of various types of carboxylic acids, such as paraffinic, terpenoid, and aromatic ones.

The second method used by Green et al. (55) is aimed at all kinds of hydroxyl compounds. These are first concentrated by nonaqueous LC on ion exchange resins (see Chapter 6 and Ref. 4 for detail) and then fractionated into compound-class fractions by HPLC as in the first method. Now the fractions are acylated to make them more volatile and amenable to GC/MS and MS measurements. Only tentative assignments could be made so far, leaving open the distinction between various isomers. Green et al. hope that unambiguous identification of the hydroxyl compounds after acylation will be possible later when spectra of model compounds for these species become available.

For both methods the first step is distillation. Green et al. (55) state:

> Distillation was not absolutely necessary for analysis of ArOH, but the level of information obtained on higher boiling ArOH was enhanced if the bulk of the phenols and indanols/tetralinols were distilled into a light (ca. 200–325°C = 390–620°F) boiling fraction.

This group is also working on improved methods for determining the total amount of OH and COOH in petroleum fractions, for example, by derivatization and IR spectrometry (53). Acylation of alcoholic and phenolic OH with trifluoroacetyl chloride converts these to esters which are unaffected by water and association effects and have higher absorption coefficients because of the trifluoro group. The carboxylic acids are esterified with trifluoroethanol to overcome dimerization effects and interference from amides.

An earlier, more general method, developed by NIPER and discussed in Chapter 6, also separates the acid fraction obtained on ion exchange resin into subclasses by employing normal phase LC on silica and solvents containing tetraalkylammonium hydroxide, but no esterification. Sturm et al. (45) used this technique to analyze a Wilmington and a Mayan VGO (650–1000°F, 345–535°C) for their acid contents. Table 10.12 shows the composition of the two distillates in terms of their acid, base, and neutrals contents, and also of their acid types. This study shows large differences in the contents of O compounds from the two sources and also in the relative amounts of their compound types. In both VGOs, the phenols made up no more than 25% of the acidic O compounds.

Table 10.12 Acid Contents of Wilmington and Mayan Gas Oils (650–1000°F, 345–540°C)

	wt% of distillate	
Results of ABN separation	Wilmington	Mayan
Polar compound groups		
Strong acids	6.0	2.6
Weak acids	4.8	1.4
Total acids	10.8	4.0
Strong bases	2.5	0.9
Weak bases	2.7	0.8
Total bases	5.2	1.7
Neutrals	80.1	94.0
Total recovered	96.1	99.7
Acidic compound classes		
N–H compounds[a]	3.7	1.5
Phenols[b]	1.3	0.4(–)
Carboxylic acids	2.9	nd[c]
Amides	1.0	1.5
Unidentified	1.9	0.6
Total	10.8	4.0

[a]Acidic N–H compounds are largely pyrrolic benzologus; indoles, carbazoles, benzcarbazoles, and so on.
[b]Determined by IR measurement of OH bands.
[c]None detected.
Source: Ref. 45.

Rose and Francisco (35) determined the acid groups in their samples, two vacuum residues and the heptane-asphaltenes of one of these, by controlled methylation using a phase-transfer technique and methyl iodide enriched either with ^{13}C or deuterium. The conversion was monitored by ^{13}C-NMR, D-NMR, and FT-IR. Examples of the shifts obtained in the spectra are presented in Figs. 10.14 and 10.15. The ^{13}C-NMR spectra clearly show the appearance of new aromatic ether CH_3 (at 55 and 60 ppm) and ester CH_3 (at 50 ppm) groups after methylation, presumably strong enough for quantitation. The FT-IR spectra illustrate the disappearance of OH (at 3615 cm^{-1}) and NH bands (at 3475 cm^{-1}) after methylation on the one hand and the appearance of ester and amide bands (1710 and 1650 cm^{-1}) on the other. Both measurements, ^{13}C-NMR and IR, can be used at least

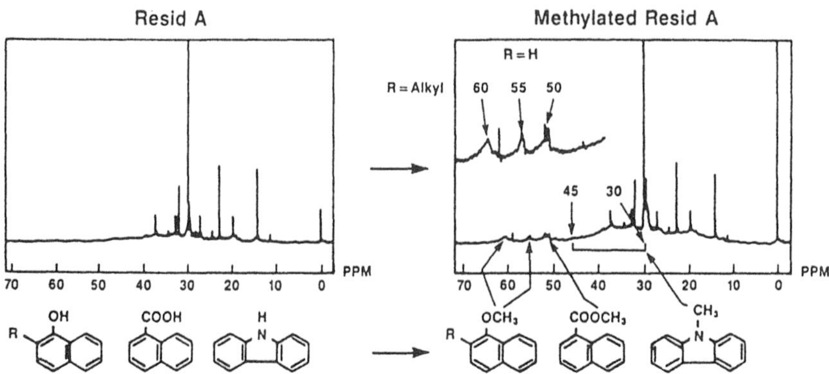

Figure 10.14 Aliphatic region of ^{13}C-NMR spectrum for a petroleum residue before and after methylation with $^{13}CH_3J$. (From Ref. 35. Reproduced with permission of the publisher.)

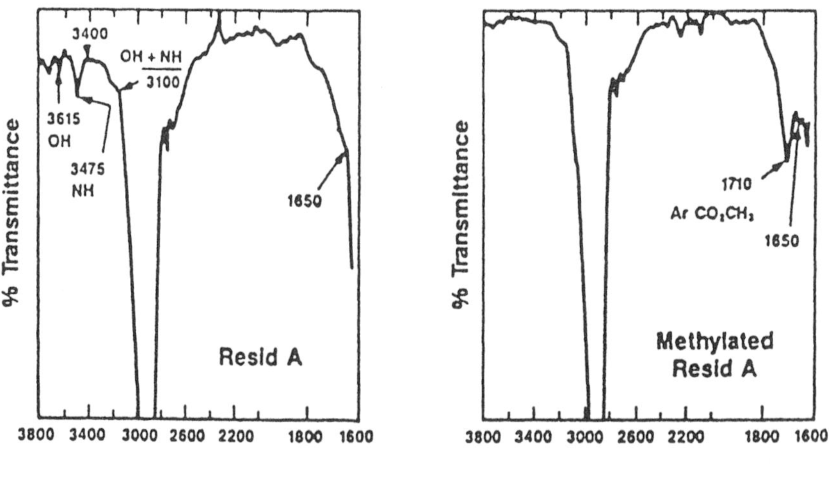

Figure 10.15 Dilute-solution FT-IR spectra of a petroleum residue before and after methylation with $^{13}CH_3J$. (From Ref. 35. Reproduced with permission of the publisher.)

to demonstrate the presence of OH and NH groups and, depending on the signal strength (concentration), even for quantitative determinations.

In summary, several new methods developed by the NIPER group have considerably advanced the analysis of O compounds in petroleum fractions. They span the range from fairly quick, cursory analyses to lengthy, very detailed ones. Nondistillable residues can be analyzed for O groups by methylation and FT-IR spectroscopy before and after the reaction. Although numerous O *compound types* have been identified in crude oils, only O-containing *functional groups* have been determined in nondistillable petroleum fractions. The most abundant ones of these seem to be carboxyl and phenolic groups. Oxygen can also occur in relatively large amounts in the form of sulfoxides and amides. We have not found in the literature of the last 20 years specific information on ketones [aside from Peterson's work on oxidized asphalts [e.g., (56)], ethers, and esters in heavy petroleum fractions, only general references or references to earlier work.

F. Multiheteroatom Compounds

So far we have described the results obtained on heterocompounds in petroleum containing only one or two heteroatoms. Nondistillable residues generally have more than one heteroatom per molecule on average; some fractions have been found to carry as many as 10 or more on average. Table 10.13 shows the heteroatom contents and other average molecular parameters in the SEF fractions of the nondistillables from five crude oils. (The numbers depend on the molecular weights which were measured by VPO in pyridine but may still be too high. Nevertheless, the general conclusions hold.) Average numbers imply that many molecules must carry an even greater number of heteroatoms, others a smaller number. Porphyrins and porphyrin derivatives, which contain four nitrogen atoms and often a metal atom, are well-known examples of multiple-heteroatom compounds in petroleum.

Many porphyrins and also some compound types with two heteroatoms can still be identified by high-resolution MS. Thus compounds with one N + O, one S + O, and one S + N have been identified in VGOs. However, this is not possible for most species containing three or more heteroatoms. These can only be characterized in general terms by elemental analysis and NMR or IR. Obviously, even HPLC cannot separate molecules with internally mixed functionalities, containing, for example, both a phenolic and a pyridinic group, into fractions with only one functionality. Most multifunctional molecules are found in the nondistillable fractions (>1300°F, >700°C AEBP).

Table 10.13 Average Molecular Parameters of Nondistillable Fractions (>1300°F, >700°C AEBP)

Fraction	Mid-AEBP (°F)	Cum. wt% from AR	Mol. wt.[a]	"Z"[b]	Average number of atoms/molecule								
					C_t	C_{ar}	H	S	N	O	V	Ni	Fe
Kern River													
SEF-1	1365	90.3	1110	−44	79	29	114	0.462	1.112	0.737	0.001	0.002	0.001
SEF-2	1945	96.0	2520	−141	179	80	218	1.141	3.957	1.888	0.009	0.019	0.013
SEF-3	2250	99.1	3620	−216	225	138	294	1.617	6.204	2.986	0.019	0.025	0.027
SEF-4	—	99.6	—	—	—	—	—	—	—	—	—	—	—
California Offshore													
SEF-1	1660	78.4	1910	−67	131	35	195	3.49	1.26	0.74	0.006	0.002	0.001
SEF-2	2305	87.9	4150	−203	282	110	362	8.25	5.18	2.18	0.076	0.017	0.012
SEF-3	2670	98.6	6070	−338	413	181	488	12.7	8.88	4.55	0.168	0.038	0.033
SEF-4	—	99.7	—	—	—	—	—	—	—	—	—	—	—
Arabian Heavy													
SEF-1	1565	87.7	1615	−60	112	36	163.7	2.745	0.461	—	0.002	0.0004	—
SEF-2	2310	95.9	3870	−224	261	134	298.7	9.155	2.460	—	0.034	0.009	—
SEF-3	2570	99.0	5060	−313	346	192	378.3	12.723	3.617	—	0.051	0.016	—
SEF-4	—	99.5	—	—	—	—	—	—	—	—	—	—	—
Maya													
SEF-1	1530	78.9	1520	−56	105	35	153	2.35	0.49	—	0.003	0.0008	—
SEF-2	2070	85.5	2970	−158	199	96	240	6.28	2.25	—	0.046	0.009	—
SEF-3	2580	98.5	5170	−315	351	192	387	11.62	4.98	—	0.188	0.029	—
SEF-4	—	99.8	—	—	—	—	—	—	—	—	—	—	—
Boscan													
SEF-1	1510	74.2	1420	−57	96	30	135.9	2.718	0.639	—	0.019	0.002	—
SEF-2	2160	87.3	3460	−175	231	103	287.0	7.590	3.806	—	0.223	0.018	—
SEF-3	2590	98.0	5510	−307	370	192	432.0	12.230	7.087	—	0.479	0.037	—
SEF-4	—	100.0	—	—	—	—	—	—	—	—	—	—	—

[a]Number-average molecular weight determined by VPO in pyridine. [b]"Z" in the general formula $C_nH_{2n+Z}X$

Let us take another look at Fig. 10.2, which shows the composition of Kern River crude oil as a function of AEBP. DISTACT Cut 1, having a mid-AEBP of 722°F (383°C), is the first heavy fraction of this crude. It consists of about 50% "aromatics" and 12% "polar heterocompound" fractions. Roughly 40% of the "aromatics" in this fraction (and, thus, 20% of Cut 1) are neutral heterocompounds (mostly containing sulfur). We may, therefore, assume that all of the heterocompounds together account for about one-third of this distillation cut. According to Table 10.14, the molecules in this cut contain about 0.1 S atoms, 0.075 N atoms, and 0.09 O atoms on average. Because the heteroatoms must reside in the nonhydrocarbons which, we assume, make up only one-third of Cut 1, these must then contain 0.3 S, 0.23 N, and 0.27 O atoms on average, together 0.8 heteroatoms. This agrees fairly well with the total heterocompound content.

Average numbers imply that about half of the heterocompound molecules contains fewer heteroatoms than average, but the other half contains more. Thus, about 13% of the molecules in this cut (33% · 0.8/2) contain one or more heteroatoms. Depending on the distribution, about half of these, say 6%, may contain two or more heteroatoms. About one-third of the heterocompounds (12 out of 32) contain a heteroatom in a "polar" or "functional" group. The fraction of molecules in Cut 1 with 2 or more polar groups amounts to about 3% (one-fourth of 12%). These (speculative) results depend to an amazingly small degree on the estimate of heterocompounds in the aromatics fraction. They would have come out almost the same with an estimate of only 10% heterocompounds in the aromatics.

The molecules in the last distillate, Cut 5, contain about 0.35 S, 0.75 N, and 0.5 O atoms on average. The polar heterocompound fractions in this cut amount to about 55% and the aromatics to about 37%. Assuming 80% of the aromatics in this cut to be neutral heterocompounds, the total amount of heterocompounds in Cut 5 is around 90%, and the average heterocompounds contain about 0.39 S, 0.83 N, and 0.55 O atoms, that is, together around 1.8 heteroatoms on average. By the same reasoning as before we arrive at the conclusion that about one-half of all the molecules in Cut 5 should contain two or more heteroatoms, and about 15%, two or more polar groups. (Our conclusion would be the same if we assumed only 50% of neutral heterocompounds in the aromatics fraction.)

The nondistillable fractions have considerably higher concentrations yet of molecules with multiple heteroatoms and multiple polar functions. For SEF-1, a pentane-soluble nondistillable fraction of 1365°F mid-AEBP, the numbers are 0.46 S, 1.11 N, 0.74 O atoms = 2.2 heteroatoms per molecule on average and, with an estimated abundance of 80% polar and close to 100% total heterocompounds, also 2.2 heteroatoms per heterocompound.

Table 10.14 Average Parameters in Kern River Fractions

Fraction	Mid-AEBP (°F)	Cum. wt% from AR	Mol. wt.[a]	"Z"[c]	Average number of atoms/molecule								
					C_t	C_{ar}	H	S	N	O	V	Ni	Fe
Distillate data													
Dist 1	<400	—	190	−1.8	13.7	—	25.6	0.020	0.001	—	—	—	—
Dist 2	≈570	—	260	−6.0	19.0	—	32.0	0.047	0.012	0.037	—	—	—
Atmospheric residue data													
Cut 1	722	21.6	335	−8.9	24.1	6.4	39.3	0.104	0.075	0.088	—	—	—
Cut 2A	825	32.8	420	−12.0	30.0	8.0	48.0	0.134	0.156	0.118	—	—	—
Cut 2B	885	42.8	470	−14.7	33.7	9.0	52.7	0.163	0.215	0.177	—	—	—
Cut 3A	955	51.4	525	−16.9	37.7	10.0	58.5	0.196	0.293	0.198	—	—	—
Cut 3B	1025	58.3	595	−19.7	42.3	12.4	64.9	0.249	0.378	0.264	—	—	—
Cut 4A	1090	64.3	680	−23.6	48.6	14.0	73.6	0.296	0.492	0.349	0.001	0.001	—
Cut 4B	1160	68.9	760	−27.0	53.8	16.1	80.6	0.316	0.615	0.415	0.001	0.002	—
Cut 5	1240	74.3	880	−31.5	62.4	17.7	93.3	0.356	0.745	0.515	0.001	0.001	—
SEF-1	1365[d]	90.3	1110[b]	−44.1	79	29.5	114	0.462	1.112	0.737	0.001	0.002	0.001
SEF-2	1945[d]	96.0	2520[b]	−140.3	179	80	218	1.141	3.957	1.888	0.009	0.019	0.013
SEF-3	2250[d]	99.1	3620[b]	−215.4	255	138	294	1.617	6.204	2.986	0.019	0.025	0.027
SEF-4	—	99.6	—	—	—	—	—	—	—	—	—	—	—

[a]Number-average molecular weight determined by VPO in toluene except as noted.
[b]Number-average molecular weight determined by VPO in pyridine.
[c]"Z" in the general formula $C_nH_{2n+Z}X$.
[d]Mid-AEBP Calculated from MW.
Source: Data adapted from Refs. 6 and 57.

Thus, not only do practically all the molecules in this fraction carry, on average, more than two heteroatoms, but 80% of them have at least one polar functionality (basic N, acidic NH and OH, SO, etc.). In addition, considering the relatively high nitrogen and oxygen content in this fraction compared to sulfur, there is a good chance that more than half of these polar compounds contain two or more polar groups.

Fraction SEF-2 (1950°F mid-AEBP and pentane insoluble but cylohexane soluble) has no more hydrocarbons. On average, its molecules contain 5.25 heteroatoms, some of these likely with different functionalities. The average heteroatom content in fraction SEF-3 (2250°F mid-AEBP, cyclohexane insoluble but toluene soluble) is even higher, namely, 10.8 per molecule. All of these molecules contain at least one polar group, and most of them several. Extrapolating the trends from lower fractions, we may assume that in SEF-2 these heterocompounds are about evenly divided into basic, pyrrolic, and acidic ones and that in the higher SEF fractions, 3 and 4, the acidic and, to a lesser extent, the pyrrolic compounds outweigh the basic ones. In Table 10.15 we have consolidated our estimates of what portion in the different cuts contain two or more heteroatoms per molecule.

The SEF fractions 2–4 are insoluble in *n*-pentane and, thus, can be viewed as representative of pentane asphaltenes. On average, SEF-2 fractions, with mid-AEBPs between about 1700 and 2200°F (930–1200°C), contain about 200 carbon atoms, 40–50% of which are aromatic. Their hydrogen deficiency (Z) is enormously large, ranging from about -150 to -200, and indicating the presence of condensed rings (see Chapter 4 for a discussion). SEF-3 fractions have even higher average MWs, aromaticities, heteroatom contents, and hydrogen deficiencies (Table 10.13). (SEF-4 fractions were too small for such analyses.) This composition makes

Table 10.15 Approximate Percentages of Molecules in Kern River Fractions Carrying Two or More Heteroatoms

	Mid-AEBP			
Fraction	°F	°C	MW	% of fraction with ≥2 heteroatoms
Cut 1	722	394	335	4
Cut 3A	955	524	525	10
Cut 5	1240	682	880	20
SEF-1	1365	740	1110	27
SEF-2	1945	1060	2520	85
SEF-3	2250	1230	3620	100

Table 10.16 Heteroatom Concentrations in Heavy Fractions of Different AEBPs from Five Crude Oils

Crude oil	Mid-AEBP	MW	Approx. % of hetero-compounds[a]	Average number of heteroatoms/heterocompound: S	N	O	Sum	Heteroatoms per 100 C of entire fraction
Kern River	722	335	20	0.52	0.38	0.44	1.34	1.1
	887	470	40	0.41	0.54	0.44	1.4	1.7
	1023	595	60	0.50	0.75	0.53	1.8	2.1
	1365	1110	80	0.58	1.39	0.92	1.9	2.6
	1945	2520	100	1.15	3.95	1.9	7.0	3.9
	2250	3620	100	1.6	6.2	3.0	10.8	4.2
Calif. Offshore	697	330						1.7
	804	400						2.0
	1026	610						2.8
	1245	955						3.7
	1660	1910	80	4.4	1.6	0.9	6.9	4.2
	2300	4150	100	8.2	5.2	2.2	15.6	5.5
	2670	6070	100	12.9	8.9	4.5	26.3	6.4
Arabian Heavy	662	305						>1.2
	796	400						>1.5

	1025	620						>2.0
	1305	940						>2.5
	1565	1620	80	3.4	0.6	—	>4.0	>2.9
	2310	3870	100	9.2	2.5	—	>11.7	>4.5
	2570	5070	100	12.7	3.6	—	>16.3	>4.7
Maya	705	315						>1.3
	813	400						>1.2
	1025	580						>1.9
	1285	950						>2.3
	1530	1520	80	2.9	0.6	—	>3.5	>2.7
	2070	2970	100	6.3	2.25	—	>8.5	>4.3
	2580	5170	100	11.6	5.0	—	>16.6	>4.8
Boscan	719	335						2.4
	816	430						2.5
	1088	675						3.4
	1197	875						3.8
	1510	1420	80	3.4	0.8	—	>4.2	3.5
	2160	3460	100	7.6	3.8	—	>11.4	4.9
	2590	5510	100	12.2	7.1	—	>17.3	5.2

[a]The numbers other than 100 are estimates.

Source: Data adapted from Refs. 6 and 43.

meaningful further subfractionation very difficult, except, possibly, for separations by molecular weight or basic and acidic character.

The numbers are greater yet in many other crudes with higher heteroatom concentrations. Table 10.16 displays this kind of data for five crude oils. Also shown in this table are the numbers of heteroatoms per 100 C atoms, which are independent of MW. These, too, increase distinctly and gradually with increasing AEBP, though more slowly than the number of heteroatoms per heterocompound (i.e., per molecule).

What are the heteroatom functionalities in these multiheteroatom compounds? IR spectroscopic measurements indicate the presence of carboxylic acid, phenolic, pyrrolic, pyridine, and amide functionalities (35,38,37,58,59) including quinolones (60). We may conclude that, to a large extent, the multiheteroatom compounds have the same functionalities as simpler petroleum constituents featuring only one or two heteroatoms, except for the fact that they contain several heteroatoms in one molecule.

G. Metal-Containing Compounds

The most important metal-containing compounds in petroleum from the refiner's point of view are those containing vanadium, nickel, and iron. Even though present only in small amounts (usually $<1\%$ by weight), they are very detrimental to petroleum processing. They cause rapid catalyst poisoning (61,62) and other problems. The metals are typically removed by a deasphalting process which also rejects a large amount of convertible material along with the metal-containing species.

The concentration of vanadium in crude oils varies from almost 1200 ppm in Boscan to 0.15 ppm in Altamont. Nickel concentrations are generally lower than vanadium except in crudes with very low V levels, ranging from about 0.3 to 100 ppm. Iron is lower yet, usually between 1 and about 30 ppm, though occasionally with somewhat higher levels.

We have already seen the distribution of vanadium and nickel in the fractions of some crude oils (Fig. 4.13 of Chapter 4). For the sake of convenience, we show that of vanadium again in Fig. 10.16. In contrast to all other element distributions, those of vanadium and nickel (and iron) are quasi-bimodal with a relatively low peak around 1200 or 1300°F, a valley at 1600°F, and a continuously rising curve from then on. The first hump in this distribution corresponds to molecular weights between 400 and 1500 daltons (super-heavy VGO) and represents the free metalloporphyrins. These are readily identifiable by the Soret bands which are absorption bands of visible light between about 400 and 570 nm. The largest one, usually observed near 408 nm, is characteristic of all porphyrins. The smaller ones near 550 nm (α, β bands) are specific for the porphyrin type

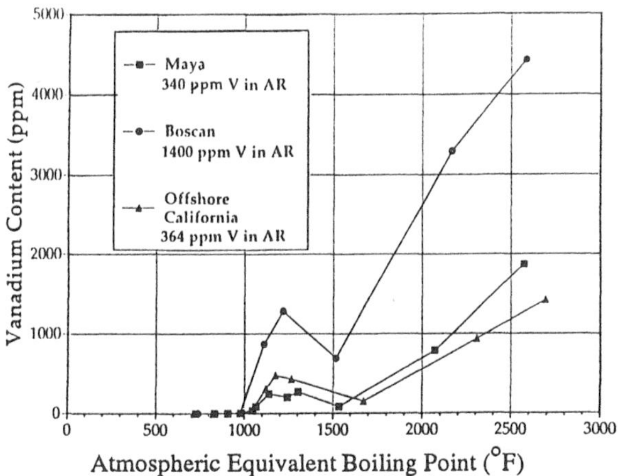

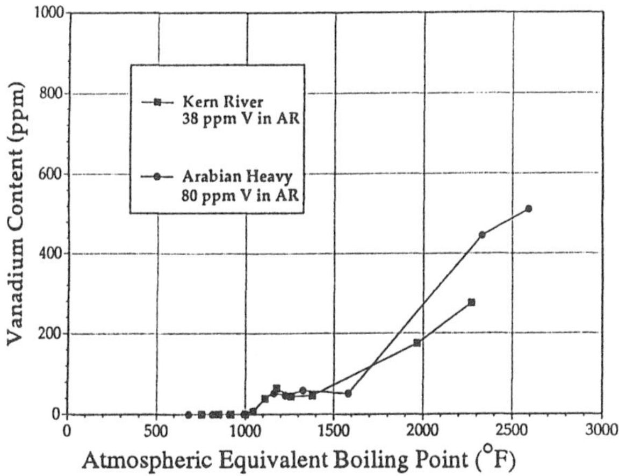

Figure 10.16 (Top and bottom) Concentration of V as a function of AEBP. (From Ref. 63. Reproduced with permission of the publisher.)

(etio, rhodo, etc.) and for the different metals. For instance, those of Ni are at 550–552 nm and those of vanadyl (VO) near 570 nm. Two examples of petroporphyrin Soret bands are shown in Fig. 10.17.

The petroporphyrins are relatively amenable to reductive (67) as well as oxidative (64) demetallation. The remainder of the metals is bound in structures which give only weak Soret bands or none at all and are much harder to break up. These are frequently called metallo-nonporphyrins.

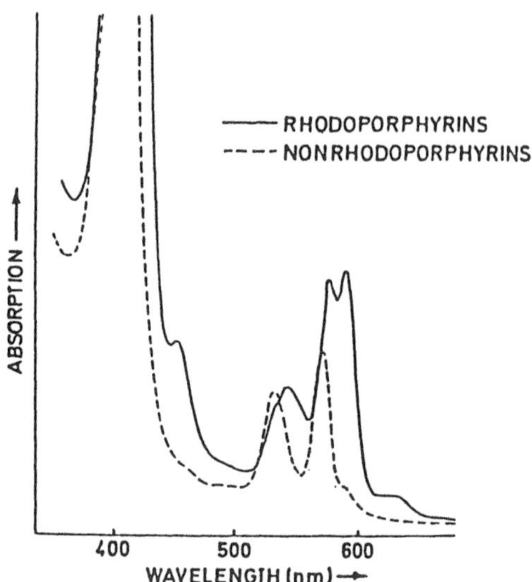

Figure 10.17 Soret bands in petroporphyrins. (From Ref. 64. Reproduced with permission of the publisher.)

Because of their combination of being resistant to processing and having such a deleterious effect on catalysts, much work was done to determine their structure. There are two schools of thought about their nature. One is represented, for instance, by Dickson et al. (65,66), Fish and co-workers (68,69), and Reynolds and co-workers (70–72). Its adherents claim that most of the metallo-nonporphyrins are distinctly different from the porphyrins. We belong to the other which is represented mainly by Blumer et al. (73,74), Goulon and co-workers (75,76), Pearson and Green (77). We believe to have good evidence (78) for the view that these metal compounds are mainly molecules in which hindered porphyrin groups held in a fairly large surrounding hydrocarbon structure either by chemical bonds or, possibly, sterically trapped.

Much of the evidence used by the first group is based on electron paramagnetic resonance spectroscopy (EPR) which seemed to indicate different types of chemical environments for the vanadyl group, VO, in the porphyrins and nonporphyrins (65,66,70–72). The applicability of EPR to this problem has recently been questioned (79). However, reversed-phase LC and size exclusion chromatography (SEC) combined with element specific detection suggested that some truly nonporphyrin metal compounds may indeed be present in several heavy petroleum samples (68,69,80). Some of these were in the MW range of 400–800 MW, that is, in the size range of free metalloporphyrins or even smaller. However, no species have been identified.

Goulon et al. (75) isolated "metallo-nonporphyrins" from Boscan asphaltenes and subjected them to two kinds of x-ray absorption spectroscopic measurements: EXAFS (extended x-ray absorption fine structure) and XANES (x-ray absorption near-edge structure). Back-to-back comparison of the "metallo-nonporphyrins" with pure model compounds clearly indicated that most or all of the metal in this sample actually was in the typical porphyrin environment. Boscan crude, however, is an exception in that all SEF fractions (>1300°F, >700°C AEBP) show a Soret band (6,43).

In our laboratories, Zhang et al. (78) also used XANES (performed at Brookhaven Lab), but they applied it to well-defined fractions of progressively higher AEBPs and combined it with other measurements. They fractionated a Mayan atmospheric residue into seven fractions of increasing AEBP by molecular distillation and by SEF (Boduszynski's sequential elution fractionation). The five fractions with AEBP > 1000°F (540°C) contained vanadium and nickel; the two with AEBP < 1000°F were metal-free.

By means of SEC with element specific detection this group measured the metal content and also the sulfur distribution in the residue and in all the fractions. X-ray absorption spectroscopy (XANES) of the metal-containing fractions proved that essentially all of the so-called metallo-nonporphyrins in the SEF fractions of their nondistillable residue had the same coordination spheres as the free metalloporphyrins in the high-boiling distillates. There was no indication of molecules with nonporphyrin coordination spheres of the kind suggested by Reynolds et al. (70,72). Also, there were no metal-containing molecules of lower MW than the free metalloporphyrins.

A paper by Pearson and Green (77) corroborates these results. In a study of the processing characteristics of Mayan and Wilmington heavy residues, these authors observed an increase of Ni- and VO-porphyrins in the weak base fractions after hydrotreating at 375 and 400°C (710 and 750°F). To judge by the Soret bands, large amounts of metallo-porphyrins had been created in these fractions. The authors excluded the possibility that porphyrins from other fractions had been converted to those in the weak base fraction. This left two other explanations. We quote:

> 1) . . . the most likely one, some of the "non-porphyrinic" metal compounds are actually porphyrinic but with external substitutions of the porphyrin ring structure of such a nature that absorption of the α and β bands in the visible region is outside the region being scanned for

> porphyrins. 2) . . . porphyrins in the polar fractions are complexed by the polar molecules so that their visible absorbance is shifted outside the region measured, via charge transfer interaction(s).

Pearson and Green cited an example of such a shift in the α band of a special VO porphyrin in piperidine caused by the formation of a piperidine adduct and remarked on the many potential complexing constituents in residues which lend substantial weight to this explanation.

At this time, the controversy has not been completely settled. But there seems to be good evidence for the view that most of the vanadium and nickel in petroleum actually is in the form of petroporphyrins, much of it hindered in still obscure ways. There appears to be some evidence for a small amount of true metallo-nonporphyrins in the low-molecular-weight fractions of petroleum residues although no such material was found in our Mayan crude oil.

III. COMPOSITION AS A FUNCTION OF AEBP

The general changes in the composition of petroleum fractions with AEBP have been outlined in the beginning of this chapter in Section I.B. Here we are trying to convey our present understanding of the composition of the major heavy petroleum fractions as it applies to most crudes. Our description is based mostly on factual knowledge but, because of limited data, some parts are also based on extrapolations from our experience. Our guesses are clearly indicated as such.

Relatively little was published on the detailed composition of heavy petroleum fractions. Most of the reported work either focussed on a particular broad fraction (for example, a VGO or residue) or on specific crude oil components (biomarkers, porphyrins, sulfur compounds, etc.). A comprehensive picture of the petroleum composition in terms of its variation with both boiling point and crude origin is not yet available.

In order to facilitate the discussion, we show in Table 10.17 compositional data of a series of progressively higher boiling fractions making up an entire (and fairly typical) heavy crude oil (22°API). Thus, the reader will get at least a general impression if not the entire picture. The AEBP ranges in the table are the cut points between adjacent fractions. These do not reveal the actual breadth of the fractions, which is greater than indicated by these numbers because of the poor separation in distillations and the resulting overlap.

A. Light and Middle Distillates (<650°F or <345°C AEBP)

Light straight-run naphthas (<300°F, <150°C AEBP) consist mainly of paraffins (see Table 10.17). Any ring compounds are restricted to alkyl-

Table 10.17 Compositional Changes with AEBP for a Heavy Crude Oil (22°API)

Crude oil fraction	Approx. AEBP range °F	Approx. yield from crude wt%	H/C ratio	S wt%	O wt%	N ppm	Compound groups (wt%) Saturates	Aromatics	Polars
Light naphtha	Start–300	8	2.05	0.1	0.01	<1	88	12	[a]
Heavy naphtha	300–400	8	1.95	0.4	0.03	2	80	20	[a]
Kerosine	400–500	5	1.85	1.0	0.10	15	75	25	[a]
AGO	500–650	12	1.8	2.0	0.12	300	65	34	1
LVGO	650–800	15	1.7	2.8	0.18	1500	45	50	5
HVGO	800–1000	10	1.6	3.3	0.24	2000	32	58	10
SHVGO	1000–1300	15	1.5	4.2	0.28	3500	15	70	15
Nondistillable residue	>1300	27	1.3	5.9	1.00	8700	2	8	90[b]

[a] A small fraction of 1 wt%.

[b] A significant fraction of "polars" is insoluble in alkane solvents.

cyclohexanes and alkylcyclopentanes and to alkylbenzenes. The sulfur content may vary from a few ppm to several hundreds ppm. The nitrogen content is so low that it is rarely measured.

Heavy naphthas (300–400°F, 150–200°C AEBP) are still mainly aliphatic, but they contain greater amounts of mono- and dinaphthenes. The aromatics may account for as much as 20 wt% and consist of alkylbenzenes. The sulfur content ranges from a few hundreds ppm to as high as 2 wt%; N content is usually <1 ppm.

Kerosine (400–500°F, 200–260°C AEBP) typically contains about 70–80% saturates and 20–30% aromatics. The saturates may be considerably more naphthenic than those in the lower-boiling fractions, depending on the crude oil. Hydrocarbons with one or two naphthenic and aromatic rings may be present. S content may vary from 0.05 to 3.0 wt%, mainly in the form of sulfides, in acyclic as well as cyclic alkanes, and of alkylthiophenes and benzothiophenes. N content is typically a few ppm but may be higher.

Atmospheric gas oils (AGO) (500–650°F, 260–345°C AEBP) may have considerable amounts of aromatics and detectable amounts of polars (see Table 10.17). Besides mono- and dinaphthenes, alkylbenzenes, and alkylnaphthalenes, they contain small amounts of higher ring-number naphthenes and aromatics. Sulfur contents are typically around 1.5–3 wt% but may occasionally reach 4.5 wt%. Besides dialkyl-, alkylaryl-, and arylarylsulfides and the alkylthiophenes and benzothiophenes, there may now also be some dibenzothiophene derivatives. Nitrogen content is usually a few hundreds ppm, but it may go up to 1500 ppm for some crudes.

Thus, we note several changes in the composition of light and middle distillates with increasing AEBP. The concentration of saturates decreases, but their content of naphthenic compounds rises slowly for most crudes. In contrast, the aromatics concentration increases with increasing boiling point. Most aromatics are alkylbenzenes and naphthalenes, some of which may contain one or several naphthenic rings. In the middle distillates and the lower-boiling fractions, we already find the five major sulfur compound classes: thiols, sulfides, disulfides, sulfoxides, and thiophene derivatives, mainly benzothiophenes. Nitrogen compounds are mostly pyrroles, indoles, pyridines, and quinolines (Table 10.9). Among the oxygen compounds, aliphatic carboxylic acids, aliphatic esters, ketones and other carbonyl derivatives as a group, dibenzofurans, and phenols were the most abundant types in the atmospheric gas oil fraction of a California crude oil (39). The concentration of polars in fractions boiling below 650°F usually does not exceed 1 wt%, but for some crudes it may be as high as a few wt%.

B. Light Vacuum Gas Oils (LVGOs) (650–800°F or 345–430°C AEBP)

Light vacuum gas oils (LVGOs) have a boiling range from about 650 to 800°F (ca. 345–430°C) and are distinctly more aromatic than AGOs. In this range, analyses begin to be challenging and are less accurate than with lower boiling fractions because of the higher MW and the increased concentration of heteroatoms.

Typically, saturates in this AEBP range account for 40–60 wt% of the fraction. They consist mainly of paraffins and alkylnaphthenes with 1–4 rings and have carbon numbers between 15 and 30. Figure 10.7 shows the MW and carbon number distribution of the tetranaphthenes in 3 short-path distillates of Kern River crude oil, the first one of which is representative of LVGOs.

The aromatic content may be as high as 50 wt% or even more as shown in Table 10.17. Most compound types in this fraction are alkylbenzenes, naphthalenes, and phenanthrenes, with or without naphthenic rings. The aromatics also contain substantial amounts of heterocompounds, primarily sulfur compounds. In addition to those present in lower boiling ranges, several types of thiophene derivatives have been reported which vary mainly by the number and type of aromatic rings to which they may be fused. Thus, we have benzo-, dibenzo-, naphthobenzo-thiophenes, and other derivatives.

The polars in LVGOs usually account for 1–10 wt%. They consist primarily of compounds containing nitrogen or oxygen or both. Snyder et al. (39) investigated the polars in the LVGO range of a California crude oil (see Table 10.9) and reported 3.4 wt% carbazoles, 0.6% indoles, 0.5% benzocarbazoles, 0.66% pyridines, 1.74% quinolines, 0.26% benzoquinolines, and 1.2% pyridones (and quinolones, etc.) (see Table 10.9). Furthermore, they found 1.74 wt% aliphatic carboxylic acids, 0.97% phenols, 1.2% benzoaromatic and higher aromatic furanes, and 0.7% aliphatic esters, ketones, and other carbonyl derivatives.

The likely carbon, heteroatom, and ring number ranges of the molecules found in this fraction are presented in Table 10.17.

C. Heavy Vacuum Gas Oils (HVGOs) (800–1000°F or 430–540°C AEBP)

Heavy vacuum gas oils (HVGOs) have a boiling range from about 800 to 1000°F (430–540°C AEBP). Here the challenge to compositional analysis is even greater than with LVGOs because of the higher MW (or carbon number) and heteroatom content. The carbon number range of this fraction is approximately 20–50.

Table 10.18 Effect of Heterocompound Type on the Carbon Number Range. Basic Fractions from 370–535°C (700–1000°F) Boiling Range of Wilmington Crude

General formula	Likely no. of ar. rings	Likely no. of naph. rings	Carbon-no. range
$C_nH_{2n-7}N$	1	2	15–30
$C_nH_{2n-9}N$	1	3	15–30
$C_nH_{2n-11}N$	2	0	15–34
$C_nH_{2n-13}N$	2	1	15–33
$C_nH_{2n-15}N$	2	2	15–35
$C_nH_{2n-17}N$	3	0	15–33
$C_nH_{2n-19}N$	3	1	15–34
$C_nH_{2n-21}N$	4	0	16–33
$C_nH_{2n-23}N$	4	0	18–33
	4	1	
$C_nH_{2n-25}N$	4	1	18–33
	4	2	
	5	0	
$C_nH_{2n-29}N$	5	0	19–33
$C_nH_{2n-17}NS$	3	1	12–29
$C_nH_{2n-19}NS$	3	1	17–31
$C_nH_{2n-12}N_2$	2	1	15–31
$C_nH_{2n-14}N_2$	2	2	15–31
$C_nH_{2n-16}N_2$	3	0	16–31
$C_nH_{2n-8}NS$	2	0	12–29
$C_nH_{2n-10}NS$	2	1	12–29

Source: Data adapted from Ref. 37.

There is a dramatic decrease of saturates and a corresponding increase in aromatics and polars. In the example in Table 10.17, the HVGO contains 32 wt% saturates, 58 wt% aromatics, and 10 wt% polars. Further analysis of compound group fractions (for example, saturates, aromatics, polars) in this boiling range is severely impaired by the limitations of analytical techniques. Even analysis by FIMs, the most suitable methodology for the analysis of aliphatic hydrocarbons, cannot distinguish, at nominal mass resolution, paraffins from heptanaphthenes. Both are likely to be present in this AEBP range.

With the aromatics and polars, the analysis becomes even more difficult. The likelihood of molecules having more than 1 heteroatom further complicates the analysis. Most aromatic and polar molecules in the HVGO AEBP range contain 1–3 aromatic rings, predominantly with naphthenic rings. Compounds with 4 and more aromatic rings are present at much lower concentrations.

Again we can assume that all the compound types found in lower-boiling fractions are present in the HVGOs, too, except that they have higher C numbers, mainly due to more naphthenic groups and to more and longer alkyl chains. In addition, heterocompound types with more aromatic rings may be found, e.g., benzonaphtho- and dinaphthothiophenes, benzo- and naphthoquinolines, aza and diaza derivatives. Here, we also find compounds with two and three heteroatoms such as amides, azaamides, sulfoxides, carboxylic acids, and hydroxycarboxylic acids. Some heterocompound types proposed by McKay et al. (37) are shown in Table 10.18 and Figures 10.10 and 10.13.

Snyder (39) identified several N and O compounds in the HVGO fraction (approximately 850–1000°F AEBP) of a California crude oil: 0.75 wt% indoles, 4.1% carbazoles, 1.3% benzocarbazoles, 1.3% pyridines, 2.0% quinolines, 1.6% benzoquinolines, 2.0% pyridones (and quinolones, etc.), 0.4% azaindoles, and so on (Table 10.9); 0.85% dibenzofuranes, 0.8% benzonaphthofuranes, 0.15% dinaphthofuranes, 0.08% dihydrobenzofuranes, 1.35% phenoles, 1.35% aliphatic carboxylic acids, and 2.3% aliphatic esters, ketones and other carbonyl derivatives.

Figure 10.18 shows the carbon number distribution of apparent *Z* series in FIMS spectra of the mono-, di-, and triaromatics from two whole VGOs (~650–1000°F, 345–540°C AEBP) of different origin. Their upper MW limits vary between 600 (for one of the triaromatics fractions) and 800 (for one of the monoaromatics fractions) which corresponds to carbon numbers of about 40 and 60. (In any given boiling range, the triaromatics have lower MWs than the monoaromatics; see Chapter 3.) Figure 10.19A presents the FIMS molecular-weight profiles of three triaromatic fractions of increasing mid-AEBP, which are representative of a HVGO boiling range. The FIMS profiles in Fig. 10.19B are those of the triaromatics in the SHVGO and SEF-1 range. The increase in the width as well as the mean of the MW range is clear. The most prominent series in these fractions is that with $Z = -22$ which most likely represents triaromatic hopane derivatives. MWs in the first three spectra (part A) range from about 250 to 800, and carbon numbers, from about C_{15} to C_{60}. The upper limit indicates close to 40 carbon atoms in side-chains of these compounds.

The effect of heterocompound type on the carbon number range is illustrated in Table 10.18 and Fig. 10.20. Heterocompounds generally have distinctly lower average and upper MWs than the aromatics from the same boiling range. This is most pronounced with pyrrolic compounds as demonstrated in Fig. 10.20. This figure also shows the structures of the four

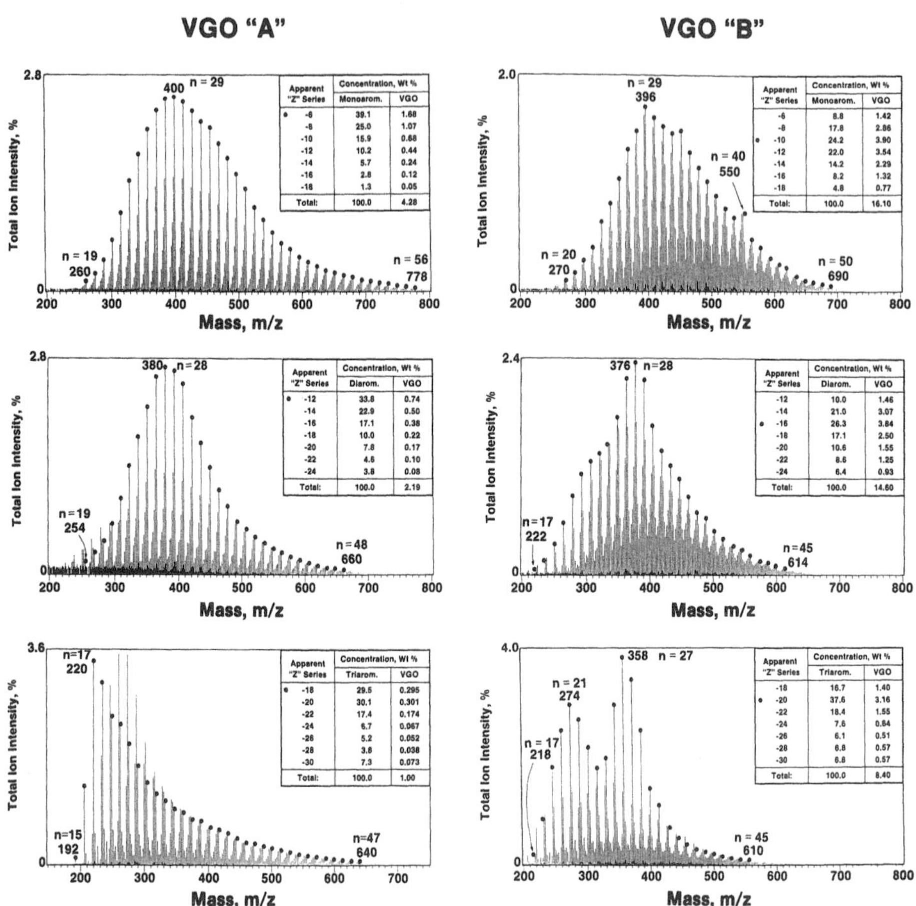

Apparent "Z" Series	Concentration, Wt %	
	Monoarom.	VGO
• -6	39.1	1.68
-8	25.0	1.07
-10	15.9	0.68
-12	10.2	0.44
-14	5.7	0.24
-16	2.8	0.12
-18	1.3	0.05
Total:	100.0	4.28

Apparent "Z" Series	Concentration, Wt %	
	Monoarom.	VGO
-6	8.8	1.42
-8	17.8	2.86
• -10	24.2	3.90
-12	22.0	3.54
-14	14.2	2.29
-16	8.2	1.32
-18	4.8	0.77
Total:	100.0	16.10

Apparent "Z" Series	Concentration, Wt %	
	Diarom.	VGO
• -12	33.8	0.74
-14	22.9	0.50
-16	17.1	0.38
-18	10.0	0.22
-20	7.8	0.17
-22	4.6	0.10
-24	3.8	0.08
Total:	100.0	2.19

Apparent "Z" Series	Concentration, Wt %	
	Diarom.	VGO
-12	10.0	1.46
-14	21.0	3.07
• -16	26.3	3.84
-18	17.1	2.50
-20	10.6	1.55
-22	8.6	1.25
-24	6.4	0.93
Total:	100.0	14.60

Apparent "Z" Series	Concentration, Wt %	
	Triarom.	VGO
• -18	29.5	0.295
-20	30.1	0.301
-22	17.4	0.174
-24	6.7	0.067
-26	5.2	0.052
-28	3.8	0.038
-30	7.3	0.073
Total:	100.0	1.00

Apparent "Z" Series	Concentration, Wt %	
	Triarom.	VGO
-18	16.7	1.40
• -20	37.6	3.16
-22	18.4	1.55
-24	7.6	0.64
-26	6.1	0.51
-28	6.8	0.57
-30	6.8	0.57
Total:	100.0	8.40

Figure 10.18 Carbon-number distribution of apparent *Z* series in FIMS spectra of mono-, di-, and triaromatics from two whole VGOs of different origin. (From Ref. 6. Reproduced with permission of the publisher.)

most prominent nitrogen compound types in the "pyrrolic" and "basic" fractions of Kern River VGO (approximately 650–1000°F AEBP). The great majority of these fractions consists of molecules with only one nitrogen. However, this does not exclude the possibility of other heteroatoms being present. Particularly, compounds containing S or O or both are likely, but high resolution MS would be needed to identify those.

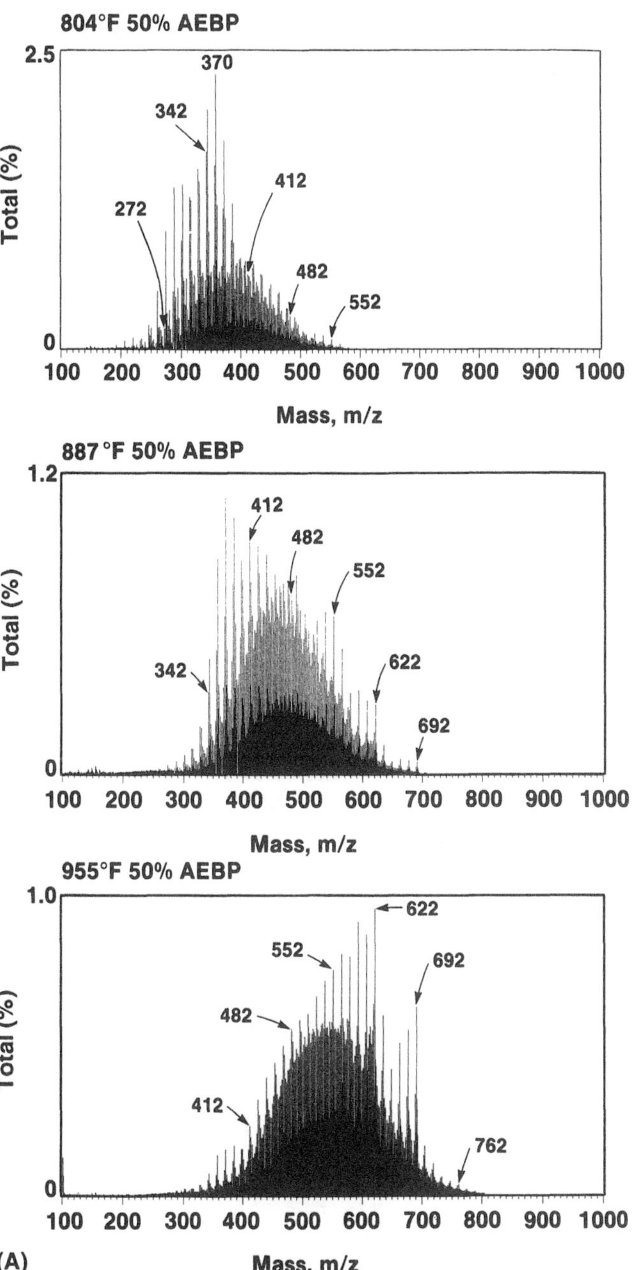

Figure 10.19 FIMS molecular weight profiles of six triaromatic fractions from offshore California crude oil. (A) HVGO range, (B) SHVGO range. (From Ref. 6. Reproduced with permission of the publisher.)

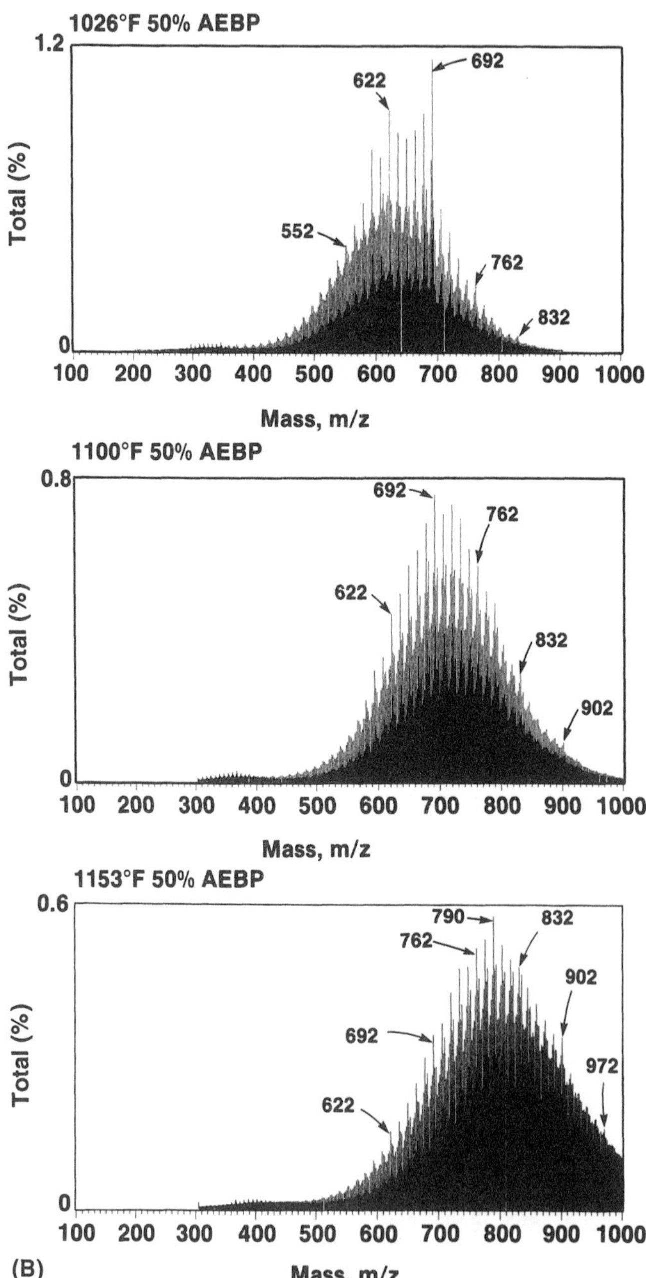

Figure 10.19 (Continued)

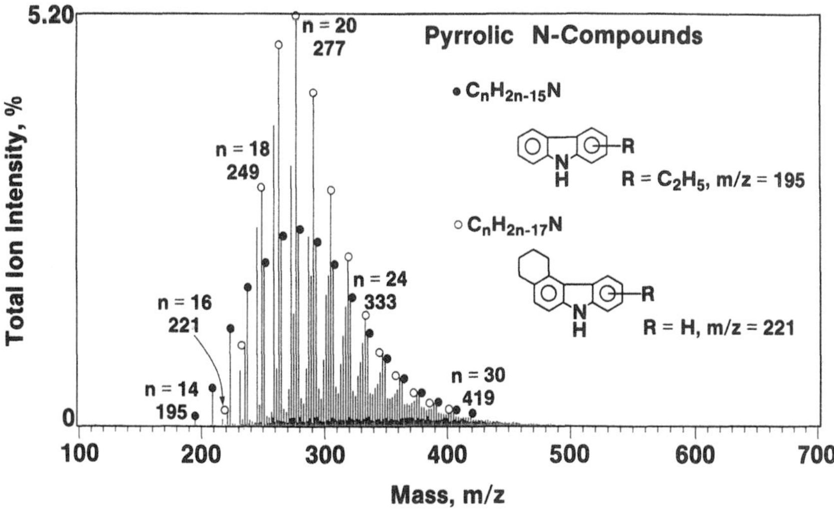

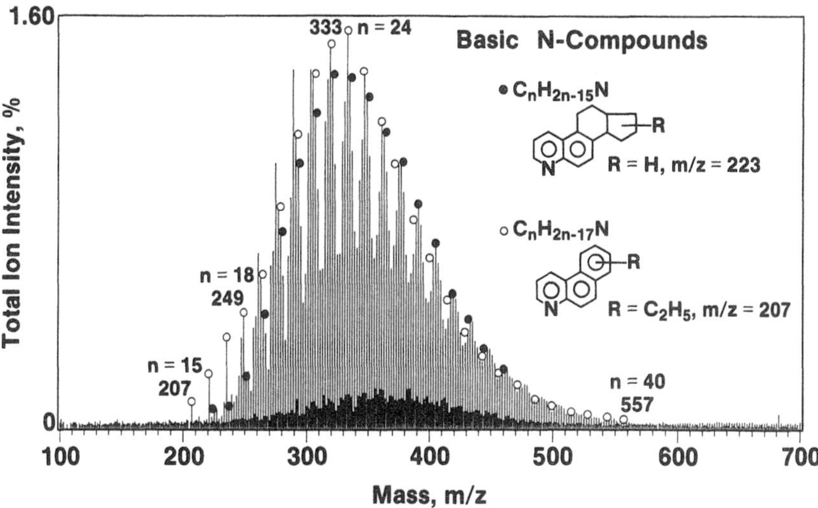

Figure 10.20 ^{13}C-corrected FIMS spectra of "pyrrolic" compounds (top) and "basic" (bottom) compounds derived from a Kern River VGO. (From Ref. 6. Reproduced with permission of the publisher.)

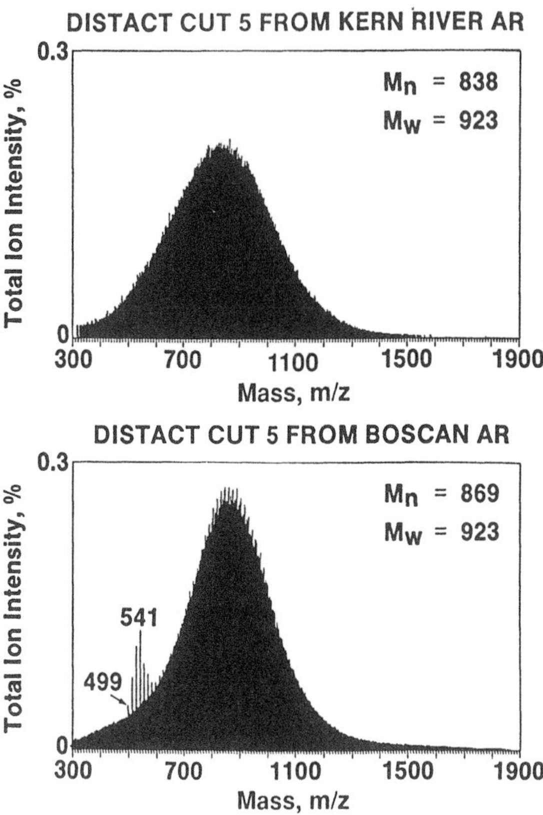

Figure 10.21 FIMS molecular-weight profiles (uncorrected for ^{13}C) of two short-path distillation cuts in the SHVGO range from Kern River Boscan crude oils. (From Ref. 6. Reproduced with permission of the publisher.)

D. Super-Heavy Vacuum Gas Oils (SHVGOs) (1000–1300°F or 540–700°C AEBP)

Super-heavy vacuum gas oils (SHVGOs), with a nominal boiling range of approximately 1000–1300°F (540–700°C) AEBP, are immensely complex. Between 50 and 80% of their molecules contain heteroatoms, mostly sulfur, nitrogen, and oxygen, but also vanadium, nickel, and iron. About half of them have two or more heteroatoms (see Section II.F and Table 10.16). Although SHVGOs can be completely volatilized during MS analysis, the resulting spectra are very complex and difficult to interpret.

Saturates in SHVGOs seldom account for more than 10–20 wt%. Aromatics typically range from 50 to over 80 wt%. However, the latter usually contain 40–80 wt% of neutral heterocompounds, which are very difficult to separate from the hydrocarbons. Polar compounds may account for 10–30 wt%. Even in this boiling range, most of the aromatics and heterocompounds contain only 1–3 aromatic rings, but they are generally highly naphthenic. A relatively small portion of molecules in these compound groups have four and more aromatic rings.

Chromatographic separations of SHVGOs into concentrates of certain groups or classes of compounds facilitate their further analysis, but adjacent fractions overlap severely. Present MS group-type methods cannot be applied to these fractions. FIMS and FDMS can provide information in terms of molar mass distribution and apparent Z series (2 to -10). Even in this boiling range, the apparent Z-numbers of saturates can be reasonably well interpreted in terms of specific Z-numbers, that is, as representing paraffins, mononaphthenes, dinaphthenes, and so on. For aromatics, and even more so the polars fractions, the analysis is much more difficult, and it would require ultra-high resolution MS, which is not readily available.

The carbon number in this range is approximately 35–100. The complexity in the SHVGOs is so great that chromatographic separations and molecular analysis of the fractions become difficult and less accurate.

The effect of increasing AEBP on separation efficiency is demonstrated in Fig. 10.22. This figure compares the HPLC-UV maps of short-path distillation cuts in the LVGO, HVGO, and SHVGO range, and a SEF-1 fraction derived from nondistillable residue. Although the first three maps allow the clear differentiation of mono-, di-, and triaromatics, this is no longer so easy in those of the SHVGO and nondistillable ranges. In the latter, we can barely recognize the main peaks of the one- to three-ring aromatics because of severe overlap.

E. Nondistillable Residues (>1300°F or >700°C AEBP)

1. General Description

The "nondistillable" residues consist of molecules of such low volatility that they cannot be distilled with present-day equipment, not even with high-vacuum short-path stills. They are the most complex of all crude oil fractions, consisting almost exclusively of heterocompounds, most of which contain several heteroatoms per molecule. They typically account for 30–50% of the total sulfur, 70–80% of the total nitrogen, and 80–90% of the total vanadium and nickel present in a crude oil. Saturates and aromatics, together, make up only a few weight percent of the "nondistillable" res-

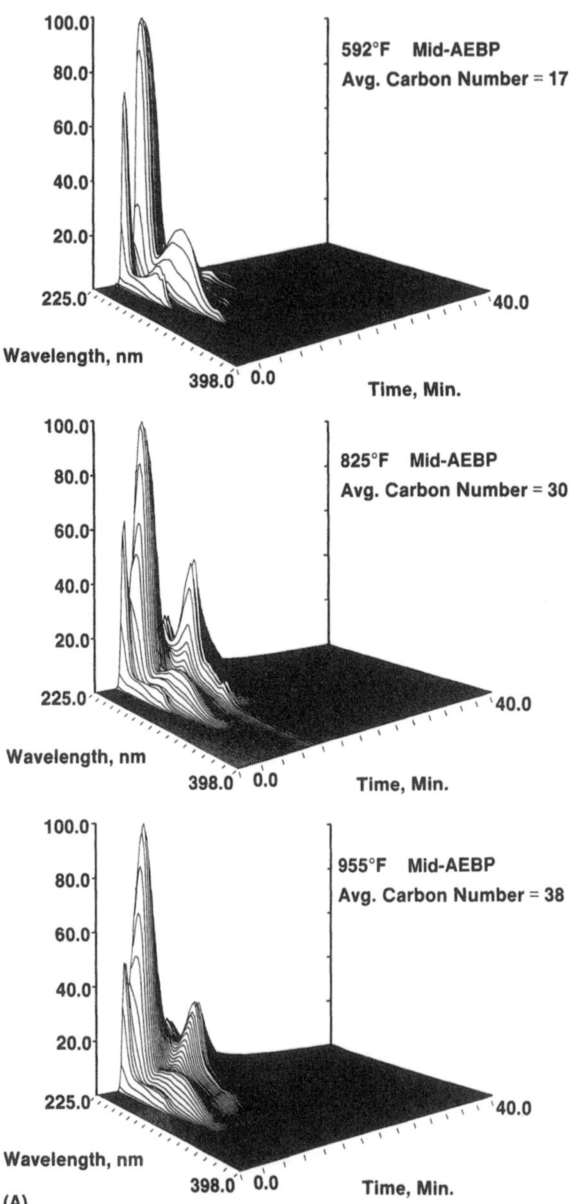

Figure 10.22 HPLC-UV maps of short-path distillation cuts in the light and heavy VGO (A) and two SHVGO ranges and a SEF-1 fraction (B). (From Ref. 6. Reproduced with permission of the publisher.)

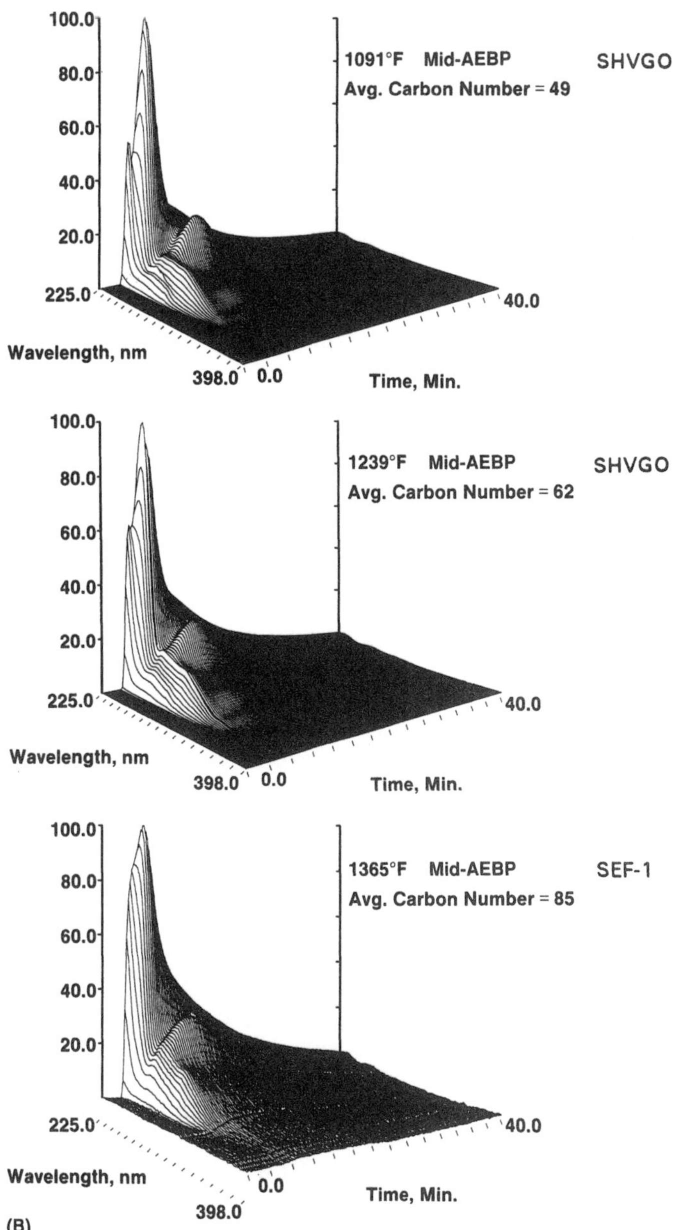
100.0
80.0
60.0
40.0
20.0
225.0
Wavelength, nm
398.0
0.0
40.0
Time, Min.
1091°F Mid-AEBP
Avg. Carbon Number = 49
SHVGO
100.0
80.0
60.0
40.0
20.0
225.0
Wavelength, nm
398.0
0.0
40.0
Time, Min.
1239°F Mid-AEBP
Avg. Carbon Number = 62
SHVGO
100.0
80.0
60.0
40.0
20.0
225.0
Wavelength, nm
398.0
0.0
40.0
Time, Min.
1365°F Mid-AEBP
Avg. Carbon Number = 85
SEF-1

(B)

idue. About 40–70 wt% of the nondistillable residue is pentane insoluble "asphaltenes."

The nondistillables can be separated into several fractions of increasing MW and, thus, of increasing AEBP. This was discussed in Chapter 3 in detail. So far we used the SEF method (sequential elution fractionation) for this purpose. SFE (supercritical fluid extraction) may be a good alternative in the future. SEF-1 fractions are pentane soluble and even susceptable to simulated distillation (SIMDIS). Figure 3.9 of Chapter 3 gave the SIMDIS "boiling curve" of a SEF-1 fraction together with those of several preceding short-path distillates as an example of the smooth progression beyond the "distillable" material. Fractions SEF-2 to 4 are distinctly different from SEF-1 and from each other in their increasing molecular weights, heteroatom contents, and hydrogen deficiencies. Table 10.13 lists the average parameters of SEF fractions from five crude oils. These data clearly illustrate the differences between consecutive SEF fractions and the consistency of these differences among diverse crudes.

The composition of the SEF fractions in molecular terms is unknown. In principle, and to judge from its chemical properties (see Tables 4.25–4.27 of Chapter 4), the composition of SEF-1 can be assumed to be similar to that of the SHVGOs except for its higher complexity: its higher MW, heteroatom content, variety of compound types, and degree of alkyl substitution. Compared with SHVGOs, the average MW of SEF-1 fractions is higher by about 50%; yet their molar mass profiles (by FIMS or FDMS) do not exceed 2000 daltons. Their aromatic carbon content is greater by 60–100% and their heteroatom content by at least as much (see Tables 4.25–4.27 of Chapter 4). The increases are even more dramatic going from SEF-1 to SEF-2 and SEF-3 (Table 10.13). Even if we eliminate the uncertainty of the MW and compare the heteroatom content per 100 C atoms (Table 10.16), the increases are distinct.

The pentane soluble fraction SEF-1 of a nondistillable residue is already so complex that further separation by chromatography is highly inefficient. This is demonstrated by the HPLC-HV map (b) in Figure 10.22. The higher AEBP fractions SEF-2, 3, and 4 are much more difficult to analyze yet.

Although we have chosen sequential elution fractionation (SEF) (discussed in Chapter 3) to divide nondistillable residues into fractions of increasing MW and, thus, AEBP, others have taken different approaches. The workers of the API-60 project and of NIPER [(1–4); see also Chapter 6] separate the nondistillable fraction on ion exchange resins into acids, bases, and neutrals. Most other groups have used precipitation by *n*-pentane or similar low-boiling alkanes to remove the insoluble "asphaltenes" from the soluble "maltenes," which were then more amenable to further separation and analysis. They apply this technique typically to vacuum residues, sometimes even to atmospheric residues. We will discuss this latter approach and our view of it in Sections III.E.5 and III.E.6. The "asphaltenes," being insoluble, are also nondistillable, no matter from what residue they were precipitated. Because practically all analytical work on nondistillable compounds was done on "asphaltenes," we must extensively refer to that work in the following sections.

What little information we have on the composition of insolubles comes from elemental analysis, from spectroscopic methods like NMR, IR, XANES, and so on, and from gentle degradation followed by analysis of the fragments. The spectroscopic methods allow the determination of some functional groups and certain aspects of the hydrocarbon skeleton in the sample. The degradative approach leads to more detailed information on the fragments but not necessarily to quantitative data on the entire molecules. (See Section III.F.2. and Chapter 9.) Some properties of the insolubles were described in Sections II.F and II.G and in Chapter 4. Other aspects will be discussed in the following sections.

2. Aromatic Ring Structures and Alkyl Substitution

In the 1960s, a "chicken wire" model with large aromatic and naphthenic pericondensed ring systems was popular for asphaltenes. Modifications of this model still show up in review articles and other papers although, generally, this view is no longer accepted. Now these insoluble molecules are seen as consisting of one or several relatively small ring clusters made up of only a few condensed aromatic and naphthenic rings, and containing polar heteroatom groups. In the larger molecules, several ring clusters, and maybe even single rings, are interconnected by paraffinic groups or chains and by sulfur linkages (7,29).

Today, the aromatic ring numbers per ring cluster in these heterocompounds are assumed to be fairly small, ranging from about 2 to 7 (81). Originally, these small numbers were based on the analysis of volatile pyrolysis or other degradation products. This could have been misleading because generally about 50% of the pyrolysis products are a solid coke-like-mass, rich in heteroatoms and of unknown ring size. Larger clusters containing a greater number of rings, if present in the original sample before pyrolysis, would likely be trapped in this mass rather than liberated into volatile products. Thus, there is no guarantee that the analysis of the volatile pyrolysis products would give us the correct ring number distribution in these clusters. But other more recent studies, not based on pyrolysis, confirm a low average ring number in the clusters. For instance, El-Mohamed et al. (82) came to this conclusion on the basis of magnetic susceptibility measurements of the asphaltenes from four heavy crude oils.

On the other hand, the high hydrogen deficiencies found in most high-boiling petroleum fractions seem to be incompatible with very small (one- or two-ring) aromatic moieties. We (83) have performed model calculations to shed some light on this problem, making use of the hydrogen deficiency and a rigorous equation relating the total number of rings, aromatic and naphthenic, R_t, to some basic compositional data:

$$R_t = C + 1 - \frac{1}{2}(C_{ar} + H - N_{pyrrolic})$$

The effects of S, N, and O are also taken into account. Our analysis, applied to asphaltene data by Ali et al. (84), in two of the three samples leads to negative numbers of paraffinic C for clusters with fewer than three aromatic rings on average unless we accept the average (!) naphthenic ring number per cluster to be at least four. Generally, three or more aromatic rings per cluster on average are required for reasonable distributions of the aliphatic carbon atoms in asphaltenes.

Strausz and co-workers (Ref. 8 and references therein, and Ref. 9) took a closer look at the composition of the asphaltenes in Athabasca bitumen. They break them up by gentle oxidation catalyzed with Ru ions, a technique they named "RICO" [for Ru ion catalyzed oxidation; see, e.g. (9,85)]. Their method, which was briefly described in Section III.B of Chapter 9, gave a wealth of new data. We feel, however, the results are subject to discussion. The following paragraphs summarize what this group found.

They attributed 9.2% of total carbon (9.2 per 100 C atoms) in the first set of volatile oxidation products to *n*-alkyl side-chains attached to aromatic rings and about 2% to methylene bridges between aromatic rings. The chain length of these *n*-alkyl side-chains ranged from 1 to 32 carbon atoms (9) with a curve steeply descending to about C_7 and then flattening out. The average length of these alkyl chains, excluding methyls, was reported as 8.7, and the total number of chains per 100 C atoms, as 1.8. More alkyl chains were found in the oxidation residue which consisted primarily of naphthenic structures with their alkyl substituents.

The average number of carbon atoms in methylene bridges connecting two aromatic rings, found in the original volatile esterified RICO products, was reported as 3.8 per 100 C. The chain length of the methylene bridges varied from 2 to 22 with a maximum at three carbons and an initially steep descent to about C_7. With 0.6 connecting chains per 100 C atoms, the average chain length here is 6.3.

The yields of benzene- di-, tri-, and tetracarboxylic acids were low (about 0.6 C per 100 C in the asphaltene sample on average), suggesting a very low content of condensed aromatic ring systems. The original aromatic carbon content of the asphaltene sample was 43 per 100 C (86). Only 1.4 of the 43 aromatic C per 100 C or 3.3% of all of the aromatic carbon atoms would have been part of a multiring cluster with the remaining 96.7% in benzene rings. Mojelski et al. (9) state that their yield of multiring carbon atoms is a lower limit rather than an average value. On the other hand,

Strausz and Lown (8) state that such a low content is in agreement with various spectroscopic measurements (87) and magnetic susceptibility studies (82) on Athabasca asphaltenes.

Based on these findings, Strausz and Lown (8) constructed a hypothetical assembly of structural units in their Athabasca asphaltene fraction which is reproduced in Fig. 10.23. This structure is a good example of how to present compositional features of a sample in a chemical structure without creating an "average molecule." It is an assembly of the main substructures found, put together more or less arbitrarily to show how they might have been linked and how the oxidation procedure would have liberated them. This "molecule" was designed for illustration only and is not to be mistaken for real ones.

Apparently, all the heterocompounds are destroyed in the oxidation process. It may be for this reason that Strausz' model does not show the

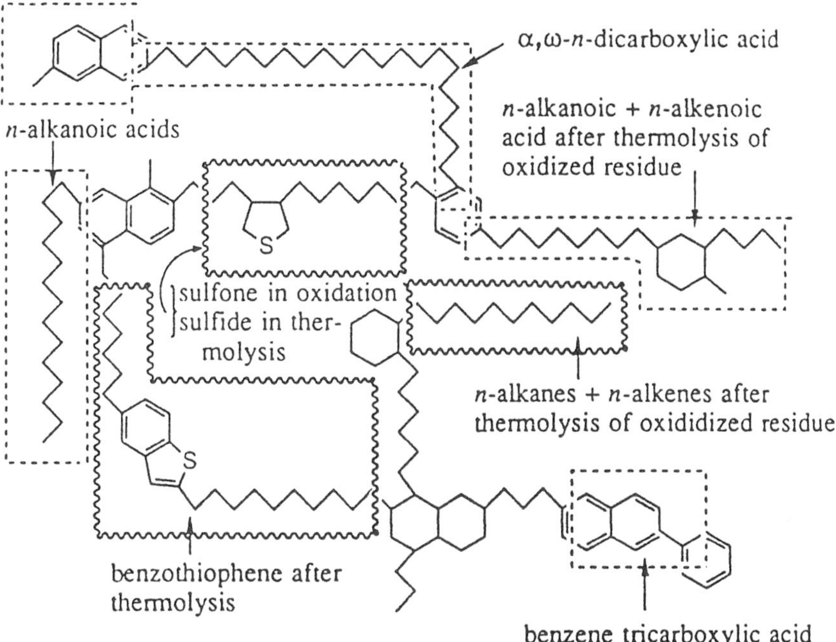

Figure 10.23 Hypothetical assembly of structural units in Athabasca asphaltenes. Units within dotted lines are recovered following Ru(VIII)-catalyzed oxidation; those within wavy lines are generated in the thermolysis of the oxidized (naphthenic) residue. (From Ref. 8. Reproduced with permission of the publisher.)

characteristics of greatest importance to the refiner, those which make pentane insolubles so intractable, the metals-, nitrogen-, and sulfur-containing groups. Taking Strausz and Lown's model at face value, one might ask, What in it makes the material it represents so hard to process? Furthermore, the model appears to us (83) incompatible with the data presented by Payzant et al. (88), in particular with the low hydrogen content. Nevertheless, the work done by Strausz and collaborators has distinctly advanced our knowledge of the most intractable fraction of petroleum. Even if we have taken a critical look at their results and interpretations, we fully recognize and appreciate the importance and potential of their approach.

3. Heteroatom Content and Functionalities

To a large extent, the nondistillables have the same functionalities as simpler petroleum constituents comprising only one or two heteroatoms, except that they have them in higher concentration. For instance, the nondistillable SEF-1 fraction (1660°F mid-AEBP) of California Offshore crude has an average MW of about 1910 and contains, on average, 131 carbon, 3.5 sulfur, 1.3 nitrogen, 0.7 oxygen, and 0.01 metal atoms per molecule. This translates to 2.7 S, 1.0 N, 0.5 O, and a total of 4.2 heteroatoms per 100 C. The SEF-2 fraction (2305°F mid-AEBP) of the same crude (Table 10.13) has an average MW of about 4100 and contains, on average, 282 carbon, 8 sulfur, 5 nitrogen, 2 oxygen, and 0.1 metal atoms per molecule, or 2.8 S, 1.8 N, and 0.7 O and a total of 5.3 heteroatoms per 100 C. For perspective, compare these values with those of lower-boiling species, for example, that of C_6-substituted dibenzothiophene (268 MW), a compound which was identified (among many similar ones) in a heavy gas oil boiling range and has a ratio of only 0.34 S per 100 C. Or look at Cut 1 (697°F mid-AEBP) of California Offshore crude which has 1.7 heteroatoms per 100 C on average. More examples are provided in Table 10.16.

If we assume an average number of three rings for our fraction SEF-2 above (2305°F mid-AEBP, 2520 MW, 80 aromatic C atoms), a cluster comprises 14 aromatic carbons on average. This means, again on average, about six such clusters per molecule, each containing at least one heteroatom (0.2 sulfur, 0.7 nitrogen, and 0.3 oxygen atoms). These numbers are obviously speculative and are not meant to describe the composition of this fraction. They are only presented to illustrate the great complexity of this type of material.

Another difference between nondistillables and heavy distillable fractions is their much higher content of polar groups. The "polar" contents of Cut 1, Cut 5, SEF-1, and SEF-2 in Kern River crude are roughly 15, 40, 80, and 100%, respectively. Thus, not only do the higher SEF fractions have more heteroatoms per molecule and per 100 C, but they have an even greater proportion (2–10 times as many) of polar functional groups such as pyrrolic and basic N, OH, and COOH.

4. What Makes Insolubles Insoluble?

What makes insolubles different from soluble petroleum components; indeed, what makes them insoluble? We will try to answer both parts of the question, first to part 1. In Chapter 4, we saw how the physical and chemical properties of petroleum fractions changed with increasing AEBP. From step to step, the change was gradual, even though the rate of change kept increasing. We discussed the need for widening the concept of atmospheric distillation to that of vacuum distillation, then short-path distillation, and finally to "equivalent distillation" to cover higher and higher "boiling" constituents. Again, going from atmospheric residue fractions to vacuum residue fractions and finally to nondistillable residue fractions (SEF fractions), the changes in composition were gradual, not abrupt.

Bunger et al. (89) reasoned that, on thermodynamic principles, the differences between asphaltenes and resins (soluble polar fraction) should be gradual rather than abrupt. Selucky et al. (90) found great similarities between the two groups when they compared the properties of asphaltenes and resins obtained from the same residue under identical conditions by means of three separation methods, namely, adsorption chromatography, chromatography on ion exchange resins, and SEC. This result, together with the great dependence of asphaltene yields on alkane type and precipitation conditions, prompted them to emphasize the gradual nature of the shift in the composition of resins and asphaltenes. An additional reason would have been the dependence of both yield and nature of asphaltenes on the material from which they are precipitated: atmospheric, vacuum, or nondistillable residue, or even whole crude oils.

Boduszynski et al. (58) demonstrated that the compounds comprising asphaltenes are part of the acid and base fractions of the residue. They separated a petroleum residue by the API-60 scheme into the respective fractions, acids, bases, neutral Lewis bases, aromatics, and saturates without prior asphaltene removal. All these fractions had VPO MWs between 1000 and 2300 daltons, in contrast to the asphaltenes which had an apparent MW > 4000 daltons (by VPO in benzene and in methylenechloride) when they were precipitated separately from the same residue. From the MW and elemental composition of the acid, base, and other fractions, there was no indication of abnormality because of the asphaltene constituents which they contained. These values were much the same as those of the same fractions obtained from the soluble polars. This is why they gave their paper the title "Asphaltenes, Where Are You?" Obviously, asphaltenes lose their identity when their acidic and basic constituents are separated from the residue without prior precipitation.

What makes asphaltenes insoluble? According to solubility theory (91), a compound is soluble in a solvent if the difference of the solubility parameters (between solute and solvent) is sufficiently small. The solubility parameter, δ, introduced by Hildebrand and Scott, is the square root of the energy of vaporization ΔE_V, over the molar volume V_m:

$$\delta = \left(\frac{\Delta E_V}{V_m}\right)^{1/2}$$

Strictly speaking, this equation is valid only for nonpolar solvents. The concept has been empirically expanded (92), and ΔE_V is viewed as arising from three contributions: hydrogen bonding, ΔE_{H}, permanent-dipole–permanent-dipole interactions, ΔE_P, and nonpolar London dispersion forces, ΔE_D. The latter arise from the hydrocarbon structure, essentially from the CH_n groups in paraffinic chains and ring structures, and, thus, depend on molecular size and shape. The energy of vaporization is the sum of these three components:

$$\Delta E_V = \Delta E_D + \Delta E_P + \Delta E_{\mathrm{H}}$$

Thus, it increases with

1. Polarity imparted by heteroatoms causing hydrogen bonding
2. Dipole interactions due to aromatic ring clusters and heteroatom groups
3. Molecular size

The molecular size effect, however, is minor for the solubility parameter, because much of it is cancelled by the molar volume which appears in the numerator. Some group contributions to partial solubility parameters are listed in Table 10.19 to give an idea about their relative magnitude.

We know from experience, and we can see from the considerations above, that highly polar species may be insoluble in *n*-pentane and other low-boiling hydrocarbons even if they have a low or moderate molecular weight and their aromatic rings are not highly condensed. Pericondensed aromatic ring clusters of only moderate size may also be insoluble if they have insufficient alkyl substitution (e.g., coronene). On the other hand, polybutenes and other polyalkenes, that is, very large molecules containing neither polar groups nor large aromatic ring systems, are soluble in low-boiling alkanes unless they are cross-linked by chemical bonds or by crystallization. Thus, contrary to general opinion, molecular weight does not have the prime importance the other two properties have for making as-

Table 10.19 Group Contributions to Partial Solubility Parameters

Functional group	Polar parameter, $V_{m\delta_{P}^{2}}$ (cal/mol)	H-bond parameter, $V_{m\delta_{H}^{2}}$ (cal/mol)	
		Aliphatic	Aromatic
>CO	390	800	400
-COOH	220	2750	2250
$-NH_2$	300	1350	2250
>NH	100	750	—
-OH	250	4650	4650

Source: Data extracted from Ref. 93.

phaltenes insoluble. Its main effect is to aid the other two in promoting insolubility.

So far we have only considered enthalpy effects.* The solubility of a compound depends on the free energy of mixing, ΔG, which is the sum of the enthalpy of mixing, ΔH, and an entropy term, ΔS:

$$\Delta G = \Delta H - T\Delta S$$

Here, $\Delta H = \Delta E$. A compound dissolves in a liquid if ΔG is negative. Ordinarily, the entropy term,

$$\Delta S = -k\,(N_1 \ln n_1 + N_2 \ln n_2)$$

is positive and quite large. N stands for the number of molecules and n for the number fraction of solvent (1) and solute (2) molecules in the mixture. In regular solutions (ideal ΔS, nonideal ΔH), $\Delta\delta$ may be 3.5–4 $(J/cm^3)^{1/2}$ without precipitation occurring because of the large entropy term. However, if we are dealing with large polar molecules, some of the polar groups may interact so strongly in poor solvents that they cause (physical) cross-linking. In Fig. 10.24, we illustrate this effect and also point out the difference between regular links and cross-links.

Cross-linking reduces the entropy term and leads to insolubility (94). We believe physical cross-linking may be another contribution to the insolubility of asphaltenes in pentane and other alkanes. In better solvents, the interactions are weak enough to cause much less or no cross-linking. To what extent, and whether at all, cross-linking is responsible for the insolubility of asphaltenes has not been established. A study of this effect

*Because of the expansion of Hildebrand's original concept to include hydrogen bonding, ΔE actually is affected by an entropic contribution. For our discussion, this is immaterial.

cross-link

cross-link

cross-link

cross-link

link

Intramolecular links and cross-links

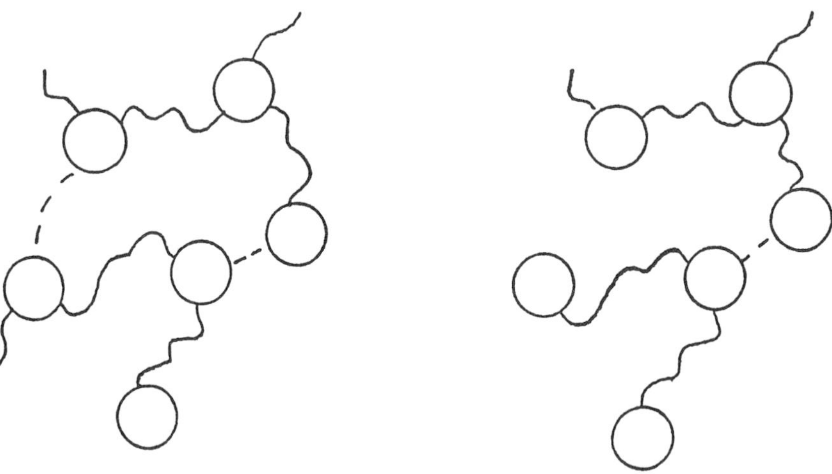

Intermolecular cross-linking

Loose aggregation
(Intermolecular linking)

Figure 10.24 Regular links, intra- and intermolecular cross-links, and loose intermolecular aggregation in large polar molecules.

would require precipitation from very dilute solutions. If it exists, it presupposes fairly large molecules of at least 700 molecular weight, in contrast to the enthalpy effects (solubility parameter) which affect the solubility behavior of any size molecules, small as well as large ones.

So far, we have considered only intramolecular cross-linking. Extensive *inter*molecular interaction, that is, aggregation, leads to even more effective cross-linking, now involving two or more molecules. Because the cross-links cause the molecules to contract, the total molar volume decreases, but the number of interacting CH and CH_2 groups increases, causing an even greater energy of mixing and, thus, loss of solubility. Loose aggregation, as we see it, for instance, with asphaltenes in benzene, just increases the effective MW (or, better, particle weight). Extensive aggregation, as with asphaltenes in pentane, causes both intra- and intermolecular cross-linking and insolubility. Figure 10.24 demonstrates the differences between intramolecular cross-linking, intermolecular cross-linking, and loose aggregation.

McKay et al. (95), expressed the view that asphaltene solubility depends on an interaction of MW, aromaticity, and polarity. Long (96,97) proposed the scheme shown in Fig. 10.25 as an illustration of asphaltene insolubility.

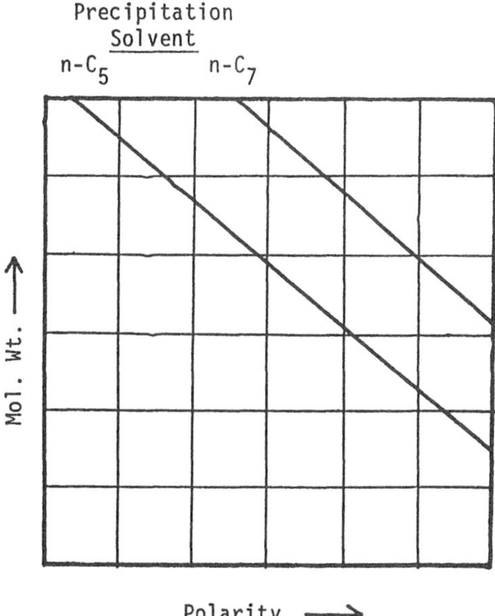

Figure 10.25 Long's scheme of asphaltene solubility. (From Ref. 96. Reproduced with permission of the publisher.)

The two lines represent the phase boundaries between soluble and insoluble structures in pentane and heptane, respectively. In this scheme, Long lumped together our variables, hydrogen bonding and dipole–dipole interaction (aromaticity), into one under the common name of "polarity."

Snape and Bartle (98) designed a three-dimensional scheme for the definition of asphaltenes in terms of three parameters: MW, acidic OH, and the number of internal aromatic carbons, C_{Int}. It is shown in Fig. 10.26. C_{Int} (which is the same as our CAI from Chapter 8) is a measure for the degree of both aromaticity and ring fusion. We present this figure to dem-

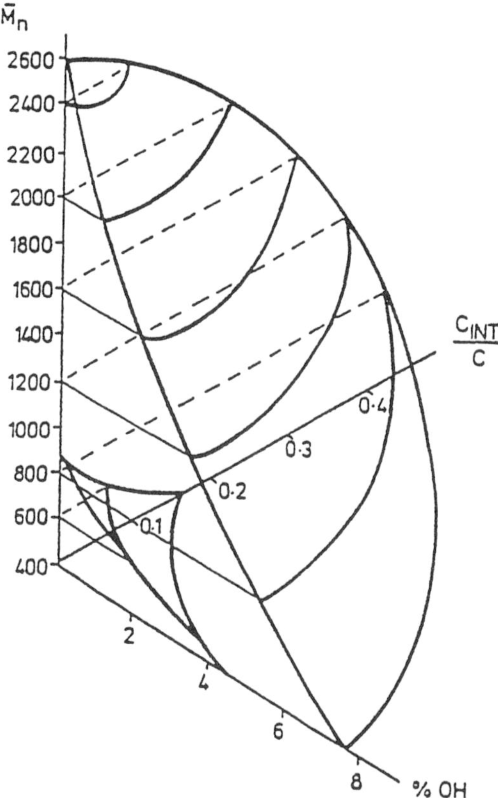

Figure 10.26 Snape and Bartle's three-dimensional scheme of asphaltene solubility. The small volume in the left lower corner of this graph defines *n*-pentane solubles, the large volume above, the asphaltenes. C_{Int}/C (internal aromatic C over total aromatic C) is a measure for the degree of aromatic condensation. (From Ref. 98. Reproduced with permission of the publisher.)

onstrate the principle, not for its detail which presumably was only estimated and not intended to be precise.

The same authors also developed an empirical equation for solubility values of asphaltenes (SV = 0.85 and 1.2) and maltenes (SV < 0.6):

$$\text{SV} = 0.75 \lg \left(\frac{\text{Mn}}{200}\right) + 0.1 \text{ (\% acidic OH)} + 1.5 \frac{C_{\text{Int}}}{C}$$

Their equation satisfied the asphaltenes and maltenes from a fair diversity, though not a large number, of sources: a (Turkish) crude oil, two bitumens, and four coal extracts. The restriction of polar groups just to acidic OH may be too severe for petroleum asphaltenes; for greater generality, some other acidic hydrogen types might also have to be included, for example, SH and possibly pyrrolic NH. But the choice of acidic groups as a measure of polarity in this case seems to be further justified by the compositional trend with increasing AEBP observed in Fig. 10.2. In exceptional cases, even crystallization of long, unbranched paraffinic side-chains may contribute to the insolubility of asphaltenes. This possibility is not part of Snape and Bartle's present equation, but it could easily be incorporated. Their equation looks like a good beginning for the quantitation of the insolubility of asphaltene or that of super-high AEBP components.

5. Molecular Weight and Aggregation

The molecular weights of the pentane soluble fraction of nondistillable residues can generally be determined by VPO without any problems. The results are the same as, or similar to, those obtained by FIMS. However, the true molecular weight (not particle weight) ranges of the insolubles are hard to determine. Most have MW distributions ranging from about 500 to a few thousand daltons (see Chapter 4). A strong tendency to aggregate makes VPO and even SEC measurements above 2000 daltons questionable. As discussed in greater detail in Chapter 4, VPO gives inflated values when performed with toluene and other nonpolar solvents, and considerably lower ones with polar solvents such as pyridine and nitrobenzene. At elevated temperatures, the molecular weights are lower yet (see Table 4.6 of Chapter 4). So far, the ultimate conditions for finding the true molecular weights of insolubles, free from aggregation and other interferences, are not known. Therefore, we do not know the true molecular weights either. The same holds for SEC fractions.

Selucky et al. (90) demonstrated a time effect on the MW of certain portions of asphaltenes in very dilute solutions. They separated Athabasca asphaltenes by size exclusion chromatography (SEC) at very low concentration (0.05%) into three main parts of different MWs. When these were redissolved in methylene chloride and rerun by SEC after different incubation times (resting periods), all three fractions exhibited major conversions to a common average MW which was slightly lower than the average MW of the lowest original fraction (no numbers given). During the resting periods, the samples were kept under nitrogen in the dark to exclude chemical changes. Thus, the drop in MW must have been caused by a physical process, presumably by deaggregation. The effect was strongest in the acid components of the asphaltenes as obtained in a separate experiment by fractionation on ion exchange resins. Brulé (99) had made similar observations on asphaltenes obtained from asphalts.

This effect calls into question any high-MW data for asphaltenes as far as they are proclaimed to be true *molecular* weights rather than *aggregate* weights. Recall Boduszynski et al.'s (58) observation of much lower MWs of acid and base components of a residue compared to those of the corresponding asphaltenes. We may assume that asphaltenes are much more strongly aggregated after precipitation (isolation) and redissolution in "good" solvents than before.

An interesting aspect of Selucky's results was the generally higher apparent MW and greater aggregate stability of the less polar fractions, that is, those eluted from the ion exchange resins with benzene compared with those eluted with benzene/methanol. Steric hindrance of the interacting groups by long alkyl chains may contribute to low aggregation as well as low dissociation rates. Or else, some of the less polar molecules only appear to be less polar because their highly polar groups are trapped in very stable polar bonds with other molecules.

Other methods for the MW measurement of these samples are also limited: FIMS by the low volatility of most asphaltenes; FDMS by experimental difficulties that make it hard to operate reliably and to quantify the amount of analyzed material; and SEC by calibration problems on top of dealing with the same aggregation effects as VPO. The advantage of FIMS/FDMS and SEC is that, at least in principle, they give the molecular-weight distribution.

Recently, Overfield et al. (100) applied small-angle neutron scattering (SANS) to asphaltenes in toluene. This technique, borrowed from polymer chemistry, can give the molecular weight and the radius of gyration, the latter being a measure of molecular size. Because of several assumptions, for instance monodispersity, the asphaltene results must be viewed as qualitative as yet. Even so, they indicate that asphaltenes in toluene are highly aggregated at room temperature and much less at 100°C (210°F). The aggregates appear to be more compact than random coils and less so than solid spheres. We have seen similar results (Chapter 4) from viscosity measurements.

Ravey et al. (101) reported, also on the basis of SANS measurements, that asphaltenes have the shape of thin sheets or flakes 6–20 nm (60–200 Å) wide and only ≤0.5 nm (5 Å) thick in various "good" solvents ranging from toluene to pyridine. Comparing an original asphaltene sample with a heart cut SEC fraction, they saw no effect of polydispersity on the results. Thus, these authors conclude that the requirement of monodispersity is unnecessary for their system. However, they found unusually high asphaltene MWs, a fact which opens all their results to questions. Their weight average "molecular" weights, Mw, of asphaltene in pyridine were around 25,000 daltons. Even with their assumption of an extraordinarily high heterogeneity factor Mw/Mn = 3–4, this would require number average molecular weights, Mn, of 6000–8000 daltons. In benzene, the MWs were higher yet by a factor of about 3. These values are about seven times higher than those customarily obtained by VPO and SEC.

The thickness of the platelets was not affected by the nature of the solvent; only the diameter increased with decreasing solvent power. The nature of the asphaltenes made little difference, the dimensions of five different asphaltene samples in THF being almost the same (13–16 nm diameter, 0.45–0.50 nm thickness).

Also by means of SANS, Ravey et al. (101) observed the behavior of asphaltenes in flocculating (poor) solvents, such as *n*-pentane and other *n*-paraffins. Here, the asphaltenes formed large, very open three-dimensional aggregates, presumably made up by agglomeration of the primary aggregates. Even though quantitatively Ravey's results seem out of line, their qualitative findings look interesting and plausible.

6. Our View of Asphaltenes

In the preceding sections, we have described the main aspects of the insoluble petroleum components. Here, we will try to explain succinctly the reasons for our reluctance to use the term asphaltenes for them.

First, "asphaltenes" differ in yield and composition depending not only on crude oil origin but also on type of residue from which they have been precipitated (atmospheric, vacuum, or truly nondistillable), the precipitant (*n*-pentane, *n*-heptane, hot heptane, etc.), and precipitation conditions (sample/precipitant ratio, addition of solvents, temperature, agitation, digestion time, etc.).

Second, "asphaltenes" are not a unique compound group or compound class. They are a strictly operationally defined solubility fraction and, as we have seen, their definition is vague. Therefore, the common phrases, "asphaltene molecules" or "asphaltene-type structures," are imprecise and can be misleading. The insolubles consist of a large variety of compounds of unknown structure and wide range of MWs. For the same reason, we contend that there are no "pure" asphaltenes. This term has been occasionally used to describe insolubles which are free from occluded soluble material.

Third, in compositional studies, the removal of "asphaltenes" by precipitation is used to facilitate the analysis of the soluble portion ("maltenes") of a residue. We believe that deep distillation, together with chromatographic separation of the fractions (including the nondistillable residue), into acids, bases, and neutrals is a better way to this end (API-60/NIPER approach, see Chapter 4). Or, as in our approach, the nondistillable residue can be subjected to "extended distillation" by sequential elution fractionation (SEF). This philosophy was discussed briefly in Section III.F.4 and at length in Chapter 3. It is worthwhile pointing out once more that our prime reason for the use of SEF is the principle of "extended distillation" and not a belief that it might be a superior or unique method. The principle of extended distillation was not invented as a consequence of our use of SEF. No, SEF was developed as a tool to help us prove our concept of extended distillation.

Our reluctance to use the term "asphaltenes" (as well as "maltenes" and some other ones) does not mean we are denying the existence of this refractory portion of a crude oil. We recognize that the insolubles, with their high concentration of heteroatoms and especially metals, are a fraction, which has the most adverse effect on crude oil refining. Knowing their amount can, indeed, greatly help in assessing crude oil processibility. Their precipitation with low-boiling hydrocarbons is a convenient shortcut for concentrating the most refractory petroleum constituents and determining their amount. It is also a very common procedure with many variations. But why not call the precipitates "insolubles" rather than "asphaltenes"?

7. Determination of Asphaltenes

Depending on precipitant, mixing ratio, agitation, contact time, and so on, the yields can vary significantly. Speight et al. (102) studied the effects of these variables and came to the following conclusions:

1. The most important variable is the precipitant (see, e.g., Fig. 10.27). *n*-Pentane is preferable when easy solvent removal is important, for example, for further fractionation of the maltenes. *n*-Heptane is preferred when an asphaltene sample is desired whose yield is less affected by the nature of the precipitant.
2. A precipitant/sample ratio of ≥30 mL/g, preferably 1 of 40, is necessary for consistent results. Thinning with benzene or other solvents before addition of the precipitant is detrimental.

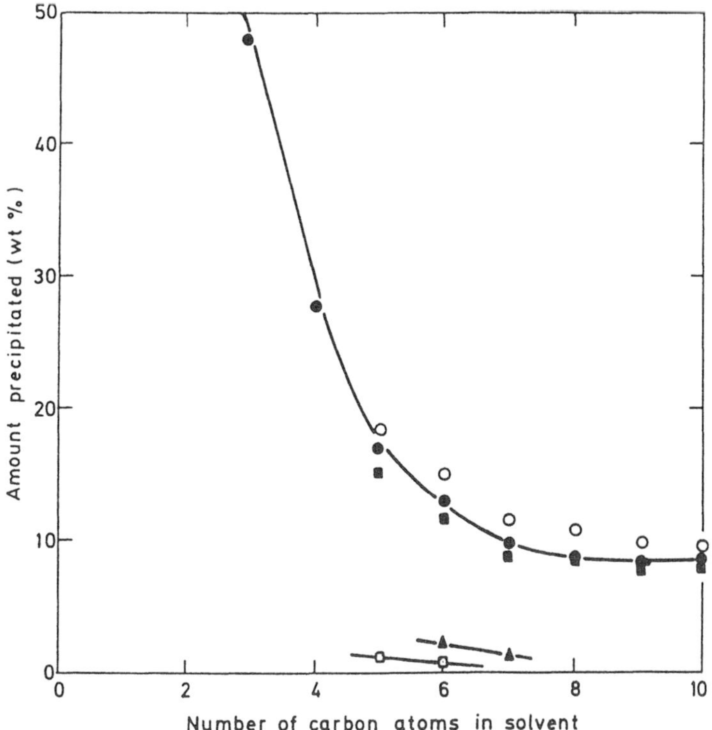

Figure 10.27 Amount of asphaltenes precipitated by various aliphatic hydrocarbons with different carbon numbers. (From Ref. 103. Reproduced with permission of the publisher.)

3. The sample–precipitant mixture should be allowed to stand for a period of 8–20 h before filtration for better reproducibility.

These suggestions helped a great deal to make asphaltene data from different laboratories comparable. Additional precautions, in those cases when the asphaltenes will be used for other measurements, include the prevention of oxidation by keeping air and light from them as much as possible.

From the refinery chemist's point of view, *n*-heptane is a better precipitant than *n*-pentane because it cuts deeper into the residue and produces an even richer concentrate of the most refractory components of crude oil.

Poirier and George (104) developed two rapid micro methods for measuring "asphaltene" contents. One uses TLC (thin-layer chromatography), the other, SEC (size exclusion chromatography). The soluble fractions are extracted from the material which stays behind in the original spot on the TLC plate. These "asphaltenes" are redissolved in toluene and measured in a colorimeter. They are, by definition, not identical with precipitated asphaltenes; however, the procedure seems to correlate well with conventional asphaltene determinations. Poirier et al. (105) report that nine samples can be run in 30 min with a standard deviation of about 3%.

Berrut and Jonathan (106) modified this method by using thin-layer rods instead of plates and evaluating them directly in an IATROSCAN which measures the "asphaltene" amount by flame ionization. Here, again, the "asphaltenes" are determined by means of chromatography rather than solubility. This technique is quite flexible and was extended by these authors to other types of measurements including the separate determination of saturates, aromatics, and polars.

For their second technique, Poirier et al. (105) use size exclusion chromatography which produces three incompletely resolved peaks. The first of these, the high MW peak, serves as a measure of the "asphaltenes." It has the advantages of speed (a run takes 30 min) and good repeatibility (1.5%), but it requires calibration with standards similar to the sample, which is a rather restrictive condition.

In these last three examples, the term "asphaltenes" is based on two or three other definitions, this time on different types of chromatographic behavior. If we consider all the variations in the original definition (by precipitation), we recognize a dozen or so different kinds of "asphaltenes" in the same sample.

IV. HOW MUCH MORE DO WE NEED TO KNOW?

In this book, we have described our present knowledge of the composition of heavy petroleum fractions and a selection of what we consider the most promising approaches and methods for their analysis. We have dealt with this subject, keeping in mind the current and future needs of the petroleum refining industry. We believe that the application of new analytical technology together with new computational tools will offer a more fundamental approach to unraveling the chemical reactions that occur in catalytic processing of complex petroleum mixtures.

We have stressed the importance of looking at the composition of petroleum in the context of the atmospheric equivalent boiling-point (AEBP) distribution. In our opinion, the description of the entire crude oil in terms of the AEBP distribution curve provides a rational basis for the comprehensive compositional analysis of crude oils and their fractions. The physical and chemical properties of various crude oil fractions as well as the results of detailed compositional analyses can then be presented as a function of AEBP. This provides a rational basis for comparing different crude oils as well as feeds, intermediates, and products from various refining processes. The continuity of changing petroleum composition as a function of AEBP is important when interpolating or extrapolating physical and chemical properties of crude oil fractions.

There are obvious gaps in our knowledge and in our analytical capabilities. How important are these gaps, how much more do we need to know, and to what extend do we need to improve our methods? That depends on the job at hand. The most important application of the detailed compositional analysis, of course, is in developing reaction networks and kinetic models of refining processes, and here we may rather quickly recognize our present limitations.

In a recent paper, Krambeck (107) stated

> Three main factors have to be considered in choosing a lumping scheme to describe chemical changes in a particular system of interest:
>
> 1. Measurability: It must be possible, at reasonable cost, to measure the composition in sufficient detail to determine the lumped composition.
> 2. Adequacy: The lumped composition must have sufficient detail to determine all product properties of interest.
> 3. Accuracy: Different feedstocks with the same lumped composition must give reaction products with the same lumped composition.

The tools presently available to us are impressive. But how much effort is justified? Will we really benefit from knowing the composition of the heavy petroleum fractions to a greater extent than we do now? Liguras and Allen (108) presented a very interesting and fairly detailed discussion of this point. In their attempt to design a pseudocomponent model for catalytic cracking, they needed an analytical method that allowed them to monitor the reactions. On the one hand, they looked for sufficient information. On the other hand, they tried to minimize the effort and cost of the analyses.

They say "One of the key issues in developing the ideal characterization and kinetic model is determining the point at which increasing the number of model compounds no longer enhances the predictions of the model." In their study, they used a carbon center approach, that is, they calculated the number of CH, CH_2, CH_3 as well as terminal and nonterminal olefinic and aromatic carbons. From the concentrations of reactants and of the products predicted by the model, they calculated:

- The concentrations of the various compound classes
- The carbon number distribution in each compound class

- The carbon center distribution for each carbon number

The main questions were: How badly are the compositional results affected before and after the model cracking reaction by

- Reducing the number of different structures considered in the calculation
- Choosing a smaller set of carbon numbers

They came to several interesting conclusions:

1. Separation of each compound class into subsets of different carbon numbers and calculation of a carbon center distribution for each carbon number was unnecessary. Equally good product concentrations resulted from the assignment of average carbon center distributions for all carbon numbers in a hypothetical distillation cut and even in the entire feed.
2. For the selection of pseudocomponents of a given carbon number within a compound class, it was possible to ignore all molecules (with different structures) of concentrations lower than 10% of the maximum pseudocomponent concentration regardless of chemical structure. Even ignoring those below 75% gave reasonable results after some reaction time (10 s rather than 1 or 2 s).
3. Lumping by structure, that is, grouping molecules of a given carbon number into compound classes and selecting only the most prominent classes (with their carbon-number distributions) for the determination of the carbon center distribution was thus permissible.
4. Lumping by carbon number, that is, not specifying the complete carbon-number (or molecular weight) spectrum in each fraction, however, caused large errors.
5. For this case, at least, the measurement of the carbon-number distribution—for instance, by mass spectrometry—was more important than the measurement of structural detail—as by ^{13}C-NMR—for each carbon number or even for a whole distillation cut.

Meanwhile, there are better approaches than that chosen by Liguras and Allen which would have led to different conclusions. Of interest in our context is the basic philosophy expressed by these authors.

The analytical technology has been advancing so rapidly that we can now measure far greater detail of structural/compositional data than the simple atomic composition or average molecular parameters versus carbon number. In his recent paper, Krambeck (107) says "Much attention has been paid in the literature to the accuracy of lumping schemes, that is, to the question of how few lumps can be used to develop kinetics that are still an accurate representation of the system."

He gives an example that "it was possible to develop accurate kinetics for the gasoline reforming process with iso and normal paraffins lumped together." However, he points out that "the octane numbers of these hydrocarbons are so different that the results of these kinetics are of limited value unless they can be subsequently delumped." He also adds: "On the other hand, it is not so easy to distinguish experimentally between these paraffin isomers over the whole boiling point range of interest."

Do we need better tools for compositional analysis than we have now? As we increase our knowledge, new insights will open new avenues, and new areas of research which, in turn, will require better and new instruments. Also, new tools have often helped to suddenly increase our knowledge, even to open new lines of research. Just as GC has revolutionized the production of gasoline, so will we some day have the means to fine-tune the conversion of high-boiling fractions and residues to light transportation fuels and lubestocks.

REFERENCES

1. McKay, J. F., Amend, P. J., Harnsberger, P. M., Cogswell, T. E., and Latham, D. R. 1981. Composition of petroleum heavy ends. 1. Separation of petroleum >675°C residues. *Fuel*, **60**:14–16.
2. McKay, J. F., Amend, P. J., Harnsberger, P. M., Erickson, R. B., Cogswell, T. E., and Latham, D. R. 1981. Composition of petroleum heavy ends. 2. Characterization of compound types in petroleum >675°C residues. *Fuel*, **60**:17–26.
3. McKay, J. F., Latham, D. R., and Haines, W. E. 1981. Composition of petroleum heavy ends. 3. Comparison of the composition of high-boiling petroleum distillates and petroleum >675°C residues. *Fuel*, **60**:27–32.
4. Green, J. B., Reynolds, J. W., and Yu, S. K-T. 1989. Liquid chromatographic separations as a basis for improving asphalt composition–physical property correlations. *Fuel Sci. Tech. Int.*, **7**(9):1327–1363.
5. Boduszynski, M. M. 1987. Composition of Heavy Petroleums. 1. Molecular weight, hydrogen deficiency, and heteroatom concentration as a function of atmospheric equivalent boiling point up to 1400°F (760°C). *Energy Fuels*, **1**:2–11.
6. Boduszynski, M. M. 1988. Composition of heavy petroleums. 2. Molecular characterization. *Energy Fuels*, **2**:597–613.
7. Ignasiak, T., Kemp-Jones, A. V., and Strausz, O. P. 1977. The molecular structure of Athabasca Asphaltene. Cleavage of the carbon–sulfur bonds by radical ion electron transfer reactions. *J. Organ. Chem.*, **42**:312–320.
8. Strausz, O. P., and Lown, E. M. 1991. Structural features of Athabasca bitumen related to upgrading performance. *Fuel Sci. Tech. Int.*, **9**(3):269–281.

9. Mojelsky, T. W., Ignasiak, T. M., Frakman, Z., McIntyre, D. D., Lown, E. M., Montgomery, D. S., and Strausz, O. P. 1992. Structural features of Alberta bitumen and heavy oil asphaltenes. *Energy Fuels*, **6**:83–96.
10. Tissot, B. P., and Welte, D. H. 1984. *Petroleum Formation and Occurrence.* Springer-Verlag, Berlin.
11. Peters, K.E., Scheuerman, G. L., Lee, C. Y., Moldowan, J. L., Reynolds, R. N., and Pena, M. M. 1992. Effects of refinery processes on biological markers. *Energy Fuels. 6*(**5**):560–577.
12. Peters, K. E., and Moldowan, J. L. 1992. *The Biomarker Guide. Interpreting Molecular Fossils in Petroleum and Ancient Sediments.* Prentice-Hall, Englewood Cliffs, NJ.
13. Johns, R. B., editor. 1986. *Biological Markers in the Sedimentary Record.* Elsevier, Amsterdam.
14. Orr, W. L., and Sinnenghe Damste, J. S. S. 1990. Geochemistry of sulfur in petroleum systems. *Geochemistry of Sulfur in Fossil Fuels*, edited by W. L. Orr and C. M. White. ACS Symposium Series 429. American Chemical Society, Washington, DC, Chap. 1.
15. Svob, V., and Jambrec, N. 1986. Contribution to characterization of the saturated part of cycloparaffinic crude oils by mass spectrometry. *Fuel*, **65**:1608–1611.
16. Sarowha, S. L. S., Dogra, P. V., Ramasvami, V., and Singh, I. D. 1988. Compositional and structural parameters of saturate fraction of Gujarat crude mix residue. *Erdoel Kohle*, **41**:124–125.
17. Payzant, J. D., Hogg, A. M., Montgomery, D. S., and Strausz, O. P. 1985. A field ionization mass spectrometric study of the maltene fraction of Athabasca bitumen. Part II—The aromatics. *AOSTRA J. Res.*, **1**:183–202.
18. Payzant, J. D., Hogg, A. M., Montgomery, D. S., and Strausz, O. P. 1985. A field ionization mass spectrometric study of the maltene fraction of Athabasca bitumen. Part III—The polars. *AOSTRA J. Res.*, **1**:203–210.
19. Singh, I. D., Ramaswami, V., Kothiyal, V., Brouwer, L., and Severin, D. 1992. Structural studies of short path distillates and supercritical fluid extract of petroleum short residue by NMR spectroscopy. *Fuel Sci. Tech. Int.*, **10**(2):267–280.
20. Strausz, O. P., Lown, E. M., and Payzant, J. D. 1990. Nature and geochemistry of sulfur-containing compounds in Alberta petroleums. *Geochemistry of Sulfur in Fossil Fuels*, edited by W. L. Orr and C. M. White. ACS Symposium Series 429. American Chemical Society, Washington, DC, Chap. 22, pp. 366–395.
21. Payzant, J. D., Montgomery, D. S., and Strausz, O. P. 1983. Novel terpenoid sulfoxides and sulfides in petroleum. *Tetrahedron Lett.*, **24**:651.
22. Valisolalao, J., Perakis, N., Chappe, B., and Albrecht, P. 1984. *Tetrahedron Lett.*, **25**:1183–1186.
23. Cyr, T. D., Payzant, J. D., Mothgomery, D. S., and Strausz, O. P. 1986. A homologous series of novel hopane sulfides in petroleum. *Organ. Geochem.*, **9**(3):139–143.

24. Waldo, G. S., Carlson, R. M. K., Moldowan, J. M., Peters, K. E., and Penner-Hahn, J. E. 1991. Sulfur speciation in heavy petroleums: Information from X-ray absorption near-edge structure. *Geochim. Cosmochim. Acta*, **55**:801–814.
25. Ali, M. F., Perzanowski, H., and Koreish, S. A. 1991. Sulfur compounds in high boiling fractions of Saudi Arabian crude oil. *Fuel Sci. Tech. Int.*, **9**(4):397–424.
26. Frakman, Z., Ignasiak, T. M., Lown, E. M., and Strausz, O. P. 1990. *Energy and Fuels*, **4**:263–270.
27. Kelemen, S. R., George, G. N., and Gorbaty, M. L. 1990. Direct determination and quantification of sulphur forms in heavy petroleum and coals. 1. The X-ray photoelectron spectroscopy (XPS) approach. *Fuel*, **69**:939–944.
28. Francisco, M. A., Rose, H. D., and Robbins, W. K. 1989. Private communication.
29. Shaw, J. E. 1989. Molecular weight reduction of petroleum asphaltenes by reaction with methyl iodide–sodium iodide. *Fuel*, **68**:1218–1220.
30. Sinnenghe Damsté, J. S., Eglington, T. I., Rijpstra, W. I. C., and de Leeuw, J. W. 1990. Characterization of organically bound sulfur in high-molecular-weight, sedimentary organic matter using flash pyrolysis and Raney Ni desulfurization. *Geochemistry of Sulfur in Fossil Fuels*, edited by W.L. Orr and C. M. White. ACS Symposium Series 429. American Chemical Society, Washington, DC, Chap. 26.
31. Nishioka, M., Campbell, R. M., Lee, M. L., and Castle, R. N. 1986. Isolation of sulfur heterocycles from petroleum- and coal-derived materials by ligand exchange chromatography. *Fuel*, **65**:270–273.
32. Nishioka, M., Whiting, D. G., Campbell, R. M., and Lee, M. L. 1986. Supercritical fluid fractionation and detailed characterization of the sulfur heterocycles in a catalytically cracked petroleum vacuum residue. *Anal. Chem.*, **58**:2251–2255.
33. George, G. N., Gobarty, M. L., and Kelemen, S. R. 1990. Sulfur K-edge X-ray absorption spectroscopy of petroleum asphaltenes and model compounds. *Geochemistry of Sulfur in Fossil Fuels*, edited by W.L. Orr and C. M. White. ACS Symposium Series 429. American Chemical Society, Washington, DC, Chap. 12.
34. Gorbaty, M. L., George, G. N., and Kelemen, S. R. 1990. Direct determination and quantification of sulphur forms in heavy petroleum and coals. 2. The sulphur K edge X-ray absorption spectroscopy approach. *Fuel*, **69**:944–949.
35. Rose, K. D., and Francisco, M. A. 1987. Characterization of acidic heteroatoms in heavy petroleum fractions by phase-transfer methylation and NMR spectroscopy. *Energy Fuels*, **1**(3):233–239.
36. Ruiz, J.-M., Carden, B. M., Lena, L. J., Vincent, E.-J., and Escalier, J.-C. 1982. Determination of sulfur in asphalts by selective oxidation and photoelectron spectroscopy for chemical analysis. *Anal. Chem.*, **54**:689–691.

37. McKay, J. F., Weber, J. H., and Latham, D. R. 1976. Characterization of nitrogen bases in high-boiling petroleum distillates. *Anal. Chem.*, **48**:891–898.
38. McKay, J. F., Cogswell, T. E., Weber, J. H. and Latham, D. R. 1975. Analysis of acids in high-boiling petroleum distillates. *Fuel*, **54**:50–61.
39. Snyder, L. R. 1969. Nitrogen and oxygen compound types in petroleum. Total analysis of a 400–700°F distillate from a California crude oil. *Anal. Chem.*, **41**:315–323.
40. Snyder, L. R. 1968. Nitrogen and oxygen compound types in petroleum. Total analysis of a 700–850°F distillate from a California crude oil. *Anal. Chem.*, **40**:1303–1317.
41. Holmes, S. A. 1986. Nitrogen functional groups in Utah tar sand bitumen and produced oils. *AOSTRA J. Res.*, **2**:167–175.
42. Sturm, G. P., Green, J. B., Tang, S. Y., Reynolds, J. W., and Yu, S. K-T. 1987. Chemistry of hydrotreating heavy crudes: II. Detailed analysis of polar compounds in Wilmington 660–1000°F distillate and hydrotreated products. *Am. Chem. Soc. Div. Petr. Chem.*, **32**(2): 369–378.
43. Boduszynski, M. M. 1986–1988. Unpublished results.
44. Green, J. B., Hoff, R. J., Woodward, P. W., and Stevens, L. L. 1984. Separation of liquid fossil fuels into acid, base and neutral concentrates. *Fuel*, **63**:1290–1301.
45. Sturm, G. P., Green, J. B., Grigsby, R. D., Tilley, F. L., Reynolds, J. W., and Yu, S. K-T. 1989. Detailed analysis of acidic compounds in Mayan gas oil and hydrotreated products. *Am. Chem. Soc. Div. Petr. Chem.*, **34**(2):367.
46. Seifert, W. K., and Howells, W. G. 1969. Interfacially active acids in a California crude oil. Isolation of carboxylic acids and phenols. *Anal. Chem.*, **41**:554–568.
47. Seifert, W. K., and Teeter, R. M. 1969. Preparative thin-layer chromatography and high resolution mass spectrometry of crude oil carboxylic acids. *Anal. Chem.*, **41**:786–795.
48. Seifert, W. K., and Howells, W. G. 1969. Interfacially active acids in a California crude oil. Isolation of carboxilic acids and phenols. *Anal. Chem*, **41**:555–568.
49. Seifert, W. K., and Teeter, R. M. 1970. Identification of polycyclic aromatic and heterocyclic crude oil carboxylic acids. *Anal. Chem.*, **42**:750–758.
50. Seifert, W. K., 1975. *Progress in the Chemistry of Organic Natural Products*, edited by W. Hertz, H. Grisebach, and G. W. Kirby. Springer-Verlag, New York, Vol. 2, pp. 1–49.
51. Moschopedis, S. E., and Speight, J. G. 1976. Oxygen functions in asphaltenes. *Fuel*, **55**:334–336.
52. Green, J. B., Treese, C. A., Yu, S. K.-T., Thomson, J.S., Renaudo, C. P., and Stierwalt, B. K. 1986. Separation and analysis of hydroxyaromatic species in liquid fuels. II. Comparison of ArOH in SRC-II Coal Liquid, Wilmington, CA, Petroleum and OSCR Shale Oil. *Am. Chem. Soc. Div. Fuel. Chem.*, **31**(2):126–143.

53. Yu, S. K.-T., and Green, J. B. 1989. Determination of total hydroxyls and carboxyls in petroleum and syncrudes after chemical derivatization by infrared spectroscopy. *Anal. Chem.*, **61**:1260–1268.
54. Green, J. B., Stierwalt, B. K., Thomson, J. S., and Treese, C. A. 1985. Rapid isolation of carboxylic acids from petroleum using high performance liquid chromatography. *Anal. Chem.*, **57**:2207.
55. Green, J. B., Thomson, J. S., Yu, S. K.-T., Treese, C. A., Stierwalt, B. K., and Renaudo, C. P. 1986. Separation and analysis of hydroxyaromatic species in liquid fuels. I. Analytical Methodology. *Am. Chem. Soc. Div. Fuel Chem.*, **31**(1):198–213.
56. Dorrence, S. M., Barbour, F. A., and Peterson, J. C. 1974. Direct evidence of ketones in oxidized asphalts. *Anal. Chem.*, **46**:2242–2244.
57. Boduszynski, M. M. 1985. Characterization of "heavy" crude components. *Am. Chem. Soc. Div. Petr. Chem.*, **30**(2):626–635.
58. Boduszynski, M. M., McKay, J. F., and Latham, D. R. 1980. Asphaltenes, where are you? *Asphalt Paving Technol.*, **49**:123–143.
59. Jacobson, J. M., and Gray, M. R. 1987. Use of I.R. spectroscopy and nitrogen titration data in structural group analysis of bitumen. *Fuel*, **66**:749–752.
60. Petersen, J. C., Barbour, R. V., Dorrence, S. M., Barbour, F. A., and Helm, R. V. 1971. Tentative identification of 2-quinolones in asphalt and their interaction with carboxilic acids present. *Anal. Chem.*, **43**:1491–1496.
61. Tamm, P. W., Harnsberger, H. F., and Bridge, A. G. 1981. Effect of feed metals on catalyst aging in hydroprocessing residuum. *Ind. Eng. Chem. Process Des. Dev.*, **20**:263–273.
62. Galiasso, R., Blanco, R., Gonzalez, C., and Quinteros, N. 1983. Deactivation of hydrodemetallization catalysts by pore plugging. *Fuel*, **62**:817.
63. Boduszynski, M. M., and Altgelt, K. H. 1992. Composition of heavy petroleums. 4. Significance of the extended atmospheric equivalent boiling point (AEBP) scale. *Energy and Fuels*, **6**:72–76.

63a. Barvise, A.J.G., and Whitehead, E.V. 1980. Characterization of vanadium porphyrins in petroleum residues. *Am. Chem. Soc. Div. Petr. Chem.*, **25**(2):268–279.

64. Gould, K. A. 1980. Oxidative demetallization of petroleum asphaltenes and residua. *Fuel*, **59**:733–736.
65. Dickson, F. E., Kunesh, C. J., McGinnis, E. L., and Petrakis, L. 1972. Use of electromagnetic resonance to characterize the vanadium(IV)-sulfur species in petroleum. *Anal. Chem.*, **42**:978.
66. Dickson, F. E., and Petrakis, L. 1974. Application of electronmagnetic resonance and electronic spectroscopy to the characterization of vanadium species in petroleum fractions. *Anal. Chem.*, **44**:1129.
67. Rankel, L. A. 1987. Degradation of metallopetroporphyrins in heavy oils before and during processing: Effects of heat, air, hydrogen, and hydrogen sulfide on petroporphyrin species. *ACS Symp. Ser.*, **344**(16):257–264.

68. Fish, R. H., and Komlenic, J. J. 1984. Molecular characterization and profile identifications of vanadyl compounds in heavy crude petroleums by liquid chromatography/graphite furnace atomic absorption spectrometry. *Anal. Chem.*, **56**:510.
69. Fish, R. H., Komlenic, J. J., and Wines, B. K. 1984. Characterization and comparison of vanadyl and nickel compounds in heavy crude petroleums and asphaltenes by reverse-phase and size exclusion liquid chromatography/graphite furnace atomic absorption spectrometry. *Anal. Chem.*, **56**:2452.
70. Reynolds, J. G., and Biggs, W. R. 1985. Characterization of vanadium compounds in selected crudes II. Electron paramagnetic resonance studies of the first coordination spheres in porphyrin and non-porphyrin fractions. *Liq. Fuels Tech.*, **3**(4):425–448.
71. Reynolds, J. G., and Biggs, W. R. 1987. Analysis of residuum desulfurization by size exclusion chromatography with element specific detection. *Amer. Chem. Soc. Div. Petr. Chem.*, **32**(2):398–405.
72. Reynolds, J. G., Gallegos, E. J., Fish, R. H., and Komlenic, J. J. 1987. Characterization of the binding sites of vanadium compounds in heavy crude petroleum extracts by electron paramagnetic resonance spectroscopy. *Energy Fuels*, **1**(1):36–44.
73. Blumer, M., and Snyder, M. 1967. *Chem. Geol.*, **2**:35.
74. Blumer, M., and Rudrum, R. 1970. High molecular weight fossil porphyrins: Evidence for monomeric and dimeric tetrapyrroles of about 1100 molecular weight. *J. Inst. Pet.*, **56**:99.
75. Goulon, J., Retournard, A., Friant, P., Goulon-Ginet, C., Berthe, C., Muller, J.-F., Poncet, J.-L., Escalier, J.-C., and Neff, B. 1984. Structural characterization by x-ray absorption spectroscopy (EXAFS/XANES) of the vanadium chemical environment in Boscan asphaltenes. *J. Chem. Soc., Dalton Trans.*, 1095–1103.
76. Goulon, J., Esselin, C., Friant, P., Berthe, C., Muller, J.-F., Poncet, J.-L., Guilard, R., Escalier, J.-C., and Neff, B. 1984. Structural characterization by x-ray absorption spectroscopy (EXAFS/XANES) of the vanadium chemical environment in various asphaltenes. *Collect. Collog. Semin. (Inst. Fr. Pet.)*, **40**:153–157.
77. Pearson, C. D., and Green, J. B. 1989. Comparison of processing characteristics of Mayan and Wilmington heavy residues. *Fuel*, **68**:465–474.
78. Zhang, G., Ziemer, J. N., Boduszynski, M. M., and Biggs, W. R. 1993. The binding of metals in Maya crude oil: Looking for the elusive "metallo-nonporphyrins." *Energy and Fuels*, manuscript submitted for publication.
79. Malhotra, R., and Buckmaster, H. A. 1985. 34GHz e.p.r. study of vanadyl complexes in various asphaltenes. *Fuel*, **64**:335–341.
80. Biggs, W. R., Fetzer, J. C., Brown, R. J., and Reynolds, J. G. 1985. Characterization of vanadium compounds in selected crudes. I. Porphyrin and non-porphyrins separation. *Liq. Fuels Tech.*, **3**(4):397–421.
81. Speight, J. G. 1986. Polynuclear aromatic systems in petroleum. *Am. Chem. Soc. Div. Petr. Chem.*, **31**(2):818–825.

82. El-Mohamed, S., Archard, M. A., Hardouin, F., and Gasparoux, G. 1986. Correlation between diamagnetic properties and structural characters of asphaltenes and other heavy petroleum products. *Fuel*, **65**:1501–1504.
83. Altgelt, K. H. 1992. Manuscript in preparation.
84. Ali, L. H., Al-Gannam, K. A., and Al-Rawi, J. M. 1990. Chemical structure of asphaltenes in heavy crude oils investigated by N.M.R. *Fuel*, **69**:519–521.
85. Mojelsky, T. W., Montgomery, D. S., and Strausz, O. P. 1985. Ruthenium (VIII) catalyzed oxidation of high molecular weight components of Athabasca oil sand bitumen. *AOSTRA J. Res.*, **2**:131–137.
86. Cyr, N., McIntyre, D. D., Toth, G., and Strausz, O. P. 1987. Hydrocarbon structural group analysis of Athabasca asphaltene and G.P.C. fractions by ^{13}C n.m.r. *Fuel*, **66**:1709–1714.
87. Ragunathan, P., Niizuma, S., and Strausz, O. P. 1991. ESR studies on Athabasca asphaltene. Manuscript in preparation.
88. Payzant, J. D., Lown, E. M., and Strausz, O. P. 1991. Structural units of Athabasca asphaltene: The aromatics with a linear carbon framework. *Energy and Fuels*, **5**:445–453.
89. Bunger, J. W., Cogswell, D. E., and Zilm, K. W. 1979. *Am. Chem. Soc. Div. Petr. Chem.*, **24**(4):1017.
90. Selucky, M. L., Kim, S. S., Skinner, F., and Strausz, O. P. 1981. Structure-related properties of Athbasca asphaltenes and resins as indicated by chromatographic separation. *The Chemistry of Asphaltenes*, edited by J.W. Bunger and N.C. Li. Advances in Chemistry Series 195. American Chemical Society, Washington, DC, pp. 83–118.
91. Hildebrand, J., and Scott, R. 1949. *Solubility of Non-Electrolytes*. Reinhold, New York.
92. Hansen, C. 1967. The three dimensional solubility parameter—key to paint component affinities: I. Solvents plasticizers, polymers, and resins. *J. Paint Technol.*, **39**:104–117.
93. Hansen, C., and Beerbower, A. 1970. Solubility parameters. *Kirk-Othmer Encyclopedia of Chemical Technology*, Suppl. Vol., 2nd ed. Interscience Publishers (John Wiley & Sons), New York, p. 869.
94. Flory, P. J. 1953. *Principles of Polymer Chemistry*. Cornell University Press, Ithaca, New York.
95. McKay, J. E., Amend, P. J., Cogswell, T. E., Harnsberger, P. M., Erickson, R. B., and Latham, D. R. 1977. Petroleum asphaltenes—chemistry and composition. *Am. Chem. Soc. Div. Petr. Chem.*, **22**:708–715.
96. Long, R. B. (1979). The concept of asphaltenes. *Am. Chem. Soc. Div. Petr. Chem.*, **24**(4):891–900.
97. Long, R. B. 1981. The concept of asphaltenes. *The Chemistry of Asphaltenes*, edited by J. W. Bunger and N. C. Li. Advances in Chemistry Series 195. American Chemical Society, Washington, DC.
98. Snape, C. E., and Bartle, K. D. 1984. Definition of fossil fuel derived asphaltenes, in terms of average structural parameters. *Fuel*, **63**:883–887.

99. Brulé, B. 1979. Characterization of bituminous compounds by gel permeation chromatography (GPC). *J. Liq. Chromatog.*, **2**:165–192.
100. Overfield, R. E., Sheu, E. Y., Sinha, S. K., and Liang, K. S. 1989. SANS study of asphaltene aggregation. *Fuel Sci. Tech. Int.*, **7**(5–6):611–624.
101. Ravey, J. C., Ducouret, G., and Espinat, D. 1988. Asphaltene macrostructure by small angle neutron scattering. *Fuel*, **67**:1560–1567.
102. Speight, J. G., Long, R. B., and Trowbridge, T. D. 1984. Factors influencing the separation of asphaltenes from heavy feedstocks. *Fuel*, **63**:616–620.
103. Mitchell, D. L., and Speight, J. G. 1972. The solubility of asphaltenes in hydrocarbon solvents. *Fuel*, **52**:149–152.
104. Poirier, M. A., and George, A. E. 1983. Thin layer chromatographic method for determination of asphaltene content in crude oils and bitumens. *Energy Sources*, **7**(2):165–176.
105. Poirier, M. A., Guiffier, N., Husson, J.-F., and Bourgognon, H. 1984. Dosage des asphaltenes dans le produits petroliers lourds par chromatographie sur couche mince (analyseur IATROSCAN) et chromatographie d'exclusion. *International Symposium, Lyon*, June 25–27, 1984, pp. 389–393.
106. Berrut, J.-B., and Jonathan, D. 1984. Application du systeme CMM-DIF a l'analyse quantitative des constituants lourds de petrole. *International Symposium, Lyon*, June 25–27, 1984, pp. 400–405.
107. Krambeck, F. J. 1992. How much more do we need to know? *Chemtech*, May, 292–299.
108. Liguras, D. K., and Allen, D. T. 1989. Structural models for catalytic cracking. 1. Model compound reactions. *Ind. Eng. Chem. Res.*, **28**:665–673.

Glossary

ABN	acid–base–neutrals
AE	atomic emission
AEBP	atmospheric equivalent boiling point
AED	atomic emission detector
AET	atmospheric equivalent temperature
API	American Petroleum Institute
AR	atmospheric residue, also aromatic ring
av.	average
c	concentration
C_α	number of C-atoms next to an aromatic ring
C_β	number of C-atoms two places removed from an aromatic ring
C_γ	number of C-atoms three places removed from an aromatic ring
C_ω	number of C-atoms at the end of a chain
CA	number of aromatic carbon atoms
CCR	Conradson carbon residue
CI	chemical ionization
CL	number of aliphatic carbon atoms

CN	number of naphthenic carbon atoms
d	density
DAO	deasphalted oil
DB	number of double bonds
DCP	direct current plasma (atomic emission)
DEPT	distortionless enhancement by polarization transfer
DISTACT	type of high-vacuum short-path still
EI-MS	electron impact mass spectrometry
f_a	aromatic to total carbon ratio
FAB	fast atom bombardment (ionization)
FCC	fluid catalytic cracking or fluid catalytic cracker
FD	field desorption
FDMS	field desorption mass spectrometry
FI	field ionization
FID	flame ionization detector
FIMS	field ionization mass spectrometry
FPD	flame photometric detector
FTIR	Fourier transform infrared (spectroscopy)
GASPE	gated spin echo
GC	gas chromatography
GC/MS	gas chromatography/mass spectrometry
GC-SIMDIS	simulated distillation by gas chromatography
GFAA	graphite furnace atomic absorption
GPC	gel permeation chromatography
H/C ratio	molar hydrogen to carbon ratio
H_α	number of H-atoms attached to carbon next to an aromatic ring
H_β	number of H-atoms attached to carbon two places removed from an aromatic ring
H_γ	number of H-atoms attached to carbon three places removed from an aromatic ring
HA	number of aromatic hydrogen atoms
HDN	hydrodenitrogenation
HECD	Hall electrolytic conductivity detector
HETCOR	heterocorrelated NMR spectroscopy
HETP	height equivalent to a theoretical plate
HL	number of aliphatic hydrogen atoms
HN	number of naphthenic hydrogen atoms
HOMCOR	homocorrelated NMR spectroscopy
HPLC	high performance liquid chromatography
ICP	inductively coupled plasma (atomic emission)
IR	infrared (spectroscopy)

LC	liquid chromatography
LEC	ligand exchange chromatography
LV-MS	low voltage mass spectrometry
LV%	liquid volume %
LVEI-MS	low-voltage electron impact MS
LVHR-MS	low-voltage high-resolution mass spectrometry
M	molar mass, also molecular weight in certain context
MBP	mid-boiling point
MCR	micro carbon residue
$\mathbf{M_n}$	number average molecular weight
MS	mass spectrometry
MTBE	methyl-tertiarybutylether
$\mathbf{M_w}$	weight average molecular weight
MW	molecular weight
MWD	molecular weight distribution
n	refractive index
N_1	mole number of solvent in a solution
N_2	mole number of solute in a solution
NF-MS	nonfragmenting mass spectrometry
NIPER	National Institute for Petroleum and Energy Research
NMR	nuclear magnetic resonance (spectroscopy)
PAC	polycyclic aromatic compounds
PAH	polycyclic aromatic hydrocarbons
PASC	polycyclic aromatic sulfur compounds
PCSE	part coupled spin-echo
PDAD	photodiode array detector
ppb	parts per billion
ppm	parts per million
R	(1) gas constant
R	(2) number of rings, same as R_T
R_A	number of aromatic rings
RCR	Ramsbottom carbon residue
R_i	refractivity intercept
RI	refractive index
RID	refractive index detector
R_N	number of naphthenic rings
R_T	total number of rings
SEC	size exclusion chromatography
SEF	sequential extraction fractionation
SEF-1 (2,3,4)	first (second, third, fourth) fraction obtained by SEF
SFC	supercritical fluid chromatography
SFE	supercritical fluid extraction

SIC	selected ion chromatogram
SIMDIS	simulated distillation
SpGr	specific gravity
T	temperature
T_b	boiling temperature
TRC	Thermodynamic Research Center at the Texas A&M University
2D NMR	two-dimensional NMR
UV	ultraviolet (spectroscopy)
VGC	viscosity gravity constant
VGO	vacuum gas oil
VPO	vapor phase osmometry (vapor pressure osmometry)
VR	vacuum residue
VTGA	vacuum thermal gravimetric analysis
VTGA-SIMDIS	simulated distillation by vacuum thermal gravimetric analysis
wt%	weight percent
XANES	x-ray absorption near-edge structure spectroscopy
XPS	x-ray photoelectron spectroscopy
"Z" value	hydrogen deficiency parameter
η	(1) kinematic viscosity (centistokes), (2) "concentration parameter" in an equation by Chung et al. (1978)
$[\eta]$	intrinsic viscosity
ν	viscosity (centipoise)
σ	degree of substitution

Index

For Product Safety Concerns and Information please contact our EU representative GPSR@taylorandfrancis.com
Taylor & Francis Verlag GmbH, Kaufingerstraße 24, 80331 München, Germany

www.ingramcontent.com/pod-product-compliance
Ingram Content Group UK Ltd.
Pitfield, Milton Keynes, MK11 3LW, UK
UKHW021114180425
457613UK00005B/79

* 9 7 8 0 8 2 4 7 8 9 4 6 6 *